Lecture Notes in Computer Science

Commenced Publication in 1973
Founding and Former Series Editors:
Gerhard Goos, Juris Hartmanis, and Jan van Leeuwen

T0238492

Paolo Atzeni   David Cheung   Sudha Ram (Eds.)

Conceptual Modeling

31st International Conference, ER 2012
Florence, Italy, October 15-18, 2012
Proceedings

 Springer

Volume Editors

Paolo Atzeni
Università Roma Tre
Dipartimento di Informatica e Automazione
Via Vasca Navale, 79
00146 Roma, Italy
E-mail: atzeni@dia.uniroma3.it

David Cheung
The University of Hong Kong
Department of Computer Science
Pok Fu Lam Road, Hong Kong, China
E-mail: dcheung@cs.hku.hk

Sudha Ram
The University of Arizona
Eller College of Management
McClelland Hall, Room 108, P.O. Box 210108
Tucson, AZ 85721-0108, USA
E-mail: ram@eller.arizona.edu

ISSN 0302-9743   e-ISSN 1611-3349
ISBN 978-3-642-34001-7   e-ISBN 978-3-642-34002-4
DOI 10.1007/978-3-642-34002-4
Springer Heidelberg Dordrecht London New York

Library of Congress Control Number: 2012948193

CR Subject Classification (1998): D.2.1-4, I.6.5, F.3.2, H.2.4, H.2.7-8, H.3.3-4, I.2.4, H.4, J.1, K.6

LNCS Sublibrary: SL 3 – Information Systems and Application, incl. Internet/Web and HCI

Typesetting: Camera-ready by author, data conversion by Scientific Publishing Services, Chennai, India

Printed on acid-free paper

Springer is part of Springer Science+Business Media (www.springer.com)

Foreword

This volume collects the papers selected for presentation at the 31st International Conference on Conceptual Modeling (ER 2012), held in Florence, Italy, during October 15–18, 2012. The ER International Conference on Conceptual Modeling is one of the top conferences in the area of information systems and database design and it gathers more than 100 well-known researchers from all around the world, involving both university and industrial research teams.

We are proud to present the proceedings of the ER 2012 conference that is a forum to exchange ideas and experiences, and to discuss current research and applications, with major focus on conceptual modeling for computer-supported and networked applications. The topics of interest include research and practice in areas such as theories of concepts and ontologies underlying conceptual modeling, methods and tools for developing and communicating conceptual models, and techniques for transforming conceptual models into effective implementations.

The host city of ER 2012 was Florence, one of the most important and historic cities in Italy, steeped in culture and atmosphere. At the time of Charlemagne, Florence was a university town. Today it includes many specialized institutes and is an international cultural center. Academies, art schools, scientific institutes, and cultural centers all contribute to the city's intense activity. The mingling and the mutual interaction of ancient and modern culture make Florence a magical, suggestive, and inspiring city that can offer modern, efficient, and impressive conference premises.

ER 2012 was the outcome of a joint effort from several sponsoring societies and organizations, plus many individuals including our colleagues, volunteers, and friends. We would like to take this opportunity to express our gratitude for their help and contributions. We would also like to thank all the authors who submitted papers, demos, tutorials, and panels and all the conference participants. We are especially grateful to the ER Steering Committee members for their support in organizing ER 2012, and to the Program Chairs, Committee, and external referees for their relentless work in reviewing submissions with expertise and patience to forge a top-quality scientific program. Many thanks are also due to the Workshop, Panel, Tutorial, Demos, and PhD Symposium Chairs and Committees for their experienced contributions and their support. The Organizing and the Sponsorship Chairs devoted plenty of time to ensure the success of the conference. Last but not least, we express our warm gratitude to Devis Bianchini and Alfio Ferrara, our Publicity Chairs and Conference Webmasters, and to Michele Melchiori, Stefano Montanelli, and Marco Patella, our Registration and Proceeding Chairs, for their excellent and timely job.

July 2012   Valeria De Antonellis
Stefano Rizzi

Preface

Conceptual modeling is a way to represent real-world objects, data, processes, systems and their inter-relationships. The International Conference on Conceptual Modeling is a premier venue for the presentation and exchange of the latest knowledge and cutting-edge research on conceptual modeling. It has been 36 years since Peter Chen's seminal paper on the Entity Relationship (ER) Model appeared in print. It profoundly impacted the world of data modeling and provided the impetus for this conference and its community. Since then the ER model has been used, modified, extended, and applied in many different fields including databases, software, business processes, and information systems. This year, we were pleased to initiate the fourth decade of the Conceptual Modeling conference. Traditionally, this long-lived conference has enabled the meeting and melding of computer science and information systems researchers. The 31st International Conference on Conceptual Modeling was appropriately held in Florence in Italy, the birthplace of the Renaissance movement.

Over the years the conference has attracted the best researchers working in the area of conceptual modeling and 2012 was no exception. This year we received a total of 176 abstracts and 141 full papers. Each paper was reviewed by at least three reviewers and based on these reviews we selected 24 regular papers, 13 short papers, and three poster papers. The acceptance rate for regular papers is 17%, for regular and short papers together it is 26.2%, for all categories it is 29.7%. In keeping with advancements in technology, this year's program reflects contemporary issues relating conceptual modeling to the social and Semantic Web. This year, the organizers decided to promote the PhD Symposium by offering to the three best contributions the opportunity to publish a full paper. Similarly, three additional papers were offered space for a poster. They are all included in this volume. The scientific program included three very interesting keynote talks by Stefano Ceri, Umeshvar Dayal, and Christian Jensen, each of which has a companion paper in these proceedings.

We sincerely thank the 83 members of the Program Committee and the external reviewers who worked very hard and provided thoughtful reviews on time. We also appreciate the diligence of the senior reviewers who facilitated the discussion and provided recommendations for a subset of the submissions. Most importantly, we thank the authors who wrote high-quality research papers and submitted them to ER 2011: without them, the conference would not have existed.

July 2012

Paolo Atzeni
David Cheung
Sudha Ram

Organization

General Conference Co-chairs

Valeria De Antonellis   University of Brescia, Italy
Stefano Rizzi   University of Bologna, Italy

Steering Committee Liaison

David W. Embley   Brigham Young University, Utah, USA

Program Committee Co-chairs

Paolo Atzeni   Università Roma Tre, Italy
David Cheung   University of Hong Kong, China
Sudha Ram   University of Arizona, USA

Organization Co-chairs

Marco Bertini   University of Florence, Italy
Fausto Rabitti   CNR, National Research Council of Pisa, Italy

Workshop Co-chairs

Silvana Castano   University of Milan, Italy
Panos Vassiliadis   University of Ioannina, Greece

Tutorial Co-chairs

Letizia Tanca   Politecnico di Milano, Italy
Gottfried Vossen   University of Münster, Germany

Panel Chair

Maurizio Lenzerini   University of Rome La Sapienza, Italy

Demonstration Co-chairs

Laks V.S. Lakshmanan   University of British Columbia, Canada
Mong Li Lee   National University of Singapore, Singapore

PhD Colloquium Co-chairs

Sonia Bergamaschi   Università di Modena e Reggio Emilia, Italy
Bernhard Thalheim   Christian Albrechts University of Kiel,
                                 Germany

Sponsorships and Exhibits Chair

Matteo Golfarelli   University of Bologna, Italy

Registration and Proceedings Co-chairs

Michele Melchiori   University of Brescia, Italy
Stefano Montanelli   University of Milan, Italy
Marco Patella   University of Bologna, Italy

Publicity Co-chairs and Conference Webmasters

Devis Bianchini   University of Brescia, Italy
Alfio Ferrara   University of Milan, Italy

Organizing Secretariat

Scaramuzzi Team   Florence, Italy

Program Committee

Jacky Akoka   CNAM & TEM, France
Akhilesh Bajaj   University of Tulsa, USA
Sonia Bergamaschi   Università di Modena e Reggio Emilia, Italy
Shawn Bowers   Gonzaga University, USA
Stephane Bressan   National University of Singapore, Singapore
Jordi Cabot   INRIA, France
Huiping Cao   New Mexico State University, USA
Sharma Chakravarty   The University of Texas at Arlington, USA
Roger Chiang   University of Cincinnati, USA
Faiz Currim   University of Arizona, USA
Alfredo Cuzzocrea   ICAR-CNR and University of Calabria, Italy
Umesh Dayal   HP Labs, USA
Olga De Troyer   Vrije Universiteit Brussel, Belgium
Lois Delcambre   Portland State University, USA
Gillian Dobbie   University of Auckland, New Zealand
David Embley   Brigham Young University, USA
Opher Etzion   IBM Research Labs Haifa, Israel

Jolita Ralyte   University of Geneva, Switzerland
Iris Reinhartz-Berger   University of Haifa, Israel
Motoshi Saeki   Tokyo Institute of Technology, Japan
Satya Sahoo   Case Western Research University, USA
Amit Sheth   Wright State University, USA
Mario Silva   University of Lisbon, Portugal
Il-Yeol Song   Drexel University, USA
Veda Storey   Georgia State University, USA
James Terwilliger   Microsoft Corporation, USA
Juan Carlos Trujillo Mondéjar   University of Alicante, Spain
Axel Van Lamsweerde   Université Catholique de Louvain, Belgium
Panos Vassiliadis   University of Ioannina, Greece
Ramesh Venkataraman   Indiana University, USA
Gerd Wagner   University of Technology Cottbus, Germany
Haixun Wang   Microsoft Research, China
Carson Woo   University of British Columbia, Canada
Masatoshi Yoshikawa   Kyoto University, Japan
Eric Yu   University of Toronto, Canada
Philip Yu   University of Illinois at Chicago, USA
DongSong Zhang   University of Maryland, Baltimore County,
                                       USA
Zhu Zhang   University of Arizona, USA
Huimin Zhao   University of Wisconsin Milwaukee, USA
Shuigeng Zhou   Fudan University, China
Xiaofang Zhou   University of Queensland, Australia

External Reviewers

Raman Adaikkalavan   David Gil
Ahmad Alaiad   Carlos A. González
Raian Ali   Francesco Guerra
Pramod Anantharam   Tilani Gunawardena
Masatoshi Arikawa   Chetan Gupta
Pierluigi Assogna   Kenji Hatano
Domenico Beneventano   Cory Henson
Marcello Buoncristiano   Mizuho Iwaihara
Yuanzhe Cai   Javier Luis Canovas Izquierdo
Yi Cai   Hideyuki Kawashima
Hesam Chiniforooshan   Sybren De Kinderen
Anthony Cleve   Jonathan Lemaitre
Soumyava Das   Henrik Leopold
Nadeem Daudpota   Tok Wang Ling
Ion-Mircea Diaconescu   An Liu
Julian Eberius   Hector Llorens
Khaled Gaaloul   Yifeng Luo

Riccardo Martoglia
Jose-Norberto Mazon
Eduardo Mena
Tadashi Ohmori
Marius Octavian Olaru
Mahsa Sadi
Tomer Sagi
Cassia Trojahn Dos Santos
Donatello Santoro
Takahiko Shintani

Mimma Sileo
Fabrizio Smith
Francesco Taglino
Chaogang Tang
Sujan Udayanga
Alexander Ulanov
Wenbo Wang
Shiting Wen
Yanwei Xu
Lamia Abo Zaid

PhD Symposium Program Committee

Jorge Cardoso	University of Coimbra, Portugal
Silvana Castano	University of Milan, Italy
Arantza Illarramendi	Universidad del País Vasco, Spain
Manfred Jeusfeld	Tilburg University, The Netherlands
Sudha Ram	University of Arizona, USA
Letizia Tanca	Politecnico di Milano, Italy
Guido Vetere	IBM Italia

Organized by

University of Bologna, Italy
University of Brescia, Italy

In cooperation with

ACM SIGMIS
ACM SIGMOD

Under the Patronage of

Comune di Firenze, Italy

Sponsors

UNIVERSITÀ DEGLI STUDI DI BRESCIA

UNIVERSITÀ
DEGLI STUDI
DI MILANO

ALMA MATER STUDIORUM
UNIVERSITÀ DI BOLOGNA

 Springer

ENTE
CASSA DI RISPARMIO
DI FIRENZE

UNIVERSITA' DEGLI STUDI DELL'AQUILA
DIP. DI INGEGNERIA ELETTRICA E DELL'INFORMAZIONE

Consiglio Nazionale
delle Ricerche

Istituto di Scienza e Tecnologie
dell'Informazione "A. Faedo"

Table of Contents

Extraction, Discovery and Clustering

Search and Documents

Process Modeling I

Process Modeling II

Data and Process Modeling

Ontology Based Approaches

Variability and Evolution

Adaptation, Preferences and Query Refinment

Queries, Matching and Topic Search

Conceptual Modeling in Action

Posters

PhD Symposium Full Papers

PhD Symposium Poster Papers

Mega-modeling for Big Data Analytics

Stefano Ceri[1], Emanuele Della Valle[1], Dino Pedreschi[2], and Roberto Trasarti[3]

[1] DEI, Politecnico di Milano, via Ponzio 34/5, 20133, Milano
[2] Dipartimento di Informatica, Università di Pisa, Largo B. Pontecorvo, 3, 56127 Pisa
[3] ISTI-CNR, Istituto di Scienze e Tecnologie dell'Informazione del CNR, Pisa

Abstract. The availability of huge amounts of data ("big data") is changing our attitude towards science, which is moving from specialized to massive experiments and from very focused to very broad research questions. Models of all kinds, from analytic to numeric, from exact to stochastic, from simulative to predictive, from behavioral to ontological, from patterns to laws, enable massive data analysis and mining, often in real time. Scientific discovery in most cases stems from complex pipelines of data analysis and data mining methods on top of "big" experimental data, confronted and contrasted with state-of-art knowledge. In this setting, we propose **mega-modelling** as a new holistic data and model management system for the acquisition, composition, integration, management, querying and mining of data and models, capable of mastering the co-evolution of data and models and of supporting the creation of what-if analyses, predictive analytics and scenario explorations.

1   Introduction

The grand challenge of modern scientific data processing is the ability to use "big data" for knowledge discovery. Progress in key areas - such as social and economic resilience, health, transportation, energy management - depends on a strategic use of data, e.g., for understanding disease spreading or economic crises, for energy distribution policies which make the best use of resources, for on-line alerting systems that take into account traffic, road conditions, hazards, and so on. Decision makers are thrilled by the possibility of anticipating the impacts of different possible decisions, i.e., exploring various future scenarios at different degrees of detail, employing a variety of predictive methods (such as multi-level and micro-macro models, patterns, simulations and what-if analyses).

As advocated by the visions of FuturICT [1], the FourthParadigm [2] and Haas et al. [3], this challenge cannot be addressed by simply deploying currently available technology; it entails a profound innovation of ICT research, by boosting a reformulation of all core information technologies in terms of a global techno-social ecosystem, where ICT opens to the challenges of supporting the complexity of scientific computational models. We need modelling capabilities that leverage on the power of big data, e.g., conceiving what-if models and scenarios that realistically portray the outcomes of possible interventions or changes onto complex socio-economic phenomena. Mastering such new scenarios requires establishing open platforms where

P. Atzeni, D. Cheung, and R. Sudha (Eds.): ER 2012, LNCS 7532, pp. 1–15, 2012.

scientists and developers can integrate and compose deep models, big-data-driven analytics and agent-based simulations and create empirically validated, computationally replicable modelling components.

But, beyond and before technological platforms, we need *a comprehensive theory blending simulation models, analytical models, ontological models and data-driven models into one picture*. Modelling, as we know it today, is required to scale up to a higher level, that we call **mega-modelling**: a comprehensive theory and technology of model construction (with an emphasis on incremental approaches), model search, model fitness evaluation, model composition, model reuse and model evolution. We need entirely new *model of models*, namely algebras of objects representing patterns, rules, laws, equations, etc., which are either mined/induced from data, or based on deep mathematical findings or agent-based reasoning – an overarching algebra of data and models that allow us to devise a new holistic system for integrated data and model acquisition, integration, querying and mining, capable of mastering the complexity of the knowledge discovery process.

This paper aims at making a first step in the above direction, by introducing the concept of **mega-modelling** and then some abstractions for mega-model composition, attempting both a top-down definition and a bottom-up recognition of these abstractions in previous projects of the authors. The remainder of the paper is organized as follows. Section 2 traces the origin of this research, by positioning it relative to model-driven research in software engineering and to inductive data mining. Section 3 provides a definition of mega-modules and a preliminary view on its composition abstractions. Section 4 introduces mega-schemas and patterns as fundamental ingredients of mega-modules. Finally, Section 5 provides some examples which trace, in existing work, the presence of mega-modules ancestors.

2 Scientific Pillars of Mega-Modelling

The roots of mega-modelling can be traced to an article appeared in 1992 on "Mega-programming" [4]; mega-programs represent large, autonomous computing systems whose interfaces are described through a data-centric approach and whose execution behaviour can be inspected. After about one decade, Bezivin et. al. in 2004 introduced the term mega-model to denote modelling large software components [5], thus paving the road to the school on Model-Driven Engineering (MDE) that developed thereafter [6]. Then, in this community, mega-modelling has been mostly used as a term for denoting the higher-order relationships between models (such as *representationOf, conformsTo, isTransformedIn*) [7], or, more recently, for tracing the dependencies between models during their evolution [8]. However, the innovative aspects of mega-modelling go beyond classical model-driven software generation, as we associate to each mega-module the potential of expressing classes of computations on top of big data, thereby highlighting the computational nature of the modules and the support of dynamic aspects related to inspection, adaptation, and integration. This is the new and enhanced meaning that we give to mega-modules, by reconsidering them from the perspective of "big" data analytics.

In the database/data mining community, the idea of a unifying approach towards data analysis and mining has been around for several years, since the seminal paper on *inductive databases* and *data mining as a querying process* by Imielinski and Mannila [9]; a fundamental aspect of the approach is the representation of data mining activities through patterns, whereas patterns can be seamlessly integrated with data and can therefore be the subject of queries; such view attempts a conceptualization and generalization of data mining. Yet, this idea has found concrete realizations only lately and partially; it is used in [10] and [11], where extracted data patterns are defined as views on top of data tables, and as such can be composed with domain-specific data representing genetic information [10] or spatio-temporal trajectories expressing human mobility [11]. Shaping computational results through regular table formats has recently found another intriguing field of application in the context of social computations, where tabular formats have been elected to drive interactions with human crowds through crowd-sourcing [12] and crowd-searching [13]; the above models hint to a possible seamless extension of scientific computation to crowd-based computations.

3   Mega-modules for Scientific Big Data Processing

In the context of this paper, a **mega-module** is a software component capable of processing "big data" for analytical purposes. Every mega-module performs a well-identified computation, which can be considered a unitary transformation from inputs to outputs. Inputs and outputs take the form of data and of patterns, where data are domain-specific both in terms of their schema end instances, while patterns are forms of data regularity or rules whose schema is domain-independent and whose content typically reflects collective or aggregated data properties; patterns may be extracted by data analysis algorithms, which may in turn be embodied within mega modules. Every mega-module can internally use data and patterns that are considered as invariant in the context of the computation, whose extension can be either local (e.g., organization-specific) or global (e.g., stored in public databases or ontologies).

Let us propose a running example inspired to mobility data mining in M-Atlas [11], where a mega-module uses positions of mobile users to infer how big masses of people move from regions to regions – thereby inferring how persons move at an aggregate level. Positions as detected by GPS systems in mobiles and reported in the territory with an associated timestamp represent the input data. Information about streets and roads can be considered as invariant. Moving points are recognized as trajectories, a known pattern featuring a starting point, an ending point, and a sequence of intermediate points. Trajectories are clustered and aggregated by the mega-module, whose output reports, in a compact way, the most significant "flock movements" of groups of people.

3.1   Phases of Mega-module Computations

Every mega-module exhibits a format that consists of three distinct phases of information processing, although such phases can vary significantly for their internal organization: data preparation, analysis, and evaluation.

- The first phase, **data preparation**, consists of the processing of input data and patterns for the purpose of assembling input objects that will be the subjects of the analysis. The distinction between data and object is of semantic nature: data preparation typically assembles several elementary data in the input to generate a single object for the purpose of analysis. The aggregative process that builds a object can be driven by a variety of purposes – abstracting irrelevant differences, recognizing common features, aggregating over elementary items which satisfy given predicates – thus, semantically interpreting and reconstructing data. The keywords for the preparation phase are: data sensing, acquisition, integration, transformation, semantic enrichment.
  In the running example, several observations of the positions of the same moving object are assembled into a single trajectory.
- The second phase, **data analysis**, consists of extracting computed objects from input data, possibly using the input patterns as references. Data analysis produces the response to a specific problem by performing the core scientific processing, and uses a variety of methods, ranging from mathematical to statistical models, from data mining to machine learning, from simulation to prediction, including crowd-sourcing as a way for asking social responses. The keywords for the analysis phase are: mining, learning, modeling, simulation, forecast.
  In the running example, trajectories are assembled and reported as movements of groups of people (flocks).
- The third phase, **data evaluation**, consists of preparing the output objects, which may in turn be presented as data and/or patterns. This phase consists of filtering or ranking computed objects based on their relevance, and possibly of a post-processing so as to observe the result in the most suitable way for the mega-module enclosing environment or user. The keywords for the evaluation phase are: quality assessment, filtering, significance measurements, presentation, delivery, visualization.
  In the running example, reported flocks have a population above a given threshold and connect specific portions of territory, e.g. recognized as regions in a map.

In **Fig. 1**, we propose (on the left) a mega-module graphical element, which visually captures the characteristics of a mega-module as described above, and (on the right) we show how to represent the running example with such a graphical element. The partitioning of mega-module activity in three phases is used in defining its compositional properties, as described next.

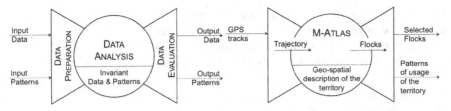

Fig. 1. Visual presentation of a generic mega-module and of the running example

The presence of the three phases allows us to define two standard inspection points within a mega-module, used for asynchronous control and feedback that mega-modules should provide to their enclosing environment. The first one, after preparation, provides a view on objects abstracted/reconstructed from data; the second one, after analysis, provides a view of the objects resulting from the analysis.

A mega-module inspection consists in extracting its controls asynchronously, during its execution; this in turn allows the enclosing environment to trace mega-module execution, to estimate completion time, and to anticipate the quality of its results. We regard the data and patterns that may be exchanged by a mega-module during its execution as the **mega-module controls**. A mega-module should expose commands to the enclosing environment that may alter its behavior, for instance by rising or by lowering confidence levels during analysis based on the quality of intermediate results or on the expected completion time. It should also be possible to suspend, resume, and terminate the mega-module computation.

3.2   Composition Abstractions

Composition abstractions are the means of combining mega-modules to the purpose of creating sophisticated analytical processes. Composition abstractions reflect the classical ways of assembling modules into higher order computations. Every abstraction induces a hierarchical decomposition, singling out an enclosing mega-module and one or more enclosed mega-modules; our goal is to describe computations over big data as top-down recursive applications of a well-designed collection of abstractions. Below, we present an initial set of abstractions; they are orthogonal, but most likely incomplete, and further investigation is needed to consolidate them.

General-Purpose Composition Abstractions. The first three abstractions apply to arbitrary mega-modules and characterize classical ways of partitioning computations, through pipelines and parallel control and by fragmenting computations.

Pipeline decomposition. The simplest form of control is a pipeline that occurs when the output of one mega-module feed the input of another mega-module. A pipeline can be abstracted by singling out the prepare phase of the first one and the evaluation phase of the last one, and, then, compounding the internal chain of analysis-evaluation-preparation-analysis as a two-step analysis phase; this generalizes to n-step pipelines.

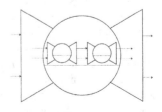

Parallel decomposition. The next form of control is a
parallel flow, which occurs when the input data after
preparation can feed two or more distinct analysis activities, each of which can be embodied by a mega-
module. Each analysis activity may itself require local
preparation and evaluation; therefore the parallel coordination abstraction entails the parallel execution of

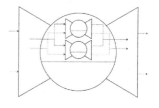

several different mega-modules, with their own preparation and evaluation, in the
context of an enclosing mega-module with one prepare and one evaluation phase.
During the evaluation phase of the enclosing mega-modules, results objects from each
internal mega-module are composed to generate global result objects, according to a
variety of possible mechanisms, e.g. they can be sorted into a list of result objects, can
be put in one-to-one correspondence and then returned identically or composed.

Map-reduce decomposition. The most important aspect
of computation over big data is the support of parallelization, that consists of solving a complex problem over
large data sets by parallel execution of less complex
problems, typically over portions of the initial data set.
The most popular parallelization paradigm, map-reduce
[14], consists of a mapping abstractions that assigns
portions of the computation to distinct processing units,

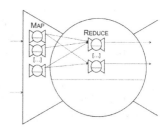

each generating a local result, and a reduction abstraction that is capable of computing
a global result from the local results. In terms of mega-modeling, the map part should
occur during data preparation and should manipulate input data and patterns into a
format that exposes input objects suitable for parallelization; the reduce part should
occur during data analysis and produce global output objects from independently
computed result objects, while the evaluation is responsible of result post-processing
and presentation to users. The ability of algorithm designers consists of expressing
computations by understanding how they can be twisted to expose both their map and
reduce components.

Specific Composition Abstractions. The following are special cases of composition
abstractions that we have recognized as typical analytical processes in existing systems.

What-if control. A classical way of mining big data is to
explore many alternative solutions that would occur for
different choices of initial setting of models and/or parameters. Essentially, this control abstraction is a form
of iteration driven by an analytical goal, allowing to
repeat a mega-module under different parameterizations
of input data and patterns, *until* a final analytical result
is obtained, which possesses a desired level of, e.g.,

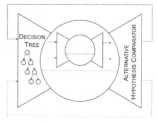

quality, precision or statistical significance; the preparation phase can be modeled by
a decision tree. Many possible instances of this "what-if" iteration control may be

envisage, pertaining to many existing alternatives for exploring a space of patterns/models studied, e.g., in machine learning, data mining, statistical physics and (agent-based) simulation.

Drift control. Many mega-module computations are based upon the validity of underlying assumptions. Thus, if the assumptions cease to be valid, the mega-module itself must be invalidated, and then either corrected or abandoned. For instance, a credit risk predictor used by a bank for granting mortgages may become obsolete as an effect of an economic crisis that impact household incomes. The phenomenon of "drifting" describes the progressive invalidation of assumptions under which a model has been learned

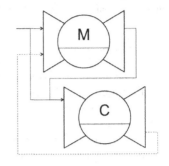

from data. A mega-module M, which is potentially subject to drifting, should be paired to an associated drift-control mega-module C, which assumes the output of M as input. The controller normally has no output, however if it perceives that the drift has occurred, then it interacts with M, by providing suitable controls.

Component-based graph decomposition. Many mega-module computations apply to input data representing (large) networks and graphs and this makes parallelization difficult; if instead a graph has modular structure of components (namely sub-networks) with high intra-module connectivity and relatively low inter-module connectivity, then a natural parallelization can be achieved, by

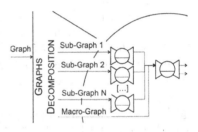

mapping of each sub-network to an internal mega-module and then integrating the results using one additional combination mega-module. Interestingly, the emerging field of network science has demonstrated that many complex networks, and notably social networks, exhibit a non-random structure with a tendency of nodes to cluster into tightly knit communities (many edges among nodes in the same community), while different communities are bridged by relatively few long-range edges: a structure that account for the small-world phenomenon (any two persons are only a few hops away from each other in the social network) and for the tendency of social triangles to form (two persons sharing common friends have a higher chance of becoming friends of each other) [18,19,20]. As a consequence, many real networks from different domains (social, web-related, biological, technological) can be partitioned into components or communities, and a wide variety of methods for community detection has flourished [21]. Treating each sub-network independently can solve a modular sub-problem in a parallelizable way; arcs connecting the sub-networks are then either disregarded or considered by a separate mega-module, which produces computed objects to be assembled and evaluated at the end. Community structure is therefore also a way of separating a micro-scale analytical process (within a community) and a meso- or macro-scale analysis (among different communities, taking each community as a node in a higher level network).

4   Data Management through Schemata and Patterns

Data-intensiveness is the characteristic feature of big data analytics, therefore it is not surprising that data and patterns are given an essential role.

4.1   Mega-schema

We define a *Scientific Data Experiment* as a unitary data research experience over a scientific big data collection, where the experiment can possibly span through years, over different laboratories, involving a huge number of data sources, each with massive amount of data. Then, we make the assumption that all the data used in the input and output schemas of the experiment must be conformant to a unique **mega-schema**.

Methodologically, the global schema design should be the first task in setting a scientific data experiment. This step should be heavily influenced by the agreed ontologies of the domain under study: schema names should reflect the agreed names as codified in the ontology. Ontology-driven schema design and annotation methods are already in use in many scientific communities, e.g. medicine and biology, and we advocate that a similar trend should take place in all cases of big data analytics.

We do not make assumption on the specific mega-schema syntax – candidates are relational, XML, and RDF – and semantics – candidates are ER, XML Schema, and OWL. Establishing a mega-schema in a context of conflicting/pre-existing data and computing resources requires the availability of a variety of data conversion tools, due to the intrinsic heterogeneous nature of those resources; we assume the a-priori establishing of one mega-schema, forcing each data source to be conformant using a "global as view" (GAV) mapping, with the strong belief that such practice will result, in the long run, beneficial also in terms of the overall data conversion complexity.

4.2   Patterns Organization

Patterns describe data regularities, which are sometimes exposed at data definition and in most times inferred as result of data analysis and mining; their format is domain-independent, therefore the "schema" of patterns reflects the underlying structure of the discovery that the scientific experiment should produce, rather than its input and output data. We are henceforth assuming that there exists a finite number of pattern structures capable of describing all the forms of regularity that are worth extracting through a given scientific data experiment. We assume patterns to describe large numbers of Items, all with the same format; Items are structured objects with a schema, and can be typed (ItemType is a label of the type). Patterns can be described by means of type constructors with Items and numerical attributes expressing their properties (either exact or approximate). **Fig. 2** lists examples of patterns from a broad spectrum of problems, although at a very shallow level of details, as each pattern should be further specialized by considering the specific data analysis experiment.

CLASSIFICATION. The computation extracts classes from a population based on some classification algorithm operating upon its property, and then computes statistics – from simple frequencies up.

    Data: Population(Item)

    Pattern: Class(Name, AggrStats)

CLUSTERING. The computation operates upon a collection and extracts its clusters, where each cluster has a name, an extent consisting of its elements, possibly a centroid element, and then statistics – from cardinalities up.

    Data: Collection(Item)

    Pattern: Cluster(Name, Extent: [Item], CentroidItem, AggrStats)

STREAMING. The computation aggregates data from a stream, by aggregating those items of a given type and within a given time interval, typically the most recent, and then computing aggregate properties.

    Data: Stream(TimeStamp, Item)

    Pattern: StreamStats(ItemType, TimeInterval, AggrStats)

STREAMING WITH WINDOWS. The computation aggregates data from a stream which is subdivided in windows, by aggregating within each window those items of a given type and then computing aggregate properties.

    Data: Stream(Window, StartTimeStamp, EndTimeStamp, Content:[Item])

    Pattern: WindowedStats(Window, ItemType, AggrStats)

TREE. Classical computations provide the descendants or ancestors of a given node, or classify a new node relative to an existing hierarchical taxonomy, e.g. by showing the path from the root to the node which is most similar to the new node.

    Data: Tree (Item, Descendent, ChildItem)

    Pattern: Descendants(Item, To: [Item])

                    Ancestors(Item, From: [Item])

                    Classify (Item, Path[Item])

GRAPH. Classical computations provide a decomposition of a graph into components by minimizing the edges which interconnect nodes of different components, or find the "friend" nodes which are at a given "nearness" from a given node.

    Data: Graph(FromItem, ToItem)

    Pattern: Components(Name, Components: [Node])

                    Friends(FromItem, NearnessLevel, To: [Item])

DISTANCE-GRAPH. If the graph includes a label expressing node distances, a classic computation find the shortest path between any two items, exptessed as a sequence of nodes connecting them and a totaldistance.

    Data: D-Graph(FromItem, ToItem, Distance)

    Pattern: ShortPath(OriginItem, DestinationItem, Path: [Item], TotalDistance)

ASSOCIATION RULES. A classical data mining problem is to find items which appear together within a basket; an association rule has an head and a body describing item sets, and then statistical properties of support and confidence defining the rule's interest.

    Data   Basket(Item)

    Pattern: Rule(Head:[Item], Body:[Item], Support, Confidence)

MOVING POINTS. When a system accumulates indications of positions from individuals, a classical computation is the reconstruction of the trajectories, i.e. the sequence of locations which are traversed by the same item from an initial location to a final location.

    Data: Point(Item, Time, Location)

    Pattern: Trajectory(Item, FromLocation, ToLocation, Steps:[Location], StepCount: Number)

FLOCKS. When a system accumulates multiple trajectories, another classical computation is the combining of trajectories together to recognize flocks, i.e. movements of groups of individuals across regions.

    Data: Trajectory(Item, FromLocation, ToLocation, Steps:[Location], StepCount: Number)

    Pattern: Flock(FlockName, FromRegion, ToRegion, Objects: [Items], ObjectCount: Number)

Fig. 2. Examples of patterns from a broad spectrum of problems

5   Examples

5.1   Parallel and Pipe

As an example of usage of parallel and pipe composition we illustrate BOTTARI [15] – an augmented reality application for personalized and localized recommendations of points of interest, experimentally deployed for recommending restaurants in the Insa-dong district of Seoul. At a first look, it may appear like other mobile apps that rec-ommend restaurants, but BOTTARI uses inductive and deductive stream reasoning [16] to continuously analyze social media streams (specifically Twitter) to understand how the social media users collectively perceive the points of interest (POIs) in a given area, e.g., Insadong's restaurants.

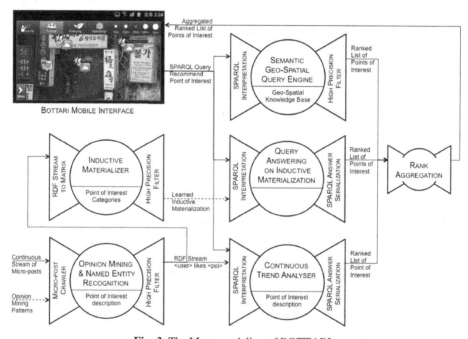

Fig. 3. The Mega-modeling of BOTTARI

BOTTARI was developed on LarKC platform [17], but the plug-able nature of LarKC and its orientation to scientific workflows, allows to cast BOTTARI in the Mega-modeling framework. In **Fig. 3**, we present a redesign of BOTTARI's interac-tion workflow that is twisted towards the mega-modules concept. The MICRO POST CRAWLER continuously streams 3.4 million tweets/day related to Seoul in the OPINION MINING & NAMED ENTITY RECOGNITION analytical sub-module that, know-ing all Insadong restaurants, identifies the subset related to the Insadong restaurants (thousands per day) and detects the users' opinions. The result is an RDF stream of positive, negative and neutral ratings of the restaurants of Insadong. It flows at an average rate of a hundred tweets/day, peaking at tens of tweets/minute. The stream is processed in real-time by the INDUCTIVE MATERIALIZER and by the CONTINUOUS TREND ANALYZER. The former transform the data in a matrix (users x restaurants) and incrementally maintain its inductive materialization using the internal knowledge about restaurant categories. The latter incrementally identifies the top restaurants in the week, month, quarter, etc. Whenever a user ask BOTTARI for a recommendation, a SPARQL query is sent to three Mega-modules: SEMANTIC GEO-SPATIAL QUERY ENGINE, which returns a list of restaurants that matches the semantic criterion requested by the user (e.g., traditional cuisine) ordered by distance from the user; the QUERY ANSWERING ON INDUCTIVE MATERIALIZATION, which uses the patterns identified by the INDUCTIVE MATERIALIZER to return a list of restaurants order by the probability that the user likes them; and the CONTINUOUS TREND ANALYZER, which return a list of restaurants ordered by number of positive ratings in a given time window (e.g., the last week). A RANK AGGREGATOR mega-module combines the lists and returns the recommendations to BOTTARI interface.

5.2   Mega-modelling of a Proactive Car-Pooling Service

A more complex analytical process is depicted in **Fig. 4**, which supports the intelligent, proactive car-pooling system of [22]. At the level of each individual mobile users, trajectories corresponding to actual trips are reconstructed, and then clustered to find the (systematic) routines of the user, e.g., her home-work-home commutes or frequent trips for bring-get activities, such as bringing kids to school. For each cluster a typical representative trip is extracted through a parallel decomposition, and a concept-drift mechanism is used to monitor when user's routine change, e.g., as a result of moving or the arrival of a new-born baby. The individual routines are, then, aggregated at collective level, according to two different purposes: to find the patterns of systematic mobility and study the city access paths (bottom right), and to find best matches among pairs of commuters with systematic routines, that can be put in contact for car-pooling. In our study over the city of Pisa, we found that there is a car-pooling potential of 67% (i.e., every routine has a 67% chance of being served from a matching routine of another user, see [22]), showing that car-pooling systems based on big data analytics have potentially a high impact in reducing systematic traffic.

5.3   Drift Control

The mega-model in Fig. 5 depicts a system that leverage the analysis of tourist routes in an area of interest for the twofold purpose of understanding aggregated touristic behavior and recommending popular routes to individual tourists. Trajectories recording tourist visits are reconstructed at the individual level and then aggregated; trajectory pattern mining is then used to discover the frequently followed routes. The discovered patterns are delivered both to the tourism authorities for analytical purposes and to the individual tourists in form of recommended popular routes. A concept-drift composition is used to monitor the validity of the discovered patterns against the incoming stream of tourist trajectories, as well as the emergence of novel patterns.

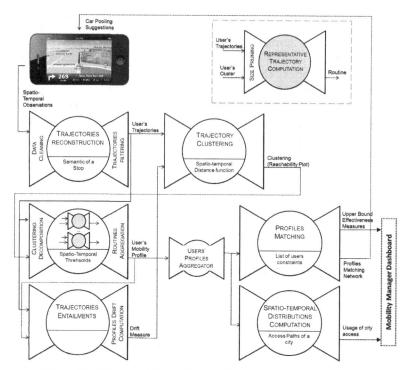

Fig. 4. The Mega-modeling of an intelligent car-pooling service

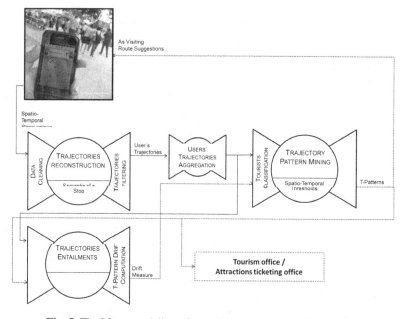

Fig. 5. The Mega-modeling of a touristic recommendation service

5.4   Component-Based Graph Decomposition

An example of component-based graph decomposition is depicted in Fig. 6, based on the idea of discovering the geographical borders delimiting the real mobility basins dictated by the big data of human mobility [23].

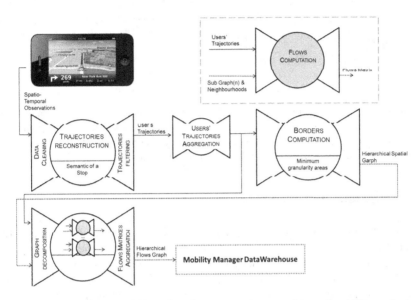

Fig. 6. The Mega-modeling of a micro-macro decomposition of mobility analysis

The key mega-module here is the *border* computation, illustrated in Fig. 7. Starting from a given zoning of the territory, tessellated into census zones, the preparation phase constructs a network whose nodes are the zones and the weighted edges between any two zones represent the number of travels originating in the first and ending in the second. The analysis phase consists in discovering densely connected sub-graphs in this network by means of a community detection method, thus highlighting groups of zones that are highly connected by many travels compared to the lower connectivity among different modules. The evaluation phase consists in mapping these modules back to geography, and drawing appropriate borders to delimit the different modules. The borders mega-module is a means to highlight a hierarchical structure in a large network, thus separating a micro level (within each separate module) and a macro-level (where each module is abstracted to a node and the links among different modules are considered). The mega-modeling of Fig. 6 then exploit this decomposition for a parallel (or map-reduce) computation over the separate modules of mobility analysis, such as origin-destination flow matrices or more sophisticated pattern mining, as in previous examples.

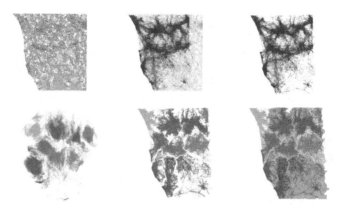

Fig. 7. The steps of the *borders* computation (from top left to bottom right): tessellation into census zones; construction of the flow network between any two zones; forgetting geography; discovery of modules (communities) in the network; mapping of communities back to geography; delimitation of modules so as to discover the borders dictated by human mobility [23]

6   Conclusions and Future Work

This paper introduces the concept of mega-modelling and some abstractions for mega-model composition; its objective is to raise the interest of the community of scientific big data processing on model composition and reuse. Our approach is very preliminary, and needs formalization and extensions, e.g. each model should be associated with a meta-model explaining the model's syntax and behaviour, and therefore mega-modelling should be associated with suitable languages for expressing meta-model aspects. In a broad vision, meta-models should be indexed and searchable, and model interoperability should be supported, so as to build a world-of-modelling platform, capable of supporting the creation of complex analytical and simulation processes by composing knowledge discovery mega-modules. Different language and visual interfaces to mega-modelling should also be devised, so as to empower different classes of users with varying levels of expertise (data scientists, social scientists, policy makers, ordinary citizens) with appropriate means for creating realistic and understandable simulations and what-if analyses. In our future work, we plan to further develop the mega-modelling abstractions in the context of large inter-disciplinary projects such as FuturICT and Genomic Computing.

References

1. Bishop, S., Helbing, D.: FuturICT Project Summary, http://www.futurict.eu
2. Hey, T., Tansley, S., Tolle, K. (eds.): The Fourth Paradigm. Data-Intensive Scientific Discovery. Microsoft Research (2009)
3. Haas, P.J., Maglio, P.P., Selinger, P.G., Tan, W.-C.: Data is Dead ...Without What-If Models. In: Proceedings of the Very Large Data Bases Endowment, PVLDB, vol. 4(12) (2011)

4. Wiederhold, G., Wegner, P., Ceri, S.: Towards Mega-Programming. ACM Communications 35, 11 (1992)
5. Bezivin, J., Journault, F., Valduriez, P.: On the need for Megamodels. In: OOPSLA 2004/GPCE Workshop
6. Favre, J.-M., Nguyen, T.: Towards a Megamodel to Model Software Evolution Through Transformations. Electr. Notes Theor. Comput. Sci. 127(3), 59–74 (2005)
7. Schmidt, D.C.: Model-Driven Engineering. IEEE Computer 39(2), 25–31 (2006)
8. Seibel, A., Neumann, S., Giese, H.: Dynamic Hierarchical Megamodels: Comprehensive Traceability and its Efficient Maintenance. Software and System Modeling 9(4), 493–528 (2010)
9. Imielinski, T., Mannila, H.: A Database Perspective on Knowledge Discovery. Communication of the ACM 39(11), 58–64 (1996)
10. Blockeel, H., Goethals, B., Calders, T., Prado, A., Fromont, E., Robardet, C.: An Inductive Database System Based on Virtual Mining Views. Data Mining & Knowledge Discovery 24(1), 247–287 (2012)
11. Giannotti, F., Nanni, M., Pedreschi, D., Pinelli, F., Renso, C., Rinzivillo, S., Trasarti, R.: Unveiling the Complexity of Human Mobility by Querying and Mining Massive Trajectory Data. The VLDB Journal 20(5), 695–719 (2011)
12. Franklin, M.J., Kossmann, D., Kraska, T., Ramesh, S., Xin, R.: CrowdDB: Answering Que-ries with Crowdsourcing. In: Proc. ACM-Sigmod, Athens (June 2011)
13. Bozzon, A., Brambilla, M., Ceri, S.: Answering Search Queries with Crowdsearcher. In: Proc. WWW 2012, Lyon (April 2012)
14. Dean, J., Ghemawat, S.: Mapreduce: Simplified data processing on large clusters. In: Operating Systems Design and Implementation (USDI 2004), pp. 137–147 (2004)
15. Celino, I., Dell'Aglio, D., Della Valle, E., Huang, Y., Lee, T., Park, S., Tresp, V.: Bottari: an Augmented Reality Mobile Application to deliver Personalized and Location-based Recommendations by Continuous Analysis of Social Media Streams. J. Web Semantics (to appear, 2012)
16. Barbieri, D.F., Braga, D., Ceri, S., Della Valle, E., Huang, Y., Tresp, V., Rettinger, A., Wermser, H.: Deductive and Inductive Stream Reasoning for Semantic Social Media Analytics. IEEE Intelligent Systems 25(6), 32–41 (2010)
17. Assel, M., Cheptsov, A., Gallizo, G., Celino, I., Dell'Aglio, D., Bradesko, L., Witbrock, M., Della Valle, E.: Large Knowledge Collider: a Service-oriented Platform for Large-scale Se-mantic Reasoning. In: Proc. WIMS 2011 (2011)
18. Watts, D.J., Strogatz, S.H.: Collective dynamics of 'small-world' networks. Nature 393, 440 (1998)
19. Barabasi, A.L., Albert, R.: Emergence of scaling in random networks. Science 286, 509 (1999)
20. Easley, D., Kleinberg, J.: Networks, Crowds, and Markets: Reasoning About a Highly Connected World. Cambridge University Press (2010)
21. Coscia, M., Giannotti, F., Pedreschi, D.: A classification for community discovery methods in complex networks. Statistical Analysis and Data Mining 4(5), 512–546 (2011)
22. Trasarti, R., Pinelli, F., Nanni, M., Giannotti, F.: Mining mobility user profiles for car pooling. In: KDD 2011, pp. 1190–1198 (2011)
23. Rinzivillo, S., Mainardi, S., Pezzoni, F., Coscia, M., Pedreschi, D., Giannotti, F.: Discovering the Geographical Borders of Human Mobility. KI - Künstliche Intelligenz (2012)

Spatial Keyword Querying

Xin Cao[1], Lisi Chen[1], Gao Cong[1], Christian S. Jensen[2], Qiang Qu[2],
Anders Skovsgaard[2], Dingming Wu[3], and Man Lung Yiu[4]

[1] Nanyang Technological University
[2] Aarhus University
[3] Hong Kong Baptist University
[4] Hong Kong Polytechnic University

Abstract. The web is increasingly being used by mobile users. In addition, it is increasingly becoming possible to accurately geo-position mobile users and web content. This development gives prominence to spatial web data management. Specifically, a spatial keyword query takes a user location and user-supplied keywords as arguments and returns web objects that are spatially and textually relevant to these arguments. This paper reviews recent results by the authors that aim to achieve spatial keyword querying functionality that is easy to use, relevant to users, and can be supported efficiently. The paper covers different kinds of functionality as well as the ideas underlying their definition.

1 Introduction

The sales of Internet-worked mobile devices such as smartphones are skyrocketing. According to the market research firm IDC, some 490 million smartphones were sold worldwide in 2011, a year-on-year increase of nearly 60%. During the first quarter of 2012, worldwide smartphone sales were 145 million, and expectations for 2012 in the range from about 650 to about 690 million units have been reported. Sales are expected to reach 1 billion by 2016, which will be about half of the mobile device market [12].

As a result of this development, the web is accessed increasingly by mobile users. In recognition of this development, Google announced in 2010 that the company would now develop services "mobile first," meaning that services are developed first for mobile devices and users and only then adapted to desktop devices and users.

Next, we are witnessing a proliferation of geo-positioning capabilities. Smartphones, navigation devices, some tablets, and other mobile devices are equipped with GPS receivers. The Galileo global satellite navigation system, which will provide more and better services than currently offered by the GPS system, is scheduled to become operational by the end of the decade.

Other available positioning technologies exploit the communication infrastructures used by mobile devices, e.g., Wi-Fi, 3G, and 2G. Stated briefly, this can be achieved by building and maintaining a so-called location fingerprint database of pairs of a ground truth location and the base stations and cell towers seen by

P. Atzeni, D. Cheung, and R. Sudha (Eds.): ER 2012, LNCS 7532, pp. 16–29, 2012.

the radios in a mobile device from that location. It is then possible to assign a location to a device when the device reports the base stations and cell towers seen from its location. This type of technology may be used for both outdoor and indoor positioning, but offers less accurate positioning than does GPS.

As a result of the developments outlined above, a spatial, or geographical, web is emerging where content and users are associated with locations that are used in a wide range of location-based services. Billions of queries are being processed by web search engines. According to one report, Google processed a daily average of 4.7 billion queries in 2011 [20]. A substantial fraction of these have local intent and target so-called spatial web objects or places, i.e., points of interest with a web presence that have locations as well as textual descriptions. One article reports that 53% of mobile searches on Bing have local intent [21]. An older, PC-centric finding from Google is that 20% of Google searches are related to location [10].

A prototypical spatial keyword query takes a user location and user-supplied keywords as arguments and returns objects that are spatially and textually relevant to these arguments. Due perhaps to the rich semantics of geographical space and the importance of geographical space to our daily lives, many different kinds of relevant spatial keyword query functionality may be envisioned. A range of contributions are already available in the literature that study different aspects of spatio-textual querying (e.g., [5, 6, 8, 14, 16, 18, 25–27]).

This paper discusses contributions by the authors that aim to improve the support currently available for web querying with local intent. In doing so, it describes the general setting of spatial web querying. It surveys fundamental spatial web querying functionality as well as examples of advanced functionality. The objective is to provide the reader with an overview so that interested readers may follow pointers to additional, in-depth coverage of the different functionalities covered. The paper purposefully does not cover implementation and performance aspects, but rather adopts a focus on semantics and aims to explain the ideas and rationale underlying the functionality it covers.

Section 2 presents the setting, covering the modeling of users and content, the modeling of the space in which users and content are embedded, notions of spatial distance and text similarity, and spatial web object ranking functions. Section 3 covers standard and moving spatial keyword querying functionality. It then covers techniques that aim to capture and exploit the benefits of co-location to compute better query results and techniques that allow users to retrieve sets of spatial web objects that jointly satisfy a query. Finally, Section 4 discusses some of the many challenges yet to be addressed in full.

2   Problem Setting

2.1   Content and Users

The formalization of the problem setting is simple. The content that is queried is a set of objects $\mathcal{D}$ that are available on the web and that we often call *spatial web objects* or *places*. A place $p \in \mathcal{D}$ has two attributes: $\langle \lambda, \psi \rangle$, where λ encodes an accurate geo-location and ψ is a text value.

Increasing volumes of places, i.e., pairs of a location and a text document, are available on the web. Websites for businesses, e.g., bars, cafes, restaurants, banks, dentists, doctors, pharmacies, tourist attractions, hotels, shops, public offices, typically list street addresses. And more and more businesses are acquiring a web presence. Further, business directories list places with descriptions that include location and text, in addition to, e.g., photos, phone numbers, and opening hours. Indeed, there are good reasons for businesses to appear in business directories. A prominent one is that they may then occur more prominently in web query results.

In addition, research is being conducted that aims to associate meaningful locations with web pages based on analyses of the text content on the web pages. This will further increase the volumes of places.

Another type of content that conforms to the assumed format is geocoded microposts, e.g., tweets posted on Twitter. While a micropost carries different semantics than what we associate with a place, spatial keyword querying can also be applied to such content.

We assume that users also have known geo-locations. We also generally assume that the users are mobile. Thus, typical users include individuals with smartphones, tablets, laptops, or similar mobile devices. In the work covered here, we assume that the users have accurate geo-locations, as can be provided by GPS. However, in other settings, it may be assumed that the location of a user is inaccurate and is known only at the granularity of a country, city, or city block. For example, the Google latitude API supports "accurate" and city-level granularities [9]. Or it may be approximated as the region associated with an internet service provider or the service region of a communication cell tower.

2.2   Spaces, Positions, and Distances

The geo-location of a place is modeled as a point location in the space that is used for the modeling of geographical space. We typically assume that geographical space is modeled as two-dimensional Euclidean space. We then typically use Euclidean distance, denoted as $\| \cdot \cdot \|$, as the distance notion. We may also use squared Euclidean distance in cases where only relative distance is important because it is simpler than Euclidean distance. These modeling choices are helpful because they are simple and allow us to abstract away the intricacies of how to actually capture geo-locations.

However, we may note that it is also possible to model space as some form of spatial network [24]. Spatial networks are often used for the modeling of transportation networks such as road networks. They are relevant because it is often reasonable to view movement as being constrained to a transportation network and because places are often reachable via such a network.

Spatial network models come in different variations. The simplest may be that of a regular undirected or directed graph where each vertex has a Euclidean point location and where each edge has a weight that captures the edge's lengths.

However, much more accurate spatial network models exist. For example, models may associate polylines, possibly at different levels of detail [11], with the edges to capture the embedding of roads into geographical space. In addition, a model may capture traffic lanes and traffic regulations [19].

In a spatial network, the position of an object is often given by an edge and a distance from the start of the edge, or it is simply assumed that objects can only be located at vertices, in which case a position is given by a vertex.

In the context of spatial network models, the relevant distance notion is that of spatial network distance. The distance from one object to another is then the length of the shortest possible path from the source object to the target object. In the context of spatial networks, travel time may also be considered as the relevant distance notion. In that case, travel times, sometimes time varying, are associated with edges. It may also be relevant to use a notion distance that is based on the assignment of eco weights to edges, thus capturing the greenhouse gas emissions of vehicles that traverse the edges.

Substantial infrastructure is available for associating places with locations and vice-versa. Internet Yellow Pages and online business directories (e.g., Bing Business Portal, Google Places, and Yahoo! Local) offer listings that associate businesses with street addresses. A geocoder is a service that typically maps a street address to a geo-location, often a latitude-longitude pair (corresponding to a two-dimensional Euclidean point location). A reverse geocoder is then the opposite: it maps a geo-location to a street address.

Finally, we observe that it is becoming increasingly relevant to also consider indoor spaces [13]. In that case, locations may be symbolic, e.g., room names or numbers, and indoor walking distance may be used as the distance notion [15].

2.3   Text Content and Relevance

The second attribute of a place is a text value or a document. We generally assume that this value is a simple, unstructured text string. As we shall see in detail later, queries also take arguments that are (shorter) text strings and that we call query keywords. Just like we need a means of quantifying the distance between two positions, we need a means of comparing the relevance of one text value (of a place) to another text value (typically of a query). To this end, we can use one of several state-of-the-art techniques.

Specifically, with p being a place and $q = \langle \lambda, \psi \rangle$ being a prototypical query that takes both a location and a text value as arguments, we let the function $tr_{q.\psi}(p.\psi)$ denote the text relevancy of $p.\psi$ to $q.\psi$. The text relevance $tr_{q.\psi}(p.\psi)$ can be computed using an information retrieval model, such as a language-based model [7] or the vector space model [3], which we review next.

The text relevance between term t and $p.\psi$ can be computed by a language model [17] as:

$$\hat{p}(t|\theta_{p.\psi}) = (1 - \lambda)\frac{tf(t, p.\psi)}{|p.\psi|} + \lambda\frac{tf(t, Coll)}{|Coll|} \tag{1}$$

Here, $tf(t, d)$ takes a term t and a document d as arguments and returns the number of occurrences of the t in d; $Coll$ is the document that consists of the collection of all documents associated with places in $\mathcal{D}$; $tf(t, d)/|d|$ is the maximum likelihood estimate of t in d; and λ is a smoothing parameter of the Jelinek-Mercer smoothing method. Finally, text relevance for the language model approach is computed as follows:

$$tr^1_{q.\psi}(p.\psi) = \prod_{t \in q.\psi} \hat{p}(t|\theta_{p.\psi}) \tag{2}$$

The text relevance can also be computed by the vector space model [28] as follows:

$$tr^{vs}_{q.\psi}(p.\psi) = \frac{\sum_{t \in q.\psi \cap p.\psi} w_{q.\psi,t} w_{p.\psi,t}}{W_{q.\psi} W_{p.\psi}}, \text{ where } w_{q.\psi,t} = \ln(1 + \frac{|Coll|}{f_t}),$$

$$w_{p.\psi,t} = 1 + \ln(tf(t, p.\psi)), \ W_{q.\psi} = \sqrt{\sum_t w^2_{q.\psi,t}}, \ W_{p.\psi} = \sqrt{\sum_t w^2_{p.\psi,t}} \tag{3}$$

Here f_t is the number of objects whose text descriptions contain the term t, and $tf(t, p.\psi)$ is the frequency of term t in $p.\psi$; $w_{p.\psi,t}$ corresponds TF and $w_{q.\psi,t}$ corresponds to IDF.

Finally, it may be observed that much text content available on the web has structure that may be exploited for computing text similarity. This then calls for new definitions of relevance.

2.4 Place Ranking

So far, we have seen that places have both a geo-location and a textual description, and we have discussed means of comparing locations with locations and textual descriptions with textual descriptions. The prototypical query $q = \langle \lambda, \psi \rangle$ takes both a location and a text argument and retrieves places that are in some sense relevant to both arguments. As a foundation for defining a variety of queries, we then need a means of determining how well a place matches a query. Intuitively, a place whose text description is more relevant to the query's text argument and whose location is closer to the query location is preferable.

In our work, we have considered two general ways of accomplishing this. First, we can rank a place with respect to a query as follows.

$$rank^1_q(p) = \alpha \frac{\|q.\lambda \ p.\lambda\|}{maxD} + (1 - \alpha) \frac{tr_{q.\psi}(p.\psi)}{maxP} \tag{4}$$

This definition builds on and combines the earlier definitions of distance and text relevance. It normalizes each by the maximum possible distance and relevance so that each becomes a value between 0 and 1. It also introduces a parameter α that makes it possible to balance the importance of the two aspects.

Second, we may use a function that weights the geographical distance by the text relevance [1].

$$rank^2_q(p) = \frac{\|q.\lambda \ p.\lambda\|}{tr_{q.\psi}(p.\psi)}, \tag{5}$$

With this function, the smaller a value of $rank_q(p)$ a place p gets, the more relevant it is to query q, and the higher it is ranked.

When setting α to 0 in Definition 4, the definition degenerates to textual relevance as used by a search engine; and setting α to 1, the definition considers only distance and degenerates to what is used in a standard nearest neighbor query. On the other hand, we are also left with the problem of providing an appropriate value for α—it is unlikely to be a good idea to ask users to set α. Definition 5 has the advantage that no parameter needs to be set, but then it of course also does not offer the control provided by the parameter.

The two definitions of ranking take into account two aspects, or signals. A commercial web search engine is likely to take into account many more signals, PageRank being a well-known one. In our setting, directions queries (users who ask for directions from a place a to a place b) can be used for the ranking of places [22]. The general idea is that a query for directions to place b is a vote that place b is interesting. As an example of further aspects of the problem, if the distance from place a to place b is long, this may be taken as an indication that place b is more interesting than if the distance is short. Another type of user-generated content, namely GPS traces, can also be used for the ranking of places [2]. It is easy to envision how additional signals can be accounted for by the introduction of additional terms in Definition 4.

3   Spatial Keyword Queries

3.1   Standard Queries

Based on the setting provided in the previous section, we proceed to describe what may be considered as standard spatial keyword queries. These all involve different conditions on the spatial and textual aspects of places. In spatial databases, the arguably most fundamental queries are range queries and k nearest neighbor queries. In text retrieval, queries may be Boolean, requiring results to contain the query keywords, or ranking-based, returning the k places that rank the highest according to a text similarity function as discussed in Section 2.3.

The *Boolean range query* $q = \langle \rho, \psi \rangle$, where ρ is a spatial region and ψ is a set of keywords, returns all places that are located in region ρ and that contain all the keywords in ψ. Variations of this query may rank the qualifying places. The *Boolean kNN query* $q = \langle \lambda, \psi, k \rangle$ takes three arguments, where λ is a point location, ψ is as above, and k is the number of places to return. The result consists of up to k places, each of which contain all the keywords in ψ, ranked in increasing spatial distance from λ.

Next, the *top-k range query* $q = \langle \rho, \psi, k \rangle$, where ρ, ψ, and k are as above, returns up to k places that are located in the query region ρ, now ranked according to their text relevance to ψ. Finally, the *top-k kNN query* takes the same arguments as the Boolean kNN query. It retrieves k objects ranked according to a score that takes into consideration spatial proximity and text relevance, as discussed in Section 2.4.

Among these queries, the latter two ones that perform textual ranking are the most similar to standard web querying, and the last one is the one that is most interesting and novel, as it integrates the spatial and textual aspects in the ranking. This query, also called the *top-k spatial keyword query*, is exemplified in Figure 1.

Figure 1(a) shows the locations of a set of objects $\mathcal{D} = \{p_1, p_2, p_3, p_4\}$. Assume that we are given the query q shown in Figure 1(a) with $q.\psi = \langle a, b \rangle$ and $q.k = 2$. The underlined number next to each object is its text relevancy to the query keywords $q.\psi$. These are computed using a text relevancy function $tr_{q.\psi}(p.\psi)$. The result of query q is $\langle p_2, p_3 \rangle$ according to function $rank_q^2(\cdot)$. The ranking values of p_2 and p_3 are 0.478 (= 0.22/0.46) and 0.54 (= 0.26/0.48), respectively.

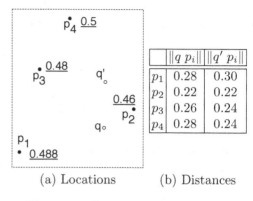

	$\|q\,p_i\|$	$\|q'\,p_i\|$
p_1	0.28	0.30
p_2	0.22	0.22
p_3	0.26	0.24
p_4	0.28	0.24

(a) Locations   (b) Distances

Fig. 1. Top-k Spatial Keyword Query

3.2   Moving Queries

The queries considered above are one-time queries: a query is issued and the result is computed and returned to the user; end of story. This is in contrast to a continuous query that is a one-time query that is registered with the system and is later deregistered. As long as the query remains registered, the system must send a new result to the user each time the result changes. In our setting, contributions have so far been assumed that set $\mathcal{D}$ is static, so no changes occur.

A moving query is akin to continuous queries. Specifically, a *moving top-k spatial keyword query* takes the normal arguments ψ and k, but then it also takes as argument a location λ that changes continuously and is intended to capture the user's location [23]. The system must now send the user a new result every time the result changes due to changes in λ.

The state-of-the-art approach to handling such queries is to utilize safe zones. When the result of a query is computed, the system also computes a safe zone, which is a region with the property that as long as λ remains inside the region, the result is guaranteed not to change. This region is sent to the user (to the part of the system on the user's mobile device) along with the result. Now the client side can monitor λ and notify the server side when λ is about to exit the safe zone, upon which the story repeats itself.

Returning to the example in Figure 1, when q moves to q', the result becomes $\langle p_2, p_4 \rangle$. The ranking value of p_2 and p_4 is 0.478 and 0.48, respectively.

Readers familiar with spatial databases may know that the safe zone for a standard nearest neighbor query is a cell in a Voronoi diagram. It turns out that if we adopt the ranking function in Equation 5, the safe zone for our problem is a cell in a kth order multiplicatively weighted Voronoi (mwV) diagram. In our setting, a place that is highly relevant to a query may belong to the query result even if it is located relatively far from the query. Thus, the safe zone for a query result consisting of such a place is a large region. In addition, the safe zone may have "holes" in it that are caused by the presence of places that are less relevant but nearer to the query.

Figure 2(a) shows an mvV diagram and a safe zone for a query q with $k = 1$. The underlined values next to the places are their text relevancies to the query. The safe zone is the region that both q and the result p_1 belong to. While mwV

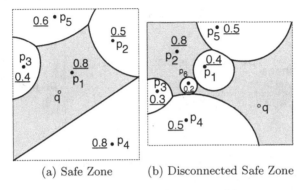

(a) Safe Zone   (b) Disconnected Safe Zone

Fig. 2. Multiplicatively Weighted Voronoi Diagrams

diagrams fit the problem perfectly, their use also raises questions. For example, Figure 2(b) shows a case where the safe zone of the result of a query q ($k = 1$; i.e., p_2) is disconnected, which means that it is impossible for the user to reach p_2 without another place becomes a better result along the way. So if users issue continuous queries for places with the intention of visiting the places, is seems becoming to explore other definitions of the ranking.

3.3 Accounting for Co-location

We often see that similar businesses locate near each other. Examples include restaurant districts (e.g., China Town, Little Italy), bar districts, shopping streets, markets and bazaars (e.g., farmer's markets, antiques markets), and regions with a concentration of car sales businesses. This phenomenon seems to suggest that businesses benefit from locating near similar businesses. A possible explanation may be that this affords consumers easy access to a larger selection of products and services. For example, if no shoes in a store are attractive to a consumer, the presence of nearby shoe stores is desirable. Thus, we may expect that concentrations of similar businesses attract proportionally more consumers than do isolated stores. Put differently: co-located stores attract more customers to such an extent that this compensates for the increased competition.

The problem is then how to integrate the benefits of co-location into the ranking of places in top-k spatial keyword queries. To solve this problem, we have developed an approach that exploits a query centric notion of prestige. Given a query, a place is assigned prestige that takes into account the presence of nearby places that are also relevant to the query [3]. Then the prestige of a place is used for the ranking.

To illustrate prestige-based object ranking, consider the query "shoes" at location q in Figure 3 ($k = 1$). Circles represent shops selling shoes or jeans, with the centers representing the locations and the areas representing the relevancies to the query. Techniques that treat objects independently (e.g., [7]) return p_5

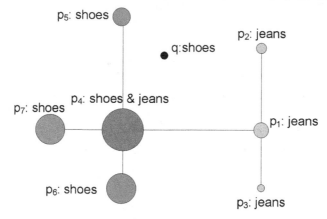

Fig. 3. Prestige Propagation

as the result, since p_5 is relevant and closest to q. However, when using prestige for ranking, p_4 becomes the result because it has more nearby shops that are relevant to the query and also is close to the query location.

We also note that the notion of prestige involves mutual reinforcement: the prestige of one place is affected by the prestige of nearby places. Therefore, we employ a PageRank-like random walk mechanism for the propagation of prestige. The graph that is used is created by connecting places with edges if the places are sufficiently close and similar.

The use of prestige also has the benefit that even if the text of a place does not contain any of the query terms, the place can still be identified as relevant. This occurs if the place has a description that matches those of nearby objects that do contain the query terms. See the places in the figure with text "jeans."

3.4   Beyond the Single-Object Result Granularity

A trend in general web querying is to produce more interesting and richer results than simply lists of blue links with search snippets. For example, a query on google.com for "Caroline Wozniacki" returns not only links, but also information on the performance of tennis star Caroline Wozniacki in the most recent tournament she participated in, a photo of her, and basic information about her such as birth date and place. It also lists names and photos of other tennis stars that people (who search for Caroline Wozniacki) also search for.

Our setting offers new opportunities for departing from the blue links where the granularity of a result is a single link or place. First, queries exist where no single, nearby place is a good result by itself, but where several places near each other in combination constitute a good result. Put differently, the places collectively meet the need expressed in the query, while none of the individual places meet need. For example, a "hotel, gym, theater" query may have its best result a pair of a theater and a hotel with a gym that are located next to each other and are close to the query.

Next, note that in the above example, a user would be happy with a hotel with both a gym and theater. In other cases, the user is directly interested in retrieving collections of places that are near each other. Specifically, the user may want to find a *region* of a specified size such that the places in the region meet the user needs in a better way than individual objects. For example, users would prefer to do shopping in a region with many shops.

We proceed to consider spatial keyword querying that targets the above two situations.

Collective Queries. To address the first situation, we consider a query q that takes a user location λ and a set of keywords ψ as arguments. Its search space is all subsets of the set of places $\mathcal{D}$. It returns a set of places such that (i) the textual descriptions of the objects collectively cover ψ, (ii) the result places are all close to λ, and (iii) the result places are close to each other.

These three conditions represent three different aspects of meeting the user's needs, and different instantiations of each condition may be preferable in different application scenarios. We have studied two instantiations of the query [4].

1. Find a group of objects χ that cover the keywords in q such that the sum of their spatial distances to the query is minimized.
2. Find a group of objects χ that cover the keywords in q such that the sum of the maximum distance between an object in χ and query q and the maximum distance between two objects in χ is minimized.

The first instantiation focuses on the distance between the query and each result object. For example, this may fit a scenario where the user is at a hotel and wants to visit nearby places that offer different services (e.g., workout in a gym, get a haircut, watch a movie) but wishes to return to the hotel in-between services. The second instantiation considers all three aspects. This may fit a scenario where the user wishes to visit all the places before returning.

Other types of instantiations exist. For example, we can relax the keyword matching requirement to allow partial matching, and we can consider other distance measures (e.g., the sum of distances) to replace the maximum distance measure. Note also that the collective query reduces to a standard spatial keyword query when individual objects can cover all the query keywords.

Region Queries. Next, we consider the situations where the user is explicitly interested in retrieving a collection of places.

Consider an example where a user wants to buy "blue jeans and fashion shoes" and wants to explore the offerings of different stores that are within walking distance. In this example, the user is interested in a region that is densely populated with stores that sell "blue jeans and fashion shoes." The following are important aspect of the query formulation: (i) the size and shape of the region, (ii) the relevance of the stores in the region to the query, (iii) the number of relevant stores in the region, and (iv) the density of the relevant stores in the region.

We call this a *relevant region query*. We instantiate the query by providing precise definitions for the four aspects. An instantiation should be well defined, should be useful from a user perspective, and should be computationally tractable.

With this in mind, we might consider square regions of fixed size (i.e., both the size and shape of returned regions are fixed). For the second aspect, we can use any information retrieval model for computing relevance scores, and with such scores available, it is straightforward to provide a definition for the third aspect. Finally, to define density, one option is to use the ratio of the area of the region to the number of relevant objects in the region, while another option is to use the average pairwise distance between relevant objects in the region.

Finally, note that the relevant region query reduces to a standard spatial keyword query when the specified region size is sufficiently small.

4   Outlook

In spatial keyword querying, Google-style keyword-based querying is enhanced to offer better support for queries with local intent by exploiting accurate locations associated with the users and content that their queries target. Here, we have surveyed recent results obtained by the authors. We hope that the coverage convinces the reader that the general research topic is interesting and that it may be possible to obtain results in this area that can have significant impact.

Research on spatial web querying has just begun, and there are many opportunities for continued research. Some were already suggested. Here, we proceed to discuss three opportunities and then three challenges that may also help direct future research efforts.

Structured Queries. While we believe that a major reason for the success of Google-style web querying is the simplicity of the interface and the ease and speed with which one can obtain useful results, it may be of interest to study ways of adding structure to spatial web querying.

For example, queries may include a transportation mode attribute that can be set to, e.g., walking, bicycling, public transport, or driving. In many cases, it may be possible to detect the value of this attribute so that it needs not be set explicitly by the user. With such an attribute, it may be possible to deliver more relevant results in a number of use cases.

Next, a scope or region-of-interest may be included that can contribute to improving query results. Opportunities also exist for inferring a setting for this attribute without involving the user: the relevant scope of a query may be inferred from the user's zoom level, or it may be inferred from the query itself. A query for a convenience store is likely to have a smaller than a query for, say, Ikea.

Amazon-Style and Social Queries. Quite generally, there is a great potential for computing very personalized results. Using the functionality of the Amazon.com website as inspiration, themes such as those illustrated next deserve further study:

Using personal history and the current location: given the places you have visited, you may be interested in the following nearby places: . . .

Using the histories of others at the current location: people who were at your location visited . . .

Using the histories of similar others and the current location: when people, who visited places that you have visited were at your current location, they visited . . .

It is also generally true that the recommendations and behaviors of individuals we know carry more significance to us than do those of strangers. Thus, there is great potential for integrating social network functionality into the above.

Feedback Mechanisms. According to Tim O'Reilly, "Figuring out how to build databases that get better the more people use them is the secret source of every web 2.0 company." This is a very important insight. In addition, services should learn and adapt to the preferences of their users. Indeed, web search engines rank results according to how people link from one page to another. And folklore has it that search engines observe which links in query results the users click. This offers a mechanism for improving. For example, if users frequently click the fifth link in the result of a particular query, this suggests that the fifth link should be placed higher in the result of that query.

Similarly, we need mechanisms that observe user actions in response to the results they receive. For example, we could observe the actual movement behaviors of users and relate these to the results we provided to the users.

Avoiding Parameter Overload. As a research topic matures and the most obvious functionality, we often see a move towards exploring more complex functionality that calls for the introduction of problem parameters.

The top-k spatial keyword query (Section 3.1) took three parameters: λ (location), ψ (keywords), and k (result cardinality), with only λ being new in comparison to standard keyword querying. Because it can be assumed that λ can be obtained automatically, this query is just as simple to formulate as a standard query. But then we introduced a parameter α for balancing the influence of spatial proximity and text relevance in the first of the two ranking functions for spatial keyword queries that we covered (Equation 4). And it is easy to imagine a more sophisticated ranking function that takes into consideration additional aspects (Recall, e.g., the relevant region query in Section 3.4.).

But then additional parameters may also be needed to control and balance the aspects.

The introduction of parameters that are exposed to the users or that are simply hard to set comes at a cost. Stated very generally, the relevance of a proposal often decreases rapidly as the number of parameters increases.

User Evaluation. A good first step when trying to improve something is to be able to reliably measure that something.

When we develop, say, a new join algorithm, the result is well defined. So in an empirical study of such an algorithm, we simply need to verify that the implementation computes the right result and can then focus on aspects such as the runtime or I/O performance that are fairly easy to measure. Furthermore, empirical studies can often be done meaningfully with synthetic data.

In the context of spatial keyword querying, things are more difficult. First, we typically do not merely aim for a correct result, but rather aim for a result that is as useful as possible for an important use case. But utility depends on the users. Thus, we need to determine how useful users will find a result. It is a complex challenge to establish a reliable ground truth for the result of a query. This is also a challenge that systems-oriented database researchers are not well equipped to address. And it is a challenge that it is costly to address. We feel that much can be gained from working with more user-focused scientists on evaluations (in addition to other aspects). Second, we observe that many problems within this research topic are difficult to address meaningfully without real data. In sum, when considering a new problem, two of the first questions to ask are how the results are to be evaluated and whether real data is available.

Tractability Versus Utility. On the one hand, we aim to support the functionality that (we think) users want. On the other hand, we need to ensure that the functionality that we provide can be supported efficiently. In spatial keyword querying, this tradeoff between tractability and utility of functionality is particularly prominent. For example, queries that return sets of places (recall Section 3.4) may benefit from considering all subsets of places and thus easily become NP hard. It is an interesting challenge to manage this tradeoff.

References

1. Aurenhammer, F., Edelsbrunner, H.: An optimal algorithm for constructing the weighted Voronoi diagram in the plane. Pattern Recognition 17(2), 51–57 (1984)
2. Cao, X., Cong, G., Jensen, C.S.: Mining significant semantic locations from GPS data. PVLDB 3(1), 1009–1020 (2010)
3. Cao, X., Cong, G., Jensen, C.S.: Retrieving top-k prestige-based relevant spatial web objects. PVLDB 3(1), 373–384 (2010)
4. Cao, X., Cong, G., Jensen, C.S., Ooi, B.C.: Collective spatial keyword querying. In: SIGMOD, pp. 373–384 (2011)
5. Chen, Y.-Y., Suel, T., Markowetz, A.: Efficient query processing in geographic web search engines. In: SIGMOD, pp. 277–288 (2006)

6. Christoforaki, M., He, J., Dimopoulos, C., Markowetz, A., Suel, T.: Text vs. space: efficient geo-search query processing. In: CIKM, pp. 423–432 (2011)
7. Cong, G., Jensen, C.S., Wu, D.: Efficient retrieval of the top-k most relevant spatial web objects. PVLDB 2(1), 337–348 (2009)
8. Felipe, I.D., Hristidis, V., Rishe, N.: Keyword search on spatial databases. In: ICDE, pp. 656–665 (2008)
9. Google: Google latitude API (2012), http://developers.google.com/latitude
10. Google: Google Places. Stats & Facts (2012), sites.google.com/a/pressatgoogle.com/googleplaces/metrics
11. Hage, C., Jensen, C.S., Pedersen, T.B., Speicys, L., Timko, I.: Integrated data management for mobile services in the real world. In: VLDB, pp. 1019–1030 (2003)
12. IDC: Smartphone statistics and market share (2012), www.email-marketing-reports.com/wireless-mobile/smartphone-statistics.htm
13. Jensen, C.S., Lu, H., Yang, B.: Graph model based indoor tracking. In: MDM, pp. 122–131 (2009)
14. Li, Z., Lee, K.C.K., Zheng, B., Lee, W.-C., Lee, D.L., Wang, X.: IR-tree: an efficient index for geographic document search. TKDE 23(4), 585–599 (2011)
15. Lu, H., Cao, X., Jensen, C.S.: A foundation for efficient indoor distance-aware query processing. In: ICDE, pp. 438–449 (2012)
16. Lu, J., Lu, Y., Cong, G.: Reverse spatial and textual k nearest neighbor search. In: SIGMOD, pp. 349–360 (2011)
17. Ponte, J.M., Croft, W.B.: A language modeling approach to information retrieval. In: SIGIR, pp. 275–281 (1998)
18. Rocha-Junior, J.B., Gkorgkas, O., Jonassen, S., Nørvåg, K.: Efficient Processing of Top-k Spatial Keyword Queries. In: Pfoser, D., Tao, Y., Mouratidis, K., Nascimento, M.A., Mokbel, M., Shekhar, S., Huang, Y. (eds.) SSTD 2011. LNCS, vol. 6849, pp. 205–222. Springer, Heidelberg (2011)
19. Speicys, L., Jensen, C.S.: Enabling location-based services—multi-graph representation of transportation networks. Geoinformatica 12(2), 219–253 (2008)
20. Statistic Brain: Google Annual Search Statistics (2012), http://www.statisticbrain.com/google-searches/
21. Search Engine Land: Microsoft: 53 percent of mobile searches have local intent (2012), searchengineland.com/microsoft-53-percent-of-mobile-searches-have-local-intent-55556
22. Venetis, P., Gonzalez, H., Jensen, C.S., Halevy, A.: Hyper-local, directions-based ranking of places. PVLDB 4(5), 290–301 (2011)
23. Wu, D., Yiu, M.L., Jensen, C.S., Cong, G.: Efficient continuously moving top-k spatial keyword query processing. In: ICDE, pp. 541–552 (2011)
24. Zeiler, M.: Modeling our World—The ESRI Guide to Geodatabase Design. ESRI Press (1999)
25. Zhang, D., Chee, Y.M., Mondal, A., Tung, A.K.H., Kitsuregawa, M.: Keyword search in spatial databases: Towards searching by document. In: ICDE, pp. 688–699 (2009)
26. Zhang, D., Ooi, B.C., Tung, A.K.H.: Locating mapped resources in web 2.0. In: ICDE, pp. 521–532 (2010)
27. Zhou, Y., Xie, X., Wang, C., Gong, Y., Ma, W.-Y.: Hybrid index structures for location-based web search. In: CIKM, pp. 155–162 (2005)
28. Zobel, J., Moffat, A.: Inverted files for text search engines. ACM Comput. Surv. 38(2) (2006)

Of Cubes, DAGs and Hierarchical Correlations: A Novel Conceptual Model for Analyzing Social Media Data

Umeshwar Dayal, Chetan Gupta, Malu Castellanos,
Song Wang, and Manolo Garcia-Solaco

Hewlett Packard Labs, USA
{Umeshwar.Dayal,Chetan.Gupta,Malu.Castellanos,
songw,Manolo.Garcia-Solaco}@hp.com

Abstract. With the advent of social media there is an ever increasing amount of unstructured data that can be analyzed to obtain insights. Two prominent examples are sentiment analysis and the discovery of correlated concepts. A convenient representation of information in such scenarios is in terms of concepts extracted from the unstructured data, and measures, such as sentiment scores, associated with these concepts. Typically, social media analysis reports these concepts and their associated measures. We argue that much richer insights can be obtained through the use of OLAP-style multidimensional analysis. It is fairly straightforward to see how to add traditional dimension hierarchies such as time and geography, and to analyze the data along these dimensions using traditional OLAP operations such as roll-up; for instance, to answer queries of the form "What was the average sentiment for X in Europe during the past month?" However, it is trickier to answer queries of the form "What was the average sentiment for concepts related to X in Europe during the past month?" We introduce a conceptual modeling framework that extends traditional multidimensional models and OLAP operators to address the new set of requirements for data extracted from social media. In this model, we organize data along both traditional dimensions (we call these metadata dimensions) and concept dimensions, which model relationships among concepts using parent-child hierarchies. Specifically: (i) we allow operations on parent-child hierarchies to be treated in a uniform way as operations on traditional dimension hierarchies; (ii) to model the rich relationships that can exist among concepts, we extend the parent-child hierarchies to be rooted level-DAGs rather than simply trees; and (iii) we introduce new equivalence classes that allow us to reason with "similar" concepts in new ways. We show that our modeling and operator framework facilitates multidimensional analysis to gain further insights from social media data than is possible with existing methods.

1 Introduction

Business Intelligence (BI) is a set of technologies and tools to support decision making. On-Line Analytical Processing (OLAP) is a key BI technology that enables answering multidimensional analytical queries. The core of an OLAP system is a data cube [6] which consists of numeric facts (measures) categorized by dimensions. Dimensions are organized into hierarchies which facilitate navigating the cube along any combination of dimensions and hierarchy levels to get the corresponding aggregated value of the

P. Atzeni, D. Cheung, and R. Sudha (Eds.): ER 2012, LNCS 7532, pp. 30–49, 2012.
© Springer-Verlag Berlin Heidelberg 2012

measures. A typical scenario is a sales cube where the fact table contains the sales transactions and the dimensions are time and geography. In this case, the hierarchy associated with time consists of year, quarter, month, day and hour, the one associated with geography consists of country, state and city, and then an analyst may want to see the aggregated value of sales by country or by month or by state and quarter. Typical OLAP operations such as rollup and cube facilitate the formulation of SQL queries to obtain these different levels of aggregation along the various dimensions.

The hierarchies associated with dimensions in the traditional OLAP cube are typically well-balanced, with equal length paths, with nodes of different types, and where facts are expressed only at the leaf level. However, with the advent of social media and the ever increasing amount of unstructured data (i.e., text) that it produces, a different set of requirements is imposed on the dimension hierarchies that support OLAP analysis on data derived/mined from text. It is common for a new generation of applications aiming at obtaining insight into what is being said in the online social channel, to mine the unstructured data and extract concepts and some measures about these concepts (e.g., sentiment scores). Typically, social media mining applications only report sentiment scores for individual concepts. We are interested in extending such applications with the ability to do OLAP-style multidimensional analysis. To perform OLAP analysis on this data, these concepts are organized into parent-child hierarchies that are typically unbalanced, with different path lengths from root to leaf nodes, with nodes of the same type, and the measures can be expressed at any level (i.e., leaf and non-leaf levels). Often there is also the need for a node in a hierarchy to have more than one parent, that is, the hierarchies are not restricted to trees anymore, instead, they become direct acyclic graphs (with some restrictions).

We illustrate these new requirements in Example 1 with a scenario from a real use case related to sentiment analysis. In fact, the work presented in this paper was motivated by our interest in extending Live Customer Intelligence (LCI) [3], a social media mining system we are developing at HP Labs, with the ability to do OLAP analysis on sentiment scores. Sentiment analysis has increasingly become a popular kind of social media analysis and it enables companies to listen to what customers think about their products, services and brands, and those of their competitors. In concept-based sentiment analysis concepts extracted from documents (e.g., reviews, tweets, blogs) are mined for their sentiment using a pipeline of operators like the one shown in Figure 1. In particular, each occurrence of a concept is assigned a sentiment score which can be positive, negative or neutral. We treat these scores as the measures that populate the fact table.

Example 1. Let us assume that a corpus of Amazon reviews about a family of IT products has been analyzed and the sentiment about different features of the products have been obtained. The sentiment scores populate the fact table and the products and features populate a dimension table to create the hierarchy shown in Figure 3. In this hierarchy we can observe paths of different length and features belonging to more than one product. Table 1 shows the fact table where we can see sentiment scores about features corresponding to leaf and non-leaf levels in the hierarchy and Figure 2 depicts visually the cube abstraction for the sentiment data.

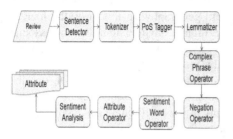

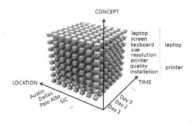

Fig. 1. Sentiment Analysis Pipeline   **Fig. 2.** Schematic of a Sentiment Cube

It is on the sentiment scores in the fact table that we want to do OLAP analysis to obtain richer insights into sentiments. Each score by itself might be insignificant but when aggregating hundreds or thousands of scores a good and meaningful signal is obtained. In particular, doing rollup on the concept hierarchy is a natural operation in analyzing sentiment, as described in Example 2.

Example 2. Given the scenario in Example 1, a common query would be to get the average sentiment for different products in the hierarchy. In particular, to get the average sentiment for "laptop" it would be necessary to roll-up the sentiment associated with all its descendants. This implies that the sentiment scores of each and every occurrence of size, resolution, screen, keyboard and laptop would be aggregated and averaged into a single sentiment score. The existence of this kind of concept hierarchy in a cube does not preclude the use of hierarchies on traditional dimensions (e.g., geography), herein called metadata dimensions. For example, we can aggregate the sentiment for "laptop" by state and month.

However, doing OLAP on cubes containing concept hierarchies with the characteristics described above is a challenge. Current OLAP systems either do not allow non-leaf level facts nor hierarchies with different length paths, or if they do, they force the (manual or automatic) insertion of artificial nodes. Even if the latter is supported, doing OLAP operations (e.g., rollup) on cubes with this kind of hierarchy is complex and cumbersome (see the related works). It is not possible to just use the rollup or cube operators as with regular hierarchies; complex queries need to be formulated and often in a non-SQL language using a separate package, product or tool coupled to a DBMS. It is clear that the need exists for mechanisms that enable a uniform and simple treatment of concept dimensions within SQL. Moreover, we also found the need for more powerful operators to extend the scope of traditional operators (e.g., selection and join) to a subset of related concepts in a single operation rather than just a single concept. We illustrate this in Example 3.

Example 3. Let us assume that the hierarchy in Figure 3 includes two nodes for HP laptops and Dell laptops, each one with the descendants shown in Figure 3 for laptop. And let us also assume that the fact table in Table 1 also includes a column for product model and reviewer id. A useful query would be to compare the sentiment scores of a given HP laptop model with a given Dell laptop model on"similar" features. Such a query becomes handy when the features on which sentiment is expressed for one laptop model are different from those of the other model. For example, we might want to compare the sentiment score for screen of one laptop model with sentiment scores for resolution, screen size, and screen itself for the other laptop model.

To extend the semantics of traditional operators in this way, we define different equivalence classes that allow specifying the scope of the operators within the hierarchies (Section 5).

In a nutshell, in this paper we tackle the limitations of traditional OLAP for social media data cubes. Specifically, our contributions can be summarized as follows:

1. We present a conceptual modeling framework to extend traditional OLAP modeling with more powerful modeling constructs:
   (a) We allow parent-child hierarchies to be treated in a uniform way as any other traditional dimensional hierarchy.
   (b) We allow measures to be associated to nodes at arbitrary levels rather than just leaf nodes and allow different path lengths from root to leaf nodes.
   (c) We further extend the parent-child hierarchies to be rooted level-DAGs rather than simply trees.
2. We extend traditional operators with new equivalence classes that enable us to expand the scope of the operators to k levels up and d levels down in a single operator. Specifically, we propose new hierarchical correlations where similarity of concepts according to their distance along a path is used to determine the scope.

The rest of the paper is organized as follows: In Section 2 we present the basic definitions and in Section 3 we discuss the concept dimension. In Section 4 we discuss the cube construction and cube operation over these hierarchies and in Section 5 we introduce the new operators. In Section 6 we discuss how to implement the new operators with SQL and existing OLAP products and finally, in Section 7 we conclude by discussing our next steps.

2   Basic Definitions

For ease of analysis, social media data is often extracted in form of tuples. The schema of such tuples can be represented by:

$$< doc\text{-}id, \{metadata\text{-}dimensions\}, concept\text{-}dimension, measure > \qquad (1)$$

For the purpose of the paper, of primary interest to us is the *concept-dimension* which is associated with a hierarchy referred to as a *concept-hierarchy*. Besides the concept dimension, a tuple has some *metadata-dimensions*. Each metadata-dimension can be associated with its own hierarchy, referred to as the *metadata-dimension-hierarchy*. Metadata dimensions are the traditional dimensions such as location, time, user-id, etc. *Measure* is some numerical quantity (for example, sentiment score in sentiment analysis) over which we compute some function of interest.

Table 1. A Sample Fact Table

S. No.	Location	Feature	Sentiment
1	Austin	Screen Size	1
2	Austin	Resolution	-1
3	Austin	Keyboard	-1
4	Dallas	Installation	-1
5	Dallas	Laptop	1
6	Palo Alto	Quality	1
7	Palo Alto	Keyboard	1
8	Palo Alto	Screen Size	-1
9	Palo Alto	Printer	-1
10	San Jose	Keyboard	1

Table 2. Concept Dimension

Level1	Level2	Level3	Level4
All	Laptop	Screen	Screen Size
All	Laptop	Screen	Resolution
All	Laptop	Keyboard	
All	Printer	Installation	
All	Printer	Quality	

Table 3. Location Dimension

All	State	City
*	TX	Austin
*	TX	Dallas
*	CA	Palo Alto
*	CA	San Jose

Example 4. To explain various concepts, we will use Table 1, Table 2, and Table 3 as running examples throughout the paper. In these tables we have shown a simplified version of sentiment tuples by considering a metadata-dimension called "Location", a concept dimension called "Feature" and a measure called "Sentiment". Table 1 is the fact table for the extracted sentiments, Table 3 is the dimension table for "Location" and Table 2 is the dimension table for "Feature". Both hierarchies are depicted pictorially in Figure 3.

To analyze social media data, we need to compute aggregates at various levels of abstraction. A systematic way of doing is through the cube abstraction [6], [7]. For our purpose, a cube is defined with the following syntax:

$$define\ cube < cube > [< dimensions >]\{measure\}$$

Where, the *dimensions* definition has the following syntax:

$$define\ dimension < dimension\ name > as(< attribute\_list >)$$

The ($< attribute\_list >$) refers to the labels of the various level of the hierarchy associated with the dimension. In our running example, the location dimension can be defined with: *define dimension* $< location > as(city, state)$ As we mentioned before, the concept hierarchy has no attribute list. One way to get around this is to construct a proxy attribute list, as presented in Table 2, where we artificially name the different levels as 'Level1', 'Level2', and so on. Then, in our running example, we can define the concept dimension as, *define dimension* $< feature > as(Level4, Level3, Level2, Level1)$.

Computing a cube implies computing the individual cuboids. For social media data that follows the schema specified in Equation 1, the number of cuboids $|Cuboids_{SC}|$ is same as that for a generic OLAP cube:

$$Cuboids_{SC} = \prod_{i=1}^{n} L_i + 1 \tag{2}$$

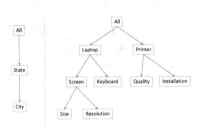

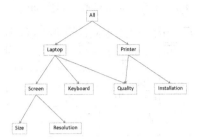

Fig. 3. Figure depicting an example location and concept hierarchy

Fig. 4. Figure depicting extended concept hierarchy

Where n is the total number of dimensions in question and L_i is the number of levels in dimension i. So, in our running example, $|L_{location} = 2|$ and $|L_{feature} = 3|$, and we need to compute $3 * 4 = 12$ cuboids to completely specify the cube. The cuboids would be: { $< City, Level4 >$, $< City, Level3 >$, $< City, Level2 >$, $< State, Level4 >$, $< State, Level3 >$, $< State, Level3 >$, $< City, Level1 >$, $< State, Level1 >$, $< All, Level3 >$, $< All, Level2 >$, $< All, Level1 >$ and $< All, Level1 >$ }.

Now that we have introduced the basic definitions, let's look at the concept dimension in greater detail.

3   Concept Dimension as a Parent Child Hierarchy

The concept dimension is unlike a traditional OLAP dimension in several ways, and presents several challenges. In a traditional OLAP hierarchy every level signifies a different resolution and the hierarchy encodes a refinement relationship. For example, in the location hierarchy ($city \rightarrow state \rightarrow country$), cities are at a finer resolution than a state, and similarly a state is a finer resolution than a nation. On the other hand, in a concept hierarchy, children typically are features of the object in the parent node. For example, for the concept hierarchy shown in Figure 3, the 'Keyboard' is a feature of a 'Laptop'. The concept hierarchy is best modeled as a parent-child hierarchy, and as mentioned earlier can be represented as a of dimension table (for example, Table 2).

Specifically, a concept hierarchy is different from an OLAP hierarchy in the following ways:

1. The hierarchy may contain root to leaf paths of different lengths. In our example concept hierarchy, the path length from root node 'All' to the leaf node 'Screen Size' is 3, whereas the path length from 'All' to the leaf node 'Installation' is 2. This is different from a traditional OLAP hierarchy where the paths from root to any leaf node are of the same length.

2. Another crucial difference is the existence of *dangling tuples*, which are fact tuples that belong to a concept that is not at a leaf level. For example, in Table 1, tuples with serial no. 5 and 9 are dangling tuples, since they express a sentiment about a 'laptop' and a 'printer' respectively which are not leaf level concepts. Specifically,

Table 4. Extended Concept Dimension

Level1	Level2	Level3	Level4
All	Laptop	Screen	Screen Size
All	Laptop	Screen	Resolution
All	Laptop	Keyboard	
All	Printer	Installation	
All	Printer	Quality	
All	Laptop	Quality	

Table 5. Cuboid $< City, Level4 >$

S. No.	Location	Feature	Count	Sum
1	Austin	Screen Size	1	1
2	Austin	Resolution	1	-1
3	Palo Alto	Screen Size	1	-1

a *dangling tuple* is a tuple in the fact table that contains a measure value for a concept that is not at the leaf node for the attribute hierarchy for that dimension. This is different from a traditional OLAP scenario, where all the facts exist at the same (lowest) level of hierarchy.

3. A traditional hierarchy can be naturally expressed in terms of "class labels". For example, in the case of location hierarchy, the labels are $(All, State, City)$. In the case of a concept hierarchy, the class labels need to be artificially defined. For example, for the concept dimension, we artificially give the labels $(Level4, Level3, Level2, Level1)$.

3.1   Extensions to the Parent-Child Hierarchy

In many practical situations, we have to model concept dimensions that are not strict hierarchies, i.e., in which a child concept may have multiple parents. Some concept dimensions are better modeled as *level-DAGs* which can be defined as:

Definition 1. *A level-DAG is an acyclic graph with a root node such that a node u can have multiple parents, and all the parents of the node u are at the same depth from the root.*

Example 5. This extended definition allows us to model hierarchies such as the one given in Figure 4. (We will add Figure 4 to our running example), where the generic 'Quality' concept belongs to both printers and laptops. (To be precise, we should have three nodes, 'Printer-quality', 'Laptop-quality', 'Quality', where 'Quality' refers to a sentiment which is not specific about which quality is being referred) Such a hierarchy can be expressed in the dimension table shown in Table 4. Notice the additional row (the last row) in the table that indicates that 'Quality' has a second parent, 'Laptop'.

For the rest of the paper we will refer to the concept hierarchy with only one parent per node (other than the root node) as a *basic concept hierarchy* and the concept hierarchy which allows for more than one parent as an *extended concept hierarchy*. In the next two sections, we discuss operations over these hierarchies.

4   Cube Construction and Operations

In the last section, we discussed the various types of concept hierarchies and now we discuss some operations over data cubes that are defined over these types of dimension hierarchies. For ease of exposition, we first discuss the semantics of cube computation for the *basic concept hierarchy* and then for the *extended concept hierarchy*. The cube is nothing but a collection of cuboids, hence computing the cube implies computing the cuboids.

4.1   Cube Construction and Roll-Up for Basic Concept Hierarchy

Just as with standard cubes, the cuboids at a higher level in our cube can be constructed by a roll-up operation on the cuboids at the lower level. Roll-up on a cube means either going up the hierarchy in a dimension or removing a dimension. Traditionally, for roll-up by going up a hierarchy, we first need to map the relevant attributes in tuples under consideration to the higher abstraction level followed by an aggregation/group-by operation. For roll-up by removing a dimension, only aggregation/ group-by is sufficient. However, we need to modify the traditional roll-up operator to obtain correct results for the concept hierarchy.

The first difference between a traditional cube and our cube is that, in a traditional data cube, the base cuboid would be the same as the fact table. But in our case, for example, as shown in Table 5, the fact table contains only a subset of tuples (where for the measure we have used a simple sum). For rolling up along metadata dimensions, the results at a higher level are obtained from the lower level cuboid as in traditional cubes, i.e., by first mapping and then aggregating.

For the concept dimension, however, the results at the higher level cannot necessarily be obtained from lower level cuboids. For going up the hierarchy in the concept dimension, we need to check if there are any leaf nodes at the higher level of abstraction and we need to check for the presence of dangling tuples. This requires augmenting the group-by operation with referring back to the fact table. This leads to a novel situation where *cuboids can contain both aggregated and unaggregated data*. In contrast, in a traditional cube, no cuboid other than the base cuboid contains unaggregated data.

Example 6. Continuing with our running example, let us construct the cuboid $< City, Level3 >$ (shown in Table 6). $< Level3 >$ contains leaf nodes, hence to construct $< City, Level3 >$ besides a group-by/roll-up on $< City, Level4 >$, we need to refer back to the fact table. In Table 6, row 1 and row 4 can be obtained from $< City, Level4 >$. Tuples for feature 'Keyboard', 'Installation', and 'Quality', (rows 2-3, rows 5-7) have to be retrieved from the fact table. Further note that, since $< Level3 >$ contains both leaf and internal nodes, the resulting table contains both aggregated and unaggregated results. Now, finally let's construct the cuboid $< City, Level2 >$ (Shown in Table 7). $< Level2 >$ does not contain any leaf nodes but contains dangling tuples, hence to construct $< City, Level2 >$ besides getting results from $< City, Level3 >$, we need to refer back to the fact table. In Table 7, row 3-4 is obtained from the fact table and the rest are obtained from $< City, Level3 >$.

In the above example, of course, *all* the results could have been obtained from the fact table but in most cases that would not be optimal. When a roll-up is performed by removing the concept dimension, similar considerations show up. Continuing with the running example, (i) when we remove $< Level4 >$ from $< City, Level4 >$

Table 6. Cuboid $< City, Level3 >$

S. No.	Location	Feature	Count	Sum
1	Austin	Screen	2	0
2	Austin	Keyboard	1	-1
3	Dallas	Installation	1	-1
4	Palo Alto	Screen	1	-1
5	Palo Alto	Quality	1	1
6	Palo Alto	Keyboard	1	1
7	San Jose	Keyboard	1	1

Table 7. Cuboid $< City, Level2 >$

S. No.	Location	Feature	Count	Sum
1	Austin	Laptop	3	-1
2	Palo Alto	Laptop	2	0
3	Dallas	Printer	1	-1
4	Palo Alto	Printer	2	0
5	San Jose	Laptop	1	1
6	Dallas	Laptop	1	1

Table 8. Cuboid $< City >$

S. No.	Location	Count	Sum
1	Austin	3	-1
3	Dallas	2	0
4	Palo Alto	4	0
5	San Jose	1	1

(Table 5) to get $< City >$ Table 8), we have to look for both dangling tuples and tuples belonging to leaf nodes in the fact table, since higher levels of the concept dimension contain leaf nodes and dangling tuples respectively; (ii) when we remove $< Level3 >$ from $< City, Level3 >$ (Table 6) to get $< City >$, we have to look for dangling tuples in the fact table, since $< Level3 >$ contains dangling tuples; (iii) when we remove $< Level2 >$ from $< City, Level2 >$ (Table 7) to get $< City >$, we don't need to refer back to the fact table since higher level abstraction in the concept dimension $< All >$ contains neither dangling tuples nor leaf nodes.

4.2 Cube Construction for Extended Concept Hierarchy

An extended hierarchy presents additional challenges due to nodes having multiple parents. To ensure correctness of results, it is important to ensure that for a node u with multiple parents, u's contribution to any ancestor is counted only once. So the cuboid $< City, Level3 >$ (shown in Table 6) remains the same as before but the cuboid $< City, Level2 >$ will need to be updated to reflect the fact that 'Laptop' is also a parent of 'Quality'. This is shown in Table 9. Compare the cuboid shown in Table 9 with the cuboid in Table 7. Notice that there is additional sentiment for the 'Palo Alto, Laptop' aggregate, reflecting the fact that 'Quality' has a new parent in 'Laptop' and the dangling tuple at 'Quality' has 'Palo Alto' for location dimension.

There are a few things to be noted. Firstly, since 'Quality' existed at $Level3$, adding another parent only impacts levels higher than $Level3$. Secondly, if we add the counts in Table 9 they add up to 11 which is greater than the size of the fact table. This is acceptable since the actual sentiment count for every concept is as it should be semantically. Going higher in the hierarchy, however, we need to ensure that if any node u is a common ancestor for both of the parents of 'Quality', then at node u, the contribution of 'Quality' should be counted only once. Thus for example, the cuboid $< City >$ should be as shown in Table 10.

Table 9. Cuboid $< City, Level2 >$

S. No.	Location	Feature	Count	Sum
1	Austin	Laptop	3	-1
2	Palo Alto	Laptop	3	1
3	Dallas	Printer	1	-1
4	Palo Alto	Printer	2	0
5	San Jose	Laptop	1	1
6	Dallas	Laptop	1	1

Table 10. Cuboid $< City >$

S. No.	Location	Count	Sum
1	Austin	3	-1
2	Dallas	2	0
3	Palo Alto	4	0
4	San Jose	1	1

Given the above discussion it should clear that to construct correct cuboids with level-DAGs where some nodes have multiple parents we need new methods. Suppose we are computing the roll-up for a node u. Then to get the correct answer we could specify the following set of steps: (i) Compute all the paths from node u to the leaf nodes in the subtree rooted at u (ii) From these paths obtain the set of all the individual nodes. It's easy to see that this procedure is correct since all the descendants of u are being counted, and counted only once. This is the method we demonstrate in the implementation section.

For large concept hierarchies the above method can be cumbersome since it would mean either computing all the descendants in real time for a node u or storing the descendants for all the nodes. To overcome this, we suggest an alternate way. We first enumerate all possible trees induced by the level-DAG. The number of trees that can be induced within a level-DAG is given by Equation 3. Let the level-DAG be H with the root r, the set of nodes be U, and all parents of a node $u \in U$ be given by $p(u)$, the total number of trees induced by H is:

$$|T_H| = \prod_{u \in \{u\}} p(u) \tag{3}$$

The correctness of this is easy to see: while constructing the trees, one can choose any of the parents of a node, and every unique combination of parents gives us one tree. Hence, the total number of trees is the product of the number of parents of a node. In a traditional hierarchy this method of construction and the formula above return one tree - the hierarchy itself. The hierarchy specified in Figure 4 gives us two trees.

Once we have enumerated all the trees induced by the level-DAG H, we are ready to compute the correct result for any node of H. First, let us introduce some notation. Let the set of all the trees induced from H be denoted by T_H. Every tree $T_i \in T_H$ induces a subtree rooted at node u in the hierarchy, for every node u in the hierarchy H. Let this set of subtrees be denoted by u_H, where for every $T_i \in T_H, u_i \prec T_i$. (The symbol $\prec$ is used to denote subtree relationship).

For a node u, we can obtain the answers for a roll-up by aggregating along any of the subtrees rooted at u. The correct answer for a roll-up is however given by the subtree that has the maximum number of nodes. Let $|u_i|$ be the size (number of nodes) of a subtree u_i. Then:

Theorem 1. *The correct result for a roll-up operation for a node $u \in H$, is given by rolling up along the subtree $u_r \in u_H$ such that $r = \arg\max_{i \in u_H}\{|u_i|\}$*

Proof. When computing the result for a roll-up for a node $u \in H$, for correctness we need to ensure that the contribution of all the nodes n in the hierarchy rooted at u is counted only once. Since we enumerate trees, and trees have no cycles, the contribution of no node can be counted twice (since there is only one path from any n to u). For the first part it is easy to see that. since we are enumerating all possible subtrees, there must be at least one subtree rooted at u that contains all the nodes. This subtree surely will have the maximum number of nodes of any subtree rooted at u.

The discussion above points to a straightforward ways of doing rollups for the case of going up the hierarchy, especially when the function computation over measures is a traditional operation such as sum. If a roll-up is performed by removing dimensions other than the concept dimension then the correct way to do roll-up is as discussed in the previous section, i.e., do a rollup as one would with a traditional OLAP hierarchy. If the roll-up is performed by removing the concept dimension then it is a bit tricky. To address this case, we need to compute the cuboid as $< dim\text{-}levels, All >$ where *dim-levels* are the traditional OLAP dimension and level pairs left once the concept dimension is removed.

Now that we have discussed cube construction and roll-ups, we discuss the other cube operations.

4.3 Other Cube Operations

The drill-down, slice and dice operators for cubes containing concepts are essentially the same as for a traditional data cube. If the cuboid at the lower dimensional level is not materialized, then to perform a drill-down, one has to go to the fact table to obtain facts that are at lower levels of abstraction.

A slice operation is essentially a selection of data from a d-dimensional materialized cuboid by selecting $d-1$ dimensions and selecting some value for the dth dimension variable. This result can be obtained through projection from the cuboid. Dice can be understood as repeated slice operations and handled similarly.

Since these operations are performed on materialized cubes, in our case, they behave similarly to the traditional data cubes.

An *iceberg query* is an important query in traditional cube computations. It computes only those cuboids which have more than a certain number of tuples in it. The BUC algorithm [2] is a standard approach to computing the iceberg cuboids. It follows a top-down approach and uses the *a-priori* property that if a cuboid C_1 at a higher abstraction level is not an iceberg cuboid (does not have sufficient number of tuples) then the lower level cuboids that can be obtained by drill down from C_1 cannot be iceberg cuboids. This property holds even in the presence of dangling tuples, uneven path lengths and nodes with multiple parents. This implies that we can compute the iceberg query using the BUC algorithm for cubes with both basic and extended concept hierarchies.

4.4 Discussion

From the sections above it is clear that dealing with the concept dimension in the presence of uneven path lengths, dangling tuples, and multiple parents creates challenges that we need to address. We have made some modeling choices and it is natural to question such choices. Here, we present some alternatives and discuss why they are insufficient.

- We could require all leaves to be at the same level of the hierarchy. This takes care of uneven path lengths but does not address the problem of dangling tuples. Moreover, since all paths share a common root and we have to make all path lengths equal, this implies adding some dummy conceptual levels, for path lengths that are shorter. This adds unnecessary levels and hence creates conceptual problems.
- We could also add a dummy leaf node(s) to ensure that all leaf nodes are at the same level. This again does not address the problem of dangling tuples, but could also unnecessarily increase the complexity of representation. For example imagine a hierarchy where only one leaf node is at level d, and the rest $n - 1$ are at depth $d - 1$. In that case we will need to extend the hierarchy for all the $n - 1$ nodes.
- We could also break up our feature dimension into multiple dimensions, such that each path becomes a separate dimension. This makes the representation sparse and cumbersome. The representation would be even sparser if each row is still to contain only one measure.

5   New Equivalence Classes and Hierarchical Correlations

In the context of concept dimension, besides the traditional cube operations, we also introduce a set of new equivalence operators. These equivalence operators aid in the construction of new equivalence classes. Given a new equivalence operator $\oplus$, the equivalence class of a node u, is the set of all nodes $n \in H$ for which $u \oplus n \equiv TRUE$.

Besides allowing us to create new aggregates over these equivalence classes, these new types of operators allow for retrieval of measure values for related concepts without actually knowing the hierarchy. We use these equivalence classes to construct hierarchical correlations.

5.1   New Equivalence Operators

We first give some notation to help with the definitions of the new equivalence relationships. Given a feature hierarchy denoted by H, let a node u's i^{th} ancestor be denoted by $A(u, i)$, where $A(u, 0) = u$. Similarly, for a node u at depth d, the set of descendants at depth $d + i$ is denoted by $\{D(u, i)\}$, with $\{D(u, 0)\} = \{u\}$. The subtree rooted at node u is denoted by $T(u)$. As previously discussed, in case of hierarchies with multiple parents, construct trees and choose the one with maximum leaf nodes.

Given the schema in Equation 1, every tuple in the fact table is associated with a concept dimension. Let the concept associated with a node u be denoted by $c(u)$ and let the node associated with a concept a be $f(a)$. Finally let the set of concepts belonging to the subtree $T(u)$ rooted at u be $c(T(u))$.

The first equivalence operator is called an *upward path equality*, $\equiv_p^k$, and defined as:

Definition 2. *For two features* $\{a, b\}$, $a \equiv_p^k b$ *iff:* $b \in \cup_{i=0}^{k} c(A(f(a), i))$

In other words, concept b is considered to be $\equiv_p^k$ to concept a, if b is one of kth ancestors of a. For example, in the concept hierarchy discussed before 'Size' $\equiv_p^2$ 'Screen', and 'Size' $\equiv_p^2$ 'Laptop'. This relation is reflexive. It is neither symmetric nor transitive. Note that this relation is valid even in the case of a concept a having multiple parents, where all of a's parents are considered.

The second equivalence operator is called an *upward sub-tree equality*, $\equiv_s^k$, and defined as:

Definition 3. *For two concepts* $\{a, b\}$, $a \equiv_s^k b$ *iff:* $b \in c(T(f(a), k))$

The second equivalence operator means that concept b is considered to be $\equiv_s^k$ to concept a, if b exists in a subtree rooted at the kth ancestor of a. For example, in the concept hierarchy discussed before 'Size' $\equiv_s^2$ 'Screen', 'Size' $\equiv_p^2$ 'Keyboard', etc. This relation is reflexive, antisymmetric and essentially a partial order, similar to a subset relationship.

The third equivalence operator is called an *downward all path equality*, $\equiv_a^k$, and defined as:

Definition 4. *For two concepts* $\{a, b\}$, $a \equiv_a^k b$ *iff:* $b \in \bigcup_{n \in \{D(f(a), k)\}} c(n)$

The third equivalence operator means that concept b is considered to be $\equiv_a^k$ to concept a, if b is upto a kth descendant of a. For example, in the concept hierarchy discussed before 'Laptop' $\equiv_a^2$ 'Screen', 'Laptop' $\equiv_p^2$ 'Resolution', etc. This relation is reflexive.

5.2 Hierarchical Correlations

We now discuss how certain useful queries that do hierarchical correlation are simplified with the use of the new equivalence operators. We add one more functionality, which we call "level", where:

Definition 5. *The operator level, denoted by* $level(f)$*, where* f *is an element of the concept hierarchy returns the depth at which* f *exists in the hierarchy tree,* T.

For example, the level of 'screen' of in our example concept hierarchy is 3. The operator $\equiv_a^k$ is useful in selecting all the tuples that are associated with the subtree rooted at a particular node of the concept hierarchy. Suppose we want to get the average sentiment for all the attributes for 'printer'. Then a simple query such as this would suffice:

```
SELECT AVG(measure)
FROM Table T1,
WHERE 'printer' ≡² T1.concept
              a
```

To compare the sentiments of a particular concept with its parents becomes easy with the use of $\equiv_p^k$. For example:

```
...
WHERE 'printer' a ≡¹ T1.concept
                p
```

To compare the sentiments of a particular concept with its other concepts at the same

Table 11. TableDimension: the Concept Hierarchy used for Implementation Examples

PathId	Level1	Level2	Level3	Level4
1	All	Laptop	Screen	ScreenSize
2	All	Laptop	Screen	Resolution
3	All	Laptop	Keyboard	
4	All	Laptop	Quality	
5	All	Printer	Installation	
6	All	Printer	Quality	

Table 12. TableConcept: the Concept-level mapping used for Implementation Examples

Concept	Level
All	Level1
Printer	Level2
Laptop	Level3
......	......

level becomes easy with the use of $\equiv_s^k$, where k specifies helps specify the subtree to which the concepts that need to be compared belong. For example to compare 'screen' with 'keyboard' :

. . .

```
WHERE 'printer' ≡¹ₛ T1.concept
AND level('printer') = level(T1.concept)
```

Similarly, to compare 'screen' with 'keyboard', 'quality' and 'installation':

. . .

```
WHERE 'printer' ≡²ₛ T1.concept
AND level('printer') = level(T1.concept)
```

Note that we don't need to actually know the hierarchy to construct any of these queries.

6   Design and Implementation

In this section, we first use a set of SQL queries to show the semantics of rolling up in the proposed level-DAG concept hierarchy and then we show how to do some of these operations with a commercial product, Microsoft SSAS.

6.1   SQL Queries

The design of the SQL queries are illustrated through examples based on the concept hierarchy (named TableDimension) shown in Table 11. We also need the concept-level mapping table (named TableConcept) shown in Table 12. The concept-level mapping table can be constructed from the concept hierarchy table. Here we assume that every concept has a unique level, i.e. there is no duplicated concept in the hierarchy.

Example 7. **Roll-Up Operation:** Assume we want to roll-up all the sentiments to the concept "Laptop" and calculate the sum score, we need a set of SQL queries to achieve this.
Q1: Get the level of corresponding concept "Laptop", assume it is "Level2"

```
SELECT level FROM TableConcept WHERE concept = "Laptop"
```

Q2: Filter out other concepts in the hierarchy.

```
SELECT * INTO TempT FROM TableDimension WHERE level2= "Laptop"
```

Q3: Using union to find all the concepts in the subtree and use DISTINCT to remove replication due to DAG structure. Here assume the maximum level in the hierarchy is "level4". The range of the levels is "level2" to "level4".

```
SELECT DISTINCT(Concept) INTO AggrTargetT FROM (
SELECT pathid, level2 AS Concept FROM TempT WHERE level2 IS
NOT NULL
UNION
SELECT pathid, level3 AS Concept FROM TempT WHERE level3 IS
NOT NULL
UNION
SELECT pathid, level4 AS Concept FROM TempT WHERE level4 IS
NOT NULL )
```

Q4: Join AggrTargetT with fact table (FactT) and do aggregation.

```
SELECT Concept, SUM(Sentiment) FROM (
SELECT * FROM AggrTargetT, FactT WHERE AggrTargetT.Concept
= FactT.Feature)
GROUP BY Concept
```

Example 8. **Upward Path Equality Operation:** Assume we want to upward 1 level from the concept "Screen" and calculate the sum score, we need a set of SQL queries to achieve this. Q1: Get the level of corresponding concept "Screen", assume it is "Level3"

```
SELECT level FROM TableConcept WHERE concept = "Screen"
```

Q2: Filter out other concepts in the hierarchy.

```
SELECT * INTO TempT FROM TableDimension WHERE level3= "Screen"
```

Q3: Using union to find all the concepts in the upward subtree and use DISTINCT to remove replication due to DAG structure. Here assume the root level "level1" is not reached. The range of the levels is "level2" to "level3".

```
SELECT DISTINCT(Concept) INTO AggrTargetT FROM (
SELECT pathid, level3 AS Concept FROM TempT WHERE level3 IS
NOT NULL
UNION
SELECT pathid, level2 AS Concept FROM TempT WHERE level2 IS
NOT NULL)
```

Q4: Join AggrTargetT with fact table (FactT) and do aggregation (omitted).

Example 9. **Downward All Path Equality Operation:** Similarly we can move downward the hierarchy. Assume we want to move downward 1 level from the concept "Laptop" and calculate the sum score, we need a set of SQL queries to achieve this. We can see here that roll-up operation is a special case of downward equality operation. Q1: Get the level of corresponding concept "Laptop", assume it is "Level2"

```
SELECT level FROM TableConcept WHERE concept = "Laptop"
```

Q2: Filter out other concepts in the hierarchy.   ·

```
SELECT * INTO TempT FROM TableDimension WHERE level2= "Laptop"
```

Q3: Using union to find all the concepts in the subtree and use DISTINCT to remove replication due to DAG structure. The range of the levels is "level2" to "level3".

```
SELECT DISTINCT(Concept) INTO AggrTargetT FROM (
SELECT pathid, level2 AS Concept FROM TempT WHERE level2 IS
```

```
NOT NULL
UNION
SELECT pathid, level3 AS Concept FROM TempT WHERE level3 IS
NOT NULL)
```
Q4: Join AggrTargetT with fact table (FactT) and do aggregation (omitted).

Example 10. **Upward Sub-tree Equality Operation:** Combining the two previous examples, we can move both upward and downward along the hierarchy. Assume we want to move upward 1 level from the concept "Screen" and downward 2 levels next and finally calculate the sum score, we need a set of SQL queries to achieve this. Q1: Get the level of corresponding concept "Screen", assume it is "Level3"
```
SELECT level FROM TableConcept WHERE concept = ".Screen"
```
Q2: Find all the subtree roots in the hierarchy. Here all subtree roots should be on the same level, "level2".
```
SELECT DISTINCT(level2) AS root INTO RootT FROM TableDimension
WHERE level3 = ''screen" AND level2 IS NOT NULL
```
Q3: Filter out other concepts in the hierarchy by the subtree roots. Here we have two subtree roots: "Laptop" and "Printer" in the RootT.
```
SELECT * INTO TempT FROM TableDimension WHERE level2= "Laptop"
OR level2= "Printer"
```
Q4: Using union to find all the concepts in the subtree and use DISTINCT to remove replication due to DAG structure. The range of the levels is "level2" to "level4".
```
SELECT DISTINCT(Concept) INTO AggrTargetT FROM (
SELECT pathid, level2 AS Concept FROM TempT WHERE level2 IS
NOT NULL
UNION
SELECT pathid, level3 AS Concept FROM TempT WHERE level3 IS
NOT NULL) UNION
SELECT pathid, level4 AS Concept FROM TempT WHERE level4 IS
NOT NULL)
```
Q5: Join AggrTargetT with fact table (FactT) and do aggregation (omitted).

6.2 Using Microsoft SSAS

Firstly for our running example the tables for facts and dimension (location and concept) are created and populated on an SQL Server instance with the concept table stored as a parent-child hierarchy. Next, a connection to from SSAS to the DB instance is established and a view, a cube and its dimensions are created. Hierarchies are created by defining relationships among the attributes of each dimension. Also, relationships between dimensions and fact tables are established at this point.

Query results (e.g., a cuboid computation) can be seen in the build-in browser, or from the SQL Server Management Studio on an MDX window, or in Excel, provided PowerPivot add-in is installed.

Next we illustrate the use of MDX by computing the cuboid for laptop that requires rolling up the sentiment scores of all its descendants. It is important to notice that our concept hierarchies are modeled as parent-child dimensions and that there is no roll-up operator on this kind of dimensions. Instead, MDX needs to be used as shown in the two following MDX queries. (In the queries below, "Concept H Parent" is the parent of a concept, "Table Concept" is the concept table and "Table Location" is the location table.

Example 11. Q1: Rollup on feature Laptop:

```
SELECT NON EMPTY {[Measures].[Sentiment]} ON COLUMNS
FROM (SELECT({[Table Concept].[Concept H Parent].&[laptop]})
ON COLUMNS FROM [ER12])
WHERE ([Table Concept].[Concept H Parent].&[laptop]
```

Q2: Rollup on feature Laptop AND city Austin:

```
SELECT NON EMPTY {[Measures].[Sentiment]} ON COLUMNS
FROM (SELECT({[Table Location].[City].&[austin]}) ON COLUMNS
FROM (SELECT({[Table Concept].[Concept H Parent].&[laptop]})
ON COLUMNS FROM [ER12]))
WHERE    ([Table    Concept].[Concept    H    Parent].&[laptop],
[Table Location].[City].&[austin])
```

6.3 Discussion

Our experience is that, it is complex and cumbersome to implement the desired semantics in a commercial system. Implementing our multi-level DAGS and operational semantics on them doesnt seem to be possible in commercial systems (for example, in SSAS the concept of multiple hierarchies for a same parent-child dimension doesnt exist). For regular dimensions, multiple hierarchies can be supported through the creation of auxiliary tables; however, it is unclear how to implement our semantics of the rollup operation on level-DAGS.

Providing direct support for the operations we have defined would make it easier to do multidimensional analysis over concept hierarchies (and even traditional hierarchies). One way of doing this would be to encode the operations in user defined functions in SQL. (The queries shown in Section 6.1 implement some of these operators.)

7   Related Work

The work presented in this paper is related to work on different topics. In this section we briefly describe some representative works and contrast them with ours.

Trying to understand the sentiment orientation or polarity (positive, negative or neutral) of a subjective element is a hard problem that has been studied by many researchers [12]. The research started from document-level sentiment classification where the goal is to predict the sentiment polarity of the whole document. More recently the focus has shifted to feature-level sentiment classification mainly due to the increasing interest in product reviews where it is not enough to discover whether the review is positive or negative but which aspects of the item being commented are positive and which are negative (products and features correspond to concepts in our sentiment cube). The goal of these approaches is to tackle the polarity problem; we could say they propose techniques to produce the sentiment score facts, but they are not concerned with their subsequent exploitation through OLAP analysis.

Zemanyi, et al. [9], [10] describe a rich conceptual model for OLAP, which includes several of the extensions we have described here: unbalanced hierarchies that allow different length paths, and nodes with multiple parents. They show how to map this conceptual model to a logical model that can be implemented using commercial RDBMSs. However, they do not show how to do roll ups and other OLAP operations over these rich dimensional models. Since our goal is to do OLAP over social media data, a key contribution is the concept dimension, and extended operators over this dimension.

We discuss two previous works that extend the traditional OLAP cube [6] to cope with text. In topic cube [15] the authors propose a topic dimension with a hierarchical topic tree (specified by the analyst) for the text field, and probabilistic distributions as the content measure of text documents in the hierarchical topic dimension. On one hand this allows the analyst to drill-down and roll-up on the text dimension and on the other, to accommodate different kinds of summaries of the content of the text documents. The traditional cube is extended to incorporate the probabilistic latent semantic analysis (PLSA) model so that a data cube carries parameters of a probabilistic model that can indicate the text content of a cell. The end goal is to enable topic modeling for OLAP. In contrast to our work, their topic hierarchies are just like standard hierarchies in the sense that they are well-balanced, with equal-length paths, and measures are associated to all leaf and non-leaf levels since each topic in the tree is mined for each cell in the original data cube (the one created from the standard dimensions). The authors propose an heuristic algorithm to materialize the cube, in particular, to mine topics from documents of one cell in the original data cube by aggregating the word distributions of topics in its subcells as a starting point and then using EM to quickly converge.

The other kind of cube is called Text Cube [8] where IR measures (TF and DF) of the aggregated text data (not stored in cells) are associated to each cell of a cuboid providing users flexibility to aggregate measures for any subset of dimensions. The challenge is how to compute the IR measures for each cell given storage constraints (i.e., how to choose the subset of cells to materialize) and aggregating cells in finer grain cuboids into a coarser grain cells without looking at the original database. The authors introduce a term hierarchy to specify the semantic levels and relationships among text terms where each leaf is a size-1 set containing only one term and the root is the set of all terms; intermediate nodes are generated by two proposed OLAP operators: pull-up that produces a higher term level and push-down (the reverse of push-down). The term hierarchy is balanced and all paths are of same length . All data except text data are considered dimensions, including price and sentiment score, thus, cells aggregate text data according to any combination of (some) dimensions and the term hierarchy allows semantic navigation of the text data. In contrast with these two types of data cubes for text analysis, we introduce the concept dimension and associated OLAP and equivalence operators suitable for rich social media analysis.

Similarity join has been studied by many researchers due to its importance in many applications. Generic similarity/distance functions were discussed in ([5], [13], [4], [1], [14]), including edit distance, set overlap, Jaccard similarity, cosine similarity,

Hamming distance and their variants. Similarity join on strings is also closely related to approximate string matching, an extensively studied topic in algorithm and pattern matching communities. We refer the reader to [11] for an excellent overview of the work as well as additional references. Different from these work on similarity join and approximate string matching, our concept equivalence operations are specifically defined by the concept hierarchy and the corresponding path information is included when considering the distance function for concepts. Moreover, the extension to level-DAGs introduces more complexity for dealing with the concept distance when moving upward or downward.

As a representative of the commercial state of the art, we investigated the use of Microsoft's SQL Server Analysis Services (SSAS). SSAS now supports two models, the original Unified Dimensional Model (UDM), now called Business Intelligence Semantic Model (BISM) Multidimensional, and the new BISM Tabular. We used the BISM Multidimensional for our experiments. It is not trivial to make the system work. SQL Server and SSAS need to be installed and properly connected. In addition, a different language needs to be used to manipulate the cube (MDX and DMX are the most used). In general, the learning curve is steep and the user has to work with two or three different environments simultaneously. BISM multidimensional models can leverage many-to-many relationships (which most OLAP engines do not offer) to present data from different perspectives and transcends the limits of traditional OLAP. These many-to-many relationships can also be used to model multiple hierarchies in a dimension. We believe that our approach is more intuitive for multidimensional analysis over concept hierarchies.

8 Conclusions

In this paper, for the purpose of social media data analysis, we have presented extensions to the traditional OLAP modeling by allowing parent-child hierarchies (included those modeled as level-DAGs) to be treated in a uniform way as any other traditional dimensional hierarchy. Furthermore, we have also proposed new operators to construct new equivalence classes that allow us to reason with "similar" concepts in novel ways.

For future work we are exploring ways to extend the hierarchies beyond level-DAGs to general DAGs, which would be the most meaningful way extension to the modeling of the concept hierarchy. We are exploring how to integrate these ideas with a column-store RDBMS such as Vertica.

References

1. Arasu, A., Ganti, V., Kaushik, R.: Efficient exact set-similarity joins. In: Proceedings of the 32nd International Conference on Very Large Data Bases, VLDB 2006, pp. 918–929 (2006)
2. Beyer, K., Ramakrishnan, R.: Bottom-up computation of sparse and iceberg cube. SIGMOD Rec. 28(2), 359–370 (1999)
3. Castellanos, M., Dayal, U., Hsu, M., Ghosh, R., Dekhil, M., Lu, Y., Zhang, L., Schreiman, M.: Lci: a social channel analysis platform for live customer intelligence. In: Proceedings of the 2011 ACM SIGMOD International Conference on Management of Data, SIGMOD 2011, pp. 1049–1058 (2011)

4. Chaudhuri, S., Ganti, V., Kaushik, R.: A primitive operator for similarity joins in data cleaning. In: Proceedings of the 22nd International Conference on Data Engineering, ICDE 2006 (2006)
5. Gravano, L., Ipeirotis, P.G., Jagadish, H.V., Koudas, N., Muthukrishnan, S., Srivastava, D.: Approximate string joins in a database (almost) for free. In: Proceedings of the 27th International Conference on Very Large Data Bases, VLDB 2001, pp. 491–500 (2001)
6. Gray, J., Chaudhuri, S., Bosworth, A., Layman, A., Reichart, D., Venkatrao, M., Pellow, F., Pirahesh, H.: Data cube: A relational aggregation operator generalizing group-by, cross-tab, and sub-totals. Data Min. Knowl. Discov. 1(1), 29–53 (1997)
7. Han, J., Kamber, M.: Data Mining: Concepts and Techniques. Morgan Kaufmann Publishers (2006)
8. Lin, C., Ding, B., Han, J., Zhu, F., Zhao, B.: Text cube: Computing ir measures for multidimensional text database analysis. In: ICDM 2008, pp. 905–910 (2008)
9. Malinowski, E., Zimányi, E.: OLAP Hierarchies: A Conceptual Perspective. In: Persson, A., Stirna, J. (eds.) CAiSE 2004. LNCS, vol. 3084, pp. 477–491. Springer, Heidelberg (2004)
10. Malinowski, E., Zimányi, E.: Hierarchies in a multidimensional model: from conceptual modeling to logical representation. Data Knowl. Eng. 59(2), 348–377 (2006)
11. Navarro, G.: A guided tour to approximate string matching. ACM Comput. Surv. 33(1), 31–88
12. Pang, B., Lee, L.: Opinion mining and sentiment analysis. Found. Trends Inf. Retr. 2(1-2), 1–135 (2008)
13. Sarawagi, S., Kirpal, A.: Efficient set joins on similarity predicates. In: Proceedings of the 2004 ACM SIGMOD International Conference on Management of Data, SIGMOD 2004, pp. 743–754 (2004)
14. Xiao, C., Wang, W., Lin, X., Yu, J.X.: Efficient similarity joins for near duplicate detection. In: Proceedings of the 17th International Conference on World Wide Web, WWW 2008, pp. 131–140 (2008)
15. Zhang, D., Zhai, C., Han, J., Srivastava, A., Oza, N.: Topic modeling for olap on multidimensional text databases: topic cube and its applications. Stat. Anal. Data Min. 2(56), 378–395 (2009)

Understanding Constraint Expressions in Large Conceptual Schemas by Automatic Filtering

Antonio Villegas, Antoni Olivé, and Maria-Ribera Sancho

Department of Service and Information System Engineering
Universitat Politécnica de Catalunya – BarcelonaTech
Barcelona, Spain
{avillegas,olive,ribera}@essi.upc.edu

Abstract. Human understanding of constraint expressions (also called schema rules) in large conceptual schemas is very difficult. This is due to the fact that the elements (entity types, attributes, relationship types) involved in an expression are defined in different places in the schema, which may be very distant from each other and embedded in an intricate web of irrelevant elements. The problem is insignificant when the conceptual schema is small, but very significant when it is large. In this paper we describe a novel method that, given a set of constraint expressions and a large conceptual schema, automatically filters the conceptual schema, obtaining a smaller one that contains the elements of interest for the understanding of the expressions. We also show the application of the method to the important case of understanding the specification of event types, whose constraint expressions consists of a set of pre and postconditions. We have evaluated the method by means of its application to a set of large conceptual schemas. The results show that the method is effective and efficient. We deal with conceptual schemas in UML/OCL, but the method can be adapted to other languages.

Keywords: Constraints, Large Conceptual Schemas, Filtering.

1   Introduction

The conceptual schema of many real-world information systems is large or very large. General ontologies (like Cyc [1]), metaschemas (like UML superstructure [2]) or information reference models (like HL7 [3]) are also very large. In general, those conceptual schemas or ontologies include many formal constraint expressions (also called schema rules), which are used for defining static or dynamic integrity constraints, derivation rules, default values, pre and postconditions of events and operations, or results of operations.

In large conceptual schemas, human understanding of such expressions is difficult. The problem is not the formal language in which they are written (logic in general, or the OCL in UML [4]), but the fact that the elements (entity types, attributes, relationship types) involved in an expression are defined in different places in the schema, which may be very distant from each other and embedded

P. Atzeni, D. Cheung, and R. Sudha (Eds.): ER 2012, LNCS 7532, pp. 50–63, 2012.

in an intricate web of irrelevant elements for the purpose at hand. The problem is insignificant when the schema is small, but very significant when it is large.

As a very simple example, consider a schema with the n-level specialization hierarchy B_n IsA ... B_i IsA ... B, the binary relationship type $R(b:B,c:C)$ and the attribute $att(C,Integer)$. Assume that in the context of B_n there is the simple constraint expression `self.c.att > 0`. Understanding such expression requires finding entity type B_n (the context) in the schema, moving upwards the n-level hierarchy until the root (entity type B), navigating towards C, and finding attribute att. When n is large and each of the entity types B_n, ... B_i, ... B, C has several attributes and participates in several relationship types, the task of localizing the relevant elements in the constraint expression, and focusing on them, becomes difficult.

The overall framework of the research is that of design science [5], where knowledge and understanding of a problem domain and its solution are achieved in the building and application of a designed artifact. The problem we try to solve is to ease the understanding of constraint expressions in large conceptual schemas. The expressions are written in a constraint language, such as the OCL. The problem is significant because:

- in a conceptual schema there may be many constraint expressions, used as –among others– invariants, derivation rules, and pre/post conditions,
- understanding such expressions is a necessity during their definition, validation, implementation, and maintenance, and
- understanding such expressions is very difficult when the conceptual schema is large, because the elements involved in those expressions are in general scattered throughout the schema, and it is not easy to navigate through the schema in the way implied by the expressions.

In this paper we describe a method that, given a set of constraint expressions and a large conceptual schema, automatically filters the conceptual schema, obtaining a smaller one that contains the elements of interest for the understanding of the expressions. We have evaluated our method by means of its application to a set of large conceptual schemas, consisting of the UML metaschema [2, 6], the Magento [7] and osCommerce [8] schemas from the e-commerce domain, and the schema of the car rental case study known as EU-Rent [9].

As far as we know, this is the first work that filters conceptual schemas with the objective of easing the understanding of constraint expressions. The method is based and extends the work reported in [10], whose aim was to filter a large schema focusing on one or more entity types and to enrich them with other schema elements according to their importance. Here, we focus on one or more constraint expressions and we obtain a schema that contains all elements that appear in them, independently of their importance. In the simple example introduced above, our method would obtain a filtered schema consisting of entity types B_n and C, the binary relationship type $R(b:B_n,c:C)$ and the attribute $att(C,Integer)$. Note that the hierarchy B_n IsA ... B_i IsA ... B does not appear in the filtered schema, and that the first participant of the relationship type R has changed to B_n.

In the literature, there are several techniques and associated tools for the visualization and comprehension of large conceptual schemas or ontologies (see [11–18]). The group of techniques more appropriate for our purposes is the one called focus+context. In these techniques, the user focuses on a single element, and the rest of the elements are presented around it, reduced in size until they reach a point that they are no longer visible. In our approach, we filter the schema with the objective of obtaining a subset of the original one that contains all relevant elements.

The rest of the paper is structured as follows. The next section introduces the basic concepts and notations of the schemas considered in this paper. Section 3 describes our filtering method. The evaluation of the method in terms of effectiveness and efficiency is presented in Sect. 4. Section 5 explains the adaptation of the method to the special case of filtering event types. Finally, Sect. 6 concludes the paper and points out to future work.

2   Conceptual Schemas

In the general case, a conceptual schema $\mathcal{CS}$ contains the following components:

- A a set of entity types $\mathcal{E}$.
- A set of event types $\mathcal{B}$.
- A set of data types and enumeration types $\mathcal{T}$.
- A set of relationship types $\mathcal{R}$. We denote by $R(p_1:C_1, ..., p_n:C_n)$ a relationship type R with participant entity or event types $C_1, ..., C_n \in \mathcal{E} \cup \mathcal{B}$ playing roles $p_1, ..., p_n$ respectively.
- A set of attributes $\mathcal{A}$. We see attributes as binary relationship types. We denote by $attr(C,T)$ an attribute owned by an entity or event type $C \in \mathcal{E} \cup \mathcal{B}$, named $attr$, and whose type is $T \in \mathcal{T}$.
- A set of generalization relationships $\mathcal{G}$ between entity or event types. We denote by $g:(C_i \text{ IsA } C_{i-1}) \in \mathcal{G}$ the generalization relationship g between C_i and C_{i-1}. $\mathcal{G}^+$ is the transitive closure of generalization relationships.
- A set of schema rules $\mathcal{S}$ including integrity constraints, derivation rules, default values, pre and postconditions of events and operations, and results of operations.

In the rest of the paper, we assume that schemas are written in UML/OCL, although the filtering method presented here could be used with schemas and formal constraint expressions written in other languages.

We illustrate the method by means of its application to the Magento[1] e-commerce system [7]. Its conceptual schema (see a snapshot in Fig. 1) contains 218 entity types and 187 event types, with 983 attributes, 319 relationship types, and 165 generalization relationships. Also, Magento's schema defines 61 data types and enumeration types. The set of schema rules includes 480 integrity constraints, 69 pre and postconditions, and 185 derivation rules.

---

[1] Magento e-commerce System http://www.magentocommerce.com/

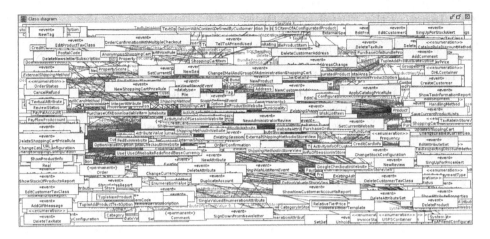

Fig. 1. Graphical representation of the conceptual schema of the Magento system

3 Filtering Conceptual Schemas

The aim of information filtering is to expose users to only information that is relevant to them. There are many filtering systems of widely varying philosophies [18], but all share the goal of automatically directing the most valuable information to users in accordance with their needs, and of helping them use their limited time and information processing capacity most optimally.

In [10] we proposed a method to deal with large conceptual schemas, in which a user focuses on one or more entity types of interest for her task, and the method filters the schema in order to obtain a reduced and self-contained view from it, consisting of the focus and a few schema elements that may be of interest for her based on the automatically computed importance of those elements.

In the method we propose here, the user focuses on a set of schema rules, which may be integrity constraints, derivation rules, or pre/post conditions, and the method obtains the smallest subset of the schema that is needed to understand those expressions. Figure 2 presents an overview of our filtering method.

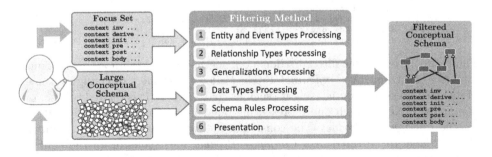

Fig. 2. Method Overview

3.1   Input of the Filtering Method

The components of the input for the filtering method are:

- **Large Conceptual Schema** ($\mathcal{CS}$): It represents the knowledge about a domain of interest. It has the components presented in Sect. 2.
- **Focus Set** ($\mathcal{FS}$): It works as the conceptual schema viewpoint of the user. Formally, the focus set $\mathcal{FS}$ contains one or more constraint expressions selected by the user.

3.2   Output of the Filtering Method

The output of our filtering method is a filtered conceptual schema $\mathcal{CS}_{\mathcal{F}} = \langle \mathcal{E}_{\mathcal{F}}, \mathcal{B}_{\mathcal{F}}, \mathcal{T}_{\mathcal{F}}, \mathcal{R}_{\mathcal{F}}, \mathcal{A}_{\mathcal{F}}, \mathcal{G}_{\mathcal{F}}, \mathcal{S}_{\mathcal{F}} \rangle$. The contents of such conceptual schema depends on the particular user information needs represented in the input of the filtering method. The properties $\mathcal{CS}_{\mathcal{F}}$ must satisfy are:

- **Minimal Subset:** The knowledge contained in $\mathcal{CS}_{\mathcal{F}}$ is the smallest subset of the knowledge of $\mathcal{CS}$ referenced by the constraint expressions of $\mathcal{FS}$. The filtering method does not create new knowledge through the filtering process. The schema elements that appear in the filtered schema come from the original large schema through a process of knowledge extraction.
- **Valid Instantiation:** The filtered conceptual schema is a valid instance of the corresponding metaschema of the original schema from which it is obtained. Thus, the filtered schema is syntactically correct since the elements included within it are concrete instances of metaclasses of that metaschema.

3.3   Stages of the Filtering Method

The filtering method is divided into six ordered stages that sequentially process the input specified by a user in order to obtain the corresponding filtered conceptual schema. As an example, we assume that the user needs to understand the following integrity constraint defined in the Magento, in the context of the entity type *ConfigurableProduct*:

```
context ConfigurableProduct
   inv: -- is associated to products with the same attributes
      self.associatedProduct->forAll(ap |
        ap.ableToRateAttribute->includesAll(self.configurableAttribute))
```

Figure 3(a) presents the filtered schema our method will obtain for the previous integrity constraint. A *ConfigurableProduct* allows customers to configure some of its attributes when purchasing it (such as color, size...). It is related to a set of *ConfiguredProducts*, each one representing one concrete available configuration for the *ConfigurableProduct*. The constraint indicates that the set of attributes that are able to be rated in a *ConfiguredProduct* must contain all the configurable attributes of its *ConfigurableProducts*. The additional constraint included in the filtered schema only references elements in $\mathcal{CS}_{\mathcal{F}}$ and indicates that the

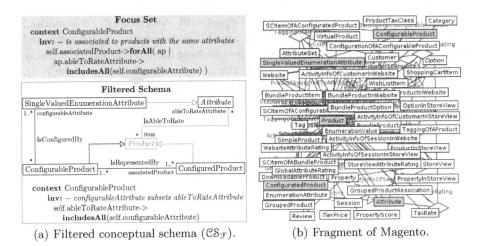

(a) Filtered conceptual schema ($\mathcal{CS}_{\mathcal{F}}$).   (b) Fragment of Magento.

Fig. 3. Comparison between (a) the filtered conceptual schema and (b) the corresponding fragment of Magento for the constraint of *ConfigurableProduct*

configurable attributes of a *ConfigurableProduct* must be included in its set of attributes that are able to be rated.

The size of the filtered schema is smaller than the fragment of the Magento concerning the elements of entity types referenced by the schema rule of focus, which is depicted in Fig. 3(b). Note that it also contains 193 attributes (not shown in Fig. 3(b)), of which 33 are owned by entity types in $\mathcal{CS}_{\mathcal{F}}$. By using our method, a user does not need to manually explore the schema in Fig. 3(b) in order to extract the required elements to understand the schema rules of focus.

Next subsections present a detailed description of the stages of the filtering method, and their application to obtain the filtered schema shown in Fig. 3(a).

Stage 1: Filtering of Entity and Event Types. The method firstly extracts the entity and event types referenced by the selected schema rules of $\mathcal{FS}$ and includes them in the resulting filtered conceptual schema. Formally,

$$\mathcal{E}' = \{e \in \mathcal{E} \mid \exists s \, (s \in \mathcal{FS} \wedge s \rightsquigarrow e)\},$$
$$\mathcal{B}' = \{b \in \mathcal{B} \mid \exists s \, (s \in \mathcal{FS} \wedge s \rightsquigarrow b)\}$$

We say that $s \rightsquigarrow x$ whenever $s \in \mathcal{FS}$ and x is (a) the context entity or event type of s, (b) an entity or event type being the target participant of a relationship type in a navigation expression of s, or (c) an attribute, entity, event, enumeration, or data type explicitly referenced in s.

Considering the integrity constraint from the running example (see Fig. 3.3), $\mathcal{E}'$ consists of the entity types *ConfigurableProduct* (context of the constraint), *ConfiguredProduct* (target participant of `self.associatedProduct` navigation), *Attribute* (target participant of `ap.ableToRateAttribute` navigation), and *SingleValuedEnumerationAttribute* (target participant of `self.configurableAttribute` navigation).

Stage 2: Filtering of Relationship Types and Attributes. The method extracts the relationship types and the attributes referenced by the selected schema rules of $\mathcal{FS}$ and includes them in the resulting filtered conceptual schema. For the case of relationships, we define the set of referenced relationship as, $\mathcal{R}' = \{R(p_1{:}C_1, ..., p_n{:}C_n) \in \mathcal{R} \mid \exists s, p_i{:}C_i \ (s \in \mathcal{FS} \wedge s \rightsquigarrow p_i{:}C_i \wedge p_i{:}C_i \in R)\}$.

Considering the running example, the method extracts the following three referenced relationship types for $\mathcal{R}'$:

```
IsRepresentedBy(configurableProduct:ConfigurableProduct,
               associatedProduct:ConfiguredProduct),
IsConfiguredBy(configurableProduct:ConfigurableProduct,
               configurableAttribute:SingleValuedEnumerationAttribute),
IsAbleToRate(item:Item,ableToRateAttribute:Attribute).
```

The next step consists in projecting the participants of the referenced relationships $\mathcal{R}'$ to the entity or event types of the filtered schema. Formally, given a participant $p{:}C$ we define,

$$projection(p{:}C) = \begin{cases} p{:}C & \text{if} \quad C \in \{\mathcal{E}' \cup \mathcal{B}'\} \\ p{:}LCA(\mathcal{D}_C) & \text{if} \quad C \notin \{\mathcal{E}' \cup \mathcal{B}'\} \end{cases},$$

where $\mathcal{D}_C = \{C_i \in \{\mathcal{E}' \cup \mathcal{B}'\} \mid C_i \text{ IsA}^+ C\}$.

The lowest common ascendant (LCA) of a set of siblings $C_1, ..., C_n \in \{\mathcal{E}' \cup \mathcal{B}'\}$ that descend from the participant C is the lowest entity or event type in the hierarchy shared by the siblings that has all $C_1, ..., C_n$ as descendants (where we allow an element to be a descendant of itself). Whenever the set of descendants $\mathcal{D}_C$ is empty, the corresponding LCA equals to C. If the LCA obtained by projecting is not a member of $\mathcal{E}' \cup \mathcal{B}'$, the method includes it as an auxiliary entity or event type in $\mathcal{E}_\mathcal{X}$ or $\mathcal{B}_\mathcal{X}$. Therefore, the final entity and event types of the filtered schema are $\mathcal{E}_\mathcal{F} = \mathcal{E}' \cup \mathcal{E}_\mathcal{X}$ and $\mathcal{B}_\mathcal{F} = \mathcal{B}' \cup \mathcal{B}_\mathcal{X}$. For the case of relationships, the final set $\mathcal{R}_\mathcal{F}$ contains the projections of the relationship types in $\mathcal{R}'$. Formally, $\mathcal{R}_\mathcal{F} = \{R(projection(p_1{:}C_1), ..., projection(p_n{:}C_n)) \mid R \in \mathcal{R}'\}$. The method performs the same filtering and projection steps for the attributes explicitly referenced by schema rules of $\mathcal{FS}$.

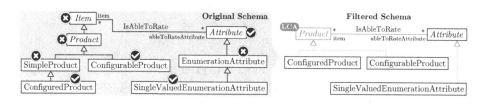

Fig. 4. Projection of the referenced relationship type *IsAbleToRate*

Figure 4 presents the referenced relationship type *IsAbleToRate* defined between *Item* and *Attribute*, and its projection to the LCA of the descendants of *Item* in the filtered schema. Note that since *Product* is the LCA of both *ConfigurableProduct* and *ConfiguredProduct*, and it was not filtered into $\mathcal{E}'$ in the previous stage, the method includes it in the auxiliary set $\mathcal{E}_\mathcal{X} \subset \mathcal{E}_\mathcal{F}$.

Considering the integrity constraint from the running example, the set of filtered attributes $\mathcal{A}_{\mathcal{F}}$ is empty since the constraint does not reference any attribute. The final filtered relationship types from the projection of the relationships of $\mathcal{R}'$ in $\mathcal{R}_{\mathcal{F}}$ are:

```
IsRepresentedBy(configurableProduct:ConfigurableProduct,
               associatedProduct:ConfiguredProduct),
IsConfiguredBy(configurableProduct:ConfigurableProduct,
               configurableAttribute:SingleValuedEnumerationAttribute),
IsAbleToRate(item:Product,ableToRateAttribute:Attribute).
```

Stage 3: Filtering of Generalization Relationships. At this point of the filtering process, the resulting filtered schema contains its entity, event, relationship types, and attributes. Now, we determine the needed generalization relationships between the filtered entity and event types. The method automatically selects the direct generalizations between filtered entity or event types, and includes them into the filtered schema. However, it is possible to have a pair of filtered entity or event types e_i, e_j so that e_i is an indirect ascendant of e_j in the original schema but they are not connected through generalization relationships in the filtered one. To avoid this inconsistency, the method creates auxiliary generalizations between those unconnected pairs. Formally, $\mathcal{G}_{\mathcal{F}} = \{g{:}(C_i \text{ IsA } C_j) \mid C_i, C_j \in \{\mathcal{E}_{\mathcal{F}} \cup \mathcal{B}_{\mathcal{F}}\} \wedge \exists g'(g'{:}(C_i \text{ IsA}^+ C_j) \in \mathcal{G}^+)\}$.

Figure 5 shows the hierarchy of *Product* referenced by the integrity constraint of the running example. Since this constraint does not reference the entity type *SimpleProduct*, it is not included into the resulting filtered schema. First, the method obtains the direct generalization relationship between *ConfigurableProduct* and *Product*. As a result, the entity type *ConfiguredProduct* that was a descendant in that hierarchy of the original schema, it is now an isolated entity type in the filtered schema. Next, in order to avoid that situation and maintain the original semantics, our method creates an auxiliary generalization in order to show that *ConfiguredProduct* is indirectly a descendant of *Product*. Similarly, the method includes an auxiliary generalization between *SingleValuedEnumerationAttribute* and *Attribute* in $\mathcal{G}_{\mathcal{F}}$.

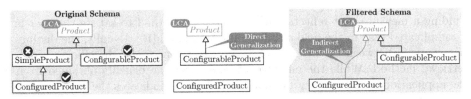

Fig. 5. Filtering of the generalization relationships from the hierarchy of *Product*

Stage 4: Filtering of Data Types. The method includes in the filtered schema those data types from the original one that are referenced in the specification of a schema rule of $\mathcal{FS}$ or define the type of a filtered attribute of $\mathcal{A}_{\mathcal{F}}$. Formally, $\mathcal{T}_{\mathcal{F}} = \{T \in \mathcal{T} \mid \exists s \ (s \in \mathcal{FS} \wedge s \rightsquigarrow T) \vee \exists a, C \ (C \in \{\mathcal{E}_{\mathcal{F}} \cup \mathcal{B}_{\mathcal{F}}\} \wedge \ a(C, T) \in \mathcal{A}_{\mathcal{F}}\}$.

In the running example, the set $\mathcal{T}_{\mathcal{F}}$ is empty since the constraint does not reference any data type.

Stage 5: Filtering of Schema Rules. The method selects the schema rules defined in the context of elements from $\mathcal{CS}_\mathcal{F}$ and only includes in $\mathcal{S}_\mathcal{F}$ those rules that are referentially-complete. A schema rule is referentially-complete whenever all the schema elements used in the specification of the rule belong to the filtered conceptual schema. All the schema rules from the focus set are referentially-complete since the schema elements they reference are all already in the filtered schema. Formally, $\forall s\,((s \in \mathcal{S} \wedge \neg \exists x (s \rightsquigarrow x \wedge x \notin \mathcal{CS}_\mathcal{F})) \Rightarrow s \in \mathcal{S}_\mathcal{F})$. The execution of this stage is performed only if the user wants in $\mathcal{S}_\mathcal{F}$ the additional schema rules.

Considering the running example, the method includes in $\mathcal{S}_\mathcal{F}$ the integrity constraint of $\mathcal{FS}$, and an additional constraint from $\mathcal{S}$ that only references the filtered elements in the former stages of the method (see Fig. 3(a) bottom).

Stage 6: Presentation of the Filtered Schema. The last step of the filtering method graphically presents to the user the elements included in the resulting filtered conceptual schema, as shown in Fig. 3(a). Note that the projected relationship types and the auxiliary entity and event types from Stage 2, and the auxiliary generalizations from Stage 3, are marked with a lighter color.

3.4 Method Correctness

It can be shown that $\mathcal{CS}_\mathcal{F}$ is minimal and a valid instance of the metaschema. The reason is that it only includes those elements from $\mathcal{CS}$ that truly participate in the specification of the schema rules of focus, apart from the additional schema rules from Stage 5 that are included only on user-demand. By projecting relationship types and attributes, and avoiding unnecessary generalization relationships, our method creates a filtered schema with standard constructions that cannot be smaller for a given input focus set. Any deletion of an element in the filtered schema produces an inconsistency in such schema with respect to the specification of the schema rules of focus.

4 Evaluation

Finding a measure that reflects the ability of our method to satisfy the user is a complicated task in the field of information retrieval [19]. Usually, the distinction is made between the evaluation of the effectiveness and the efficiency of a retrieval method. While the effectiveness measures the benefits obtained from the application of the filtering method, the efficiency indicates the time interval between the request being made and the answer being given.

We have implemented the filtering method described in the previous section as an extension of the prototype tool described in [20]. We have then evaluated the efficiency and effectiveness of the method by using four distinct case studies: the schema of Magento [7], the UML metaschema [2, 6], the schema of osCommerce [8], and the EU-Rent car rental schema [9]. Table 1 summarizes the components of these four schemas. We have applied our filtering method for each schema rule specified in these conceptual schemas. In the following, we present the results we have obtained from the analysis of the resulting data.

Table 1. Components of the conceptual schemas used in the evaluation

CS	$|\mathcal{E}|$	$|\mathcal{B}|$	$|\mathcal{A}|$	$|\mathcal{R}|$	$|\mathcal{G}|$	$|\mathcal{J}|$	$|\mathcal{S}|$
Magento	218	187	983	319	165	61	734
UML metaschema	293	0	93	377	355	13	170
osCommerce	84	262	209	183	393	17	457
EU-Rent	65	120	85	152	207	7	283

4.1 Effectiveness Analysis

Our method produces a filtered conceptual schema of small size that helps understanding the schema rules of a large schema. We compare the final size of the filtered schema with the size of the context constraint schema, i.e, the portion of the large schema the user needs to manually explore in order to cover the elements referenced by the formal specification of the schema rules of focus.

For a set of schema rules of focus $\mathcal{FS}$, we define the components of the constraint context schema $\mathcal{CS_C} = \langle \mathcal{E_C}, \mathcal{B_C}, \mathcal{R_C}, \mathcal{A_C}, \mathcal{G_C} \rangle$ as:

$$\mathcal{E_C} = \mathcal{E_S} \cup \mathcal{E_G} \cup \mathcal{E_R},$$
$$\mathcal{B_C} = \mathcal{B_S} \cup \mathcal{B_G} \cup \mathcal{B_R},$$
$$\mathcal{R_C} = \{R(p_1:C_1, ..., p_n:C_n) \in \mathcal{R} \mid \exists C_i, p_i \ (C_i \in \{\mathcal{E_C} \cup \mathcal{B_C}\} \wedge p_i:C_i \in R)\},$$
$$\mathcal{A_C} = \{a \in \mathcal{A} \mid \exists C, T \ (C \in \{\mathcal{E_C} \cup \mathcal{B_C}\} \wedge T \in \mathcal{J} \wedge a(C,T))\},$$
$$\mathcal{G_C} = \{g \in \mathcal{G} \mid \exists C_i, C_j \ (C_i, C_j \in \{\mathcal{E_C} \cup \mathcal{B_C}\} \wedge g:(C_i \ IsA \ C_j))\},$$

where $\mathcal{E_S}$ and $\mathcal{B_S}$ are the entity and event types referenced by the schema rules of focus. For the example of Fig. 4, $\mathcal{E_S} = \{ConfigurableProduct, ConfiguredProduct, Attribute, SingleValuedEnumerationAttribute\}$. $\mathcal{E_G}$ and $\mathcal{B_G}$ contain the entity and event types that are intermediate members of the paths of generalization relationships between members of $\mathcal{E_S}$ and $\mathcal{B_S}$. For the example of Fig. 5, $\mathcal{E_G} = \{SimpleProduct\}$. $\mathcal{E_R}$ and $\mathcal{B_R}$ are the participant entity and event types in relationship types and attributes referenced by schema rules from the focus set, without applying projection. For the example of Fig. 4, $\mathcal{E_R} = \{Item, Attribute\}$.

Therefore, we define the filtering utility factor between the filtered schema $\mathcal{CS_F}$ and the context constraint schema $\mathcal{CS_C}$ as follows:

$$\text{Filtering Utility Factor: } \Delta = 1 - \frac{\Sigma(\mathcal{CS_F})}{\Sigma(\mathcal{CS_C})}, \text{ where}$$
$$\Sigma(\mathcal{CS_F}) = |\mathcal{E_F}| + |\mathcal{B_F}| + |\mathcal{R_F}| + |\mathcal{A_F}| + |\mathcal{G_F}|, \text{ and}$$
$$\Sigma(\mathcal{CS_C}) = |\mathcal{E_C}| + |\mathcal{B_C}| + |\mathcal{R_C}| + |\mathcal{A_C}| + |\mathcal{G_C}|.$$

Figure 6(a) presents a box plot with the resulting values for the filtering utility factor applied to each of the 1644 schema rules of the case studies. For each schema, the plot indicates the smallest observation (sample minimum), lower quartile (Q1), median (Q2), upper quartile (Q3), and largest observation (sample maximum). Also, the black diamonds indicate the mean of each sample. The bottom and top of each box (Q1 and Q3) are the 5th and 95th percentiles, which means that the box contains the 90% of the samples.

We observe that the mean value of the filtering utility factor exceeds 0.7 in any case, which indicates a size reduction greater than 70% using filtered schemas instead of working manually. It implies a significant reduction of the cognitive effort a user has to face when understanding the schema rules of a large schema.

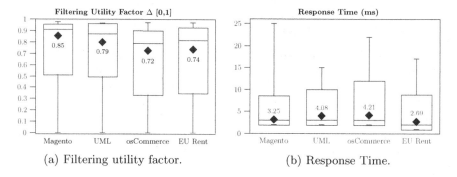

(a) Filtering utility factor.   (b) Response Time.

Fig. 6. Effectiveness and efficiency analysis

The smallest observations of the filtering utility factor in any schema ($\Delta = 0$) indicate that the size of the filtered schema equals the size of the context constraint schema ($\Sigma(\mathcal{CS}_\mathcal{F}) = \Sigma(\mathcal{CS}_\mathcal{C})$). This situation occurs whenever the schema rule of focus only references all the attributes of a single entity or event type without relationships to other elements, as in the case of primary key constraints of isolated types. In our experimentation, the schema rules that cause this represent less than a 2% of the total schema rules analyzed by the method.

4.2   Efficiency Analysis

It is clear that a good method does not only need to be useful, but it also needs to obtain the results in an acceptable time according to the user's expectations. To find the time spent by our method it is only necessary to record the time lapse between the request of knowledge, i.e. once a focus set $\mathcal{FS}$ containing schema rules has been indicated by the user, and the obtainment of the filtered schema.

Figure 6(b) presents a box plot with the resulting values for the response time (in milliseconds) obtained by an Intel Core 2 Duo 3GHz processor with 4GB of DDR2 RAM. The mean value of the response time is less than 5 milliseconds, which indicates that the time a user expends waiting for the resulting filtered schema is negligible. Furthermore, the largest observations of the response time in any schema are below 40 milliseconds. It is expected that as the number of projections of relationship types and subsumed generalization relationships to process increases, the response time will increase linearly. However, the resulting times for all the schema rules of the case studies are short enough for our purpose.

5   Application to Event Types

Our method can be used to help understanding the specification of event types in large conceptual schemas. We see this as an important application of the method. An event describes a nonempty set of changes in the population of entity or relationship types in the domain of the conceptual schema. We assume event types are represented as entity types with the stereotype «event». This allows one to define relationships between events and other entities, integrity constraints, derivation rules, etc. in a way very similar to that for ordinary entity types [21].

We define an *effect()* operation in each event type, whose purpose is to specify the effect of the event in the domain. The postcondition of this operation will be exactly the postcondition of the corresponding event. We can use a constraint language like OCL to specify these postconditions. Then, we can directly apply our filtering method to the a set of postconditions of a particular event type in order to obtain its corresponding filtered schema.

Figure 7 (left) presents the two postconditions –*addProduct* and *decrease-Quantity*– of the *AddProductToShoppingCart* event, and the resulting filtered schema (right) that our method automatically obtains ($\Delta = 0.8$ in 3ms). Such

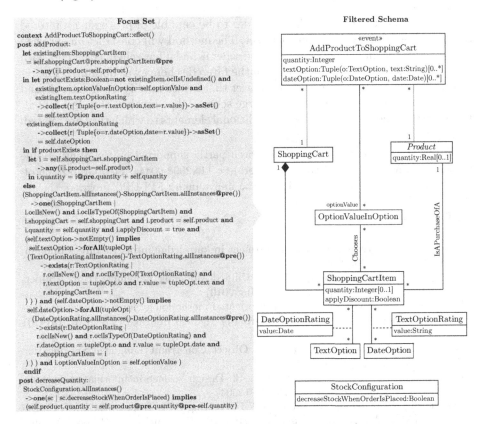

Fig. 7. Application of the filtering method to the event *AddProductToShoppingCart*

schema can be seen as an Effect Correspondence Diagram (ECD) [22] that shows all the entities affected by a given event type.

6   Conclusions

We have focused on the problem of understanding constraint expressions in large conceptual schemas, in which the elements referenced by the expressions may be very distant from each other and embedded in an intricate web of irrelevant elements for this purpose.

We have proposed a filtering method in which a user focuses on a set of schema rules and the method obtains a filtered schema that includes the smallest subset of the original schema that is needed to understand those expressions. We have implemented our method in a prototype tool and we have evaluated it by means of its application to four large conceptual schemas. The results show that our method achieves a size reduction greater than 70% in the number of schema elements to explore when understanding a schema rule by using filtered schemas instead of working manually, with an average time per request that is short enough for the purpose at hand. We have also shown the application of the method to help understanding the specification of event types.

We plan to extend our method in order to be capable of processing a filtered conceptual schema as the input focus set. The method will construct an enriched schema with the knowledge of the filtered schema and a few schema elements with relation to the schema rules from which the filtered schema was obtained. The resulting schema may be of interest for a user that has to modify a schema rule and needs to explore the fragment of the large schema that involves the elements affected by the schema rule and a set of additional elements to which they are related.

Acknowledgements. This work has been partly supported by the Ministerio de Ciencia y Tecnologia and FEDER under project TIN2008-00444/TIN, Grupo Consolidado, and by Universitat Politècnica de Catalunya under FPI-UPC program.

References

1. Lenat, D.B.: Cyc: a large-scale investment in knowledge infrastructure. Commun. ACM 38(11), 33–38 (1995)
2. Object Management Group (OMG): Unified Modeling Language (UML) Superstructure Specification, version 2.3 (May 2010)
3. Beeler, G.: HL7 Version 3–An object-oriented methodology for collaborative standards development. International Journal of Medical Informatics 48(1-3), 151–161 (1998)
4. Object Management Group (OMG): Object Constraint Language Specification (OCL), version 2.0 (February 2010)
5. Hevner, A., March, S., Park, J., Ram, S.: Design science in information systems research. Mis Quarterly 28(1), 75–105 (2004)
6. Bauerdick, H., Gogolla, M., Gutsche, F.: Detecting OCL Traps in the UML 2.0 Superstructure: An Experience Report. In: Baar, T., Strohmeier, A., Moreira, A., Mellor, S.J. (eds.) UML 2004. LNCS, vol. 3273, pp. 188–196. Springer, Heidelberg (2004)

7. Ramirez, A.: Esquema conceptual de Magento, un sistema de comerç electrónic. Technical report. Universitat Politécnica de Catalunya (2011), http://hdl.handle.net/2099.1/12294

8. Tort, A., Olivé, A.: The osCommerce conceptual schema. Technical report, Universitat Politécnica de Catalunya (2007), http://hdl.handle.net/2099.1/5301

9. Frias, L., Queralt, A., Olivé, A.: EU-Rent car rentals specification. Technical report, Universitat Politécnica de Catalunya (2003), http://www.lsi.upc.edu/~techreps/files/R03-59.zip

10. Villegas, A., Olivé, A.: A Method for Filtering Large Conceptual Schemas. In: Parsons, J., Saeki, M., Shoval, P., Woo, C., Wand, Y. (eds.) ER 2010. LNCS, vol. 6412, pp. 247–260. Springer, Heidelberg (2010)

11. Tzitzikas, Y., Hainaut, J.-L.: How to Tame a Very Large ER Diagram (Using Link Analysis and Force-Directed Drawing Algorithms). In: Delcambre, L.M.L., Kop, C., Mayr, H.C., Mylopoulos, J., Pastor, Ó. (eds.) ER 2005. LNCS, vol. 3716, pp. 144–159. Springer, Heidelberg (2005)

12. Auddino, A., Dennebouy, Y., Dupont, Y., Fontana, E., Spaccapietra, S., Tari, Z.: SUPER – Visual interaction with an object-based ER model. In: Pernul, G., Tjoa, A.M. (eds.) ER 1992. LNCS, vol. 645, pp. 340–356. Springer, Heidelberg (1992)

13. Lanzenberger, M., Sampson, J., Rester, M.: Visualization in ontology tools. In: Intl. Conf. on Complex, Intelligent and Software Intensive Systems, pp. 705–711. IEEE Computer Society (2009)

14. Shoval, P., Danoch, R., Balabam, M.: Hierarchical entity-relationship diagrams: the model, method of creation and experimental evaluation. Requirements Engineering 9(4), 217–228 (2004)

15. Moody, D.L., Flitman, A.: A Methodology for Clustering Entity Relationship Models - A Human Information Processing Approach. In: Akoka, J., Bouzeghoub, M., Comyn-Wattiau, I., Métais, E. (eds.) ER 1999. LNCS, vol. 1728, pp. 114–130. Springer, Heidelberg (1999)

16. Campbell, L.J., Halpin, T.A., Proper, H.A.: Conceptual schemas with abstractions making flat conceptual schemas more comprehensible. Data & Knowledge Engineering 20(1), 39–85 (1996)

17. Kuflik, T., Boger, Z., Shoval, P.: Filtering search results using an optimal set of terms identified by an artificial neural network. Information Processing & Management 42(2), 469–483 (2006)

18. Hanani, U., Shapira, B., Shoval, P.: Information filtering: Overview of issues, research and systems. User Modeling and User-Adapted Interaction 11(3), 203–259 (2001)

19. Belkin, N.J., Croft, W.B.: Information filtering and information retrieval: two sides of the same coin? Commun. ACM 35, 29–38 (1992)

20. Villegas, A., Sancho, M.-R., Olivé, A.: A Tool for Filtering Large Conceptual Schemas. In: De Troyer, O., Bauzer Medeiros, C., Billen, R., Hallot, P., Simitsis, A., Van Mingroot, H. (eds.) ER Workshops 2011. LNCS, vol. 6999, pp. 353–356. Springer, Heidelberg (2011)

21. Olivé, À.: Definition of Events and Their Effects in Object-Oriented Conceptual Modeling Languages. In: Atzeni, P., Chu, W., Lu, H., Zhou, S., Ling, T.-W. (eds.) ER 2004. LNCS, vol. 3288, pp. 136–149. Springer, Heidelberg (2004)

22. Downs, E., Clare, P., Coe, I.: Structured Systems Analysis and Design Method: Application and Context, 2nd edn. Prentice-Hall (1992)

Understanding Understandability of Conceptual Models – What Are We Actually Talking about?

Constantin Houy, Peter Fettke, and Peter Loos

Institute for Information Systems (IWi)
at the German Research Center for Artificial Intelligence (DFKI) and
Saarland University, Campus, Building D3$_2$
66123 Saarbrücken, Germany
{Constantin.Houy,Peter.Fettke,Peter.Loos}@iwi.dfki.de

Abstract. Investigating and improving the quality of conceptual models has gained tremendous importance in the past years. In general, model understandability is regarded one of the most important model quality goals and criteria. A considerable amount of empirical studies, especially experiments, have been conducted in order to investigate factors influencing the understandability of conceptual models. However, a thorough review and reconstruction of 42 experiments on conceptual model understandability conducted in this research shows that there is a variety of different understandings and conceptualizations of the term *model understandability*. As a consequence, this term remains ambiguous, research results on model understandability are hardly comparable and partly imprecise, which shows the necessity of clarification what the conceptual modeling community is actually talking about when the term *model understandability* is used. In order to overcome this shortcoming, our research classifies the different observed dimensions of model understandability in a reference framework. Moreover, implications of the findings are presented and discussed and some guidelines for future model understandability research are given.

Keywords: conceptual modeling, model understandability, model comprehensibility, model quality, experimental research.

1 Introduction

Conceptual models are considered a capable tool supporting many different tasks in the context of designing technical as well as organizational aspects of information systems (IS). They are practically used as methodical instruments for initial IS development and for managing IS in all the following phases of their life cycle [1]. Furthermore, conceptual models offer application potential in fields such as database design and management, business process management as well as for the choice, implementation and customization of standard software.

However, conceptual models can only fulfill their function and purpose if they possess an appropriate quality, which makes the topic of conceptual model quality highly relevant for research and practice. In this context, the understandability of

P. Atzeni, D. Cheung, and R. Sudha (Eds.): ER 2012, LNCS 7532, pp. 64–77, 2012.
© Springer-Verlag Berlin Heidelberg 2012

conceptual models is considered an important quality criterion [2]. A high understandability is especially important when conceptual models are used in order to support the communication about and a collective understanding of the functionality and structure of IS during the analysis, design and usage, which is one of the main purposes of conceptual models [3]. Against this background, research on model understandability and its influencing factors has a long tradition within the conceptual modeling community and there are quite a number of research contributions focusing on model understandability, predominantly experimental works.[1]

In order to successfully investigate and understand factors influencing the understandability of conceptual models in experimental research, it is necessary to clearly conceptualize and operationalize these factors (*independent variables*) and, furthermore – even more important – to have a clear understanding of the dependent variable *understandability* and what is constitutive of it. Therefore, it is of major importance to clearly conceptualize and operationalize *model understandability*. In general, without a clear conceptualization and operationalization of variables, it remains unclear what is actually investigated and measured in experimental research. Thus, besides the two important criteria *objectivity* (the measurement's independence of the person using the measurement instrument) and *reliability* (ability of a measurement instrument to reproduce results), the criterion of *validity* (what does the measurement instrument measure and does it really measure what it intends to measure?) is of highest importance. So far, no in-depth research concentrating on the *validity* aspect and putting into question *what* is actually measured in model understandability experiments has been presented.

To investigate the latter aspect is the *goal* of this article. Our research reconstructs and analyzes the underlying understanding of model understandability in experimental contributions in order to clarify what the research community is actually talking about when the term *model understandability* is used. Furthermore, our research aims at systemizing the different identified understandings of model understandability in a reference framework and, moreover, at developing appropriate guidelines in order to support future experimental research in this context.

The *research approach* chosen in this article is based on a thorough review of systematically retrieved experimental research contributions [4] and a methodical reconstruction [5] of the understanding, conceptualization, operationalization and measurement of conceptual model understandability. The reconstruction results are analyzed, classified into a structured reference framework and later on discussed.

The article at hand has the following *structure*: after this introduction, the conceptual and theoretical background of our research is introduced in section two and three. Section four clarifies the applied research approach in detail before section five presents the results of the reconstruction of the underlying understanding, conceptualization and operationalization of conceptual model understandability. Then, section six discusses the findings and presents implications as well as our guidelines before section seven summarizes and concludes the article.

---

[1] 42 out of the 48 research articles treating model understandability (87.5%) which have been systematically retrieved in the context of this research use experimental research approaches.

2   Conceptual Models and Understandability Research

Conceptual models are an instrument to express and clarify the meaning and relationship of terms and concepts in a domain in order to support communication and to avoid problems with different interpretations of these terms and concepts during the development of information systems [6]. Conceptual models are important artefacts for the design and maintenance of IS as they aim at clearly documenting and communicating their functionalities and structure [7].

Against this background, it is critical that conceptual models are easily understandable by all users and stakeholders [8]. If a user or stakeholder does not correctly understand a conceptual model, communication about the model content or the verification of whether important requirements concerning IS are effectively met is not possible [2]. Therefore, it is important to identify potential principles, characteristics or relationships influencing the understandability of conceptual models in order to improve the success of conceptual modeling. This is the reason why research on conceptual model understandability and especially experimental research in this context have a long tradition in IS research and the conceptual modeling community.

Understandability can be interpreted as a kind of pragmatic quality of conceptual models [9]. It is related to the ease of use, respectively the effort for *reading* and correctly *interpreting* a conceptual model, which is a cognitive process of assigning meaning to the different parts of a conceptual model [10]. In contrast to things which are easily and objectively observable (*observable constructs*), like the number of elements of a conceptual model or if it is raining at the moment, it is a crucial question *how* such cognitive processes can be observed, if they are objectively accessible at all. Mental processes are connected to emotions, intentions or wishes and can therefore not easily be accessed. Such phenomena are commonly considered *theoretical constructs* which have to be accessed by means of an adequate observable construct which is able to represent a theoretical construct [10]. Identifying such an adequate observable construct is highly problematic and actually the crux of the matter which our article treats. In experimental research, it is often persons' behavior which is observed and measured in order to examine the theoretical construct *understandability*. In this context, the fundamentals of measurement theory and its requirements have to be considered in order to develop a valuable scientific experiment.

3   Measurement Theory and Understandability Research

As already mentioned, it is important to verbally capture investigated objects and to make theoretical constructs measurable [11]. Several requirements concerning the definition, conceptualization and operationalization of theoretical constructs have to be met in order to clarify what is actually investigated respectively measured. The following general requirements are of special importance in this context [12]:

R.1: *Explicit definition of the basic understanding of a theoretical construct*
In the first place, it is necessary to clearly define and describe the basic and overall understanding of a theoretical construct – an abstraction which describes a phenomenon of theoretical interest [13], such as *model understandability*. In this context, a circle-free definition of a theoretical construct should be given for clarification.

R.2: *Clear conceptualization of a theoretical construct*

Secondly, it is fundamental to clearly conceptualize a theoretical construct. *Conceptualization* means that relevant aspects and dimensions of a theoretical construct are clearly elaborated. Theoretical constructs can consist of one or more relevant dimensions. If there are different important aspects or dimensions of a theoretical construct, all of these dimensions have to be identified and clearly defined in order to specify what the theoretical construct exactly consists of.

R.3: *Clear operationalization of a theoretical construct*

As the different dimensions of the conceptualization of a theoretical construct have to be measured, adequate indicators – observable, quantifiable scores which can be gathered by empirical means [14] – measurement methods and adequate measurement scales for each dimension have to be elaborated. An operationalization gives an exact specification of what is going to be observed and, thus, turns an abstract thing into a measurable thing. In this context, reflexive and formative operationalizations can be distinguished based on the consideration whether the omission of an indicator changes the conceptualization of a construct. Several different formative indicators induce ("form") one certain construct. Thus, a formatively operationalized theoretical construct is measured as a composite of specific indicators. In contrast to that, in the context of reflexive operationalization each indicator of its own is a manifestation of the theoretical construct [13]. A clearly defined relationship between theoretical constructs and indicators is of paramount importance as it assures a clear and meaningful mapping of theoretical constructs onto empirical phenomena, which helps to avoid ambiguity of research results [15]. This is equally valid in the context of conceptual model understandability research as the operationalization of the different model understandability dimensions determines a meaningful access to the cognitive process of understanding a conceptual model.

In the following, we systematically investigate the validity of experimental research on conceptual model understandability, especially the construct validity, by analyzing the fulfillment of these requirements. We systematically review and reconstruct the understanding, conceptualization and operationalization of model understandability in experimental research conducted by the conceptual model community. Furthermore, we group and classify the different understandings in a structured conceptual model understandability reference framework.

4   Research Approach

In our research, a systematic literature review of experimental conceptual model understandability research is conducted [4]. First of all, relevant research articles investigating conceptual model understandability have been systematically retrieved. In order to limit the retrieval complexity and to maximize the spectrum of research contributions, three leading literature databases have been interrogated: (1) Science Citation Index (Thomson Scientific), (2) Scopus (Elsevier) and (3) Business Source Premier (EBSCO). In order to identify relevant literature, the following search terms have been used for the forward search: "understand*", "comprehens*", "perceivab*" and "conceptual model*", "process model*", data model*". Moreover, further contributions have been identified by means of a backward search and any relevant literature known to the authors which has not been retrieved by the above described search has been added to the population of articles. Thus, in total 48 research articles using different research methods have been identified. 42 of these articles describe experimental research results and have been investigated in our research.

In order to analyze the fulfillment of the requirements **R.1** to **R.3,** the basic understanding, conceptualization and operationalization of model understandability within these experiments have been captured by means of a reconstruction of the contributions' underlying research designs, the total amount of investigated variables as well as the measurement instruments [5]. Furthermore, similar model understandability conceptualizations have been grouped and classified in order to inductively create a structured reference framework based on all identified understandability conceptualizations. Due to the high demands which are made on the detailed operationalization of variables in experiments, the underlying understanding of conceptual model understandability could be precisely specified and reconstructed. In the following, we present the results of our research.

5   Model Understandability in Experimental Research

The analysis of the 42 experiments showed very interesting results and also a lot of different basic understandings of model understandability. Particularly, in contrast to the requirement **R.1**, many of the investigated contributions do not explicitly define the term *model understandability* and use the term as if its meaning was totally clear and as if there was no need to explain a basic understanding (a detailed overview is given in table 2). Most of the works indirectly define *model understandability* for the first time in their methodical sections by means of their own operationalization of the different dimensions of understandability. Only 14 of the 42 contributions explicitly define their basic understanding of *understandability* or synonyms like *comprehensibility*. Eight of these contributions also conduct a terminological discussion concerning possible definitions of model understandability, e. g. [16, 17]. In some cases, even if the basic understanding of understandability is defined, however, given definitions are not very illuminative as they are not circle-free, e.g. the definitions of understandability in [18] ("the degree to which information contained in a [...] model can be easily understood"), in [19] ("the ease with which concepts and structures in the [...] model can be understood") or in [20] ("the ability to understand the [...] model"). As a consequence, it can be stated that the requirement **R.1** is not fulfilled by many experiments on model understandability.

In order to get a more detailed understanding of model understandability and to investigate the fulfillment of **R.2** and **R.3,** every experiment's research design has been examined and documented. In this context, all investigated constructs, their operationalization and the used measurement instruments for model understandability were analyzed. Table 1 shows an excerpt taken from the overview of this documentation.

Table 1. Overview of investigated constructs and their operationalization (excerpt)

Reference	Research design	N	Independent variables	Dependent variables	Measurement instrument for understandability
1. Agarwal et al. 1999 [21]	Laboratory experiment + replication, two groups, participants randomly assigned	36 + 35	*Modeling approach:* 1. Usage of object-oriented models (structure) 2. Usage of process-oriented models (behaviour)	*Understandability:* 1. Accuracy of model comprehension	1. comprehension test: comprehension score rating participants' answers (7-point-Likert scale) on eight comprehension questions.
2. Bavota et al. 2011 [22]	Quasi-experiment + two replications	37 + 52 + 67	*Modeling notation:* 1. UML class diagrams 2. ER diagrams *Modeling experience*	*Understandability:* 1. model comprehension level 2. subjectively perceived understanding (preference for notation)	1. comprehension test: number of correctly answered multiple choice questions out of 10 (one or more correct answers per question) 2. Qualitative assessment of notation preference
3. Bodart et al. 2001 [23]	Three laboratory experiments, mixed designs, randomly assigned participants	52 + 52 + 96	*Representational complexity:* 1. Mandatory properties representation 2. Optional properties representation	*Understandability:* 1. surface-level understanding 2. deeper-level understanding (response accuracy and problem-solving)	1. Seven measures for recall accuracy: total number of correctly recalled construct instances (entities, relationships, attributes, attributes recalled and typed correctly, relationships recalled with correct cardinalities etc.) 2. Response accuracy: 10 comprehension questions, response time (in seconds), normalized accuracy (accuracy score divided by time score) and three measures for problem-solving performance concerning 9 questions (the number of correct answers based upon information recalled in the conceptual model; (b) the number of correct answers provided by a participant based upon extra-model knowledge; and (c) the number of incorrect answers provided by the participant.)
4. Burton-Jones et al. 1999 [24]	Laboratory experiment, 2x2 mixed design, randomly assigned participants	67	*Ontological model clarity* (relationships in ERM with / without attributes) *Domain knowledge of users*	*Understandability:* 1. problem-solving performance 2. perceived ease of understanding	1. problem-solving measure: number of acceptable answers to six problem-solving questions (PST), number of answers coming from aspects represented differently in the two groups (PSD) 2. PEU: Six items from ease of use-instruments
5. Burton-Jones et al. 2006 [17]	Laboratory experiment, 1*3 between-groups design, randomly assigned participants + replication	57 + 66	*Model decomposition*	*Understandability:* 1. Actual understanding (comprehension, problem-solving) 2. perceived understanding	1. problem-solving test: number of acceptable answers to problem-solving questions and cloze test (participant's ability to complete a narrative of the domain, number of filled blanks) 2. Four items to measure perceived ease-of-use
6. ...	...	...	...	...	...

This reconstruction shows in detail that the term model understandability is conceptualized and operationalized very differently within the 42 experiments. Different researchers define several different dimensions which either reflexively or in other cases also formatively operationalize model understandability. The different identified dimensions are also used in different combinations. In contrast to the work of Agarwal et al. [21] which only considers one dimension of understandability (*accuracy of model comprehension*) and measures it by means of comprehension questions concerning the model content, the work of Bodart et al. [23] defines different depths of understanding which formatively operationalize model understanding, viz. *surface-level* ("correctly recalling model parts") and *deeper-level understanding* ("correctly answering questions concerning the model content and problem solving"). Moreover, many different instruments for measuring model understandability have been used, e.g. in Agarwal et al. [21] the answers to the comprehension test were documented and their correctness was subjectively evaluated by the conductor of the experiment using a 7-point Likert scale. In contrast, Bodart et al. [23] counted the number of correctly recalled model parts respectively successfully solved problems. As a result, it can be stated that not only is model understandability conceptualized very differently, but also its measurement significantly varies in experimental research contributions. Although all of the investigated experiments fulfill both the requirements **R.2** and **R.3**, it becomes clear that the conceptual modeling community is far from having a consensus about *how* model understandability should be conceptualized and operationalized.

For the inductive development of our *reference framework* (figure 1) we have grouped and systemized all the different identified dimensions of model understandability. This framework is also the basis for our further analysis and assessment of all 42 investigated articles. In total, six different dimensions have been identified which comprise objectively measurable dimensions (dimension Nr. 1-5) and one subjective understandability dimension (dimension Nr. 6, "perceived ease of understanding"). Furthermore, we have classified the six dimensions differentiating their focus concerning understandability effectiveness (dimensions Nr. 1-4+6) and understandability efficiency (dimension Nr. 5), "time needed to understand a model").

Conceptual model understandability					
Objectively measurable dimensions of understandability					Subjective dimension of understandability
effectiveness				efficiency	effectiveness
1. Recalling model content	2. Correctly answering questions about model content	3. Problem-solving based on the model content	4. Verification of model content	5. Time needed to understand a model	6. Perceived ease of understanding a model

Fig. 1. Conceptual model understandability reference framework

This inductively developed reference framework covers all identified dimensions of conceptual model understandability which have been observed in different combinations in our reconstruction results [5]. In table 2 on the following page this framework is used to further analyze and assess the applied understandability dimensions in the investigated experiments and gives an overview of their usage.

Table 2. Investigated dimensions of understandability (references in chronological order)

| References | basic definition given: y / n | Objectively measurable dimensions | | | | | Subjective dimension |
| | | effectiveness | | | | efficiency | effectiveness |
		1. Recalling model content	2. Correctly answering questions about model content	3. Problem-solving based on the model content	4. Verification of model content	5. Time needed to understand a model	6. Perceived ease of understanding a model
1. Juhn and Naumann 1985, [25]	n		●	●			
2. Palvia et al. 1992, [26]	n		●			●	
3. Shoval et al. 1994, [27]	y		●				
4. Hardgrave and Dalal 1995, [28]	y		●			●	●
5. Kim et al. 1995, [29]	n		●		●		●
6. Shanks 1997, [30]	y						●
7. Agarwal et al. 1999, [21]	n		●				
8. Burton-Jones et al. 1999, [24]	n			●			●
9. Nordbotten et al. 1999, [31]	n		●				
10. Bodart et al. 2001, [23]	n	●	●	●			
11. Moody 2002, [32]	n		●		●	●	
12. Purchase et al. 2002, [33]	n		●			●	●
13. Parsons 2003, [34]	n		●			●	
14. Moody 2004, [35]	n		●		●	●	
15. Serrano et al. 2004, [36]	n					●	
16. Gemino et al. 2005, [37]	n		●	●		●	●
17. Poels et al. 2005, [38]	n		●				
18. Sarshar et al. 2005, [39]	n		●				●
19. Burton-Jones et al. 2006, [17]	y		●	●			●
20. Khatri et al. 2006, [16]	y		●	●			
21. Cruz-Lemus et al. 2007, [40]	y	●	●	●			
22. Mendling et al. 2007, [41]	y		●				●
23. Recker et al. 2007, [42]	n		●	●		●	
24. Serrano et al. 2007, [43]	n		●			●	
25. Burton-Jones et al. 2008, [44]	y		●	●			●
26. De Lucia et al. 2008, [45]	n		●				
27. Genero et al. 2008, [46]	n		●			●	●
28. Mendling et al. 2008, [47]	y		●				
29. Patig 2008, [10]	y		●			●	
30. Reijers et al. 2008, [48]	n		●				
31. Vanderfeesten et al. 2008, [49]	n		●				
32. Fuller et al. 2010, [50]	n		●				
33. Sánchez-González et al. 2010, [51]	y		●			●	
34. Bavota et al. 2011, [22]	n		●				
35. Figl and Laue 2011, [52]	y		●				●
36. Ottensooser et al. 2011, [53]	n		●	●			
37. Parsons 2011, [54]	n		●				●
38. Recker et al. 2011, [55]	y		●	●		●	
39. Reijers et al. 2011a, [56]	y		●				
40. Reijers et al. 2011b, [57]	n		●				
41. Sánchez-González et al. 2011, [58]	n		●			●	
42. Schalles et al. 2011, [59]	n	●	●			●	
●: understandability dimension has been observed in this contribution							

6 Discussion, Implications and Some Guidelines

Our research showed that there are a lot of different basic understandings of model understandability. Many of the investigated contributions do not fulfill the requirement **R.1**. Furthermore, also based on the reconstruction of the underlying conceptualizations and the operationalization of model understandability, it becomes clear that no consensus about an exact basic understanding of model understandability exists in experimental research. It shows that the term has been very differently defined, interpreted, conceptualized and measured. The requirements **R.2** and **R.3** are usually effectively met, but it shows that understanding a conceptual model comprises different dimensions within the different investigated articles. According to our reference framework in figure 1, model understandability has subjective and objective as well as effectiveness- and efficiency-related dimensions which are combined in different constellations in the different experiments. *Correctly answering questions about the model content* seems to be a very important observable construct representing model understandability which is used in 39 of the 42 experiments. Furthermore, *problem solving based on the model content, perceived ease of understanding a model* or *time needed to understand a model* are also considered relevant dimensions while the other two dimensions seem to be of minor importance.

All the investigated contributions are concerned with conceptual model understandability and present research results based on their different understandings. However, if the underlying conceptualization and operationalization of model understandability is not explicitly considered, research results can be ambiguous and cannot be adequately compared [5]. If, e.g. two experiments compare the understandability of *UML class diagrams* and *entity relationship models* and one of them investigates the dimension *perceived ease of understanding* while the other analyzes the *time needed to understand a model* and both come to the conclusion that UML class diagrams are easier to understand, these final statements are hardly comparable.

The consideration of model understandability conceptualizations and their operationalization are particularly necessary since otherwise, scientific results in all likelihood are interpreted ambiguously. This is especially the case if the readers of a contribution do not keenly regard the details on conceptualization and operationalization in the methodical section, but focus on the summary of results which commonly do not contain details on *how* understandability has been defined and understood. This circumstance not only impedes the development of a more precise research but also the expedient accumulation of the conceptual modeling community's research results, such as reliable empirical relationships for the development of dedicated theories. Against this background, we formulate our first guideline for future experimental research on model understandability as follows:

Guideline 1: *Experimental research on model understandability should clearly define the basic understanding of understandability and clearly conceptualize the term in order to avoid ambiguity and make research result comparable.*

We expect that by following this guideline, a more consistent understanding and use of terminology concerning model understandability can be established which supports an easier accumulation of comparable research results in this area. Moreover, our reference framework can contribute to a further clarification of model understandability and its dimensions, support the communication between researchers and serve for the preparation of future experiments.

An interesting observation of our analysis, which earlier research has also shown [5], is that *the understanding and conceptualization of model understandability in IS research has not been further differentiated during the last years (1985-2011)*. It could be intuitively expected that in almost thirty years of research on conceptual model understandability, a more differentiated understanding of the topic should have evolved. However, it shows that this is not the case. The contributions which we have analyzed in this research often cite related work; but conceptualizations and operationalization used in former work are seldom adopted in detail and then further developed. From the perspective of cumulative research, this can be considered unfavorable as existing research results on model understandability are not always taken into account and in some cases experimental research in this area seems to begin again "from scratch". Therefore, we formulate our second guideline for future experimental research, which can also be supported by the content of our reference framework, as follows:

Guideline 2: *Experimental research on model understandability should put a stronger focus on the pool of different understandability dimensions identified in related work and use them in order to further our understanding of model understandability regarding all its possible dimensions.*

The tables 1 and 2 show that especially the exact operationalization of model understandability is far from being consistent. Different combinations of formative and reflexive operationalization are used, which can also have a strong influence on the comparability of results. In this context, mistaken validity tests which do not provide valid information about the experiments validity are not seldom [14]. Therefore our third guideline is as follows:

Guideline 3: *Experimental research on model understandability should especially clarify which measurement instruments are appropriate for the chosen understandability operationalization and take care of the right combination and treatment of formative and reflexive operationalization.*

Our research also has some *limitations*: our approach aims at understanding conceptual model understandability by means of the reconstruction of existing experimental work. In this context, it cannot be excluded that further relevant dimensions of understandability exist and have not yet been discovered and captured in existing research. Thus, it is possible that our research results and conclusions are not exhaustive. However, the systematic retrieval improves the comprehensiveness of investigated sources and thus contributes to the minimization of this limitation. Moreover, it could be noted that further work on model understandability not using experimental research methods exists which has not been considered in our research. Thus, it is possible that potentially interesting aspects of model understandability have not been considered in our review and reconstruction process. However, we chose to reconstruct the conceptualization and operationalization of model understandability based on experiments because this reconstruction can be objectively and transparently executed. This is mainly related to the circumstance that all researchers performing experiments have to meet the quality requirements of experimental research regarding conceptualization and operationalization of investigated constructs. The high quality of underlying contributions and the full accessibility of construct conceptualizations and operationalization provided good preconditions for a proper reconstruction. Thus, we think that our research provides valuable insights into the concept of understandability.

7   Conclusions and Outlook

In this article, we have investigated the understanding, conceptualization and operationalization of conceptual model understandability by reviewing and reconstructing the approaches and results of relevant experimental research. We found that there is so far no consensus about the central aspects and dimensions of model understandability within the conceptual modeling community. Conceptual model understandability is, in fact, understood very differently. We found that, based on this fact, research results on model understandability are likely to be ambiguous if the underlying conceptualization and operationalization are not considered in detail during the interpretation of results. As a more consistent understanding of concepts and use of terminology is of essential relevance, we have developed a structured reference framework comprising all the different dimensions of conceptual model understandability used in experimental research based on our reconstruction. This reference framework can contribute to the further clarification and creation of a consistent and consensual understanding of conceptual model understandability and, thus, support the future conceptualization and operationalization of the dependent variable. With our research we hope to contribute to a fruitful discussion on the topic which is an important research stream and which can significantly contribute to the success of conceptual modeling.

Acknowledgements. The research described in this paper was supported by a grant from the German Research Foundation (DFG), project name: "Pluralistische Beurteilung der Qualität von Unternehmensmodellen - Qualitätsdiskurse und Diskursqualität innerhalb der Wirtschaftsinformatik (PluralistiQue)", support code LO 752/4-1. The authors would also like to thank the anonymous reviewers for their valuable comments which helped to improve this paper.

References

1. Germonprez, M., Hovorka, D., Gal, U.: Secondary Design: A Case of Behavioral Design Science Research. Journal of the AIS 12, 662–683 (2011)
2. Lindland, O.I., Sindre, G., Sølvberg, A.: Understanding Quality in Conceptual Modeling. IEEE Software 11, 42–49 (1994)
3. Krogstie, J.: Modelling of the People, by the People, for the People. In: Krogstie, J., Opdahl, A.L., Brinkkemper, S. (eds.) Conceptual Modelling in Information Systems Engineering, pp. 305–318. Springer, Berlin (2007)
4. Cooper, H., Hedges, L.V.: Research Synthesis As a Scientific Enterprise. In: Cooper, H., Hedges, L.V. (eds.) The Handbook of Research Synthesis, pp. 3–14. Russell Sage Foundation, New York (1994)
5. Fettke, P., Houy, C., Vella, A.-L., Loos, P.: Towards the Reconstruction and Evaluation of Conceptual Model Quality Discourses – Methodical Framework and Application in the Context of Model Understandability. In: Bider, I., Halpin, T., Krogstie, J., Nurcan, S., Proper, E., Schmidt, R., Soffer, P., Wrycza, S. (eds.) EMMSAD 2012 and BPMDS 2012. LNBIP, vol. 113, pp. 406–421. Springer, Heidelberg (2012)

6. Wand, Y., Storey, V.C., Weber, R.: Analyzing the Meaning of a Relationship. ACM Trans. Database Systems 24, 494–528 (1999)
7. Wand, Y., Weber, R.: Research Commentary: Information Systems and Conceptual Modeling - A Research Agenda. Information Systems Research 13, 363–377 (2002)
8. Moody, D.L.: Cognitive Load Effects on End User Understanding of Conceptual Models: An Experimental Analysis. In: Benczúr, A.A., Demetrovics, J., Gottlob, G. (eds.) ADBIS 2004. LNCS, vol. 3255, pp. 129–143. Springer, Heidelberg (2004)
9. Moody, D.L., Sindre, G., Brasethvik, T., Sølvberg, A.: Evaluating the Quality of Process Models: Empirical Testing of a Quality Framework. In: Spaccapietra, S., March, S.T., Kambayashi, Y. (eds.) ER 2002. LNCS, vol. 2503, pp. 380–396. Springer, Heidelberg (2002)
10. Patig, S.: A practical guide to testing the understandability of notations. In: Proceedings of the Fifth Asia-Pacific Conference on Conceptual Modelling (APCCM 2008), Wollongong, Australia (2008)
11. Lazarsfeld, P.F., Pasanella, A.K., Rosenberg, M. (eds.): Continuities in the Language of Social Research. The Free Press, New York (1972)
12. Babbie, E.R., Mouton, J.: The Practice of Social Research. Oxford University Press, Cape Town (2001)
13. Edwards, J.R., Bagozzi, R.P.: On the Nature and Direction of Relationships Between Constructs and Measures. Psychological Methods 5, 155–174 (2000)
14. Petter, S., Straub, D., Rai, A.: Specifying Formative Constructs in Information Systems Research. MIS Quarterly 31, 623–656 (2007)
15. Costner, H.L.: Theory, deduction, and the rules of correspondence. American Journal of Sociology 75, 245–263 (1969)
16. Khatri, V., Vessey, I., Ramesh, V., Clay, P., Park, S.-J.: Understanding conceptual schemas: Exploring the role of application and IS domain knowledge. Information Systems Research 17, 81–99 (2006)
17. Burton-Jones, A., Meso, P.N.: Conceptualizing Systems for Understanding: An Empirical Test of Decomposition Principles in Object-Oriented Analysis. Information System Research 17, 38–60 (2006)
18. Reijers, H.A., Mendling, J.: A Study Into the Factors That Influence the Understandability of Business Process Models. IEEE Transactions on Systems, Man, and Cybernetics Part A: Systems and Humans 41, 449–462 (2011)
19. Shanks, G.: Conceptual Data Modelling. An Empirical Study of Expert and Novice Data Modellers. Australian Journal of Information Systems 4, 63–73 (1997)
20. Hardgrave, B.C., Dalal, N.P.: Comparing Object-Oriented and Extended-Entity Relationship Data Models. Journal of Database Management 6, 15–21 (1995)
21. Agarwal, R., De, P., Sinha, A.: Comprehending object and process models: An empirical study. IEEE Trans. Softw. Eng. 25, 541–556 (1999)
22. Bavota, G., Gravino, C., Oliveto, R., De Lucia, A., Tortora, G., Genero, M., Cruz-Lemus, J.A.: Identifying the Weaknesses of UML Class Diagrams during Data Model Comprehension. In: Whittle, J., Clark, T., Kühne, T. (eds.) MODELS 2011. LNCS, vol. 6981, pp. 168–182. Springer, Heidelberg (2011)
23. Bodart, F., Patel, A., Sim, M., Weber, R.: Should Optional Properties Be Used in Conceptual Modelling? A Theory and Three Empirical Tests. Information Systems Research 12, 384–405 (2001)
24. Burton-Jones, A., Weber, R.: Understanding relationships with attributes in entity-relationship diagrams. In: 20th Annual International Conference on Information Systems (ICIS 1999), Charlotte, NC, USA, pp. 214–228 (1999)
25. Juhn, S., Naumann, J.: The Effectiveness of Data Representation Characteristics on User Validation. In: 6th International Conference on Information Systems, pp. 212–226 (1985)
26. Palvia, T., Lio, C., To, P.: The Impact of Conceptual Data Models on End User Performance. Journal of Database Management 3, 4–14 (1992)

27. Shoval, P., Fruerman, I.: OO and EER Schemas: A Comparison of User Comprehension. Journal of Database Management 5, 28–38 (1994)
28. Hardgrave, B.C., Dalal, N.P.: Comparing Object Oriented and Extended Entity Relationship Models. Journal of Database Management 6, 15–22 (1995)
29. Kim, Y.-G., March, S.T.: Comparing Data Modelling Formalisms. Communications of the ACM 38, 103–115 (1995)
30. Shanks, G.: Conceptual Data Modelling: An Empirical Study of Expert and Novice Data Modellers. Australasian Journal of Information Systems 4, 63–73 (1997)
31. Nordbotten, J.C., Crosby, M.E.: The Effect of Graphic Style on Data Model Interpretation. Information Systems Journal 9, 139–155 (1999)
32. Moody, D.L.: Complexity Effects on End User Understanding of Data Models: An Experimental Comparison of Large Data Model Representation Methods. In: Wrycza, S. (ed.) 10th European Conference on Information Systems (ECIS 2002), Gdansk, Poland, pp. 482–496 (2002)
33. Purchase, H.C., Colpoys, L., McGill, M., Carrington, D.: UML collaboration diagram syntax: an empirical study of comprehension. In: First International Workshop on Visualizing Software for Understanding and Analysis (2002)
34. Parsons, J.: Effects of local versus global schema diagrams on verification and communication in conceptual data modeling. Journal of MIS 19, 155–184 (2003)
35. Moody, D.L.: Cognitive Load Effects on End User Understanding of Conceptual Models: An Experimental Analysis. In: Benczúr, A.A., Demetrovics, J., Gottlob, G. (eds.) ADBIS 2004. LNCS, vol. 3255, pp. 129–143. Springer, Heidelberg (2004)
36. Serrano, M., Calero, C., Trujillo, J., Lujan, S., Piattini, M.: Empirical Validation of Metrics for Conceptual Models of Data Warehouses. In: Persson, A., Stirna, J. (eds.) CAiSE 2004. LNCS, vol. 3084, pp. 506–520. Springer, Heidelberg (2004)
37. Gemino, A., Wand, Y.: Complexity and Clarity in Conceptual Modeling: Comparison of Mandatory and Optional Properties. Data and Knowledge Engineering 55, 301–326 (2005)
38. Poels, G., Gailly, F., Maes, A., Paemeleire, R.: Object Class or Association Class? Testing the User Effect on Cardinality Interpretation. In: Akoka, J., Liddle, S.W., Song, I.-Y., Bertolotto, M., Comyn-Wattiau, I., van den Heuvel, W.-J., Kolp, M., Trujillo, J., Kop, C., Mayr, H.C. (eds.) ER Workshops 2005. LNCS, vol. 3770, pp. 33–42. Springer, Heidelberg (2005)
39. Sarshar, K., Loos, P.: Comparing the Control-Flow of EPC and Petri Net from the End-User Perspective. In: van der Aalst, W.M.P., Benatallah, B., Casati, F., Curbera, F. (eds.) BPM 2005. LNCS, vol. 3649, pp. 434–439. Springer, Heidelberg (2005)
40. Cruz-Lemus, J.A., Genero, M., Morasca, S., Piattini, M.: Using Practitioners for Assessing the Understandability of UML Statechart Diagrams with Composite States. In: Hainaut, J.-L., Rundensteiner, E.A., Kirchberg, M., Bertolotto, M., Brochhausen, M., Chen, Y.-P.P., Cherfi, S.S.-S., Doerr, M., Han, H., Hartmann, S., Parsons, J., Poels, G., Rolland, C., Trujillo, J., Yu, E., Zimányie, E. (eds.) ER Workshops 2007. LNCS, vol. 4802, pp. 213–222. Springer, Heidelberg (2007)
41. Mendling, J., Reijers, H.A., Cardoso, J.: What Makes Process Models Understandable? In: Alonso, G., Dadam, P., Rosemann, M. (eds.) BPM 2007. LNCS, vol. 4714, pp. 48–63. Springer, Heidelberg (2007)
42. Recker, J., Dreiling, A.: Does it matter which process modelling language we teach or use? An experimental study on understanding process modelling languages without formal education. In: Toleman, M., Cater-Steel, A., Roberts, D. (eds.) Proceedings of the 18th ACIS 2007, Toowoomba, Australia, pp. 356–366 (2007)
43. Serrano, M., Trujillo, J., Calero, C., Piattini, M.: Metrics for data warehouse conceptual models understandability. Inf. Softw. Technol. 49, 851–870 (2007)

44. Burton-Jones, A., Meso, P.N.: The Effects of Decomposition Quality and Multiple Forms of Information on Novices' Understanding of a Domain from a Conceptual Model. Journal of the AIS 9, 748–802 (2008)
45. De Lucia, A., Gravino, C., Oliveto, R., Tortora, G.: Data model comprehension an empirical comparison of ER and UML class diagrams. In: IEEE International Conference on Program Comprehension, Amsterdam, pp. 93–102 (2008)
46. Genero, M., Poels, G., Piattini, M.: Defining and validating metrics for assessing the under-standability of entity-relationship diagrams. Data and Knowledge Engineering 64, 534–557 (2008)
47. Mendling, J., Strembeck, M.: Influence factors of understanding business process models. In: Abramowicz, W., Fensel, D. (eds.) BIS 2008. LNBIP, vol. 7, pp. 142–153. Springer, Berlin (2008)
48. Reijers, H.A., Mendling, J.: Modularity in Process Models: Review and Effects. In: Dumas, M., Reichert, M., Shan, M.-C. (eds.) BPM 2008. LNCS, vol. 5240, pp. 20–35. Springer, Heidelberg (2008)
49. Vanderfeesten, I., Reijers, H.A., Mendling, J., Van Der Aalst, W.M.P., Cardoso, J.: On a Quest for Good Process Models: The Cross-Connectivity Metric. In: Bellahsène, Z., Léonard, M. (eds.) CAiSE 2008. LNCS, vol. 5074, pp. 480–494. Springer, Heidelberg (2008)
50. Fuller, R.M., Murthy, U., Schafer, B.A.: The effects of data model representation method on task performance. Information & Management 47, 208–218 (2010)
51. Sánchez-González, L., García, F., Mendling, J., Ruiz, F., Piattini, M.: Prediction of Business Process Model Quality Based on Structural Metrics. In: Parsons, J., Saeki, M., Shoval, P., Woo, C., Wand, Y. (eds.) ER 2010. LNCS, vol. 6412, pp. 458–463. Springer, Heidelberg (2010)
52. Figl, K., Laue, R.: Cognitive Complexity in Business Process Modeling. In: Mouratidis, H., Rolland, C. (eds.) CAiSE 2011. LNCS, vol. 6741, pp. 452–466. Springer, Heidelberg (2011)
53. Ottensooser, A., Fekete, A., Reijers, H.A., Mendling, J., Menictas, C.: Making sense of business process descriptions: An experimental comparison of graphical and textual notations. Journal of Systems and Software 85, 596–606 (2012)
54. Parsons, J.: An Experimental Study of the Effects of Representing Property Precedence on the Comprehension of Conceptual Schemas. Journal of the AIS 12, 441–462 (2011)
55. Recker, J., Dreiling, A.: The Effects of Content Presentation Format and User Characteristics on Novice Developers Understanding of Process Models. Communications of the AIS 28, 65–84 (2011)
56. Reijers, H.A., Mendling, J.: A Study Into the Factors That Influence the Understandability of Business Process Models. IEEE Transactions on Systems, Man, and Cybernetics Part A: Systems and Humans 41, 449–462 (2011)
57. Reijers, H.A., Mendling, J., Dijkman, R.M.: Human and automatic modularizations of process models to enhance their comprehension. Information Systems 36, 881–897 (2011)
58. Sánchez-González, L., Ruiz, F., García, F., Cardoso, J.: Towards thresholds of control flow complexity measures for BPMN models. In: Proceedings of the ACM Symposium on Applied Computing, SAC 2011, TaiChung, pp. 1445–1450 (2011)
59. Schalles, C., Creagh, J., Rebstock, M.: Usability of Modelling Languages for Model Interpretation: An Empirical Research Report. In: Bernstein, A., Schwabe, G. (eds.) 10th International Conference on Wirtschaftsinformatik, Zurich, Switzerland, pp. 787–796 (2011)

Rules from Cognition for Conceptual Modelling

Jeannette Stark and Werner Esswein

Technische Universitt Dresden

Abstract. Conceptual Modelling is a cognitive intensive process. Prior research has acknowledged the importance of cognitive theories and their implications for Conceptual Modelling. Several authors have developed hypotheses to give modellers a hint how to improve their models. Although much effort has been made, researchers and practitioners cannot easily apply or broaden these hypotheses. Yet, they are forced to spend a lot of review work, as a comprehensive overview about past research is missing. With this paper we give a review of hypotheses developed from Cognition for Conceptual Modelling.

1 Introduction

Understanding a conceptual model can be supported by adjusting the process of modelling, the modelling tool as well as the modelling grammar in such a way that fits human cognition [1]. Yet, modellers usually do not receive instructions how to fit their models to human cognition. Instead, how they use the grammar to create the model is based on intuition and experience and hence, is likely to result in a distortion of information. The resulting model might look good to them, but does not necessarily communicate effectively [2]. This is why prior research has used Cognitive Theories to develop hypotheses for the conception of cognitive effective models.

Two problems prevent from applying these hypotheses for Conceptual Modeling. First, accessing these hypotheses is not yet easy. It requires a lot of review work to find relevant studies and consolidation work to make its statements applicable. Second, several dependent variables of these hypotheses are used without relating them to those of other studies. Thus, readers have to cope with a multiplicity of dependent variables without knowing how these variables interact. That way, readability is used without any relation to domain understanding. But how does it differ from domain understanding?

Rockwell and Bajaj have focused on the first problem by reviewing prior research and have concluded their findings within a propositional framework they called COGEVAL [3]. This framework relates independent variables such as chunking, fragmentation and relationship information to the dependent variables modelling performance and readability. In addition, they fitted past empirical work to some of their hypotheses, which is why COGEVAL is an important work to make past studies applicable. However, COGEVAL is limited to the dependent variables modeling performance and readability and independent variables related to the design of a modeling grammar. Yet, we can find other dependent

P. Atzeni, D. Cheung, and R. Sudha (Eds.): ER 2012, LNCS 7532, pp. 78–87, 2012.

variables such as understanding [4][5] and learning performance [6] and further independent variables concerning the modelling process and the modeller.

In this study we address the first problem by reviewing relevant hypotheses for Conceptual Modeling founded in Cognition. We do not limit our study to the design of modelling grammars. For a comprehensive overview we summarize hypotheses within a framework proposed by Gemino and Wand [7]. We address the second problem by relating dependent variables of the study. This research makes contributions for researchers and practitioners. Researchers get an overview of hypotheses from Cognition for Conceptual Modeling. A classification of independent and dependent variables leads to future conceptual and empirical research ideas. The identification of relations between dependent variables allows a systematic identification of dependent variables for future empirical investigations. For practitioners independent variables serve as rules on how to design models and modelling grammars that fit human cognition.

The reminder of the study is organized as follows. In the next section we present variables of prior research, their origin and empirical status. In section 3 we derive implications for future research. Section 4 will give a short conclusion.

2 Presentation of Variables

During the last years IS researchers have found several approaches to develop hypotheses for Conceptual Modelling from Cognition. Some researchers tested their hypotheses (Section 2.1). Others developed frameworks of hypotheses for further empirical investigations (Section 2.2).

2.1 Empirically Tested Variables

Theories from Cognition that have served to derive hypotheses include Classification Theory, Cognitive Fit, Cognitive Load Theory (CLT), Cognitive Theory of Multimedia Learning (CTML), Diagrammatic Reasoning, Semantic Network and Spreading Activation (see Table 1). Each hypothesis is tested in a laboratory experiment. While the method is the same, variables differ among the studies.

Weber has studied the effect of the distinction between entities and properties [15]. If the distinction is made one might question if properties should be mandatory or optional [4][14]. These independent variables are tested for domain understanding as well as deep and surface-level understanding. If the resulting model consists of several diagrams it is of interest which effect fragmentation of information across several diagrams [6] and its conceptual [12][6] and perceptual integration [13][9] has. A further variable addressing the modelling grammar is the 'number of concepts of a modelling language' which is tested for the dependent variable readability [16].

We have found further variables that address the presentation of the resulting model such as 'if a conceptual model is presented with a narrated instead of a written explanation' and 'if it is animated'. These variables are tested for domain understanding [4].

Table 1. Past Empirical Research for Conceptual Modeling from Cognition

Theory	Dependent Variable	Independ. Variable	Result	Lit.
Classification Theory	Local information in local diagrams	Domain understanding	supported	[8]
	Conceptual integr. for complex situations		supported	
CLT	Conceptual integr. Perceptual integr.	Surface-level understanding	part. supp.	[9]
		Deep understanding	part. supp.	
Cognitive Fit	Problem solving skills match task and representation	Problem solving performance	supported	[10]
			supported	
	Problem representation match task	Modelling effectiveness	part. supp.	[11]
	Domain knowledge	Domain understanding	supported	[5]
CTML	Narrated explanation Animated Models	Domain understanding	supported not supp.	[12]
	Modelling language experience	Learning performance	no supp.	[6]
	Mandatory properties	Domain understanding	supported	[4]
Diagrammatic Reasoning	Conceptual integr. Perceptual integr.	Domain understanding	supported supported	[13]
Semantic Network	Mandatory and optional properties	Deep understanding surface-level understanding	supported supported	[14]
Semantic Network Spreading Activation	Disctinction between entities and properties	Domain understanding	part. supp.	[15]
not stated	Number of concepts	Readability	partially supp.	[16]
	Conceptual integr.	Modelling, perceived ease of use, perceived usefulness, intention to use	supported supported partially supp. partially supp.	[17]

Another category of variables represents user characteristics such as domain experience [5] and modelling language experience [6] (tested for the dependent variables learning performance and conceptual schema undertanding). Variables resulting from Cognitive Fit include problem solving skills match task and problem representation [10] as well as problem representation matches task [11], which are tested for problem solving performance and modelling effectiveness.

2.2 Variables Not Subject to an Empirical Validation

We have not only focussed on empirical studies, because hypotheses that have not yet been subject for empirical investigations might be a starting point for future research. We focus on three studies that give a framework of hypotheses. Rockwell and Bajaj contributed eight hypotheses with their COGEVAL-Framework [3], which they derived from different theories such as Capacity Theory [18], Theories of Short Term Memory (STM) [19], Levels-of-Processing Theory [20] and Readability Theory [21].

Furthermore Moody developed a Theory of Visual Notations he called 'Physics of Notations' [22]. Based at the top level on a specialization on Communication Theory [23] he synthesized nine hypotheses from a wide range of fields such as 'Semiology of Graphics' [24], 'Human Problem Solving' [25] and Goodman's Theory of Symbols [26].

Hypotheses so far presented mainly focus on how to improve the modeling grammar. In contrast, Parsons and Wand focus on the modelling process itself. They used Theories of Classification to derive rules on how to guide classification [27]. We have included these rules as Parsons and Wand make the assumption that adhering to these rules effects cognitive economy of the resulting model. You can find the above presented hypotheses in Table 2 .

2.3 Integrated View on Variables

We have selected 16 studies, whose authors have developed hypotheses. We identified 13 studies in which hypotheses were subject to an empirical investigation. For a profound summary we apply variables of these hypotheses to a framework for categorizing empirical investigations of conceptual modelling techniques Gemino and Wand have developed [7]. We have used most of their classes to insert variables found during the review presented in Fig. 1.

Independent Variables. For independent variables Gemino and Wand proposed eight classes of which we have used six classes: Grammar's constructs, Use of grammar(s), Medium of content delivery as well as User and Task characteristics.

Most research has so far been done within the class Grammar's Constructs. In this class we found 21 variables which are partly tested. We have distinguished variables of this class in an inner and an inter view subclass. Variables of both subclasses provide relations to each other. For a modeling language designer it is an important question if properties are optional or mandatory [4][14].

Table 2. Hypotheses for Conceptual Modeling from Cognition (Not Tested)

Theory	Dependent Variable	Independent Variable	Lit
Capacity Theory STM	**H1:** Degree of chunking supported by a modelling language; **H2:** Number of simultaneous items required over seven to create chunking;	Modelling effectiveness and efficiency	[3]
Semantic Organization Levels-of-Processing Pattern Recognition Language Compr. Readability	**H3:** Similarity of a model's constructs to the concept node - relationship arc depiction of information; **H4:** Required amount of semantic processing to create a model; **H5:** Lack of relationship information; **H6:** Degree of chunking allowed by the constructs; **H7:** Degree of requirements' context provided by the constructs; **H8:** Degree of fragmentation of requirements by the constructs;	Readability effectiveness and efficiency	
Graphical Communication	**H1: Semantic Clarity:** 1:1 correpondence between semantic constructs and graphical symbols; **H2: Perceptual Discriminability:** Symbols are clearly to distinguish from one another; **H3: Semantic Transparency:** Symbols let suggest their meaning; **H4: Complexity Management:** Explicit mechanisms to deal with complexity; **H5: Visual expressiveness:** Full range and capacities of visual variales are used; **H6: Dual Coding:** Text complements graphics; **H7: Graphic Economy:** Number of different graphical symbols is congnitively manageable; **H8: Cognitive Fit:** Different visual dialects for different tasks and audiences;	Cognitive effectiveness	[22]
Classification Theory	**H1:** A proper subset of the properties of a potential class sufficient to determine that an instance belongs to that class muss exist; **H2:** A candidate subclass possesses qualifying properties for the subclasses; **H3:** A candidate subclass of a class has at least two additional properties; **H4:** A subclass of multiple superclasses permits identifying its instances using fewer properties than required for identifying it as instance of all superclasses; **H5:** A potential subclass of multiple superclasses possesses properties in addition to all superclasses; **H6:** A potential subclass based on restricting the value of properties of a superclass provides a qualifying set for the original class; **H7:** A potential subclass based on restricting the value of properties of a superclass possesses at least one additional property;	Cognitive economy	[27]

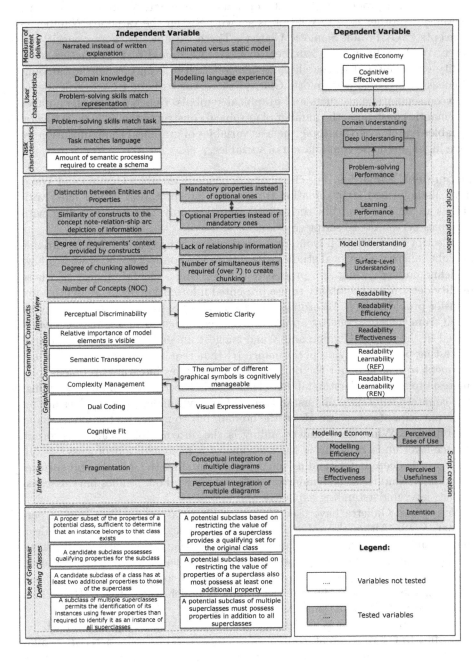

Fig. 1. Variables from Cognition for Conceptual Modeling

If, however, the designer does not distinguish between entities and properties the question, if properties should be optional or mandatory, does not have any relevance. Another type of relation is shown between the variables 'Number of Concepts (NOC)' and 'Semiotic Clarity'. These variables have an oppositional effect. While reducing the Numbers of Concepts should result in better readability efficiency and learnability, 'Semiotic Clarity' requires a certain amount of constructs. To gain 'Semiotic Clarity' there has to be a 1:1 correspondence between semantic constructs and graphical symbols. Depending on the number of semantic constructs, the number of graphical constructs can be high. If several variables match a certain topic, such as variables of graphical communication we have highlighted the combination of variables.

Dependent Variables. For classifying dependent variables Gemino and Wand have distinguished script creation from script interpretation. While most variables can be found within script interpretation only three studies deal with variables pertaining to script creation.

In script interpretation Domain understanding is used as dependent variable in eight different empirical studies. Staying in the context of the studies allows to define understanding as a process of constructing a mental model [28] whose quality determines the ability to use presented information of the model to solve transfer problems [29]. Accordingly, understanding is characterized not only by recall from memory, but by application of knowledge to problems not clearly solved in the model. Gemino and Wand suggest understanding to be measured by testing problem solving performance [7] that requires participants of an experiment to reason about the domain [12]. Six studies integrated problem solving tests to assess understandability. These studies focus on deep understanding. The counterpart of deep understanding, surface-level understanding, does not require participants to reason about the domain but to memorize the model under examination. While deep understanding is the outcome of a learning process, surface-level understanding is the outcome of a memorization process [29]. In other words deep understanding means understanding the domain and surface-level understanding means understanding the model. Four studies have assessed surface-level understanding explicitly.

Vessey and Galletta have used Cognitive Fit to assess problem solving performance [10]. Although problem solving performance is used to measure deep understanding we treat this variable as a seperate one for two reasons. First, we did not found problem solving performance as the only test to determine deep understanding, but a combination with comprehension tests. Second, problem solving is the outcome variable in the standard model of cognitive fit [10], which should therefore be maintained as separate variable.

Learning performance is used as dependent variable resulting from CTML [6]. According to CTML content, content presentation and user characteristics influence knowledge construction, that in turn is a cognitive learning outcome and can be measured by using learning performance indicators [29]. Mayer also specified deep understanding as the outcome of the learning process and hence, provides a relation between learning outcome, measured by learning performance, and deep understanding. Another fact for the relation between learning performance and deep understanding is the operationalization Recker and Dreiling used. They used model comprehension, problem solving, as well as cloze-tests. These tests are generally used for domain understanding. Therefore we apply deep understanding as input variable for learning performance.

Another dependent variable is readability of a modelling method which is defined as indicator how easy the resulting model is to read and the underlying domain reality to reconstruct. Bajaj proposed to measure readability effectiveness by the percentage of correct answers when asked questions about the domain and readability efficiency by the inverse of the time it takes to answer these questions [3]. Bajaj has operationalized readability with a comprehension test which resembles the operationalization of surface-level understanding. We relate these two variables since readability and surface-level understanding focus on model comprehension. To summarize, dependent variables so far presented can be distinguished in two categories: Understanding the model (surface-level understanding and readability) and understanding the domain (deep understanding, problem solving performance and learning performance).

A further category represents cognitive economy and effectiveness. Cognitive effectiveness can be maximized within a solution space. The solution space comprises perceptual and cognitive processing [22]. In perceptual processing retinal variables such as colour, shape and size of model constructs are detected [30] and separated from the background. Structure and relationships among constructs are perceived [31]. This is what Moody summarized with 'seeing'. What we have seen is then used within cognitive processing. It is brought into working memory, which is a temporary storage area for what is actively processed in this moment. For understanding, information of the working memory has to be integrated with knowledge stored in long-term memory. Moody summarized cognitive processing as understanding [22]. This is why we decide to relate the dependent variable understanding to cognitive effectiveness.

For script creation we found five variables. Modelling effectiveness is defined as the degree to which a conceptual modeller can correctly create a model and modelling efficiency as the amount of effort spent to create the model [3]. Moody tested how modelling efficiency and effectiveness effect perception-based variables such as perceived ease of use, perceived usefulness and intention. He argued that perception-based measurements play a more important role than objective measurements for e. g. deciding for a certain modelling grammar [17].

3   Implication for Future Research

By classifying variables into the framework of Gemino and Wand we could identify that several variables of the inner view of Grammar's Constructs as well as of Use of Grammar have not been tested. We could furthermore identify gaps within the framework. That way, it can be of interest how the application of colour or brightness effects dependent variables of model understanding such as readability or surface-level understanding (Grammar's Constructs). Another question might be how many elements within a chunk communicate most effectively (Grammar's Constructs and Use of Grammar). In Addition, one might ask if modelling on paper or with a computer might affect modelling effectiveness and efficiency (Medium of Content Delivery). Within User characteristics questions for further empirical investigations might be if a modelling grammar expert or domain expert is more effective and efficient in the usage of a higher NOC grammar than a novice of a certain modelling grammar or modelling domain.

Dependent variables for model application range from readability and comprehension to understanding. Different authors examine most of these variables independently. However, as we have shown in section 2 these variables are not independent but have a close relationship to each other. Since our definition of and relation between dependent variables is based on the studies we found, more conceptual work is needed to define the relations thoroughly.

4   Conclusion

This review comprises 13 empirical studies and three conceptual studies of hypotheses developed from Cognition for Conceptual Modelling, which we have classified and whose variables we have related to each other. We have further investigated the relationship between dependent variables, since several authors have used the same tests for different variables. This investigation showed that different authors have used different names for the same variables. That way documentation effectiveness and efficiency in [17] resembles modelling effectiveness and efficiency in [3]. Nontheless giving a profound classification and providing relations between dependent variables needs more conceptual work, which should not be limited to studies presented within this review.

References

1. Larkin, J., Simon, H.: Why a diagram is (sometimes) worth ten thousand words. Cognitive Science 11(1), 65–100 (1987)
2. Moody, D.: What makes a good diagram? improving the cognitive effectiveness of diagrams in is development. ADBIS 492, 481–492 (2007)
3. Rockwell, S., Bajaj, A.: Cogeval: Applying cognitive theories to evaluate conceptual models. Advanced Topics in Database Research 4, 255–282 (2004)
4. Gemino, A., Wand, Y.: Complexity and clarity in conceptual modeling: Comparison of mandatory and optional properties. Data & Knowledge Engineering 55(3), 301–326 (2005)
5. Khatri, V., Vessey, I., Ramesh, V., Clay, P., Park, S.: Understanding conceptual schemas: Exploring the role of application and IS domain knowledge. ISR 17(1), 81–99 (2006)
6. Recker, J., Dreiling, A.: Does it matter which process modelling language we teach or use? an experimental study on understanding process modelling languages without formal education. In: ACIS 2007, pp. 356–366 (2007)
7. Gemino, A., Wand, Y.: A framework for empirical evaluation of conceptual modeling techniques. Requirements Engineering 9(4), 248–260 (2004)

8. Parsons, J.: Effects of local versus global schema diagrams on verification and communication in conceptual data modeling. JMIS 19(3), 155–183 (2002)
9. Moody, D.L.: Cognitive Load Effects on End User Understanding of Conceptual Models: An Experimental Analysis. In: Benczúr, A.A., Demetrovics, J., Gottlob, G. (eds.) ADBIS 2004. LNCS, vol. 3255, pp. 129–143. Springer, Heidelberg (2004)
10. Vessey, I., Galletta, D.: Cognitive fit: An empirical study of information acquisition. ISR 2(1), 63–84 (1991)
11. Agarwal, R., Sinha, A., Tanniru, M.: Cognitive fit in requirements modeling: A study of object and process methodologies. JMIS 13(2), 137–162 (1996)
12. Gemino, A.: Empirical comparisons of animation and narration in requirements validation. Requirements Engineering 9(3), 153–168 (2004)
13. Kim, J., Hahn, J., Hahn, H.: How do we understand a system with (so) many diagrams? cognitive integration processes in diagrammatic reasoning. ISR 11(3), 284–303 (2000)
14. Bodart, F., Patel, A., Sim, M., Weber, R.: Should the optional property construct be used in conceptual modeling? ISR 12(4), 384–405 (2001)
15. Weber, R.: Are attributes entities? a study of database designers' memory structures. ISR 7(2), 137–162 (1996)
16. Bajaj, A.: The effect of the number of concepts on the readability of schemas: an empirical study with data models. Requirements Engineering 9(4), 261–270 (2004)
17. Moody, D.: Comparative Evaluation of Large Data Model Representation Methods: The Analyst's Perspective. In: Spaccapietra, S., March, S.T., Kambayashi, Y. (eds.) ER 2002. LNCS, vol. 2503, p. 214. Springer, Heidelberg (2002)
18. Kahneman, D.: Attention and effort. Prentice-Hall, Englewood Cliffs, NJ (1973)
19. Miller, G.: The magical number seven, plus or minus two: some limits on our capacity for processing information. Psychological Review 63(2), 81 (1956)
20. Craik, F., Lockhart, R.: Levels of processing: A framework for memory research. J. of Verbal Learning and Verbal Behavior 11(6), 671–684 (1972)
21. Kintsch, W., Vipond, D.: Reading comprehension and readability in educational practice and psychological theory, Erlbaum, Hillsdale, NJ (1979)
22. Moody, D.: The physics of notations: Toward a scientific basis for constructing visual notations in software engineering. IEEE Transactions on Software Engineering 35(6), 756–779 (2009)
23. Shannon, C., Weaver, W.: The mathematical theory of communication, vol. 19. University of Illinois Press Urban (1962)
24. Bertin, J.: Semiology of graphics: Diagrams, networks, maps. Wisconsin Press (1983)
25. Newell, A., Simon, H.: Human problem solving. Prentice-Hall (1972)
26. Goodman, N.: Languages of Art: An Approach to a Theory of Symbols. Bobbs-Merrill Co. (1968)
27. Parsons, J., Wand, Y.: Using cognitive principles to guide classification in information systems modeling. MISQ 32(4), 839–868 (2008)
28. Hegarty, M., Just, M.: Constructing mental models of machines from text and diagrams. J. of Memory and Language 32, 717–742 (1993)
29. Mayer, R.: Models for understanding. Rev. of Educat. Research 59(1), 43–64 (1989)
30. Lohse, G.: A cognitive model for understanding graphical perception. Human-Computer Interaction 8(4), 353–388 (1993)
31. Palmer, S., Rock, I.: Rethinking perceptual organization: The role of uniform connectness. Psychonomic Bulletin & Review 1(1), 29–55 (1994)

Using Domain Ontologies
as Semantic Dimensions in Data Warehouses[*]

Stefan Anderlik, Bernd Neumayr, and Michael Schrefl

Department of Business Informatics - Data & Knowledge Engineering,
Johannes Kepler University Linz, Austria
{anderlik,neumayr,schrefl}@dke.uni-linz.ac.at
http://www.dke.jku.at/

Abstract. More and more transaction systems collect records that reference concepts of domain ontologies in so-called semantic attributes. This paper investigates how such semantic attributes together with their referenced domain ontology can be best exploited for data analysis in data warehouses. This gives rise to two challenges: first, exploit the rich knowledge represented in domain ontologies for targeted analysis and, second, aggregate data along the subsumption hierarchy of concepts and ensuring summarizability in the absence of predefined aggregation levels. To meet these challenges, the paper extends OLAP (OnLine Analytical Processing) by integrating concept expressions and proper level definitions over domain ontologies into OLAP operations. A prototype demonstrates the feasibility of the approach.

Keywords: Business Intelligence, Ontologies, OLAP.

1   Introduction

Domain ontologies or terminologies are typically developed within large-scale, collaborative and long-term projects to formally define a shared vocabulary for a domain. With the advancement and common acceptance of domain ontologies, more and more transaction systems collect data records that reference concepts of ontologies in so-called *semantic attributes*. The Snomed Clinical Terminology (SNOMED CT [25]) is such a domain ontology which has been developed over the last 30 years in a multi-national effort, contains about 300000 concepts and is widely employed in the medical domain. For example, semantic attribute *disease* of a medical record may hold a reference to a concept defined by SNOMED CT. In our project[*] we are confronted with the need of our project partners from health insurance industry to exploit such semantic attributes for explorative data analysis.

[*] This work is funded by the Austrian Ministry of Transport, Innovation, and Technology in program FIT-IT Semantic Systems and Services under grant FFG-829594 (*Semantic Cockpit:* an ontology-driven, interactive business intelligence tool for comparative data analysis).

P. Atzeni, D. Cheung, and R. Sudha (Eds.): ER 2012, LNCS 7532, pp. 88–101, 2012.

A *data warehouse* is a special kind of database dedicated to complex data analysis, also referred to as OnLine Analytical Processing (OLAP), and is organized—at least conceptually—according to the multi-dimensional data model (for the purpose of this paper we make simplifications and appropriate assumptions): A data warehouse consists of *facts* collected in *cubes*. A cube is defined by a set of *dimensions* and a set of numeric *measures*. A dimension is a partial order of *nodes* (potentially described by *attributes*) representing a *roll-up hierarchy* where every node belongs to a certain *level* and every *leaf node* belongs to a single bottom. level and—to ensure *summarizability* [12]—rolls-up to exactly one node at each level of the dimension. A cube consists of fact instances (or simply: *facts*); each fact is identified by a multi-dimensional *point* (a node for each dimension) and quantified by *measure values*. The *granularity* of a point is the tuple of levels of its nodes. Facts are only asserted for points at the base granularity (i.e., points defined by a leaf node for each dimension). Data in data warehouses is analyzed and aggregated using OLAP operations, most notably: *dice* selects facts that roll up to a given point at a higher granularity; *slice* selects facts that roll up to points with nodes fulfilling some given predicates (over attributes of nodes); *roll-up* groups facts according to points at a given higher granularity and aggregates measure values according to given *aggregation functions*.

In order to exploit semantic attributes for OLAP-style analysis one may transform the domain ontology to a common dimension (i.e., non-semantic dimension) or map semantic attribute values from complex and un-leveled ontologies (such as SNOMED-CT) to leveled terminologies (such as ICD-10) which can easily be used as dimensions [4]. In both cases one loses valuable knowledge. The third approach, which we will investigate in this paper, is to use semantic attributes and domain ontologies as-is without much mapping or transformation. This last approach is favorable especially for explorative data analysis where one needs to fully exploit the rich knowledge encoded in ontologies.

The first major contribution of this paper is to show how semantic attributes together with the referenced ontologies may be seamlessly accommodated in the multidimensional data model as *semantic dimensions*: First, one picks a concept of interest from the referenced ontology. Second, the subsumption hierarchy rooted in this concept serves as core of the roll-up hierarchy of the semantic dimension. Third, one has to consider that semantic attributes refer to concepts at different levels of abstraction, e.g., the value of semantic attribute *disease* in one medical record is *'Pneumonia'* while in another medical record it is *'Viral Pneumonia'*. To cope with this situation, one adds for each concept a leaf node, e.g., *'Pneumonia (without further information)'*, representing direct references to that concept and ensures that asserted facts only refer to such leaf nodes.

The second major contribution of this paper is to adapt and extend OLAP operations *dice*, *slice* and *roll-up* to cubes with semantic dimensions. Due to our seamless integration of semantic attributes in the multidimensional data model, operation *dice* can be used as before. The operation selects facts that refer to a concept which is subsumed by the given concept. Dice operations can be evaluated by consulting the pre-coordinated subsumption hierarchy. The *slice* operation is extended to semantic dimensions by allowing to give a *concept expression* for each semantic dimension. The operation selects facts that refer to a concept which is subsumed by the given concept expression. In order to evaluate slice operations one needs to consult an ontology reasoner to determine concepts subsumed by the concept expression. To aggregate data along semantic dimensions without pre-defined levels we extend the *roll-up* operation to take a *semantic level* definition as input; in its simplest form, a semantic level is defined as an enumeration of concepts; to ensure summarizability we require level definitions to be partitioning with regard to the node in the dice point.

The feasibility of our approach is demonstrated by a proof-of-concept prototype extending relational OLAP with OWL-based semantic dimensions. The prototype is implemented in Java on top of PostgreSQL and employing the OWL API [8] together with Description Logic (DL) reasoner HermiT [26].

The remainder of the paper is organized as follows: Section 2 introduces a simple multi-dimensional data model and Section 3 extends it with semantic dimensions. Section 4 describes how to deal with semantic dimensions within OLAP operations dice, slice and roll-up. Section 5 describes a proof-of-concept prototype implementing OWL-extended relational OLAP. Related work is discussed in Section 6. We conclude the paper in Section 7.

2   Data Warehouses with Common Dimensions

In this section we introduce a conceptual multi-dimensional data model as employed by data warehouses. For the purposes of this paper we make simplifying and appropriate assumptions. In Section 3 we will extend this multi-dimensional data model to accommodate semantic dimensions.

Business records in operational systems usually have a set of descriptive fields and a set of measures. In business intelligence, business records are analyzed along hierarchies (called dimensions) associated with descriptive fields by "rolling-up" measure values through aggregation.

A *dimension* is a rooted directed acyclic graph of nodes, with $\top$ as root node, describing a roll-up hierarchy. Typically, dimensions (i) are homogenously leveled in that every node has a level and for any two levels l, l' and any two nodes o_1, o_2 of one level l they either both directly roll-up to a level l' or none, and vice versa, and (ii) provide for summarizability in that, next to (i), for any node o of some level l it directly or indirectly rolls-up to at most one node o' of some other level l'. If nodes of level l directly or indirectly roll-up to level l' we say l rolls-up to l', we write $l \uparrow l'$. Optionally, nodes may be described by attribute values, where nodes at common levels have common attributes. A *dimensional predicate* defined for level l is a predicate over attributes common to nodes at level l or any l' where $l \uparrow l'$, and selects every node of level l satisfying the predicate over the node and its ancestors.

Let D be the nodes of a dimension and l a level of the dimension we denote by D^l the set of nodes of D at level l. The sub-dimension $D_{/o}$ is the part of dimension D which is rooted in o, i.e. $D_{/o} = \{o' \in D | o' \uparrow o\}$ and $D_{/o}^l$ is the set of

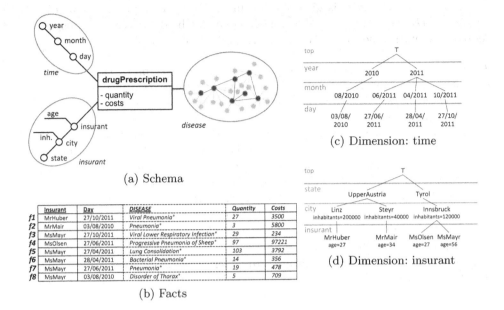

(a) Schema

(c) Dimension: time

(d) Dimension: insurant

	Insurant	Day	DISEASE	Quantity	Costs
f1	MrHuber	27/10/2011	Viral Pneumonia*	27	3500
f2	MrMair	03/08/2010	Pneumonia*	3	5800
f3	MsMayr	27/10/2011	Viral Lower Respiratory Infection*	29	234
f4	MsOlsen	27/06/2011	Progressive Pneumonia of Sheep*	97	97221
f5	MsMayr	27/04/2011	Lung Consolidation*	103	3792
f6	MsMayr	28/04/2011	Bacterial Pneumonia*	14	356
f7	MsMayr	27/06/2011	Pneumonia*	19	478
f8	MsMayr	03/08/2010	Disorder of Thorax*	5	709

(b) Facts

Fig. 1. Example Cube *drugPrescription*

nodes at level l in sub-dimension $D_{/o}$. The bottom or base level of a dimension is denoted as $\bot$ with $D^{\bot}$ being the set of leaf nodes of dimension D. The top level of a dimension is denoted as $\top$ with $D^{\top} = \{\top\}$.

A *fact* consist of an n-tuple $\langle o_1, \ldots, o_n \rangle$ of dimension nodes, the *multi-dimensional point*, and an m-tuple $[v_1, \ldots, v_m]$ of measure values drawn from value domains $V_1, \ldots, V_m$; the *granularity* of the fact is an n-tuple of levels, $\langle l_1, \ldots, l_n \rangle$ with l_i being the level of node o_i (with $i=1..n$).

We say a multidimensional point $p = \langle o_1, \ldots, o_n \rangle$ rolls up to another multidimensional point $p' = \langle o'_1, \ldots, o'_n \rangle$, denoted as $p \uparrow p'$, iff each coordinate o_i rolls-up to the corresponding coordinate o'_i (with $i = 1..n$).

Let f be a fact over an n-dimensional point p then $f\langle\rangle$ selects p. $f\langle i \rangle$ selects the i-th coordinate and $f[j]$ the j-th measure value.

We say a fact f rolls-up to or is subordinate to a multidimensional-point p, written $f \uparrow p$, if the multidimensional-point $f\langle\rangle$ rolls-up to p.

A *cube* is a homogenous set of facts, i.e., all facts within a cube refer to the same dimensions at the same level, i.e., they have the same granularity, and they possess the same set of measures. The granularity of a cube is the granularity of its facts. The dimensions of an n-dimensional cube span a dimensional space $D_1 \times \cdots \times D_n$ with each fact describing a point $\langle o_1, \ldots, o_n \rangle$ in this space. A *base cube* is at base granularity, $\langle \bot, \ldots, \bot \rangle$, referencing only leaf nodes of dimensions. A granularity $g = \langle l_1, \ldots, l_n \rangle$ rolls up to a granularity $g' = \langle l'_1, \ldots, l'_n \rangle$, denoted as $g \uparrow g'$, iff each level l_i of g rolls up to the corresponding level l'_i of g' (with $i=1..n$).

The *domain* of a cube is given by its dimension space and its granularity, denoted as $D_1^{l_1} \times \cdots \times D_n^{l_n}$. The domain of a cube may be restricted to a subspace rooted in a multi-dimensional point $\langle o_1, \ldots, o_n \rangle$, denoted as $D_{1/o_1}^{l_1} \times \cdots \times D_{n/o_n}^{l_n}$.

Example 1 (Running Example). A base cube *drugPrescription* (Fig. 1) in a data warehouse from health insurance industry contains base facts such as *f7* = $\langle MsMayr,27/06/2011 \rangle [19,478]$ referring to leaf nodes of dimensions *insurant* and *time* (we ignore dimension *disease* for the moment) and being described by values of measures *quantity* and *costs*. The domain of the cube is insurant^{insurant} × time^{day}. Dimensions consist of a roll up hierarchy of nodes, e.g., *Innsbruck* ↑ *Tyrol*, and a roll-up hierarchy of levels, e.g., *city* ↑ *state*. Points roll up to other points, e.g., $\langle MsMayr,06/2011 \rangle$ ↑ $\langle Tyrol,2011 \rangle$. Facts roll up to points, e.g., *f7* ↑ $\langle MsMayr,06/2011 \rangle$. Granularities roll up to other granularities, e.g., $\langle insurant,month \rangle$ ↑ $\langle state,year \rangle$. Sub-dimension *time*$_{/2011}$ is the part of dimension *time* rooted in node *2011*. The space of the domain of the cube may be restricted to sub-dimensions, e.g., insurant$_{/Innsbruck}^{insurant}$ × time$_{/2011}^{day}$ only covers points that roll up to point $\langle Innsbruck, 2011 \rangle$. Dimensional predicate *(city.inh > 100000 ∧ insurant.age > 32)* defined for level *insurant* selects all nodes at level *insurant* satisfying the logical expression.

3   Extending Data Warehouses with Semantic Dimensions

In this section we first review the basics of domain ontologies and clarify the meaning of semantic attributes. We then show how semantic attributes together with their referenced domain ontologies can be used as semantic dimensions. We introduce levels in semantic dimensions and discuss summarizable consistency for these levels.

A *domain ontology* consists of concepts and roles, together with axioms defining their meaning. Each concept represents a set of objects, its members or instances. The definition of a concept are the criteria which objects must fulfill to be considered members of the concept. Concepts are organized in a subsumption hierarchy, this subsumption hierarchy may be asserted by the ontology designer or be derived by an automatic reasoner based on the definitions of the concepts. A concept is *subsumed by* (or 'is subconcept of' or 'is-a') another concept, if each of its members is—in every possible world—also member of the subsuming concept. A concept may be directly subsumed by multiple other concepts. In addition to *pre-coordinated* concepts (named concepts described in the ontology) one may later introduce *post-coordinated* concepts (complex concepts that are composed from simple concepts [24]) in terms of *concept expressions*. The automatic reasoner may determine all concepts subsumed by such a concept expression. While our approach is widely agnostic towards the employed ontology formalism, in this paper we use Description Logics (DL) [1].

Data records reference—via a semantic attribute—concepts of (some part of) a domain ontology. We assume the domain of a semantic attribute is given by a concept and all its subconcepts. A reference to a concept represents an instance

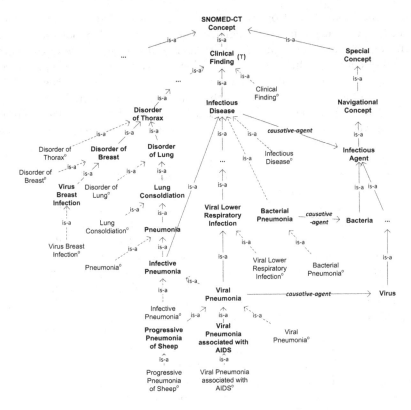

Fig. 2. Example Semantic Dimension *Disease*

of the referenced concept (whereby this instance is as such neither represented in the ontology nor in the data record). A reference to a concept with subconcepts represents a case that can be classified as an instance of exactly the concept but none of its subconcepts, i.e., a reference to concept o is treated as instance of 'o (*without further information*)', we also say a reference to o represents a *direct instance* of o.

We now describe how to use a domain ontology as a *semantic dimension* in data warehousing:

1. We identify the relevant part of the domain ontology by picking a concept from the ontology (i.e., the domain of the semantic attribute).
2. We use the subsumption hierarchy rooted in this concept as initial roll-up hierarchy. A concept o rolls up to a concept o', $o \uparrow o'$, iff o is subsumed by o'.
3. We add to each concept o a leaf node $o°$, with $o° \uparrow o$, representing all cases of 'o (*without further information*)'. Facts of base cubes refer to such leaf nodes only.

Semantic dimensions differ from common dimensions in that they do not have pre-defined levels (apart from a singleton top level $\top$ and a bottom-level $\bot$ consisting of all leaf nodes) and in that their nodes have richer descriptions in the form of concept definitions. These richer descriptions can be exploited by using concept expressions as dimensional predicates. A concept (together with its subconcepts and leaf nodes) *satisfies* such a dimensional predicate if the concept is subsumed by the concept expression. The flexibility of the approach may be further improved by using post-coordinated concepts, i.e., concept expressions, as nodes in semantic dimensions.

Example 2 (Semantic Dimension). To derive semantic dimension *disease* (see Figure 2) we first pick '*Clinical Finding*' from ontology SNOMED-CT. The subsumption hierarchy rooted in '*Clinical Finding*' is augmented by leaf nodes such as '*Pneumonia°*' meaning '*Pneumonia (without further information)*'. Facts from base cube *drugPrescription* (see Figure 1) refer to such leaf nodes only, e.g., fact f7 refers to '*Pneumonia°*'. Looking at the rich knowledge in semantic dimensions, concept '*Viral Pneumonia*'—which is defined as (using DL syntax) '*Pneumonia* $\sqcap \exists causative\text{-}agent.Virus$' (in Figure 2 this existential quantification is represented as link between '*Viral Pneumonia*' and '*Virus*')—satisfies dimensional predicate '$\exists causative\text{-}agent.(Virus \sqcup Bacteria)$'.

For roll-up operations we need to define levels, either a priori or inline with OLAP operations. In this paper we only discuss the simple case where a level is defined by explicitly enumerating the set of concept and leaf nodes at this level. In contrast to common dimensions, a node in a semantic dimension may belong to multiple levels. As levels cannot be easily globally superimposed over subsumption hierarchies we define levels locally, relative to some concept. A level l defined locally to a concept o, denoted as $^o l$, contains only nodes that roll up to o.

Aggregation along subsumption hierarchies need to be summarizable consistent. A level $^o l$ is *summarizable consistent* with regard to the bottom level, written as $\bot \, \mathord{\updownarrow} \, ^o l$, iff it partitions the set of leaf nodes that roll up to o, i.e., $\forall o' \in D_{/o}^{\bot} : \exists_{=1} o'' \in {}^o l : o' \uparrow o''$ (note that $\exists_{=1}$ denotes an unique existential quantification meaning 'there exists exactly one'). A level $^o l$ in subdimension $D_{/\hat{o}}$, denoted as $D_{/\hat{o}}^{o l}$, is *summarizable consistent*, iff it partitions the set of leaf nodes that roll up to $\hat{o}$; this is the case if $\hat{o}$ rolls-up to o and $^o l$ is summarizable consistent with regard to the bottom level, i.e., $\hat{o} \uparrow o \wedge \bot \, \mathord{\updownarrow} \, ^o l$. A level $^{\bar{o}}\bar{l}$ is *summarizable consistent* with regard to $^o l$, written as $^o l \, \mathord{\updownarrow} \, ^{\bar{o}}\bar{l}$, iff $\forall o' \in D_{/\hat{o}}^{o l} : \exists_{=1} o'' \in {}^{\bar{o}}\bar{l} : o' \uparrow o''$. As with common dimension hierarchies, our definitions guarantee for semantic dimensions summarizability with respect to the bottom level if aggregations are performed iteratively along subsumption hierarchies, i.e., $\bot \, \mathord{\updownarrow} \, ^o l \wedge {}^o l \, \mathord{\updownarrow} \, ^{\bar{o}}\bar{l} \Rightarrow \bot \, \mathord{\updownarrow} \, ^{\bar{o}}\bar{l}$.

Example 3 (Levels in Semantic Dimensions). We define a global level $^\top clinical = \{$'*Disorder of Thorax°*', '*Disorder of Breast°*', '*Virus Breast Infection*', '*Disorder of Lung°*', '*Lung Consolidation*', '*Infectious Disease°*', '*Viral Lower Respiratory*

Infection°', *'Bacterial Pneumonia°'*, *'Clinical Finding°'*}. Level $^\top clinical$ is summarizable consistent with regard to the bottom level of *disease* since every leaf node is subsumed by exactly one of the elements of $^\top clinical$. We define a level locally to concept *'Disorder of Thorax'* as $^{'Disorder\ of\ Thorax'}lung = \{$*'Disorder of Thorax°'*, *'Disorder of Breast°'*, *'Disorder of Lung'*} whereby every leaf node subsumed by *'Disorder of Thorax'* rolls up to one of the nodes of the level. Level $^{'Disorder\ of\ Thorax'}lung$ is summarizable consistent with regard to level $^\top clinical$ since every node at level $^\top clinical$ that is subsumed by concept *'Disorder of Thorax'* rolls up to exactly one node at level $^{'Disorder\ of\ Thorax'}lung$.

4   Extending OLAP to Semantic Dimensions

In this Section we revisit the common OLAP operations over cubes (dice, slice, and roll-up), and show how they, adequately defined, seamlessly apply to common as well as semantic dimensions due to our careful and diligent design of the semantic dimensions and its levels.

4.1   Dice

The *dice* operator δ applied to a cube c of domain $D^{l_1}_{1/o_1} \times \cdots \times D^{l_n}_{n/o_n}$ selects the sub-cube of the cube whose facts are subordinate to a given multidimensional point $p = \langle \bar{o}_1, \ldots, \bar{o}_n \rangle$, which is called the dice point. The application of the dice operator is written as $\delta_p(c)$ and is valid if $p \uparrow \langle o_1, \ldots, o_n \rangle$, and $\delta_p(c) = \{f \in c | f \uparrow p\}$. The operation reduces the dimensional space of the domain of the input cube but not its granularity yielding a result cube with domain $D^{l_1}_{1/\bar{o}_1} \times \cdots \times D^{l_n}_{n/\bar{o}_n}$.

Example 4 (Dice). The dice operation (consider Fig. 1 and Fig 2) *drugPres1* =

$$\delta_{\langle \text{Innsbruck}, 2011, \text{'Disorder of Thorax'}\rangle}(drugPrescription)$$

results in a cube *drugPres1* consisting of all *drug prescriptions* for *insurants* in *Innsbruck* in *2011* and referring to concepts subsumed by *'Disorder of Thorax'*. Its domain is $insurant^{insurant}_{/Innsbruck} \times time^{day}_{/2011} \times disease^{\bot}_{/\text{'Disorder of Thorax'}}$.

4.2   Slice

The *slice* operator σ applied to a cube c selects the facts of the cube whose multidimensional points satisfy given dimensional predicates. The application of the slice operator is written as $\sigma_{\langle e_1, \ldots, e_n \rangle}(c)$ where $e_1, \ldots, e_n$ are dimensional predicates; and $\sigma_{\langle e_1, \ldots, e_n \rangle}(c) = \{f \in c | f\langle i \rangle$ satisfies $e_i, i = 1..n\}$. Note that in the definition of the slice operator we did not indicate a domain of the cube since the slice operator does not alter the domain.

The definition of the slice operator seamlessly applies to common and semantic dimensions by using concept expressions in Description Logics as dimensional predicates (see Section 3).

Example 5 (Slice). The slice operation $drugPres2 =$

$$\sigma_{\langle \text{city.inh}>100000 \,\wedge\, \text{insurant.age}>32,\text{true},\exists\text{causative}-\text{agent.}(\text{Virus} \,\sqcup\, \text{Bacteria})\rangle}(drugPres1)$$

results in a sub-cube *drugPres2* consisting of facts referring to insurants older than *32* and living in *cities* with more than *100000 inhabitants*, and referring to *diseases* caused by some *'Virus'* or some *'Bacteria'*.

4.3   Rollup

The *roll-up* operator ρ applied to a cube c of domain $D^{l_1}_{1/o_1} \times \cdots \times D^{l_n}_{n/o_n}$ with m measures rolls-up all facts of the cube to a given granularity $\langle \bar{l}_1, \ldots, \bar{l}_n \rangle$ by applying indicated aggregation functions $\Sigma_1, \ldots, \Sigma_m$ to its measures. The application of the roll-up operator is written as $\rho_{\langle \bar{l}_1,\ldots,\bar{l}_n \rangle \langle \Sigma_1,\ldots,\Sigma_m \rangle}(c)$ and is *valid* if the granularity of the cube rolls-up to $\langle \bar{l}_1, \ldots, \bar{l}_n \rangle$, and $\rho_{\langle \bar{l}_1,\ldots,\bar{l}_n \rangle \langle \Sigma_1,\ldots,\Sigma_m \rangle}(c) = \{ \langle \hat{o}_1, \ldots, \hat{o}_n, v_1, \ldots, v_m \rangle \in D^{\bar{l}_1}_{1/o_1} \times \cdots \times D^{\bar{l}_n}_{n/o_n} \times V_1 \ldots \times V_m | \exists f \in c : f \uparrow \langle \hat{o}_1, \ldots, \hat{o}_n \rangle \wedge v_i = \sum_{i \atop f\uparrow\langle \hat{o}_1,\ldots,\hat{o}_n \rangle} f[i], i = 1..m \}$. The operation changes the granularity of the domain of the input cube but not its dimensional space yielding a result cube with domain $D^{\bar{l}_1}_{1/o_1} \times \cdots \times D^{\bar{l}_n}_{n/o_n}$.

The roll-up operator can now be seamlessly applied to common and semantic dimensions by allowing pre-defined and in-line-defined levels for rolling up along semantic dimensions. The summarizability conditions as introduced in Section 3 need to hold. A roll-up operation with level $^{\bar{o}}\bar{l}$ as the i-th coordinate of the given granularity applied to an input cube with $D^{ol}_{/\hat{o}}$ as the i-th coordinate of the input cube's domain is *summarizable consistent*, if $\hat{o} \uparrow \bar{o} \wedge {}^{o}l \,\updownarrow\, {}^{\bar{o}}\bar{l}$. This ensures that $D^{\bar{o}\bar{l}}_{/\hat{o}}$ (the i-th coordinate of the domain of the resulting cube) is summarizable consistent.

Example 6 (Roll-Up). Consider level $^{\top}clinical$ (Example 3) and cube *drugPres2* (Example 5) with domain $insurant^{insurant}_{/Innsbruck} \times time^{day}_{/2011} \times disease^{\perp}_{/\text{'Disorder of Thorax'}}$. The roll-up operation $drugPres3 =$

$$\rho_{\langle \text{city}, \top, {}^{\top}\text{clinical}\rangle \langle \text{SUM},\text{SUM}\rangle}(drugPres2).$$

results in a roll-up cube *drugPres3* with domain $insurant^{city}_{/Innsbruck} \times time^{\top}_{/2011} \times disease^{{}^{\top}clinical}_{/\text{'Disorder of Thorax'}}$.

The operation

$$\rho_{\langle \top, \top, {}^{\text{'Disorder of Thorax'}}\text{lung}\rangle \langle \text{SUM},\text{SUM}\rangle}(drugPres3).$$

further rolls up cube *drugPres3* to level $^{\text{'Disorder of Thorax'}}lung$. This is summarizable consistent since *drugPres3* is at level $^{\top}clinical$ in the sub-dimension of dimension *disease* rooted in *'Disorder of Thorax'*, denoted as $disease^{{}^{\top}clinical}_{/\text{'Disorder of Thorax'}}$ and since $^{\top}clinical \,\updownarrow\, {}^{\text{'Disorder of Thorax'}}lung$ (see Example 3).

5   Prototype

In this Section we exemplify how relational OLAP can be extended with OWL-based semantic dimensions. We implemented a simple *Semantic OLAP Engine* in the Java programming language using PostgreSQL9.0 [20], OWL reasoner HermiT [26], and the OWL API [8] for the interaction with the OWL reasoner. Figure 3 shows the architecture of our prototype and the steps (0-3) involved. In the remainder of this section we will look at these steps in more detail.

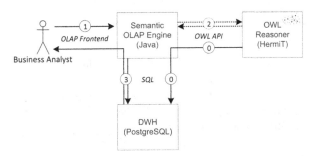

Fig. 3. Prototype Architecture

In a precoordination step 0—in order to keep expensive reasoner interaction during query time as low as possible—the reasoner infers for each semantic dimension the subsumption hierarchy of concepts. This concept hierarchy is augmented by °-concepts (as described in Section 2.2) and then stored in the DWH. A table (as exemplified in left part of Fig. 4) holds all concepts of the semantic dimension. Another table (middle part of Fig. 4) holds for each concept its equivalent concepts and direct subconcepts. For convenience of querying a view is defined for the transitive-reflexive closure of the subsumption hierarchy (right part of Fig. 4). For performance reasons this view might be materialized.

The business analyst may then, in step 1, apply Semantic-OLAP operations consecutively using a (graphical) OLAP frontend. Consider, for example, the following combination of dice, slice, and rollup operations:

$$\rho_{\langle \text{city}, \top, \{\text{`Disorder of Thorax}^{\circ}\text{'},\text{`Disorder of Breast}^{\circ}\text{'},\text{`Disorder of Lung'}\}\rangle}\langle\text{SUM},\text{AVG}\rangle$$
$$\left(\sigma_{\langle \text{city.inh}>100000 \ \land \ \text{insurant.age}>32,\text{true},\exists\text{causative}-\text{agent.}(\text{Virus} \sqcup \text{Bacteria})\rangle}\right)$$
$$\left(\delta_{\langle\top,\top,\text{`Disorder of Thorax'}\rangle}(\text{drugPrescription})\right)$$

If a slice operation contains a concept expression over a semantic dimension, the semantic OLAP engine asks, in step 2, the OWL reasoner for pre-coordinated concepts that are subsumed by this concept expression. This reasoner interaction during query time is also referred to as *post-coordination*. Since the subsumption hierarchy of pre-coordinated concepts is already stored in the DWH, reasoner interaction is minimized to only retrieving direct subconcepts of the concept expression.

In step 3, the OLAP operations are translated to SQL and evaluated by the DWH. The combination of dice, slice, and rollup operations is translated to a concepts_disease

id	label
1	Clinical Finding
2	Clinical Finding^o
3	Disorder of Thorax
4	Disorder of Thorax^o
5	Disorder of Lung
6	Disorder of Lung^o
...	...

sub_disease

sup	sub
1	2
1	3
3	4
3	5
5	6
...	...

```
CREATE VIEW h_disease AS
    WITH RECURSIVE hi(sup, sub) AS (
        SELECT sup, sub FROM sub_disease
    UNION
        SELECT hi.sup, h.sub
        FROM hi, sub_disease h
        WHERE hi.sub = h.sup
    )
SELECT sup, c.label AS sup_label, sub
FROM hi JOIN concepts_disease c ON sup = c.id
UNION
SELECT id AS sup, label AS sup_label, id AS sub
FROM concepts_disease
```

Fig. 4. Pre-coordinated Subsumption Hierarchy of Semantic Dimension *Disease*

single SQL query as exemplified in Fig. 5 (see explanations below) with the SQL-mapping of a single operation being similar, just ignoring the mappings of the other operations. We assume that the DWH employs a star or snowflake schema [27] where facts are stored in a *fact table* similar to Figure 1b but with (semantic) dimensions having numeric IDs.

```
1:  SELECT i.city, d.sup, SUM(f.quantity), AVG(f.costs)
2:  FROM drugPrescription f
3:      JOIN h_disease d ON f.disease = d.sub
4:      JOIN h_insurant i ON f.insurant = i.insurant
5:  WHERE f.disease IN (SELECT sub FROM h_disease
6:                      WHERE sup_label = 'Disorder of Thorax')
7:  AND f.insurant IN (SELECT insurant FROM h_insurant
8:                     WHERE city_inh > 100000 AND insurant_age > 32)
9:  AND f.disease IN (SELECT sub FROM h_disease
10:                     WHERE sup_label IN ('Viral Pneumonia', 'Bacterial Pneumonia'))
11: AND d.sup_label IN ('Disorder of Thorax^0',
12:                     'Disorder of Breast^0', 'Disorder of Lung')
13: GROUP BY i.city, d.sup
```

Fig. 5. Combination of Dice, Slice and Rollup translated to SQL

The *dice* operation is translated—for each semantic dimension—by a subquery that retrieves all subconcepts (including equivalent concepts, direct and indirect subconcepts) of the given concept and a WHERE-clause that selects facts that refer to such a subconcept (lines 5-6 in Fig. 5). The *slice* operation is translated—for each semantic dimension—by a subquery that retrieves all subconcepts of a concept directly subsumed by the concept expression (as inferred by the reasoner in step 2) and a WHERE-clause that selects facts that refer to such a subconcept (lines 9-10). The *rollup* operation is translated—for each semantic dimension—as follows, the fact table is joined with the transitive-reflexive closure of the subsumption hierarchy (line 3) to determine for each fact the concepts directly or indirectly referred to, restricted to concept that appear in the dynamic level (line 11-12), facts are then grouped to that concepts (line 13) and the aggregation operations are applied in the SELECT-clause (line 1).

We have tested the combined application of operations dice, slice, and rollup of Fig. 5 with data (1.7 million facts) stemming from Austrian insurance companies and a materialized semantic dimension *disease* containing 5,728 concepts from SNOMED-CT. This application was evaluated in *3.2 seconds* at a standard workstation (hardware configuration: *Intel Core i7 M640, DDR800, 4GB*). We consider this as reasonable performance for our use cases. Further optimization together with detailed performance studies are subject of future work.

6 Related Work

Querying over semantic attributes in relational databases is discussed by Das et al. [6] and implemented in *Oracle Database 11g Semantic Technologies*. In addition to their approach, we also allow to use concept expressions as selection criteria (post-coordination) and provide for a seamless integration in data warehousing and OLAP. Martinenghi and Torlone [15] extend relational algebra to querying leveled taxonomies.

Defining levels local to a sub-dimension has previously been investigated by Lehner et al. [11] and by Neumayr et al. [17]. Summarizability in heterogeneous dimensions is investigated by Hurtado and Mendelzon [9]. Malinowski and Zimányi [14] give an overview of different kinds of hierarchies other than our simple hierarchies (in Section 2), some of these are supported by commercial OLAP systems of Microsoft and Oracle. None of these works and commercial systems support exploiting the rich knowledge represented in domain ontologies and aggregating data along subsumption hierarchies as discussed in this paper.

Domain ontologies and taxonomies play a major role in medical information systems: concepts from complex ontologies like SNOMED-CT [25] are used to encode diagnosis and other information in medical records [3], leveled taxonomies like ICD-10 [28] are used for healthcare statistics. Already Pedersen and Jensen [19] highlighted the importance of using such 'advanced classification structures' as dimensions in data warehouses. Eder and Koncilia [7] discuss how to use ICD-10 for data warehousing. Lieberman et al [13] show the usefulness of complex ontologies like SNOMED-CT for querying medical data warehouses but they do not provide tools or query language extensions for this purpose.

Formalization of data warehousing topics such as aggregation and summarizability has been discussed in the framework of Description Logics. Baader and Sattler [2] and Calvanese et al. [5] investigate aggregation functions in Description Logics and their complexity bounds. Rizzi et al. [22] highlight the possibilities of analyzing semantic web data encoded according to formal ontologies. Nebot et al. [16] propose *Multi-dimensional Integrated Ontologies* as basis for the multi-dimensional analysis of semantic web data.

Using ontologies in the data warehouse design process has received considerable attention. Pardillo and Mazón [18] identify several shortcomings of traditional data warehouse design and show how using ontologies in the design process can overcome them. Khouri and Bellatreche [10] introduce a methodology for designing data warehouses from ontology-based operational databases. Romero and Abelló [23] present a methodology in which the multidimensional

DWH schema is derived from a domain ontology. Rizzi et al. [22] mention the use of domain ontologies for data mart integration. In contrast to these works we do not use ontologies for the data warehouse design process, but instead used them as part of the data and for OLAP-style querying.

7 Conclusion

In this paper we showed how semantic attributes referring to pre-coordinated concepts in domain ontologies can be seamlessly integrated in data warehousing and OLAP by using subsumption hierarchies of pre-coordinated concepts as roll-up hierarchy. As a prerequisite for summarizability, we added to each concept a leaf node representing direct instances of the concept and ensured that facts only refer to such leaf nodes. We defined summarizable consistent levels by enumeration of pre-coordinated concepts and leaf nodes. We exploited the rich knowledge in domain ontologies by using post-coordinated concepts (i.e., concept expressions) as dimensional predicates in slice operations. We demonstrated the feasibility of the approach by a prototype based on off-the-shelf technology, extending relational OLAP with OWL-based semantic dimensions.

References

1. Baader, F., Nutt, W.: Basic description logics. Description Logic Handbook, 43–95 (2003)
2. Baader, F., Sattler, U.: Description logics with aggregates and concrete domains. Inf. Syst. 28(8), 979–1004 (2003)
3. Benson, T.: Principles of health interoperability HL7 and SNOMED. Springer (2010)
4. Berndt, D., Hevner, A., Studnicki, J.: The catch data warehouse: support for community health care decision-making. Decision Support Systems 35(3), 367–384 (2003)
5. Calvanese, D., Kharlamov, E., Nutt, W., Thorne, C.: Aggregate queries over ontologies. In: ONISW 2008: Proceeding of the 2nd International Workshop on Ontologies and Information Systems for the Semantic Web, pp. 97–104. ACM, New York (2008)
6. Das, S., Chong, E.I., Eadon, G., Srinivasan, J.: Supporting ontology-based semantic matching in rdbms. In: VLDB, pp. 1054–1065 (2004)
7. Eder, J., Koncilia, C.: Incorporating icd-9 and icd-10 data in a warehouse. In: CBMS, pp. 91–96. IEEE Computer Society (2002)
8. Horridge, M., Bechhofer, S.: The owl api: a java api for working with owl 2 ontologies. In: Proc. OWLED 2009 Workshop on OWL: Experiences and Directions. CEUR Workshop Proceedings, vol. 529 (2009)
9. Hurtado, C.A., Mendelzon, A.O.: Reasoning about Summarizability in Heterogeneous Multidimensional Schemas. In: Van den Bussche, J., Vianu, V. (eds.) ICDT 2001. LNCS, vol. 1973, pp. 375–389. Springer, Heidelberg (2000)
10. Khouri, S., Bellatreche, L.: A methodology and tool for conceptual designing a data warehouse from ontology-based sources. In: DOLAP, pp. 19–24 (2010)

11. Lehner, W., Albrecht, J., Wedekind, H.: Normal forms for multidimensional databases. In: Rafanelli, Jarke [21], pp. 63–72.
12. Lenz, H.J., Shoshani, A.: Summarizability in olap and statistical data bases. In: SSDBM, pp. 132–143 (1997)
13. Lieberman, M.I., Ricciardi, T.N., Masarie, F., Spackman, K.A.: The use of SNOMED CT simplifies querying of a clinical data warehouse. In: AMIA Annu. Symp. Proc. (2003)
14. Malinowski, E., Zimányi, E.: Hierarchies in a multidimensional model: From conceptual modeling to logical representation. Data Knowl. Eng. 59(2), 348–377 (2006)
15. Martinenghi, D., Torlone, R.: Querying Databases with Taxonomies. In: Parsons, J., Saeki, M., Shoval, P., Woo, C., Wand, Y. (eds.) ER 2010. LNCS, vol. 6412, pp. 377–390. Springer, Heidelberg (2010)
16. Nebot, V., Llavori, R.B., Pérez-Martínez, J.M., Aramburu, M.J., Pedersen, T.B.: Multidimensional Integrated Ontologies: A Framework for Designing Semantic Data Warehouses. In: Spaccapietra, S., Zimányi, E., Song, I.-Y. (eds.) Journal on Data Semantics XIII. LNCS, vol. 5530, pp. 1–36. Springer, Heidelberg (2009)
17. Neumayr, B., Schrefl, M., Thalheim, B.: Hetero-homogeneous hierarchies in data warehouses. In: Link, S., Ghose, A. (eds.) APCCM. CRPIT, vol. 110, pp. 61–70. Australian Computer Society (2010)
18. Pardillo, J., Mazón, J.N.: Using ontologies for the design of data warehouses. International Journal of Database Management Systems (IJDMS) 3(2), 73–87 (2011)
19. Pedersen, T.B., Jensen, C.S.: Research issues in clinical data warehousing. In: Rafanelli, Jarke [21], pp. 43–52.
20. The PostgreSQL Global Development Group: PostgreSQL 9.0.7 Documentation (2010), http://www.postgresql.org/docs/9.0/static/index.html (viewed March 11, 2012)
21. Rafanelli, M., Jarke, M. (eds.): Proceedings of 10th International Conference on Scientific and Statistical Database Management, July 1-3. IEEE Computer Society, Capri (1998)
22. Rizzi, S., Abelló, A., Lechtenbörger, J., Trujillo, J.: Research in data warehouse modeling and design: dead or alive? In: Proceedings of the 9th ACM International Workshop on Data Warehousing and OLAP, pp. 3–10. ACM (2006)
23. Romero, O., Abelló, A.: A framework for multidimensional design of data warehouses from ontologies. Data Knowl. Eng. 69(11), 1138–1157 (2010)
24. Rosenbloom, S., Miller, R., Johnson, K., Elkin, P., Brown, S.: Interface terminologies: facilitating direct entry of clinical data into electronic health record systems. Journal of the American Medical Informatics Association 13(3), 277–288 (2006)
25. Schulz, S., Cornet, R.: Snomed cts ontological commitment. In: International Conference on Biomedical Ontology (2009)
26. Shearer, R., Motik, B., Horrocks, I.: Hermit: A highly-efficient owl reasoner. In: Proceedings of the 5th International Workshop on OWL: Experiences and Directions (OWLED 2008), pp. 26–27 (2008)
27. Vassiliadis, P., Sellis, T.: A survey of logical models for olap databases. ACM SIGMOD Record 28(4), 64–69 (1999)
28. World Health Organization: The icd-10 classification of mental and behavioural disorders: diagnostic criteria for research (1993)

Sliced Column-Store (SCS): Ontological Foundations and Practical Implications

Yoones A. Sekhavat[1] and Jeffrey Parsons[2]

[1] Department of Computer Science, Memorial University of Newfoundland,
St. John's, Canada
[2] Faculty of Business Administration, Memorial University of Newfoundland,
St. John's, Canada
{yoonesas,jeffreyp}@mun.ca

Abstract. Advances in business intelligence systems based on processing large data volumes are driving efforts toward read-optimized databases. Recently, the use of column-store approaches as a solution for such databases has become quite popular. The main idea behind the column-store approach is reducing I/O requirements through vertical partitioning of data in which only those attributes that are required to answer a query are read. This paper offers two contributions to column-store data models. First, we show that such models can be grounded in ontological foundations that provide a theoretical basis for column-store databases based on representational adequacy. Second, we use these ontological foundations as the basis to propose an extended model of the column-store model called Sliced Column Store (SCS), and show that this model outperforms column-store models for read-oriented queries.

Keywords: Column-store, data partitioning, ontology, query processing.

1 Introduction

The performance of query-intensive systems is strongly dependent on the performance of underlying databases and query processing engines. These systems require read-optimized database engines in which efficiently answering read-oriented ad-hoc queries has priority over write-oriented queries. Relational databases are optimized for Online Transaction Processing (OLTP) in which handling a large number of small inserts and updates in an acceptable time is more important than reducing query answering time for complex read-oriented queries. These databases are implemented based on the row-store approach in which records (rows of tables) are contiguously stored in memory such that a single disk write is usually enough to write all fields of a record on a disk, and provides acceptable query performance in sequential access to the data. However, relational databases are not optimized for read-oriented tasks such as querying data warehouses. In terms of read-oriented queries, such architecture usually requires a full table scan where many data columns are projected out since not all attributes are required to answer a query. Consequently, many irrelevant properties are read, even though they are not necessary to process a query. As a result, relational

P. Atzeni, D. Cheung, and R. Sudha (Eds.): ER 2012, LNCS 7532, pp. 102–115, 2012.

systems are not I/O efficient because they use I/O bandwidth for reading unnecessary data. Based on this fact and the simple idea of reading only those attributes that are required to answer a query, the Column Store (CS) approach has been proposed. Unlike the row-store model in which the properties of a record are stored contiguously in memory, in the column-store approach, the values of each property for different records are stored contiguously. Recent years have witnessed the introduction of many database systems based on the column-store approach [1-4].

In addition to improving the performance of read-oriented queries, some research has also shown the potential of the column-store approach in addressing other problems in database management. Vertical table partitioning that uses the column store approach, is one of the major techniques used in database management to address difficulties in managing large database systems. As discussed in [5], table partitioning can increase the manageability of database systems by allowing parallel access to different properties, easier backup, and a fine grained access control through providing facilities to enforce different access-right policies for different partitions. In addition to query efficiency and database manageability, the column-store approach provides other advantages such as supporting multi-value attributes, handling null values and efficient handling of wide tables with sparse data.

Based on these promising advantages of the column-store model, in this paper we explore a theoretical foundation that provides a basis for the column-store approach. While prior research treats the model only as an ad-hoc approach used for physical data storage, we examine the question whether it is possible to provide a theoretical foundation that can further improve the performance of query processing. Answering this question is crucial since the answers can direct further efforts to improve data storage models based on the column-store approach.

We first discuss ontological foundations behind the column-store approach. We show that, unlike the row-store approach, the column-store approach is more compatible with a particular ontological view of the nature of reality represented in information systems. In particular, we show how the assumption of *inherent classification* [6] has permeated the design of row-oriented model. We show the column-store model represents one step forward towards addressing this problem by vertical partitioning of data. Using these theoretical foundations, we show the column-store model can be improved by column slicing. We suggest two different query independent column slicing technique for nominal and string attributes. We argue that Sliced Column Store (SCS), which is a step towards full data partitioning, is more compatible with the nature of data. The main advantage of column slicing techniques is that only those values specified in query predicates are read. We show how column slicing results in reducing the cost of selection operations and consequently speeding up join operations in comparison to the pure column store model.

The horizontal reorganization of data in column-store approach is already considered in terms of database cracking [7, 8]. This approach is a type of partial sorting in which each request for a particular result set (through posing queries) is an advice for partitioning columns to smaller parts. The rationale behind this technique is that future queries are somehow similar to previous queries that are already posed by users, and as a result, they are more likely to have similar query predicates. We argue that reorganizing physical storage of data based on input queries has some drawbacks. First, business intelligence applications usually involve many new ad-hoc queries that are not based on previous queries. For such queries, reorganization of columns may even worsen the query performance because of the extra effort required for reorganization of data, while future queries may not take advantages of this reorganization. Second, reorganizing before or during query answering not only requires extra time that negatively affects query answering, but also provides many concurrency and consistency problems that require additional consideration. Moreover, inserting new data requires reorganization of data in columns by shifting data. The proposed column slicing technique in this paper is an effort to address these issues in which instead of sorting, the list of instances possessing the same value of a property are physically stored in different slices. Our experiments show the effectiveness of using column slicing techniques to improve query performance as well as to address many problems of database cracking and column sorting.

2   SCS: Ontological Foundations and Practical Implications

Although reducing I/O overhead to improve query performance is considered as the main technical motivation behind column-oriented databases, to our knowledge no theoretical foundation has been proposed for this model. In this section, we propose a theoretical foundation behind the column store model and show why this model is appropriate for processing read-oriented queries. We also aim to find whether it is possible to improve query performance by refining the column store model by adopting a suitable theoretical foundation.

2.1   From Row Store to Column Store

Ontological Foundations. In order to explore the theoretical foundations behind the column store approach, we turn to formal ontology, the branch of philosophy that deals with the order and structure of reality in the broadest way possible. Ontology has been widely used as a theoretical foundation for conceptual modeling, both in theoretical analyses [6] and in empirical studies [9]. The rationale behind using ontology is that it provides a meta-model of existing things in the real world, and database systems represent knowledge or facts about the real world. As a result, understanding the actual components and relations between things in the real world helps in designing databases that better reflect this reality. More specifically, we show how the row-oriented approach has been the consequence of the *assumption of inherent classification* in information system modeling [6]. As discussed in [6], many difficulties in information system management, such as schema integration, schema evolution and ability to exchange information between heterogeneous data sources, can be attributed to this assumption. According to the assumption of inherent classification, everything that is modeled in a domain of interest in an information system is treated as an instance of a class in an object-oriented model (or an entity belonging to an entity type in the Entity Relationship model). Contrary to this assumption, although classification is one of the rudimentary abilities of humans in understanding the things of interest in any domain, real world objects do not inherently belong to classes; rather, classification is a consequence of an effort to organize knowledge about existing things. Inherent classification is incompatible with ontological assumptions about the nature of things in the real world where things and their properties exist prior to and independent of their classification [6].

From this perspective, the row-store model can be viewed as a consequence of the assumption of inherent classification that pervades database design. An implementation consequence of this assumption is that information about instances that belong to the same class is stored contiguously in memory. However, based on ontological foundations, data is not inherently classified; rather, classification is an outcome of humans' efforts to organize information. Thus, there is no fundamental reason why instances belonging to the same class should be stored contiguously in memory. The row-store model (i.e., based on contiguously storing records that are in the same class) adds complexity to data extraction that consequently affects query answering time. This complexity is the result of extra operations required to filter data based on query predicates. We argue that query answering in a database is nothing but a set of data partitioning and data combining operations based on query predicates. For example, in the relational model, data partitioning is provided through selection, projection, intersection, difference, and division operations. On the other hand, combinations are provided through operations such as join, union, and Cartesian product. In the row-store model, the properties of instances belonging to the same class (tuples in a relation) are stored contiguously in memory. However, to answer many queries, there is no need for simultaneously extracting these properties. Therefore, extra selections and projections are required to answer these queries.

To address this problem of relational databases, [6] proposed an Instance-Based Data Model (IBDM) that separates the definition of instances (records) from the definition of classes. IBDM has a two-layer architecture, in which the first (instance) layer specifies the existing things in a domain by defining their properties or attributes independent of their membership in a class. The second (class) layer specifies which properties belong to each class, where one property may belong to more than one class. From an implementation point of view, instances are represented through unique identifiers plus pointers to intrinsic properties (those depending on only one instance, such as 'gender' of a patient), and mutual properties (those depending on two or more instances, such as 'date of surgery' depending on a patient and a doctor). In particular, intrinsic properties can be stored as a set of (InstanceID, Value) pairs, and mutual properties as a set of (InstanceId1, InstanceId2, Value) triples. In this model, each intrinsic property can be represented as a binary table where the name of the table is the unique identifier of that property. The first column of this table is the unique identifier of the instances (InstanceID) that possess this property, and the second column is the value of that property for each instance (value). Accordingly, a mutual property can be represented through a three-column table in which the first and second columns are unique identifiers of the instances that jointly possess this mutual property, and the third column is the value of that mutual property. From this description, IBDM can be viewed as a column-store model originally proposed as a solution for the problem of inherent classification.

Practical Implications. As [10] argues, "There is in fact something fundamental about the design of column-store systems that makes them better suited to data-warehousing workloads." From the implementation point of view, relational databases are based on the row-store approach in which data is stored in two dimensional relations (tables) including a set of properties (columns) for each record of data (row). Since data rows are stored contiguously in memory, this approach wastes I/O band-width as all attributes of a table are read even if not all of them are required to answer a query. The column-store approach is proposed as a solution to address this problem by limiting the number of attributes to those required to answer a given query. As shown in Fig. 1, regarding a query that only address property A and property B in its query predicate, the whole table is scanned for the row-store model while columns C and D are not scanned in the column-store model. As discussed in [11] reducing the amount of data read has a significant effect on reducing the time required to execute queries where I/O is the main bottleneck.

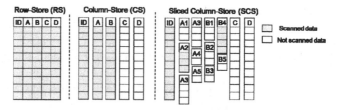

Fig. 1. Scanned data in row-store, column-store and sliced column-store for a query including particular values of A and B

In the architecture of column-store databases, every n-ary relation is represented by a group of binary relations that are stored in form of two-column tables. The tuples of each binary relation are stored physically adjacent to speed up scanning data in these tables. Recently, there has been considerable effort devoted to improving the column-store based database systems [1-4]. C-store [1] is one of the leading column-store projects in which data is stored in groups of correlated columns called projections. In this architecture, the same property may appear in different projections with different sorting orders. Each projection is stored in a different physical storage structure. In this model, relations between different properties of the same record (data row) are provided through implementing join indexes. These join indexes are necessary to reconstruct the original table from existing projections.

2.2   From Column Store to Sliced Column Store (SCS)

Ontological Foundations. From the ontological point of view, characteristics of existing things in the real world are represented by 'properties,' which are the basic constructs in data models [12]. In order to study the relations between properties, we use the property precedence notion of Bunge's ontology [13]. According to this notion, property P_1 precedes property P_2 if and only if the set of things possessing P_2 is a subset of things possessing P_1. For example, the property of 'having color' precedes the property of 'having red color' since the set of instances that are red is a subset of the set of instances that have color. We focus on property precedence since it has special importance in the context of classification. A class in a data model is represented in terms of set of generic properties while instances of that class possess specific properties implying those generic properties [14]. In other words, possessing a specific property manifests possessing a generic property. For example, 'gender=male' and 'gender=female' are two specific properties of a generic property 'having gender'.

According to Bunge's ontology [13], things and their properties exist prior to and independent of any classification. However, in the relational model (and consequently the row-store model based on it), an instance should be member of a class before insertion to the database. In this paper, we distinguish between the classification based on generic properties (that constitutes the definition of classes in a data model) and classification based on specific properties (different manifestations of a generic property). Classification of instances based on generic properties prior to existence of things and their properties results in the problem of inherent classification. However, classification of instances based on specific properties after insertion to the database is not in contradiction with ontological foundations. In addition, this type of classification is more compatible with the nature of read-oriented queries. Query answering in read-oriented queries is nothing but selecting instances from a database based on specific properties (manifestations). As a result, if these different manifestations are classified and stored contiguously on the memory, a query processor needs to read only those manifestations that are indicated in query predicates. The notion of classification based on manifestations constitutes the main idea behind the sliced column-store model (SCS). Although the column-store approach has addressed the problem of inherent classification by reading only the relevant properties, the query processor still needs to read all values of these relevant properties even if only particular values (that are indicated in query predicates) are required to answer a query.

In particular, the row-store model is the consequence of the classification of instances based on a set of generic properties where each set constitutes a class. This model prescribes instances that belong to the same class should be stored contiguously in memory. In the CS model, all values of a specific property are stored contiguously for the instances that possess that property. In SCS, classification in the manifestation layer propagates to the storage where identifications of instances possessing a specific manifestation are stored contiguously on the memory.

Practical Implementation. Sliced Column Store (SCS) model is an extended version of the column-store model that exploits the advantages of this model while providing better query performance by narrowing the search space in query processing. Column-store techniques reduce the amount of data read by considering only those attributes that are necessary to answer a query. SCS goes beyond this idea, and not only ignores irrelevant attributes, but also reads only those particular values that are specified in query predicates (selection criteria). This is achieved through slicing property tables based on different values. For example, a gender column that stores gender of people in a column in the column store model is sliced and stored in two columns *gender_male* and *gender_female* such that each slice stores the ID of instances that possess each particular property value. In SCS, the value of properties is implicit in the name of properties. As shown in Fig 1, if the CS model is an effort to reduce I/O by vertically narrowing search space, SCS narrows this space by further partitioning of the search space. Note that unlike the common horizontal data partitioning in which the main table is horizontally partitioned (all partitions have the same data schema), in SCS partitioning is performed on binary tables that are already partitioned vertically according to the column-store model.

In spite of the advantages discussed earlier, the column store model has some limitations. The main problem is the cost of materialization [15]. Column store databases store data in a set of binary tables, while users request data in form of row-style tuples that requires merging existing column (i.e. called materialization). Materialization is an important issue in column-store databases since directly affects query processing. Two common techniques in materialization are Early Materialization (EM), that involves forming intermediate tuples from set of columns as they are accessed, and Late Materialization, in which intermediate tuples are not formed until after some part of query is performed based on query predicates. For example, for a query over three columns A, B and C with selection operations $\sigma1$, $\sigma2$ and $\sigma3$, EM technique reads a block of A, B and C, and creates a row-style tuple (A, B, C). Then, it applies selection operations on these tuples. However, in LM, first the query processor reads tuples satisfying $\sigma1$ from A, then reads B and C based on $\sigma2$ and $\sigma3$ respectively. Finally, these items are stitched together. According to [16], the total cost of joins among many small partitions with few properties is much less than a join between two tables with many properties. Although simple joins are required because joins are between two-column tables, the cost of these join operations are not negligible. As a result, any step towards reducing the cost of materialization in column store databases can have an important effect on the overall query answering performance. The column slicing technique diminishes the cost of materialization by reducing the cost of joins. This is achieved by eliminating the cost of selection operations in joins since IDs of instances possessing the same value of a particular attribute are stored in the same slice.

3   Column Slicing Techniques

The main contribution of this paper is the basic idea of *column slicing* that can be implemented with any appropriate column slicing techniques. To implement this idea, in this paper we propose two different slicing techniques for nominal and string data types, and we defer slicing techniques for numerical and other data types for future work. In the following, the details of these partitioning techniques are elaborated. An example of column slicing on a sample column store database is shown in Fig. 2.

In the case of categorical properties, in accordance with the concept of classification based on manifestations, each binary table in the column-store model for a nominal property (with domain of n possible values) is split to n single-column slices where each slice stores the unique identifiers of instances possessing a particular value of that property. The name of each slice implicitly shows the value of that property for those instances that are stored in that slice (i.e. property_value). Unlike the column store approach in which the values of properties are stored as data, in the column slicing approach, values of properties are considered as a metadata (i.e. the name of slice) representing the class of objects that possess a particular value of a property. For example, as shown in Fig. 2 the nominal property *A* with the domain of values *a1*, *a2*, *a3*, *a4* and *a5* is split to five single-column slices *A_a1*, *A_a2*, *A_a3*, *A_a4*, *A_a5* that stores ID of instances possessing each particular value of *A*.

Compression, which is used widely in column store databases to reduce the size of databases and speeding up the query answering [1, 17], is implicit inside the column slicing approach for nominal properties since each particular value of a property is stored one time as a metadata in the name of the single-column slice, and only ID of instances possessing this particular value are stored in that slice.

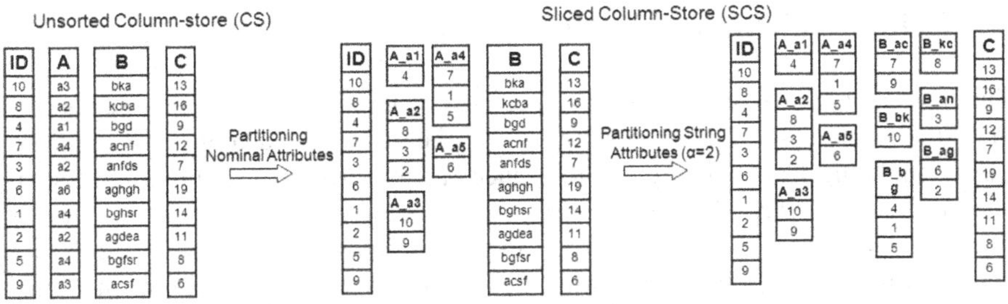

Fig. 2. An example of partitioning technique for nominal (A) and string (B) attributes

In the case of string properties, we propose an α-level slicing technique based on the concept of *trie* data structure [18]. A trie is a tree-based data structure that is useful for handling strings over alphabet. In the trie data structure, information about the content of each node is stored in the path from the root to the node, rather than the node itself. We use this notion to classify the values of string properties based on their characters starting from the first character. More specifically, the first level includes 26 classes (a-z) where each class represents a class of strings starting with a particular English letter. In the case of lower- and upper-case letters, the number of classes will be 52 (a..z, A...Z) and in the case of all ASCII characters, the total number of classes in the first level is 128). At each level, nodes are further classified based on the next character. An example of this classification for data of Fig. 2 is shown in Fig. 3.

A path from the root to each leaf node (i.e. an order of characters) represents the name of the slice that stores the IDs of instances possessing string values starting with that path. Note that the concept of the trie data structure is used only for the purpose of explaining the slicing and classifying string properties, and there is no need to store this structure in the database since the name of each slice implicitly indicates the value of string properties starting with that name. For example, *B_be* represents a slice of property *B* that stores instances that their value for *B* starts with *be*. Since not all slices are created in the beginning, and slices are created in the record insertion time, there is no slice without a value. That way, we control the number of slices by avoiding empty slices. However, the number of slices can growth exponentially with increasing α. The effect of applying different values for α on query answering and database size is discussed in the evaluation section.

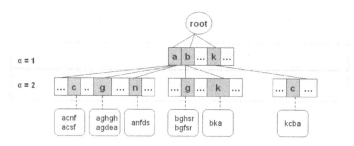

Fig. 3. An example of slicing a string proeprty (property B in Fig. 2)

4   Column Slicing vs. Database Cracking and Sorting

Column-stores are heavily optimized to perform materialization as tuple reconstruction is the main cost of column-store query plans. Having unaligned columns is always something to avoid in order to reduce random accesses during tuple reconstruction. This is why C-store [1] uses column-store projections where data is replicated, sorted and then compressed. Similarly, database cracking [7] replicates columns in multiple orders but performs the sorting partially (i.e., only for the range of values referenced in query predicates). In this section we discuss how these techniques are different from column slicing.

Database cracking [7, 8] is a dynamic partitioning technique for numerical properties such that the physical organization of data is continuously updated based on input queries. More specifically, based on the query predicate of input queries, those values of a numerical column that satisfy the query predicate are partially sorted. The main idea behind dynamic reorganization of data based on input queries is that the way that users request data in the future is similar to the way they requested data in the past. We argue that database cracking has three main drawbacks. First, Business Intelligence applications usually deal with many new ad-hoc queries in which the behavior of users in query building is not necessarily the function of their behavior in the past. As a result, dynamic reorganization can worsen query performance of queries in which there is no overlap between the query predicates of past and future queries. However, the proposed column slicing technique in this paper is a query independent technique in which the same manifestations of a property are classified and stored contiguously on the memory. Such a query independent technique is more appropriate to answer new ad-hoc queries. Second, the cost of reorganization of data in database cracking during query processing is an additional overhead that is added to the cost of query answering time. However, in the column slicing technique, physical reorganization and query processing are two independent processes performed in two different phases, and physical reorganization of data does not negatively affect query processing time. Third, in database cracking, inserting new data to the database may result in full reorganization of data since it will require shifting and physical reorganization of data. This would not be a problem for our column slicing technique as new records are stored in predefined slices that there is no need for reorganization of data.

Column sorting (in which binary tables are sorted) is a more general solution in comparison to database cracking (i.e. partially sorting column based on query predicates). Although column sorting can significantly reduce the cost of selection operations and consequently the cost of record materialization by reducing the cost of join operations, we argue that column slicing works better than column sorting. Assume a binary table including n rows of (ID, value) for a property. To answer a query including a selection operation on a specific property, a binary search with the cost of $O(\log n)$ is required to select data from a sorted column. However, in the case of full data partitioning, the list of instances satisfying the selection operation already exist in a single-column slice where the name of each slice implicitly indicates its contents.

5   Experiments

In this Section, we evaluate the effect of applying the column slicing technique on query answering; as well, we explore the side effects of this technique. SQL Server 2012 [4] is used as a database engine. The new added feature in this database engine (called column store index) is a pure column-store system in which data for different columns are stored on separate pages. As a result, performing the column index on a column ensures physically creating a column store database. The hardware setting of the employed database server is AMD Athlon, 64 X2 Dual Core 2.71 GH, 1GB RAM, and 280 GB HD. To gain an understanding of column slicing techniques, we conducted experiments in two phases with two different datasets. In the first phase, a dataset generated by our simple data generator (MUN-DGen) is used to explore the pure effect of slicing on query operations. In the second phase, TPC-H benchmark [19] with data that typically found in data warehousing and business intelligence applications is used.

5.1   Phase1: The Pure Effect of Slicing on Query Operations

In the first phase, we have conducted a set of experiments on a simple database. The purpose of using this simple database for exploring the pure effect of slicing is controlling over the number of tables, number of properties, types of the properties and number of categorical values. We implemented a data generator (MUN-Dgen) that automatically generates column store and sliced column store datasets for a big table. This big table A includes m sting properties A_SP, n integer properties A_IP, and k categorical (represented by a integer data type.) properties A_CP. In the case of categorical properties in SCS, $A\_CP_{i\_j}$ is the value j of the categorical property A_CP_i.

Simple Selections. In the case of column slicing, we expect better performance in comparison to the pure column store technique for simple selection queries with high selectivity query predicates because they access only few records in comparison to the whole column scanning in column store. To explore this effect, we take into account a query $Q_S = \pi_{InId}(\sigma_{A\_CP1 = 'CP1-1'}(A))$ that selects instances from a single table based the value of a categorical property (A_CP_1 is the referenced categorical property, and CP_{1-1}

is a specific value of this property). Let r the number of records in table A and s the average size of allocated space by a nominal property (that can be shown by an integer or a string data type). The amount of data scanned in CS is rs this amount is rs/t for SCS where t is the average number of categorical values. In SCS, the selection criterion is implicit in the name of table, and there is no need for explicit filtering of data. The experiment supporting this idea is shown in Fig. 4. With increasing the number of categorical values, the execution time of Q_S on SCS is reduced because of increasing the selectivity of the query predicate.

Aggregation Functions. In order to explore the effect of slicing on queries including aggregate functions, we consider the query $Q_A = \pi_{A\_CP1, Count(A\text{-}ID)}(A)$ that adds an aggregation operator on top of the selection query. Since the list of instances possessing each specific value of a categorical property is stored in different slices, aggregate functions access those data directly. However, since all records must be scanned in both CS and SCS, there is no considerable improvement on aggregate functions running on SCS. The experiments regarding Q_A are shown in Fig. 4.

Join Operations. We expect a considerable improvement in performance of queries including join operations because of reducing the cost of materialization (reconstructing row-style tuples from a set of columns) in SCS. Reduction in the cost of the materialization is the consequence of reducing the cost of selections operations on columns since the list of instances are pre-categorized in a set of column slices. In the case of joining multiple columns, all relevant columns (including columns that will be materialized and columns that are referenced in the query predicates) participate in the join operation. Since only pre-selected values of columns are entered into join operations, we expect considerable improvement for SCS. Recall that according to [16], the join between two tables with $m>2$ properties is more expensive than joins between m tables with two properties. To study the effect of slicing technique on the join operations, we used the query Q_J: $\pi_{InId, A\_CP1, A\_SP1, A\_1P1} [\sigma_{A\_CP2= \text{'}CP2\text{-}1\text{'}} (A)]$. As shown in Fig. 4, as the number of categorical values increases, the execution time of Q_J decreases.

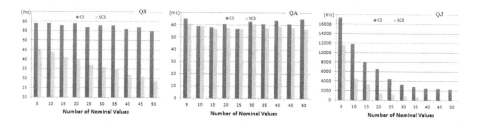

Fig. 4. Execution time of Q_S, Q_A and Q_J for different number of nominal values

5.2   Phase2: TPC-H Benchmark

In the second phase of experiments, in order to evaluate the effect of column slicing on real business intelligence and data warehousing dataset, an instance of TPC-H at scale 1 is generated using the TPC-H data generator where the total database size is around 1GB. The database generator program in TPC-H was modified such that creates equivalent databases in three different states: 1) slicing only nominal attributes, 2) slicing only string attributes, and 3) slicing both nominal and string attributes. This categorization allows to gain an understanding about both pure and mutual effects of different column slicing techniques on the performance of query processing.

Among 22 Queries in TPC-H benchmark [19], queries Q3, Q10, Q12 were selected to explore the pure effect of partitioning nominal attributes because at least one nominal attribute exists in their predicates while there are no string attributes. These queries were performed on an instance of the TPC-H dataset in which only columns regarding nominal attributes were sliced. The query answering time is compared with running these queries on original instances of the generated database (Table 1).

In order to explore the effect of slicing string properties, Queries Q2, Q5, Q7, Q11, Q17 and Q20 were selected from TPC-H benchmark queries as they include at least one string property in their query predicates with no nominal property. Three different instances of TPC-H dataset (regarding $\alpha = 1, 2, 3$) in which all string properties are sliced were created. The results of these experiments regarding these queries are shown in Table 1. We do not suggest more than three levels of slicing for TPC-H dataset as exponentially increasing the number of slices negatively affects the advantages of column slicing. The effect of increasing slices and consequently increasing the database size is discussed in Section 5.3.

Among TPC-H queries, Q8, Q19 and Q21 are queries that both nominal and string attributes are addressed in their query predicates. To explore the effect of slicing nominal properties in conjunction with string properties, instances of TPC-H dataset are generated in which all nominal and string properties were sliced. As shown in Table 1, our experiments demonstrate that the proposed techniques for slicing nominal and string properties can increase the performance of read-oriented queries.

Table 1. Execution time of TPC-H queries on CS and SCS (ms)

Category	Nominal -No string			String-No Nominal						Nominal and string		
Query	Q3	Q10	Q12	Q2	Q5	Q7	Q11	Q17	Q20	Q8	Q19	Q21
CS	1086	6103	1573	196	2376	1116	246	4090	1186	876	1913	2716
SCS (α=1)	724	2878	953	109	1033	791	161	2763	878	515	865	1364
SCS (α=2)	-	-	-	91	865	721	124	2235	687	406	723	1148
SCS (α=3)	-	-	-	85	791	702	113	2103	654	389	631	1007

5.3   The Side Effects of Column Slicing Approach

Increasing the number of slices by increasing the diversity of property values is the main side effect of column slicing approach. In the case of string attributes, at most 128^n slices in level n can be created. One obvious consequence of applying full column slicing on a property with a large domain of values would be increasing a large number of slices while each slice stores only a few items. The result of increasing the number of slices will be increasing the metadata required to store these slices. With increasing the number of slices, we may reach a point where the advantage of slicing is offset by the extra cost to store and manage metadata regarding large number of slices. In other words, column slicing in SCS is acceptable only to the extent that the amount of growth for database is not a big issue in an application. As shown in Table 1, there is only slight improvement in performance of SCS(α=2) to SCS(α=3) that shows the negative effect of increasing the number of slices. Unlike column sorting and column cracking technique in which data insertion is expensive and requires re-organization of physical storage, insertion is not problematic in SCS as identification of new instances possessing a particular value of a property must be saved in one of the fixed and predefined slices.

6   Conclusion and Future Work

In this paper, we discussed the ontological foundations behind the column-store data model and we showed that how this model can be a solution for the problem of inherent classification in the row-store model. Unlike the row-store model, which prescribes contiguously storing properties of instances that are in the same class, in the column store model the problem of inherent classification is alleviated by allocating a separate property table for each property and contiguously storing all values of each property. In this paper, we go beyond this approach and contended that from ontological point of view, there is no rationale behind storing all values of a property in the same column where extra selection operations are required to access a particular value of that property. We proposed the Sliced Column Store model in which property columns are horizontally partitioned to some slices, where each slice stores the identification of instances that possess a particular value of that property. We suggested a full slicing technique for nominal properties, and the α-level slicing technique for string properties. We conducted a set of experiments to study the effect of column slicing on the performance of query processing. The pure effect of each partitioning technique and the mutual effect of them were explored using both a simple dataset (generated by MUN-DGEN data generator) and the TPC-H benchmark [18]. Unlike the column-store approach in which the whole property table is scanned to find instances possessing a specific value, the major consequence of column slicing in SCS is eliminating this effort by directly reading only those instances that possess this property. We argued that this virtue can result in significant improvement in query processing from two points of view. First, improving query performance by reducing the cost of selection operations through reducing I/O amount; this is achieved by reading only those property values that are indicated in query predicates since instances possessing particular values are stored in separate slices. Second, by reducing the cost of join operations through decreasing the cost of selection operations, resulting in simple joins (i.e. joins between columns having smaller records). These improvements are achievable with a slight cost of increasing the size of database.

In this paper, we sketched a research landscape with a large number of opportunities that slicing can contribute to database design, and we suggested two slicing techniques for string and nominal properties. For future works, we focus on efficient slicing techniques for other data types. We also aim to implement an SQL-independent prototype of SCS to study the scalability this method to handle big data.

References

1. Stonebraker, M., Abadi, D.J., Batkin, A., Chen, X., Cherniack, M., Ferreira, M., Lau, M., Lin, A., Madden, S., O'Neil, E., O'Neil, P., Zdonik, S.: C-store: a Column-Oriented DBMS. In: International Conference on Very Large Data Bases. VLDB Endowment, pp. 553–564 (2005)
2. Boncz, P.A., Zukowski, M., Nes, N.: MonetDB/X100 Hyper-Pipelining Query Execution. In: Conference on Innovative Data Systems Research, pp. 225–237 (2005)
3. MacNicol, R., French, B.: Sybase IQ multiplex - designed for analytics. In: International Conference on Very Large Data Bases. VLDB Endowment, pp. 1227–1230 (2004)
4. Larson, P., Clinciu, C., Hanson, E.N., Oks, A., Price, S.L., Rangarajan, S., Surna, A., Zhou, O.: SQL Server Column Store Indexes. In: ACM SIGMOD International Conference on Management of Data, pp. 1177–1184. ACM, New York (2011)
5. Herodotou, H., Borisov, N., Babu, S.: Query Optimization Techniques for Partitioned Tables. In: ACM SIGMOD International Conference on Management of Data, pp. 49–60. ACM (2011)
6. Parsons, J., Wand, Y.: Emancipating Instances from the Tyranny of Classes in Information Modeling. ACM Trans. Database Syst. 25, 228–268 (2000)
7. Idreos, S., Kersten, M., Manegold, S.: Database Cracking. In: 3rd Biennial Conference on Innovative Data Systems Research (CIDR), pp. 68–78 (2007)
8. Idreos, S., Manegold, S., Kuno, H., Graefe, G.: Merging What's Cracked, Cracking What's Merged: Adaptive Indexing in Main-Memory Column-Stores. In: VLDB Endowment, pp. 586–597 (2011)
9. Gemino, A., Wand, Y.: A Framework for Empirical Evaluation of Conceptual Modeling Techniques. Requirements Engineering Journal 9(4), 248–260 (2004)
10. Abadi, D.J., Madden, S.R., Hachem, N.: Column-Stores vs. Row-stores: How Different are They Really? In: ACM SIGMOD International Conference on Management of Data, pp. 967–980. ACM, New York (2008)
11. Harizopoulos, S., Liang, V., Abadi, D.J., Madden, S.: Performance Tradeoffs in Read-Optimized Databases. In: International Conference on Very Large Data Base. VLDB Endowment, pp. 487–498 (2006)
12. Wand, Y., Storey, V., Weber, R.: An Ontological Analysis of the Relationship Construct in Conceptual Modeling. ACM Trans. Database Sys. 24, 494–528 (1999)
13. Bunge, M.: Treatise on Basic Philosophy: Ontology I: The Furniture of the World, vol. 3. D. Reidel Publishing Co., New York (1977)
14. Parsons, J., Wand, Y.: Attribute-Based Semantic Reconciliation of Multiple Data Sources. In: Spaccapietra, S., March, S., Aberer, K. (eds.) Journal on Data Semantics I. LNCS, vol. 2800, pp. 21–47. Springer, Heidelberg (2003)
15. Abadi, D.J., Myers, D.S., DeWitt, D.J., Madden, S.R.: Materialization Strategies in a Column-Oriented DBMS. In: IEEE International Conference on Data Engineering, pp. 466–475 (2007)
16. Agrawal, S., Narasayya, V., Yang, B.: Integrating Vertical and Horizontal Partitioning into Automated Physical Database Design. In: SIGMOD International Conference on Management of Data, pp. 359–370. ACM, New York (2004)
17. Lemke, C., Sattler, K.-U., Faerber, F., Zeier, A.: Speeding Up Queries in Column Stores. In: Bach Pedersen, T., Mohania, M.K., Tjoa, A.M. (eds.) DAWAK 2010. LNCS, vol. 6263, pp. 117–129. Springer, Heidelberg (2010)
18. Heinz, S., Zobel, J., Williams, H.E.: Burst Tries: A Fast, Efficient Data Structure for String Keys. ACM Trans. Inf. Syst. 20, 192–223 (2002)
19. TPC-H. Benchmark Specification, http://www.tpc.org

Schema Decryption for Large Extract-Transform-Load Systems

Alexander Albrecht and Felix Naumann

Hasso Plattner Institute for Software Systems Engineering,
Prof.-Dr.-Helmert-Straße 2-3, 14482 Potsdam, Germany
{alexander.albrecht,felix.naumann}@hpi.uni-potsdam.de

Abstract. Extract-Transform-Load (ETL) tools are used for the creation, maintenance, and evolution of data warehouses, data marts, and operational data stores. ETL workflows populate those systems with data from various data sources by specifying and executing a DAG of transformations. Over time, hundreds of individual workflows evolve as new sources and new requirements are integrated into the system. The maintenance and evolution of large-scale ETL systems requires much time and manual effort. A key problem is to understand the meaning of unfamiliar attribute labels in source and target databases and ETL transformations. Hard-to-read attribute labels in schemata lead to frustration and time spent to develop and understand ETL workflows.

We present a schema decryption technique to support ETL developers in understanding cryptic schemata of sources, targets, and ETL transformations. For a given ETL system, our recommender-like approach leverages the large number of mapped attribute labels in existing ETL workflows to produce good and meaningful decryptions. In this way we are able to decrypt attribute labels consisting of a number of unfamiliar few-letter abbreviations, such as UNP_PEN_INT, which we decrypt to UNPAID_PENALTY_INTEREST. We evaluate our schema decryption approach on three real-world repositories of ETL workflows and show that our approach is able to suggest high-quality decryptions for cryptic attribute labels in a given schema.

Keywords: ETL, Data Warehouse and Repository, Data Integration.

1 Introduction

ETL systems are visual programming tools that allow the definition of complex workflows to extract, transform, and load heterogeneous data from one or more sources into a target database. Designing and maintaining ETL workflows requires significant manual work, e.g., the effort is up to 70% of the development cost in a typical data warehouse environment [8]. ETL workflows are stored in repositories to be executed periodically, e.g., daily or once a week. In the course of a complex data warehousing project up to several hundred ETL workflows are created by different individuals [1] and stored in such repositories. Moreover, the created ETL workflows get larger and more complex over time. Cryptic

P. Atzeni, D. Cheung, and R. Sudha (Eds.): ER 2012, LNCS 7532, pp. 116–125, 2012.

schemata are a well-known problem in the context of data warehousing. The main reason for cryptic schemata is the tendency to assign compact attribute labels consisting of a number of domain-specific abbreviations and acronyms.

Example 1 (Cryptic Attribute Labels). Consider a repository of ETL workflows to extract, transform, and load data of an OLTP system with attribute labels from the well-known TPC-E schema. With the to-be-generated decryption pairs ⟨CO ≈ COMPANY⟩ and ⟨SP ≈ STANDARD, POOR⟩, it would be easier for a developer who is unfamiliar with this schema to identify the semantics of attribute labels, such as CO_SP_RATE.

Manually finding decryption pairs is ineffective and time consuming. To illustrate this problem, consider the attribute label CO_SP_RATE from the previous example. As this attribute label is too specific to have a mapped attribute label as decryption in the given ETL repository, the developer has to look up all pairs of mapped attribute labels that give a hint on an appropriate decryption of tokens CO and SP. With over ten thousand of pairs of mapped attribute labels in the evaluated ETL repositories, manual schema decryptions becomes infeasable. Readers are referred to our technical report [2] for a comprehensive overview of schema and ETL workflow characteristics in the given real-world ETL repositories.

In this paper, we regard ETL workflows as transformation graphs of the well-known model introduced by Cui and Widom [7]. This model is generally applicable to ETL workflows from common ETL tools: An ETL workflow is a directed acyclic transformation graph (DAG) and the topologically ordered graph structure determines the execution order of the connected transformations. In ETL, most transformations are a generalization of relational operators supporting multiple inputs and outputs. Two transformations are connected in the graph if one transformation is applied to the output obtained by the other transformation. Accordingly, attributes in the output schema of a transformation are connected to the corresponding attributes in the input schema of the subsequent transformations. We leverage these *connected* attribute labels in the existing ETL workflows as valuable source of information for automated schema decryption. We have observed that *connected* attributes with different labels often contain reasonable decryptions – often not for the entire label but for tokens within the labels. As cryptic attribute labels are often too specific to have a connected attribute label as decryption in the given ETL repository, the problem is to pair portions of the cryptic attribute label with portions of more descriptive attribute labels to produce reasonable decryptions.

Example 2 (Connected Attribute Labels). Consider the ETL repository from Example 1. Within some ETL workflows, extracted source attributes were renamed in the succeeding transformation to provide a better readability. For example, the attribute label CP_COMP_CO_ID was renamed to COMPETITOR_COMPANY_ID and CO_CEO to COMPANY_CEO. Thus, labels CO_CEO and COMPANY_CEO and labels CP_COMP_CO_ID and COMPETITOR_COMPANY_ID are connected, respectively.

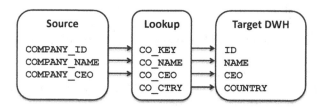

Fig. 1. An exemplary ETL workflow

As ETL tools allow the developer to drag-and-drop attribute labels from output to input schemata, there are many connected attributes with equivalent labels. But in large ETL repositories there is also a large number of connected attributes having different labels. There are several reasons for this, such as (1) source-, lookup-, and target-schemata used in an ETL workflow are often created independently and thus contain different attribute labels; (2) a data warehouse schema based on instances of cryptic source schemata uses attributes consisting of more descriptive tokens to provide a better readability; (3) copy-and-paste of entire transformations is a common practice in ETL development, which results in ETL sub-workflows connected to intermediate attributes with different labels.

In this paper, we will focus on attribute pairs between connected transformations for schema decryption. We ignore the connections among attribute labels within a single transformation, because a developer usually avoids using synonyms within a single transformation. Our approach overcomes the weaknesses in existing approaches, such as string and schema matching techniques. These methods lead to poor decryption results due to the use of domain-specific abbreviations, acronyms, and tokens in ETL schemata. Moreover, it is infeasible to exploit data redundancies between different schemata to find pairs of corresponding attribute labels: The data created in the intermediate ETL processing steps is not persisted and we lack this helpful information. Re-executing and storing data from intermediate processing steps is an unrealistic assumption in a typical data warehouse scenario.

2 Using Connected Attributes for Decryption

To illustrate our approach upfront we introduce a toy example of an ETL workflow in Fig. 1. The ETL workflow loads company data into a dimension table of a data warehouse. The extracted source data is the input of a lookup transformation. There, a company record is assigned a country from a lookup table using the company identifier as lookup key. Finally, the data is loaded into the data warehouse (DWH).

We observe that (1) attributes can be tokenized based on special characters. We also observe that (2) no two connected attributes have the same label. This is a typical situation if source, lookup, and target schemata were developed independently or for different purposes. Finally, we observe that (3) some attributes use abbreviations that appear in extended form in connected attribute labels. These observations were made repeatedly in our analysis of three real-world ETL repositories, each with up to several hundred ETL workflows containing thousands of connected attribute pairs with different labels.

Our decryption approach finds reasonable decryptions within the given ETL repository by making use of all three observations: For each ETL workflow in the ETL repository, we first break all labels into tokens, based on case-change or non-alphabetical separators. In the example, we tokenize using the underscore as separator. The second observation allows us to identify attribute labels with same or similar semantics. If data from an attribute in the source or preceding transformation is used as input for some attribute in the target or subsequent transformation, it is reasonable to assume that their two labels are semantically related – in most cases they are semantically equivalent. For instance, CO_CTRY and COUNTRY in Fig. 1 are such *connected* attribute labels. Finally, using the third observation, we realize that the tokens CO and COMPANY co-occur in multiple pairs of connected attribute labels, leading us to believe that they are synonymous (and not for instance CO and COUNTRY).

With the identified decryptions from the ETL repository, we can suggest decryptions for cryptic attribute labels of a given schema. For instance, given a schema with the cryptic attribute label CO_ID, it is decrypted to COMPANY_ID using the decryption ⟨CO ≈ COMPANY⟩ derived from ETL workflow in Fig. 1.

3   Schema Decryption

Our goal is to suggest "decryption pairs" to provide developers with a better understanding of cryptic schemata and ETL workflows.

Definition 1 (ETL Workflow). *An ETL workflow comprises a set of transformations T with input and output schemata, interconnected with each other forming a DAG. Let $W = (V, E)$ be a DAG representing an ETL workflow consisting of a set of vertices V representing the involved transformations. The edges $e \in E \subseteq V \times V$ connect the output schema of one transformation with the input schema of another transformation, i.e., e represents an ordered pair of transformations.*

In this section we explain how to find decryptions for cryptic schemata leveraging the large number of connections among attribute labels in the given ETL repository.

Definition 2 (Connected Attribute Labels). *Two attribute labels are connected if there exists at least one ETL workflow in which a direct link is established between the corresponding attributes in the output and input schemata of two connected transformations.*

Table 1. Sample results for a Spanish ETL repository from the finance industry domain

Input Schema	DEBT_RT, RESID_VAL_AT_RISK, UNP_PEN_INT
Top-5 Decryption Pairs	⟨UNP ≈ UNPAID⟩, ⟨INT ≈ INTEREST⟩, ⟨PEN ≈ PENALTY⟩, ⟨RT ≈ RATE⟩, ⟨RESID, VAL ≈ RESIDUAL, VALUE⟩

3.1 Our Schema Decryption Approach

For a given schema consisting of a set of cryptic attribute labels, our algorithm returns a ranked list of decryptions in descending order of their frequency of occurrence in the given ETL repository. We regard an attribute label as a set of tokens and represent a decryption as a pair of corresponding token sets that appear to be used synonymously within the ETL repository. The algorithm iterates over all attribute labels l from the given schema and returns the set of all applicable decryptions to decrypt l. Thus, for each attribute label l, we create all possible decryptions leveraging the large number of connected attribute labels in the given ETL repository (see Sec. 3.2). Each decryption is then added to the result. Finally, the algorithm returns a compact list of decryptions ranked in descending order of their frequency of occurrence in the ETL repository.

Let T_i and T_j be disjoint sets of tokens, i.e., $T_i \cap T_j = \emptyset$. We define a decryption pair $\langle T_i \approx T_j \rangle$, where T_i and T_j are synonyms. We regard token sets and not single tokens, because a decryption often applies to multiple tokens or even contains multiple tokens. Table 1 shows a sample schema decryption in which individual tokens but also token sets are decrypted. A decryption pair is applicable to an attribute label only if either all tokens from T_i, or all tokens from T_j occur in the (tokenized) attribute label. Tokens from T_i or T_j may occur in any order in the attribute label.

Definition 3 (Decryption Pair). *Let T_i and T_j be disjoint sets of tokens. We call $\langle T_i \approx T_j \rangle$ a decryption pair if the token set denoted by T_i is synonymous to the token set denoted by T_j.*

Finally, to suggest a compact list of decryption pairs, we remove all subsumed decryption pairs from the result list, retaining only maximal decryption pairs.

Example 3 (Maximal Decryption Pair). Consider the three created decryption pairs ⟨SP ≈ STANDARD⟩, ⟨SP ≈ POOR⟩ ⟨SP ≈ STANDARD, POOR⟩ derived from the same pairs of connected attribute labels. We only suggest ⟨SP ≈ STANDARD, POOR⟩ and remove the other two subsumed decryption pairs from the result list.

Definition 4 (Maximal Decryption Pair). *Let $L = \{(l_m, l_n)\}$ be the set of pairs of connected attribute labels containing decryption pair $\langle T_i \approx T_j \rangle$. We call $\langle T_i \approx T_j \rangle$ a maximal decryption pair if there is no decryption pair $\langle T_i' \approx T_j' \rangle$ for every $\{(l_m, l_n)\} \in L$ with $T_i \subseteq T_i'$ and $T_j \subseteq T_j'$.*

3.2   Finding Decryption Pairs

We now describe how we identify decryption pairs $\langle T_i \approx T_j \rangle$: Given an attribute label and some contained tokens T_i, we want to find all applicable decryption pairs for T_i. To this end, we search among all connected attribute labels in the given ETL repository for those that contain T_i. More formally, we consider all attribute labels l that contain T_i and are connected to some attribute label containing no subset of T_i. Using these pairs of connected attribute labels, we choose candidate decryption pairs $\langle T_i \approx T_j \rangle$, where T_j is some subset of tokens from the other attribute label. A candidate decryption is added to the result if all three of the following observations hold.

Our first observation is that connected attribute labels often share tokens, i.e., such tokens appear in both connected attribute labels. In Example 2 in Sec. 1, connected attribute labels CP_COMP_CO_ID and COMPETITOR_COMPANY_ID share token ID and connected attribute labels CO_CEO and COMPANY_CEO share token CEO. Considering shared tokens for decryption makes no sense, since their counterpart is the same token in the other label. Thus, we do not create decryption pairs containing a shared token. In the example we would not create a decryption pair such as $\langle CO \approx CEO \rangle$; the token CEO is already 'taken' by its counterpart CEO in the other attribute label.

Our second observation (and assumption) is that synonymous token sets are never used together in a single attribute label, as it would be useless to label a single attribute with synonyms. That is, if tokens x and y appear together in one attribute label, there is no decryption pair $\langle T_i \approx T_j \rangle$ with $x \in T_i$ and $y \in T_j$ or vice versa. Considering the attribute labels in Example 2 in Sec. 1, we do not create decryption pair $\langle CO \approx COMP \rangle$ from a corresponding pair of connected attribute labels, because both tokens appear together in the attribute label CP_COMP_CO_ID and thus are very unlikely to represent synonyms.

Our last observation is that a decryption is consistently used between two connected transformations. To determine the consistency of a decryption pair derived from a pair of connected attribute labels, we determine its correctness (*confidence*) and frequency of occurrence (*hit-ratio*) throughout the corresponding schemata of the two connected transformations: Let $L_{T_i} = \{(l_m, l_n)\}$ be the set of pairs of connected attribute labels in which all tokens of T_i appear either in l_m or l_n (but not both, as these are the trivial cases). These pairs represent the *positive class* for the decryption of T_i. Further, let L_{T_i, T_j} be the set of pairs of connected attribute labels in which T_i appears in one label and T_j in the other label. These pairs represent the *true positive class* for the decryption. Note that $L_{T_i, T_j} \subseteq L_{T_i}$. Then we define confidence as

$$confidence_{T_i, T_j} = \frac{|L_{T_i, T_j}|}{|L_{T_i}|}$$

and we define the hit-ratio for decryption pair $\langle T_i \approx T_j \rangle$ as

$$hit\text{-}ratio_{T_i, T_j} = \frac{|L_{T_i, T_j}|}{|L_{T_j}|}.$$

Note that both confidence and hit-ratio have to be considered. A high hit-ratio may result in a poor confidence, i.e., the decryption from T_i to T_j may occur frequently, but T_i also occurs frequently with other tokens in connected attribute labels. Similarly, a high confidence, e.g., achieved by returning only correct decryptions producing no false positives, may result in a poor hit-ratio.

Example 4 (Quality of Decryption Pairs). Consider the connected attribute labels from Example 2 in Sec. 1. Decrypting CO to COMPETITOR might have a high hit-ratio in the corresponding schemata of the two connected transformations, if COMPETITOR often co-occurs with CO. As CO also occurs frequently with tokens different from COMPETITOR, such as COMPANY, decrypting CO to COMPETITOR results in a poor confidence. On the other hand, decrypting COMP to COMPANY might have a high confidence: labels with the token COMP are almost always connected to labels containing COMPANY, but labels containing COMPANY might also often be connected with labels containing CO (but not COMP). Thus the decryption of COMP to COMPANY has a low hit-ratio.

As the quality of a decryption pair depends on both measures, we choose the harmonic mean of confidence and hit-ratio to determine the quality of a decryption pair. The harmonic mean is a typical way to aggregate measures:

$$harmonicMean = \frac{2 \cdot confidence \cdot hit\text{-}ratio}{confidence + hit\text{-}ratio}$$

As the reverse decryption of T_j to T_i results in the same harmonic mean value, we can ignore order. In our experiments, we choose a threshold value of 80% for the harmonic mean to suggest consistent decryptions from pairs of connected attribute labels.

4   Experimental Study

We evaluated our schema decryption approach on three real-world ETL repositories. These repositories were created separately by different departments of a banking organization in Switzerland (CH), Germany (DE), and Spain (ES) using Informatica PowerCenter[1]. Informatica provides ETL workflow specifications in a proprietary XML format, which our schema decryption algorithm takes as input. Schemata and connections between attribute labels are pre-indexed offline and are used to compute schema decryptions in an online fashion. Our algorithm operates efficiently and typically returns a ranked list of decryption pairs for a given schema in under a second.

4.1   Evaluation Technique

We have successfully tested our schema decryption approach on all ETL workflows from the three given ETL repositories. To evaluate the accuracy of our

---

[1] www.informatica.com

Table 2. Calculating average precision for top-3 decryption pairs

Rank	Decryption Pair	$rel(i)$	Precision
1	⟨UNP ≈ UNPAID⟩	1	1
2	⟨INT ≈ OUTPUT⟩	0	1/2
3	⟨PEN ≈ PENALTY⟩	1	2/3

schema decryption, we randomly selected three schemata consisting of at least 20 attribute labels from each repository and use schema decryption to generate ranked lists of decryption pairs for the selected nine schemata.

In our evaluation we consider the top-k decryption pairs p_i in the ranked list. Let i be the position of p_i in the ranked list, i.e., $i \leq k$. Then, we manually determine whether p_i is relevant/correct or not for the given schema, i.e., we set $rel(i)$ to 0 or 1, respectively. We consider a decryption pair to be accurate if it helps to understand the underlying semantic domain of the original attribute label. Then, we calculate the average precision measure for the top-k decryption pairs. The average precision is the average of the precision values for the seen accurate decryption pairs. Average precision is a widely-used evaluation measure in information retrieval to indicate ranking accuracy [4]:

Definition 5 (Average Precision). *Let $P(i)$ be the precision of the first i suggested decryption pairs. Then, the average precision at position k is*

$$AvP_k = \frac{\sum_{i=1}^{k} P(i) \cdot rel(i)}{\sum_{i=1}^{k} rel(i)}$$

where precision is defined as $P(i) = \frac{\sum_{j=1}^{i} rel(j)}{i}$

Example 5 (Average Precision). Table 2 shows an illustrative top-3 example of ranked decryption pairs. The examples are from the ETL repository from Spain (ES). The precision values after each new accurate decryption is observed are 1 and $\frac{2}{3}$. Thus, the *average precision* of the top 3 results (with two seen accurate decryptions) is given by $(1 + \frac{2}{3}) / 2 = 83\%$.

4.2   Results

Fig. 2 shows the accuracy of our schema decryption approach. We measure the mean average precision for each of the experiments and show the top-5, top-10, top-15, and top-20 results. For all three repositories the algorithm achieves an accuracy of above 90%. For the CH repository the algorithm provides the best accuracy. This is expected, because if there is a pair of connected attributes with different labels in the CH repository, it often contains an accurate decryption. The experiments demonstrate the advantages of identifying decryption pairs based on tokens and based on their characteristics. Additional experiments confirmed that our approach results in a significantly lower number of incorrect,

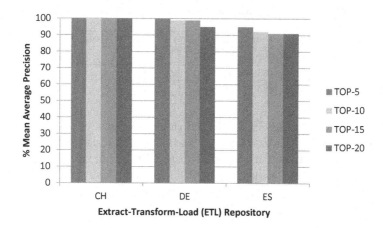

Fig. 2. Schema Decryption Accuracy for ETL Repositories (CH), (DE), (ES)

conflicting and redundant decrpytion pairs compared to other approaches. We compared our approach to a straightforward alternative of choosing entire labels of two connected attributes as decryption pair. In addition, we compared our approach against different string-similarity measures, such as Levenshtein distance [9].

5   Related Work

Our work is related to research on schema normalization in the field of data integration [10], attribute-synonym finding for relational tables and spreadsheet data in web pages [6] and string and schema matching techniques [9,3,5].

Sorrentino et al. present a semi-automatic technique for schema normalization and motivate the importance of incorporating individual examples in the process of schema normalization [10]. In contrast, ours is the first work that incorporates connected attribute labels from *complementing schemata* as source of information for fully-automated schema decryption.

The authors of [6] point out that distance metrics and global dictionaries, as often used for string and schema matching, are not appropriate to automatically find synonyms for arbitrary attribute labels. This observation is supported by our experimental results: Common distance metrics result in poor decryptions and global dictionaries lead to a relatively poor coverage of domain-specific abbreviations, acronyms, and tokens. The authors of [6] propose a large-scale discovery method on 125 million relational tables extracted from 14.1 billion HTML tables. Their approach is based on pairs of attribute labels co-occurring in tables with same context attributes. As already pointed out in the introduction, those data-driven approaches are infeasible for ETL systems. Furthermore, with our approach we can identify accurate decryptions from a substantially smaller corpus of examples compared to data-driven approaches relying on a large set of web-scale example data [3,6].

6 Conclusion

With this paper we presented a fully-automated schema decryption method leveraging the large number of mapped attribute labels in a given ETL repository. Our work is motivated by observing the need of easy-to-understand schemata during ETL development and maintenance. Cryptic schemata significantly increase the amount of time to understand unfamiliar data, as many readers might have experienced themselves. We demonstrated that our schema decryption approach provides helpful suggestions for three different real world ETL repositories. An ETL developer is now able to quickly grasp the underlying semantics of data records in cryptic schemata.

Acknowledgment. This research was funded by InfoDyn AG; we want to thank Benjamin Böhm and Dieter Radler of InfoDyn AG for supplying real-world data.

References

1. Agrawal, H., Chafle, G., Goyal, S., Mittal, S., Mukherjea, S.: An Enhanced Extract-Transform-Load System for Migrating Data in Telecom Billing. In: Proceedings of the International Conference on Data Engineering (ICDE), Cancún, México (2008)
2. Albrecht, A., Naumann, F.: Understanding Cryptic Schemata in Large Extract-Transform-Load Systems. Tech. rep., Hasso Plattner Institute for Software Systems Engineering (October 2012)
3. Arasu, A., Chaudhuri, S., Kaushik, R.: Learning String Transformations from Examples. In: Proceedings of the International Conference on Very Large Databases (VLDB), Lyon, France (2009)
4. Baeza-Yates, R.A., Ribeiro-Neto, B.: Modern Information Retrieval. Addison-Wesley, Boston (1999)
5. Bernstein, P.A., Madhavan, J., Rahm, E.: Generic Schema Matching, Ten Years Later. VLDB Journal 4 (2011)
6. Cafarella, M.J., Halevy, A., Wang, D.Z., Wu, E., Zhang, Y.: WebTables: Exploring the Power of Tables on the Web. In: Proceedings of the International Conference on Very Large Databases (VLDB), Auckland, New Zealand (2008)
7. Cui, Y., Widom, J.: Lineage Tracing for General Data Warehouse Transformations. VLDB Journal 12(1) (2003)
8. Dayal, U., Castellanos, M., Simitsis, A., Wilkinson, K.: Data Integration Flows for Business Intelligence. In: Proceedings of the International Conference on Extending Database Technology (EDBT). Saint Petersburg, Russia (2009)
9. Levenshtein, V.I.: Binary codes capable of correcting deletions, insertions and reversals. Soviet Physics Doklady 10(8) (1966)
10. Sorrentino, S., Bergamaschi, S., Gawinecki, M., Po, L.: Schema Normalization for Improving Schema Matching. In: Laender, A.H.F., Castano, S., Dayal, U., Casati, F., de Oliveira, J.P.M. (eds.) ER 2009. LNCS, vol. 5829, pp. 280–293. Springer, Heidelberg (2009)

Fast Group Recommendations by Applying User Clustering

Eirini Ntoutsi[1], Kostas Stefanidis[2], Kjetil Nørvåg[2], and Hans-Peter Kriegel[1]

[1] Institute for Informatics, Ludwig Maximilian University, Munich
{ntoutsi,kriegel}@dbs.ifi.lmu.de
[2] Department of Computer and Information Science, Norwegian University of Science and Technology, Trondheim
{kstef,Kjetil.Norvag}@idi.ntnu.no

Abstract. Recommendation systems have received significant attention, with most of the proposed methods focusing on personal recommendations. However, there are contexts in which the items to be suggested are not intended for a single user but for a group of people. For example, assume a group of friends or a family that is planning to watch a movie or visit a restaurant. In this paper, we propose an extensive model for group recommendations that exploits recommendations for items that similar users to the group members liked in the past. We do not exhaustively search for similar users in the whole user base, but we pre-partition users into clusters of similar ones and use the cluster members for recommendations. We efficiently aggregate the single user recommendations into group recommendations by leveraging the power of a top-k algorithm. We evaluate our approach in a real dataset of movie ratings.

1 Introduction

Recommendation systems provide users with suggestions about products, movies, videos and many other items. Many systems, such as Amazon, NetFlix and MovieLens, are very popular. Collaborative recommendation systems (e.g., [14,7]) try to predict the utility of items for a particular user based on the items previously rated by other users, that is, users similar to a target user are identified, and then items are recommended based on the preferences of the similar users. Users are considered similar if there is an overlap in the items consumed. The two types of entities that are dealt in recommendation systems, i.e., users and items, are represented as sets of ratings, preferences or features. Assume, for example, a restaurant recommendation application (e.g., ZAGAT.com). Users initially rate a subset of restaurants that they have already visited. Ratings are expressed in the form of preference scores. A recommendation engine estimates preference scores for the items, e.g., restaurants, that are not rated by a user and offers appropriate recommendations. Once the unknown scores are computed, the k items with the highest scores are recommended to the user.

Since recommendations are typically personalized, different users are presented with different suggestions. However, there are cases where a group of people participates in a single activity. For instance, visiting a restaurant or a tourist attraction,

P. Atzeni, D. Cheung, and R. Sudha (Eds.): ER 2012, LNCS 7532, pp. 126–140, 2012.

watching a movie or a TV program and selecting a holiday destination are examples of recommendations well suited for groups of people. For this reason, recently, there are methods for group recommendations, trying to satisfy the preferences of all the group members. These methods can be classified into two approaches [12]. The first approach creates a joint profile for all users in the group and provides the group with recommendations computed with respect to this joint profile (e.g., [27]). The second approach aggregates the recommendations of all users in the group into a single recommendation list (e.g., [3,5]). Our work follows the second approach, since it is more flexible [12,19] and offers opportunities for improvements in terms of efficiency.

In this paper, we propose a model for group recommendations following the collaborative filtering approach. We are mainly motivated by the observation that similarity can be used to cluster users in small groups with strong similarity [21]. This way, our framework applies user clustering for organizing users into groups of users with similar preferences. To do this, we employ a hierarchical agglomerative clustering algorithm. Thereafter, we propose the use of these clusters to efficiently locate similar users to a given one; recommendations for users are produced with respect to the preferences of their cluster members without extensively searching for similar users in the whole database. The k most prominent items for the group are identified by exploiting a top-k algorithm. Group recommendations are presented to users along with explanations about the reasons that the particular items are being suggested. Finally, to deal with the sparsity of the explicitly defined user preferences, we introduce the notion of support in recommendations to model how confident the recommendation of an item for a user is.

To summarize, our contributions are as follows:

- We enhance recommendations with the notion of support to model the confidence of recommendations.
- We formulate the top-k group recommendations problem as a top-k query problem and leverage the power of a top-k algorithm to efficiently derive the most prominent items for the whole group.
- We present recommendations along with explanations about why the specific recommendations appear in the top-k list.
- We introduce user clustering for partitioning users into clusters and use these clusters to efficiently compute personal recommendations. Personal recommendations are aggregated to produce group recommendations.

We have also evaluated both the efficiency and effectiveness of our approach using a real dataset of movie ratings.

The rest of the paper is organized as follows. Section 2 introduces the personal and group recommendation model, and clarifies the problem statement. Section 3 proposes methods for locating users similar to a target one and identifying the top-k recommendations for a group. Section 4 contains extensive experimentation to evaluate the efficiency and effectiveness of our approach, while Section 5 describes related work. Finally, Section 6 concludes the paper with a summary of our findings.

2   Recommendation Model

Assume a set of items $\mathcal{I}$ and a set of users $\mathcal{U}$ interacting with a recommendation application. Each user $u \in \mathcal{U}$ may express a preference for an item $i \in \mathcal{I}$, $preference(u, i)$, in the range $[0.0, 1.0]$. We use $\mathcal{P}_i$ to denote the set of users in $\mathcal{U}$ that have expressed a preference for item i. The cardinality of the items set $\mathcal{I}$ is usually high and typically users rate only a few of these items. For the items unrated by the users, we estimate a *relevance score*, denoted as $relevance(u, i)$, where $u \in \mathcal{U}$, $i \in \mathcal{I}$. To do this, a recommendation strategy is invoked. We distinguish between personal recommendations referring to a single user (Section 2.1) and group recommendations referring to a set of users (Section 2.2).

2.1   Personal Recommendations

There are different ways to estimate the relevance of an item for a user. In general, the recommendation methods are categorized into: (i) *content-based*, that recommend items similar to those the user has preferred in the past (e.g., [17]), (ii) *collaborative filtering*, that recommend items that similar users have liked in the past (e.g., [14,7]) and (iii) *hybrid*, that combine content-based and collaborative ones (e.g., [4]). Our work falls into the *collaborative filtering* category. The key concept of collaborative filtering is to use preferences of other users that exhibit the most similar behavior to a given user in order to produce relevance scores for unrated items. Similar users are located via a *similarity function* $simU(u, u')$ that evaluates the proximity between u and u'.

We use $\mathcal{F}_u$ to denote the set of the most similar users to u. We refer to such users as the *friends* of u.

Definition 1 (Friends). *Let $\mathcal{U}$ be a set of users. The friends $\mathcal{F}_u$, $\mathcal{F}_u \subseteq \mathcal{U}$, of a user $u \in \mathcal{U}$ is a set of users, such that, $\forall u' \in \mathcal{F}_u$, $simU(u, u') \geq \delta$ and $\forall u'' \in \mathcal{U} \backslash \mathcal{F}_u$, $simU(u, u'') < \delta$, where δ is a threshold similarity value.*

Clearly, one could argue for other ways of selecting $\mathcal{F}_u$, for instance, by taking the k most similar users to u. Our main motivation is that we opt for selecting only highly connected users even if the resulting set of users $\mathcal{F}_u$ is small.

Given a user u and his friends $\mathcal{F}_u$, if u has expressed no preference for an item i, the relevance of i for u is estimated as follows:

$$relevance(u, i) = \frac{\sum_{u' \in (\mathcal{F}_u \cap \mathcal{P}_i)} simU(u, u')preference(u', i)}{\sum_{u' \in (\mathcal{F}_u \cap \mathcal{P}_i)} simU(u, u')}$$

However, since the number of items is huge and usually users rate only a few items, the following question usually arises: *How confident are the relevance scores associated with the recommended items?* Towards this direction, we introduce the notion of *support* for each candidate item i for user u, which defines the percentage of friends of u that have expressed preferences for i. Formally:

$$support(u, i) = |\mathcal{F}_u \cap \mathcal{P}_i| / |\mathcal{F}_u|$$

To estimate the worthiness of an item recommendation for a user, we propose to combine the *relevance* and *support* scores in terms of a *value* function.

Definition 2 (Personal Value). *Let $\mathcal{U}$ be a set of users and $\mathcal{I}$ be a set of items. Let $w_1, w_2 \geq 0 : w_1 + w_2 = 1$. The personal value of an item $i \in \mathcal{I}$ for a user $u \in \mathcal{U}$ with friends $\mathcal{F}_u$, such that, $\nexists preference(u, i)$, is:*

$$value_{\mathcal{F}_u}(u, i) = w_1 \times relevance(u, i) + w_2 \times support(u, i)$$

Although more sophisticated functions can be designed, the weighted summation of the relevance and support scores is simple and intuitive. Moreover, when $w_2 = 0$, *value* maps to *relevance*, the typically used recommendation score.

2.2   Group Recommendations

The majority of recommendation systems are designed to make personal recommendations, i.e., recommendations for individual users. However, there are cases in which the items to be selected are not intended for personal usage but for a group of users. For example, assume a group of friends or a family that is planning to watch a movie, visit a restaurant or select a holiday destination. For this reason, some recent works have addressed the problem of identifying recommendations for a group of users, trying to satisfy the preferences of all the group members (e.g., [3,5,6]).

In our approach, we first compute the *personal value* scores for the unrated items for each user in the group, and then, based on these predictions, we compute the aggregated value scores for the group.

Definition 3 (Group Value). *Let $\mathcal{U}$ be a set of users and $\mathcal{I}$ be a set of items. Given a group of users $\mathcal{G}$, $\mathcal{G} \subseteq \mathcal{U}$, the group value of an item $i \in \mathcal{I}$ for $\mathcal{G}$, such that, $\forall u \in \mathcal{G}$, $\nexists preference(u, i)$, is:*

$$value(\mathcal{G}, i) = Aggr_{u \in \mathcal{G}}(value_{\mathcal{F}_u}(u, i))$$

We employ three different designs regarding the aggregation method *Aggr*, each one carrying different semantics: (i) the *least misery design*, capturing cases where strong user preferences act as a veto (e.g., do not recommend steakhouses to a group when a vegetarian belongs to the group), (ii) the *fair design*, capturing more democratic cases where the majority of the group members is satisfied, and (iii) the *most optimistic design*, capturing cases where the more satisfied member of the group acts as the most influential member (e.g., recommend a movie to a group when a member is highly interested in it and the remaining members have reasonable satisfaction). In the least misery (respectively, most optimistic) design, the predicted value score of an item for the group is equal to the minimum (respectively, maximum) value score of the item scores of the members of the group, while the fair design, that assumes equal importance among all group members, returns the average score. Table 1 summarizes the aggregation methods.

Table 1. Aggregation methods

Design	Aggregation Method
Least misery	$value(\mathcal{G}, i) = \min_{u \in \mathcal{G}}(value_{\mathcal{F}_u}(u, i))$
Fair	$value(\mathcal{G}, i) = \left(\sum_{u \in \mathcal{G}} value_{\mathcal{F}_u}(u, i)\right) / |\mathcal{G}|$
Most optimistic	$value(\mathcal{G}, i) = \max_{u \in \mathcal{G}}(value_{\mathcal{F}_u}(u, i))$

2.3 Problem Statement

Given a group of users and a restriction k on the number of the recommended items, we would like to provide k suggestions for items that are highly relevant to the preferences of all the group members and, also, exhibit high support.

Definition 4. (TOP-K GROUP RECOMMENDATIONS). *Let $\mathcal{U}$ be a set of users and $\mathcal{I}$ be a set of items. Given a group of users $\mathcal{G}$, $\mathcal{G} \subseteq \mathcal{U}$, and an aggregation method Aggr, recommend to $\mathcal{G}$ a list of items $\mathcal{I}_\mathcal{G} = <i_1, \ldots, i_k>$, $\mathcal{I}_\mathcal{G} \subseteq \mathcal{I}$, such that:*
(i) $\forall i_j \in \mathcal{I}_\mathcal{G}, u \in \mathcal{G}, \nexists preference(u, i_j)$,
(ii) $value(\mathcal{G}, i_j) \geq value(\mathcal{G}, i_{j+1})$, $1 \leq j \leq k - 1$, $\forall i_j \in \mathcal{I}_\mathcal{G}$, and
(iii) $value(\mathcal{G}, i_j) \geq value(\mathcal{G}, x_y)$, $\forall i_j \in \mathcal{I}_\mathcal{G}$, $x_y \in \mathcal{I} \backslash \mathcal{I}_\mathcal{G}$.

The first condition ensures that the suggested items do not include already evaluated items by some users in the group (e.g., do not recommend a movie that a group member has already watched). The second condition ensures the descending ordering of the items with respect to their group value, while the third condition defines that every item in the result set has group value greater than or equal to the group value of any of the remaining items.

2.4 Group Recommendations Explanations

Recently, it has been shown that the success of recommendations relies on explaining the cause behind them [25]. To this end, except for the suggested items, we also provide the group with an explanation for each suggested item, i.e., why the specific item appears in the top-k list.

Although our explanations depend on the employed design, they have the following general form: "ITEM i HAS GROUP VALUE SCORE $value(\mathcal{G}, i)$ BECAUSE OF USER(S) $\{u_x, \ldots, u_y\}$". For instance, for a movie recommendation system, an example explanation is: "Movie *Dracula* has group value score 0.9 because of user Jeffrey". More specifically, for the least misery design, we report with each suggested item its group value score and the member of the group with the minimum personal value score for the item, i.e., the member that is responsible for this selection. Similarly, for the most optimistic design, we report the member of the group with the maximum personal value score for the item. Finally, for the fair design, we report with each item the members of the group with personal value scores for the item close to its group value score (up to a distance p), i.e., the members that are highly satisfied, and hence, direct towards this selection.

3   Group Recommendations Computation

Our approach for suggesting items for a group of users $\mathcal{G}$, consists of the following steps: (i) locate the set of users $\mathcal{F}_u$, $\forall u \in \mathcal{G}$, (ii) compute the personal value scores $value_{\mathcal{F}_u}(u, i)$, $\forall u \in \mathcal{G}$, $i \in \mathcal{I}$, (iii) combine the independent scores according to an aggregation method $Aggr$ to derive the group value scores $value(\mathcal{G}, i)$, $\forall i \in \mathcal{I}$, and (iv) present the k items with the highest group value scores along with explanations.

For computing the scores of all items for the group, a solution that involves no pre-computation is to first find the friends of each user in $\mathcal{G}$, by computing the similarity measures between each user in $\mathcal{G}$ and each user in $\mathcal{U}$, then produce the value scores of the candidate items and finally rank them based on these scores. We refer to this approach as the *baseline approach*. Performance can be improved by performing some preprocessing steps offline. In particular, in this work, we propose building clusters of similar users, considering as similar those users that have similar preferences. Ideally, the friends of each user are the members of the cluster that the user belongs to. Based on the preferences of these cluster members, the group value scores are computed. We refer to this approach as the *user clustering approach*.

Next, we present our method for locating the friends of a user in the baseline and user clustering approach (Section 3.1). Then, we focus on how to identify the top-k group recommendations (Section 3.2).

3.1   Finding User Friends

Baseline Approach. To find the friends set $\mathcal{F}_u$ for a specific user $u \in \mathcal{U}$, the baseline approach would need to calculate all similarity measures $simU(u, u')$, $\forall u' \in \mathcal{U}$ and select those with $simU(u, u') \geq \delta$. We refer to this algorithm, as the *Baseline Friends Finder* algorithm. Such an approach though, would be extremely inefficient in large systems, since it requires the online computation of the set of friends for each user of the query group.

User Clustering Approach. Since the baseline approach is expensive for a real recommendation application where the response time is a critical parameter, we propose to organize the users into groups of users with similar preferences and employ these pre-computed groups to speed up the recommendation process.

We use clustering for partitioning users into groups[1]. In particular, we use a bottom up hierarchical agglomerative clustering algorithm. Initially, the *Clustering Friends Finder* algorithm (Algorithm 1) places each user in a cluster of his own. Then, at each step, it merges the two most similar clusters. The similarity between two clusters is defined as the minimum similarity between any two users that belong to these clusters (max linkage). The algorithm terminates when the similarity of the closest pair of clusters violates the user similarity threshold δ.

---

[1] Hereafter, we refer to the groups of users with similar preferences as *clusters*, to remove any ambiguity with the term *group* referring to the group of users asking for recommendations (query group).

Algorithm 1. Clustering Friends Finder Algorithm

Input: A set of user $\mathcal{U}$ and a threshold similarity value δ.
Output: A set of clusters of users with similar preferences.

1: create a cluster for each user $u \in \mathcal{U}$;
2: repeat
3:   **if** the max similarity between any pair of clusters is greater or equal to δ **then**
4:     merge these two clusters;
5:   **else** end loop;

Property 1. Let δ be a threshold similarity value. Each cluster produced by the Clustering Friends Finder algorithm contains users, such that, for each pair of users u, u' in the cluster $simU(u, u') \geq \delta$.

Proof: From lines $3 - 5$ of the Clustering Friends Finder algorithm, we merge the two most similar clusters, if their similarity is greater than or equal to δ. This similarity score represents the minimum similarity between a user u of the first cluster and a user u' of the second cluster. Therefore, any two users u, u' that belong to the same cluster, have similarity $simU(u, u') \geq \delta$.   □

This means that, if for a user u, we employ for estimating personal value scores the users in the cluster of u, we may lose some users similar to u, but we never consider as similar, users that are not, and so, our method does not result in some form of false positives.

In general, following this clustering approach, we consider as friends of each user u the members of the cluster $\mathcal{C}_u$ that the user u belongs to. We use $value_{\mathcal{C}_u}(u, i)$ to denote the personal value score of item i for u computed taking into account the users in $\mathcal{C}_u$. It is easy to show that:

Property 2. Let $\mathcal{G}$ be a group of users and i be an item in $\mathcal{I}$.
(i) If the resulting score of both $\max_{u \in \mathcal{G}}(value_{\mathcal{F}_u}(u, i))$ and $\max_{u \in \mathcal{G}}(value_{\mathcal{C}_u}(u, i))$ is associated with a user $u' \in \mathcal{G}$, and the most optimistic design is applied, both the baseline and the user clustering approaches employ the personal value score of u' for determining $value(\mathcal{G}, i)$.
(ii) If the resulting score of both $\min_{u \in \mathcal{G}}(value_{\mathcal{F}_u}(u, i))$ and $\min_{u \in \mathcal{G}}(value_{\mathcal{C}_u}(u, i))$ is associated with a user $u' \in \mathcal{G}$, and the least misery design is applied, both the baseline and the user clustering approaches employ the personal value score of u' for determining $value(\mathcal{G}, i)$.

Note that we cannot have a counterpart observation for the fair design, since in the fair design the personal value scores of sets of users are taken into account for computing group value scores, in both approaches.

3.2   Identifying Top-k Group Recommendations

Having established the methodology for finding the friends of a single user, we focus next on how to generate *valued recommendations* for a group of users. The first step towards this direction is to compute the personal value scores of each item for each user in the group. The next step is to combine the scores of each item in order to select, based on the group value scores, the items to be suggested.

Given a user $u \in \mathcal{U}$ and his similar users $\mathcal{C}_u$, the procedure for estimating the personal value score of each item i in $\mathcal{I}$ for u requires the computation of its *relevance* and *support*. Pairs of the form $(i, value_{\mathcal{C}_u}(u, i))$ are maintained in a set $\mathcal{V}_u$. As a post-processing step, we rank all pairs in $\mathcal{V}_u$ on the basis of their personal value scores. The set $\mathcal{C}_u$ refers to the clustering approach. For the baseline approach, we use, for each user u, the set $\mathcal{F}_u$, instead of $\mathcal{C}_u$.

For a group of users $\mathcal{G} = \{u_1, \ldots, u_n\}$, a common way to provide k recommendations to $\mathcal{G}$ is assigning to all items their group value scores and reporting the k items with the highest scores. Instead of following the naive approach of computing group value scores of items and ranking them based on these scores, we employ a variation of a threshold-based algorithm, called TA, first proposed in [10]. Given a group $\mathcal{G} = \{u_1, \ldots, u_n\}$, the *Top-k Group Recommendations* algorithm uses as input the ranked sets $\mathcal{V}_{u_1}, \ldots, \mathcal{V}_{u_n}$ and it considers two types of available item accesses: the *sorted access* and the *random access*. Sorted access enables item retrieval in a descending order of their scores, while random access enables retrieving the score of a specific item in one access.

The algorithm's main steps are (Algorithm 2):

(i) Do sorted access to each ranked set $\mathcal{V}_{u_j}$. For each item seen, do random accesses to the other ranked sets to retrieve the missing item personal value scores.
(ii) Compute the group value score of each item that has been seen. Rank the items based on their group value scores and select the top-k ones.
(iii) Stop to do sorted accesses when the group value scores of the k items are

---

Algorithm 2. Top-k Group Recommendations Algorithm

Input: A group of users $\mathcal{G} = \{u_1, \ldots, u_n\}$ with ranked sets $\mathcal{V}_{u_1}, \ldots, \mathcal{V}_{u_n}$ and an aggregation method $Aggr$.
Output: k pairs $(i_j, value(\mathcal{G}, i_j))$.

1: $topRecs = \emptyset$;
2: $scoreK = 0$;
3: $thresholdK = \infty$;
4: $t = 1$;
5: **while** scoreK < thresholdK **do**
6:      retrieve the score of the tth item from $\mathcal{V}_{u_1}, \ldots, \mathcal{V}_{u_n}$, $value_{\mathcal{C}_{u_1}}(u_1, i_1), \ldots,$ $value_{\mathcal{C}_{u_n}}(u_n, i_n)$;
7:      $thresholdK = Aggr(value_{\mathcal{C}_{u_1}}(u_1, i_1), \ldots, value_{\mathcal{C}_{u_n}}(u_n, i_n))$;
8:      retrieve the missing personal value scores for $i_1, \ldots, i_n$ from $\mathcal{V}_{u_1}, \ldots, \mathcal{V}_{u_n}$;
9:      compute the group value scores for $i_1, \ldots, i_n$;
10:     add, in decreasing order, the pairs of the form $(i_j, value(\mathcal{G}, i_j))$ to $topRecs$;
11:     make $scoreK$ equal to the kth group value score in $topRecs$;
12:     $t{+}{+}$;
13: **end while**
14: **return** the k first pairs $(i_j, value(\mathcal{G}, i_j))$ in $topRecs$;

at least equal to a threshold value that is defined as the aggregation score of the scores of the last items seen in each ranked set.

The *Top-k Group Recommendations* algorithm is correct when the item group value scores are obtained by combining their individual scores using a monotone function [10]. In our approach, aggregations are performed in a monotonic fashion, hence the applicability of the algorithm is straightforward.

4   Experiments

For the evaluation, we used the MovieLens dataset [1], which consists of 100,000 user ratings given by 1,000 users for 1,700 items. We show that the user clustering method is both efficient and effective comparing to the baseline approach.

To illustrate the efficiency of the user clustering approach versus the baseline alternative, we measure the execution time for computing the personal recommendations of the members of a query group. We omit the aggregation time for computing the top-k recommendations, since this time is the same in both cases.

We also evaluate the quality of our approach against the baseline method. We employ the *recommendations agreements* as a quality measure, which is defined as the number of items recommended by both approaches. That is, we compute the top-k group recommendations for both approaches and we find their intersection, which stands for the common suggested items. We denote this measure by *commonRecs*. The recommendations agreement measure computes the common items in both recommendation lists, but does not consider the actual ordering of the recommendations. To this end, we also employ a variation of the Kendall tau distance for partial rankings that computes the distance between two partial rankings based on the number of pairwise disagreements between them [9]. We denote this measure by *rankRecsDist*.

We use distance instead of user similarity. We define the distance between two users as the Euclidean distance over the items rated by both. Let $u, u' \in \mathcal{U}$ be two users, $\mathcal{I}_u$ be the set of items for which $\exists preference(u,i)$, $i \in \mathcal{I}_u$, and $\mathcal{I}_{u'}$ be the set of items for which $\exists preference(u',i)$, $i \in \mathcal{I}_{u'}$. We denote by $\mathcal{I}_u \cap \mathcal{I}_{u'}$ the set of items for which both users have expressed preferences. Then, the distance between u, u' is defined as:

$$distU(u,u') = \left(\sqrt{\sum_{i \in \mathcal{I}_u \cap \mathcal{I}_{u'}} (preference(u,i) - preference(u',i))^2} \right) / |\mathcal{I}_u \cap \mathcal{I}_{u'}|.$$

To set up a query group, we randomly select the members of the group from the user base. Group recommendations are extracted using both approaches and the execution time, the *commonRecs* and the *rankRecsDist* scores are computed. We run each experiment 100 times and report the average values.

4.1   Performance Evaluation

The main difference between the two approaches lies on the friends set computation for the users of the query group. In the baseline approach, for each user of the query group, we compute his distance to all database users and the users within distance δ are selected. So, for each user of the query group, a total number of $|\mathcal{U}|$ distance computations is required, where $|\mathcal{U}|$ is the number of users in the database. If $|\mathcal{G}|$ is the number of users in the query group, the total number of distance computations is $|\mathcal{G}| \times |\mathcal{U}|$. In the user clustering approach, the friends of a user are taken directly from his cluster, so no further computations are required. There is though an initialization cost for creating the clusters ($\approx 2h$ for $\delta = 0.15$, $\approx 2.5h$ for $\delta = 0.3$), however this is done once and it can be reused afterwards for answering different query groups requests.

In Figure 1, we display the time complexity for the two approaches for different query group sizes under different distance thresholds δ. The user clustering approach requires around 25% of the time required by the baseline approach. As $|\mathcal{G}|$ increases, the reduction becomes more evident, since the larger the query group size, the more the distance computations of the baseline approach. Note that for bigger threshold values, the execution time increases in both approaches, since each user of the query group has more friends/cluster members.

If the support is not taken into account (i.e., $w_1 = 1, w_2 = 0$), the running time slightly reduces for both approaches because less computations are required.

4.2 Quality Evaluation

In this set of experiments, we demonstrate the effectiveness of the user clustering approach. We compute the *commonRecs* and the *rankRecsDist* scores varying the query group size $|\mathcal{G}|$, the number k of the recommended items and the distance threshold δ.

Figure 2 depicts the *commonRecs* score for $\delta = 0.15$ and $\delta = 0.3$, for the three available designs, while Figure 3 depicts the corresponding *rankRecsDist* scores. $G1$ stands for groups with 1 member. Similarly, $G3$, $G5$ and $G7$ stand for groups with 3, 5 and 7 members, respectively. As the query group size $|\mathcal{G}|$ increases, the *commonRecs* score decreases, since the group recommendations rely in a more diverse set of users and personal values. Based on the same rationale, *rankRecsDist* increases as $|\mathcal{G}|$ increases. Also, we experimentally confirm that *commonRecs* increases with k, while *rankRecsDist* decreases.

Regarding the different designs, we observe that the fair and the least misery designs achieve better results when compared to the most optimistic design.

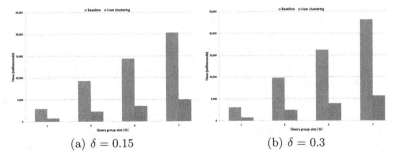

(a) $\delta = 0.15$ (b) $\delta = 0.3$

Fig. 1. Time complexity for the fair design with $w_1 = 0.5$, $w_2 = 0.5$

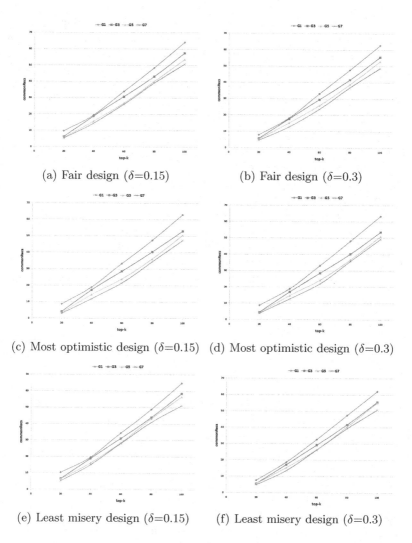

(a) Fair design (δ=0.15)   (b) Fair design (δ=0.3)

(c) Most optimistic design (δ=0.15) (d) Most optimistic design (δ=0.3)

(e) Least misery design (δ=0.15)   (f) Least misery design (δ=0.3)

Fig. 2. *commonRecs* for $\delta = 0.15$, $\delta = 0.3$ with $w_1 = 0.5$, $w_2 = 0.5$

In our scenario, where the members of the query group are selected randomly, this is expected, since it is more difficult to find agreements for max personal values. When δ increases, the *commonRecs* decreases for the fair and least misery designs and slightly increases for the most optimistic design. Corresponding findings also hold for *rankRecsDist*.

Finally, we consider the case where only relevance is taken into consideration, that is, $w_1 = 1$ and $w_2 = 0$. In Figure 4, we show the results for the fair design and $\delta = 0.3$. *commonRecs* and *rankRecsDist* behave worst comparing to the equal importance case, showing that support improves the quality of recommendations.

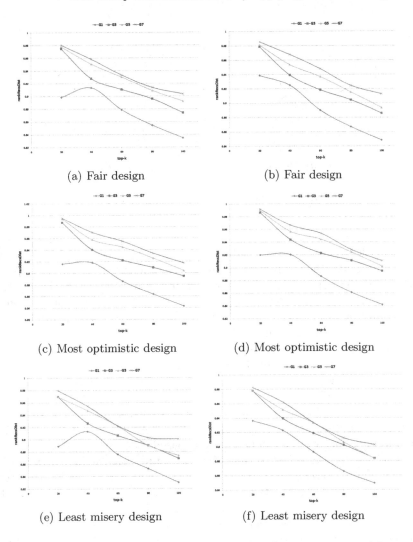

(a) Fair design

(b) Fair design

(c) Most optimistic design

(d) Most optimistic design

(e) Least misery design

(f) Least misery design

Fig. 3. $rankRecsDist$ for $\delta = 0.15$, $\delta = 0.3$ with $w_1 = 0.5$, $w_2 = 0.5$

5   Related Work

The research literature on recommendations is extensive. Typically, recommendation approaches are distinguished between: content-based, that recommend items similar to those the user previously preferred [17], collaborative filtering, that recommend items that users with similar preferences liked [14,7] and hybrid ones [4]. Several extensions have been proposed, such as employing multi-criteria ratings [2] or further contextual information [20], and providing time-aware recommendations [26,23]. Recently, there are also approaches focusing on extending database queries with recommendations [15,22].

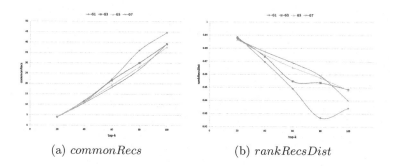

(a) *commonRecs*   (b) *rankRecsDist*

Fig. 4. Fair design for $\delta = 0.3$ with $w_1 = 1$, $w_2 = 0$

Moreover, there has been some recent work on group recommendations. PolyLens [19], a group recommendation system for movies, performed a user study to evaluate the usefulness of group recommendations. This approach employs the least misery design to aggregate personal recommendations. [3] produces group recommendations by aggregating the personal ones using a consensus function that takes into account both the relevance of the items to the users and the level at which users disagree with each other. [11] proposes a group recommendation method that employs both the social and content interests of the users of a group, while [13] provides a two-step approach: first recommendations for groups are generated and then items are filtered to increase the individual users satisfaction. [5] compares the effectiveness of personal and group recommendations. To our knowledge, none of the proposed approaches perform user clustering for finding similar users. As part of our effort, we have also designed a user interface for group recommendations based on clustering [18].

A topic related to group recommendations is *rank aggregation*, that is, given a set of different rankings of items, produce a single ranking for the items. An instance of rank aggregation known as social choice studies the problem of determining the ranking of alternatives that is best for a group given the individual opinions of its members. Social choice has been studied extensively in economics, politics, sociology, and mathematics (e.g., [24]). Rank aggregation is also studied in the web, for example, in meta-search, where ranked lists of web pages produced by different search engines need to be combined into a single one (e.g., [8]). In database middleware, a related problem refers to finding the most preferred items when there are multiple rankings based on preferences defined on different attributes or dimensions of these items (e.g., [10]). [16] reviews different aggregation strategies or functions for group modeling.

6   Conclusions

We proposed an efficient framework for group recommendations, by organizing users into clusters of users with similar preferences. These clusters are used to efficiently locate similar users to a given one; this way, recommendations for users are produced with respect to the preferences of their cluster members without extensively searching for similar users in the whole database. Top-k group recommendations are computed by aggregating the personal recommendations of the individual users, while they are presented along with explanations on the reasons that the particular items are being suggested to the group. Our results show that employing user clustering considerably improves the execution time, while preserves a satisfactory quality of recommendations.

References

1. Movielens data sets (November 2011), http://www.grouplens.org/node/12
2. Adomavicius, G., Kwon, Y.: New recommendation techniques for multicriteria rating systems. IEEE Intelligent Systems 22(3), 48–55 (2007)
3. Amer-Yahia, S., Roy, S.B., Chawla, A., Das, G., Yu, C.: Group recommendation: Semantics and efficiency. PVLDB 2(1), 754–765 (2009)
4. Balabanovic, M., Shoham, Y.: Content-based, collaborative recommendation. Commun. ACM 40(3), 66–72 (1997)
5. Baltrunas, L., Makcinskas, T., Ricci, F.: Group recommendations with rank aggregation and collaborative filtering. In: RecSys., pp. 119–126 (2010)
6. Berkovsky, S., Freyne, J.: Group-based recipe recommendations: analysis of data aggregation strategies. In: RecSys, pp. 111–118 (2010)
7. Breese, J.S., Heckerman, D., Kadie, C.M.: Empirical analysis of predictive algorithms for collaborative filtering. In: UAI, pp. 43–52 (1998)
8. Dwork, C., Kumar, R., Naor, M., Sivakumar, D.: Rank aggregation methods for the web. In: WWW10 (2001)
9. Fagin, R., Kumar, R., Sivakumar, D., Sivakumar, D.: Comparing top k lists. In: SODA, pp. 28–36 (2003)
10. Fagin, R., Lotem, A., Naor, M.: Optimal aggregation algorithms for middleware. In: PODS (2001)
11. Gartrell, M., Xing, X., Lv, Q., Beach, A., Han, R., Mishra, S., Seada, K.: Enhancing group recommendation by incorporating social relationship interactions. In: GROUP, pp. 97–106 (2010)
12. Jameson, A., Smyth, B.: Recommendation to groups. In: The Adaptive Web, pp. 596–627 (2007)
13. Kim, J.K., Kim, H.K., Oh, H.Y., Ryu, Y.U.: A group recommendation system for online communities. International Journal of Information Management 30(3), 212–219 (2010)
14. Konstan, J.A., Miller, B.N., Maltz, D., Herlocker, J.L., Gordon, L.R., Riedl, J.: Grouplens: Applying collaborative filtering to usenet news. Commun. ACM 40(3), 77–87 (1997)
15. Koutrika, G., Bercovitz, B., Garcia-Molina, H.: Flexrecs: expressing and combining flexible recommendations. In: SIGMOD Conference, pp. 745–758 (2009)
16. Masthoff, J.: Group modeling: Selecting a sequence of television items to suit a group of viewers. User Modeling and User-Adapted Interaction 14(1), 37–85 (2004)
17. Mooney, R.J., Roy, L.: Content-based book recommending using learning for text categorization. In: ACM DL, pp. 195–204 (2000)
18. Ntoutsi, I., Stefanidis, K., Nørvåg, K., Kriegel, H.P.: gRecs: A Group Recommendation System Based on User Clustering. In: Lee, S.-g., Peng, Z., Zhou, X., Moon, Y.-S., Unland, R., Yoo, J. (eds.) DASFAA 2012, Part II. LNCS, vol. 7239, pp. 299–303. Springer, Heidelberg (2012)

19. O'Connor, M., Cosley, D., Konstan, J.A., Riedl, J.: Polylens: A recommender system for groups of user. In: ECSCW, pp. 199–218 (2001)
20. Palmisano, C., Tuzhilin, A., Gorgoglione, M.: Using context to improve predictive modeling of customers in personalization applications. IEEE Trans. Knowl. Data Eng. 20(11), 1535–1549 (2008)
21. Rajaraman, A., Ullman, J.D.: Mining of massive datasets. Lecture Notes for Stanford CS345A Web Mining, p. 328 (2010), http://infolab.stanford.edu/~ullman/mmds.html
22. Stefanidis, K., Drosou, M., Pitoura, E.: You May Also Like results in relational database. In: PersDB, pp. 37–42 (2009)
23. Stefanidis, K., Ntoutsi, I., Nørvåg, K., Kriegel, H.P.: A framework for time-aware recommendations. In: DEXA (2012)
24. Taylor, A.: Mathematics and politics: Strategy, voting, power and proof. Springer, New York (1995)
25. Tintarev, N., Masthoff, J.: Designing and evaluating explanations for recommender systems. In: Recommender Systems Handbook, pp. 479–510. Springer (2011)
26. Xiang, L., Yuan, Q., Zhao, S., Chen, L., Zhang, X., Yang, Q., Sun, J., Sun, J.: Temporal recommendation on graphs via long- and short-term preference fusion. In: KDD, pp. 723–732 (2010)
27. Yu, Z., Zhou, X., Hao, Y., Gu, J.: Tv program recommendation for multiple viewers based on user profile merging. User-Adapt. Interact 16(1), 63–82 (2006)

Understanding Tables on the Web

Jingjing Wang[1,*], Haixun Wang[2], Zhongyuan Wang[2], and Kenny Q. Zhu[3,**]

[1] University of Washington
[2] Microsoft Research Asia
[3] Shanghai Jiao Tong University

Abstract. The Web contains a wealth of information, and a key challenge is to make this information machine processable. In this paper, we study how to "understand" HTML tables on the Web, which is one step further from finding the schemas of tables. From 0.3 billion Web documents, we obtain 1.95 billion tables, and 0.5-1% of these contain information of various entities and their properties. We argue that in order for computers to understand these tables, computers must first have a brain – a general purpose knowledge taxonomy that is comprehensive enough to cover the concepts (of worldly facts) in a human mind. Second, we argue that the process of understanding a table is the process of finding the right position for the table in the knowledge taxonomy. Once a table is associated with a concept in the knowledge taxonomy, it will be automatically linked to all other tables that are associated with the same concept, as well as tables associated with concepts related to this concept. In other words, understanding occurs when computers will understand the semantics of the tables through the interconnections of concepts in the knowledge base. In this paper, we illustrate a two phase process. Our experimental results show that the approach is feasible and it may benefit many useful applications such as web search.

1 Introduction

The World Wide Web contains a wealth of information. Unfortunately, most of this information is understood only by humans but not by machines. A key challenge is thus to make such information machine accessible and processable.

In this paper, we focus on enabling machines to understand information on the Web. We are particularly interested in HTML tables, for the following two reasons. First, there are billions of tables on the Web, and many of them contain valuable information. Second, tables are well structured and relatively easier to understand, whereas converting text into structured data using natural language processing techniques is very costly and impractical for large web corpus.

Informally, we define the problem of *understanding* a web table as the problem of associating the table with one or more *semantic concepts* in a general purpose knowledge base we have created. In particular, after the association is established, each row of the table will describe the attributes of a particular *entity* of that concept (In this work, we assume tables are horizontal). For example, consider the tables in Figure 1. Our goal

---

* The work was done at Microsoft Research Asia.
** Kenny Q. Zhu was partially supported by NSFC grants No. 61033002 and 61100050.

P. Atzeni, D. Cheung, and R. Sudha (Eds.): ER 2012, LNCS 7532, pp. 141–155, 2012.
© Springer-Verlag Berlin Heidelberg 2012

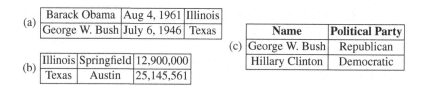

(a)	Barack Obama	Aug 4, 1961	Illinois
	George W. Bush	July 6, 1946	Texas

(b)	Illinois	Springfield	12,900,000
	Texas	Austin	25,145,561

(c)	Name	Political Party
	George W. Bush	Republican
	Hillary Clinton	Democratic

Fig. 1. Three Web Tables

is to associate, Table (a) to such concepts as *US presidents* in the knowledge base. This means the knowledge base contains *US presidents* as a concept, as well as at least some of its instances (e.g., *George W. Bush*) and some of its attributes or properties (e.g., *date of birth*). Similarly, we need to associate Table (b) to such concepts as *US states*, which contain attributes such as *capitals* and *populations* and Table (c) to such concepts as *US politicians*, which contain attributes such as *political parties*. Note that the knowledge base may not contain all the information in the table including instances, attributes or values, but the knowledge base need to contain enough information for the machines to "guess" what the table is about. Note that this is different from schema detection. For example, in Table (c), the schema is given, but without associating the table to concepts such as *US politicians*, or *politicians*, it is not possible to understand the meaning of the data.

Once we understand tables, information contained in them can be used to support applications such as semantic search. The following examples show some "smart" answers to semantic queries (assuming Figure 1 contains all the tables for input).

If a user asks for "*Obama birthday*," the system returns the following table (For this work, we use tables to answer any queries. Of course, answers can be provided in any form, including in natural languages):

President	Date of Birth
Barack Obama	Aug 4, 1961

This shows that the system correctly identified the first column of Table(a) is about *presidents* and the second column about their birthdays.

Second, for query "*Illinois*," it returns the following table about the state of Illinois:

State	Capital City	Population
Illinois	Springfield	12,900,000

Note that i) the result has an inferred schema (a header), and ii) even though the word *Illinois* appears in Table (a) as well, our system realizes Table (a) is more about presidents than states, and the query is looking for information about Illinois, the state.

Finally, when asked for "*politicians*", the machine returns the following table by combining Table (a) and (b) and adding a header:

Politician	Date of Birth	State	Political Party
Barack Obama	Aug 4, 1961	Illinois	-
George W. Bush	July 6, 1946	Texas	Republican
Hillary Clinton	Oct 26, 1947	-	Democratic

What is interesting about this answer is that the system not only knows that Table (a) is about US presidents and Table (b) is about politicians, but it also knows that presidents are politicians, the two tables can be combined for the query. Such semantic association and reasoning between tables is significantly more intelligent than relational schema induction techniques which merely generate a "data type" for each column of a table.

Knowing the structure of the data does not mean knowing the semantics of the data, or understanding its content. For example, given Figure 1(a), how do we know *Barack Obama* is a president, and *Aug 4, 1961* is not just a date, but also Obama's birth date? Furthermore, how do we know that the first row is about Obama, the president, and not Illinois, the state? And finally, how do we know we can merge Figure 1(a) and (c)?

As human beings, to answer these questions, we need certain background knowledge. If we know Obama and Bush are US presidents, then immediately we can add title *"President"* to the first column of Figure 1(a). Next, we recognize that the second column is dates. Since our knowledge tells us presidents are persons and one of the most important dates about a person is his or her date of birth, we can add the *"Date of Birth"* header to the second column. If our knowledge also knows that Illinois and Texas are both US states, then it is not difficult to tag the third column with *"State"*. Now, given that this table has three column titles: *President*, *Date of Birth* and *State*, our common sense tells us that every row of this table is more about a president than about a state, because a president has a birth date and a home state as two of his important attributes, whereas a state does not have a date of birth and a president for sure!

We can see that the key for understanding the tables in the above example is to answer these two related questions:

1. *What is the most likely concept that contains a set of given entities?*
2. *What is the most likely concept that has a set of given attributes?*

We call the process to answer these two questions *conceptualization*. In this paper, we utilize a general-purpose taxonomy called Probase [1,4,2] as our knowledge to conceptualize the information in web tables and achieve understanding. Probase is made up of worldly facts automatically constructed from 50 Terabytes Web corpus and other data. One dramatic difference between Probase and any manually built taxonomy (e.g. Open Mind Common Sense [5], Freebase [6]) is that Probase has over 2.7 million concepts, which cover most if not all concepts about worldly facts in human minds. Each concept contains a set of entities ranked by their popularity or other scores, and also a set of attributes used to describe entities in that concept. Given the rich concepts and probabilistic scores of Probase, we have a better chance to understand web tables than using other manually built taxonomies, since web tables also contain diverse and probabilistic information, which is hard to be well captured manually.

Figure 2 shows a snippet of the Probase taxonomy. Oval boxes denote concepts and their attributes, rectangles denote entities, while arrows denote *isA* relations. More details about Probase is given in Section 2.

Figure 3 illustrates the overall process for understanding web tables, which is detailed in Section 3. We first try to detect the header in a table (Section 3.2). If no header can be detected, we generate one using the table content (Section 3.3). Then we identify the column that contains entities (others are attribute columns) (Section 3.4). If we

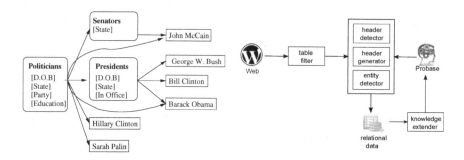

Fig. 2. A Snippet of the Probase Taxonomy   **Fig. 3.** The flowchart for understanding tables

fail to derive the header or the entity column, or the result quality is considered too low, we discard the table. Otherwise, the recognized content is *attached* to the Probase concept that best describes the entity column. This essentially *extends* the Probase. We evaluate our prototype system in Section 4, and discuss some of the most related work in Section 5 before concluding in Section 6.

2 Building a Knowledge Taxonomy

To conceptualize any web table, it is essential for our knowledge to contain large amount of concepts, their attributes and their representative entities. In this paper, we use Probase [1] to understand tables. Probase is a research prototype that aims at building a unified taxonomy of worldly facts from web data and search logs. The backbone of Probase is constructed by the Hearst patterns [7], or "SUCH AS" like patterns. For example, a sentence containing "... politicians such as Barack Obama ..." can be considered as an evidence that *politicians* is a hypernym of *Barack Obama*.

Despite their popularity in concept-entity extraction, Hearst patterns are not powerful enough for extracting attributes and values. As a result, the core Probase taxonomy is not rich in the attribute/value space. To enrich Probase with more attributes, we use the following linguistic pattern to discover seed attributes for a concept **C**:

What is the A of I?

Here, A is the seed attributes we want to discover, and I is an entity in concept **C** obtained from Probase. For example, assume we want to find seed attributes for concept *countries*. From Probase, we know *China* is a country. Then, we use the pattern "What

Table 1. Probase Notations

$\mathcal{C}, \mathcal{I}, \mathcal{A}$ concepts, instances, and attributes in the taxonomy	
$E(c)$ entities of concept c	$p_c(i)$ plausibility of entity i for class c
$A(c)$ attributes of concept c	$a_c(i)$ ambiguity of entity i for class c
$C(i)$ concepts entity i belongs to	

is the * of China?" to search the Web. A match "What is the capital of China?" indicates *capital* is a candidate seed attribute for *countries*. The seed attributes thus extracted are good enough to detect many table schemas. Below, we address several important issues in attributes discovery. The notations used in our discussion are summarized in Table 1.

1. Why use the rigid pattern "What is the A of I"? We deliberately choose a rigid pattern to increase the precision. Because i) we are only interested in seed attributes in this phase; ii) we use multiple entities to find attributes for each concept, and iii) we are matching the pattern against a huge Web corpus (50 Terabytes), which means we can always find good candidate attributes. Note that the above pattern is a skeleton which represents a set of concrete patterns. For example, the word *What* can be replaced by *Where* or *Who*, and *is* can be replaced by *are* or *was*.

2. What entities should we use as the **I** *in the pattern?* For each concept, Probase returns a list of entities associated with a set of scores. Two essential scores are *plausibility* and *ambiguity*. Plausibility measures how likely an entity is really a member of a given concept, and ambiguity measures how likely an entity is also a member of other unrelated concepts. For example, both *Microsoft* and *Apple* are companies, but *Apple* is more ambiguous because it is also a fruit. If we use *Apple*, we may mistake *taste* as an attribute of *companies*. So we want entities with high plausibility and low ambiguity. For how the scores are derived, we refer readers to [1]. Specifically, we find $ES(c)$, the eligible entities of class c:

$$ES(c) = \{i | i \in E(c), p_c(i) > \delta_p, a_c(i) < \delta_a\},$$

where δ_p and δ_a denote the thresholds for plausibility and ambiguity, respectively.

3. How to rank candidate seed attributes to obtain final seeds? For each concept c, we score and merge candidate seed attributes derived from $ES(c)$ to obtain final ones. If attribute a is derived with entity i, then we add a weight equals to $p_c(i)$ to a. We then aggregate the weights of all the candidate seed attributes, and choose top-ranked attributes as the final seed attributes of c. Specifically, let $D(a)$ denote the set of entities in $ES(c)$ having attribute a. The weight of a candidate attribute a with respect to c is:

$$w_c(a) = \sum_{i \in D(a)} p_c(i) / \sum_{i \in ES(c)} p_c(i). \tag{1}$$

Note that Eq (1) considers both plausibility and ambiguity, because $i \in ES(c)$.

One problem of using linguistic patterns to find attributes from the Web is noises, or wrong attributes. However, for different entities I, the pattern typically returns different noises, hence the ranking scheme effectively removes such noises.

Besides the above pattern, we also get attributes from DBPedia. We use exact matchings when locating Probase's entities in it.

In total, we obtain 10.5 million raw seed attributes (0.21 million distinct) for about 1 million classes. These are enough to bootstrap the process of understanding tables. Once we identified table schema, they can be used to expand the current set of attributes and essentially enrich Probase. In an evaluation of 30 concepts and their top 20 seed attributes, we found our approach reaches an average precision of 0.96. This provides a strong foundation for further understanding process.

3   Understanding Tables

From 0.3 billion web documents, we extract 1.95 billion raw HTML tables. Many of them do not contain useful or relational information (e.g., be used for page layout purpose); others have structures that are too complicated for machines to understand. We use a rule-based filtering method to acquire 65.5 million tables (3.4% of all the raw tables) that contain potentially useful information.

In this section, we describe the process of *understanding* a table. With the help of "enriched" Probase, we identify entities (of a single concept/class), attributes, and values in a table. Then, we use the information in the table to enrich Probase.

3.1   Knowledge APIs for Schema Extraction

We first introduce two functions $\kappa_A(A)$ and $\kappa_E(E)$, which are part of the knowledge API provided by Probase. Probase contains rich information about concepts, entities, and attributes. For a given concept, we may use Probase to find its entities and its attributes. Likewise, for a given set of entities or attributes, we may use Probase to find the set of concepts that they may belong to. The two functions serve this purpose. Given a set of attributes A and a set of entities E, functions $\kappa_A(A)$ and $\kappa_E(E)$ find the most likely concept to which A or E belongs [1]. More specifically, we have

- $\kappa_A(A)$: for a set of attributes A, $\kappa_A(A)$ returns a list of triples $\cdots, (c_i, A_i, sa_i), \cdots$ ordered by score sa_i, where c_i is a likely concept for A, $A_i \subseteq A$ are attributes of concept c_i, and sa_i is the score indicating the confidence of c_i given A.
- $\kappa_E(E)$: for a set of entities E, $\kappa_E(E)$ returns a list of triples $\cdots, (c_i, E_i, se_i), \cdots$ ordered by score se_i, where c_i is a likely concept for E, $E_i \subseteq E$ are entities of concept c_i, and se_i is the score indicating the confidence of c_i given E.

Table 2. A running example for table understanding

Name	Birthdate	Political Party	Assumed Office	Height
Barack Obama	4 Aug 1961	Democratic	2009	6'1
Arnold Schwarzenegger	30 Jul 1947	Republican	2003	6'2
Hillary Clinton	26 Oct 1947	Democratic	2009	5'8

The two functions may help us discover the schema of a table. Let us take Table 2 as an example. Suppose we want to know if the first row is the head of the table. As we know, a head usually contains a set of attributes of a concept. So by applying $\kappa_A()$ on $A = \{Name, Birthdate, Political Party, Assumed Office, Height\}$, we may expect to get some concepts of high confidence. Assume Probase returns the following for $\kappa_A(A)$:

(*US presidents, {Birthdate, Political Party, Assumed Office}, 0.90*)
(*politicians, {Birthdate, Political Party, Assumed Office}, 0.88*)
(*NBA players, {Birthdate, Height}, 0.65*)

 . . .

---

[1] The implementation of the two functions is discussed in [1,8].

The result ranks *US presidents* higher than *politicians*. This is probable because we are only looking at the header, and have not studied the table content yet.

On the other hand, if we compute $\kappa_A(A)$ on the second row $A = \{$*Barack Obama, 4 Aug 1961, Democratic, 2009, 6'1*$\}$, we may get empty results or results with very low confidence. This indicates that $\kappa_A()$ is useful in table header detection. We discuss this in more detail in Section 3.2.

However, many tables do not have headers. Function $\kappa_E()$ is useful in this case. Consider the first column $E = \{$*Name, Barack Obama, Arnold Schwarzenegger, Hillary Clinton*$\}$. Applying $\kappa_E()$ on E may get:

(*politicians, {Barack Obama, Arnold Schwarzenegger, Hillary Clinton}, 0.95}*)
(*actors, {Arnold Schwarzenegger}, 0.5}*)

. . .

In this case, $\kappa_E()$ identifies *politicians* as the top match. If we apply $\kappa_E()$ on other columns, we may recover more possible attributes. Thus, there is a possibility we can generate a header for the table. We discuss this in more detail in Section 3.3.

In summary, combining the outcomes of $\kappa_A()$ and $\kappa_E()$, we may get a clue regarding what the table is about. The scores sa_i and se_i returned by $\kappa_A()$ and $\kappa_E()$ play an important role in reaching the conclusion. One important issue is how sa_i and se_i are calculated. Intuitively, the more matching attributes/entities we find, the higher score of a concept should be. Specifically, assuming (c_i, E_i, se_i) is in the output of $\kappa_E(E)$ and (c_i, A_i, sa_i) is in the output of $\kappa_A(A)$, a straightforward way to define se_i and sa_i is:

$$se_i = \frac{1}{|E|} \sum_{e_i \in E_i} p_{c_i}(e_i), \qquad sa_i = \sum_{a_i \in A_i} w_{c_i}(a_i) \qquad (2)$$

where $p_{c_i}(e_i)$ is the plausibility score of e_i given c_i, and $w_{c_i}(a_i)$ is the confidence score of a_i being the attribute of c_i in Probase.

3.2 Header Detector

The first step in understanding a table is locating the header. The header of a table contains a set of attributes, which form the schema of the table. It also indicates the orientation, depending on whether it appears as a row or column. In our discussion, we assume all tables are horizontal, e.g. in Table 2, the header is the top row.

Let P denote the set of possible headers for a table T. In one extreme, we may consider every row of T, that is, P contains all rows of T. In the other extreme, we may just consider the top row, which is the mostly header if T has one. We use function $\kappa_A()$ to evaluate each possibility and generate a set of candidate schema.

$$candidate\_schema = \{x \mid p \in P, x \in \kappa_A(p), x.sa + \alpha(p, T) > \gamma_h\} \qquad (3)$$

Eq 3 filters out candidates whose score is below a threshold γ_h. [2] The score consists of two parts, the raw sa score returned by $\kappa_A()$, and an adjustment $\alpha(p, T)$, which evaluates whether a row is likely a header based on the syntax. The reason we need an adjustment is that the header usually has some syntactic characteristics that set it apart from the rest of the table. To name a few:

---

[2] γ_h was picked based on our observation in practice.

- HTML <th> tag.
- Other syntactic clues, including i) cells of p end with a colon ':', and ii) cells of p has a formatting (e.g., **bold** font is used) different from other rows in table T, etc.
- Data type differences[3]: The fact that a cell in p is of different data type from other cells in its column may indicate p is a header. For example, a cell in p has string type (e.g., *Birthdate* or *Age*), while other cells in that column are dates (*4 Aug 1961*) or numbers (e.g., *61*).

If *candidate_schema* returned by Eq 3 is empty, then we fail to detect any header. This is possible because tables may not come with a header. In this case, we generate a header using the method elaborated in Section 3.3.

As for the example of Table 2, a properly set threshold will find the first row as the header, with two candidate schema that are shown below. For simplicity, we assume $\alpha(p, T) = 0$ here, that is, the table provides no syntactic clues for header identification.

(*US presidents, {Birthdate, Political Party, Assumed Office}, 0.90*)

(*politicians, {Birthdate, Political Party, Assumed Office}, 0.88*)

In Section 3.4 we will discuss how to further narrow the above two candidates down to a single interpretation.

3.3   Header Generator

If no candidate header is found in the header detection phase, then we assume that the table does not come with a header. In this case, we try to generate a header for the table.

For each column L_i, we find its most likely concept. For example, from the 3rd column of Table 2, we may derive *political parties*. Then, we check if the concepts of all columns together describe some other concepts. Specifically, for each column L_i, we first find its best matched top-k candidate concepts by calling $\kappa_E(L_i)$. We denote the top-k candidate concepts of L_i as $c^1(L_i), \cdots, c^k(L_i)$. Then we generate P, a set of possible headers by selecting one concept from each of the top-k candidates, that is

$$P = \{(a_1, \ldots, a_i, \ldots, a_n) | a_i \in \{c^1(L_i), \cdots, c^k(L_i)\}\}$$

where n is the number of columns. Finally, we derive *candidate_schema* using P and Eq 3. Note that in this case, $\alpha(p, T)$ is 0, as the table does not have syntactic clues.

If *candidate_schema* is again empty, which means the table does not come with a header and we cannot generate one for it, then we drop it from further consideration.

3.4   Entity Detector

The entity detector tries to accomplish two tasks. First, it detects the *entity column* of the table. For example, for Table 2, the entity column is the 1st column instead of the 3rd column, as the table is about politicians, not political parties. Second, we have derived a set of candidate schemata previously, and we need to narrow them down to a final interpretation. The method presented here solves the two problems at the same time.

---

[3]   Although data types are not an entirely syntactic issue, we included it here since they can be detected syntactically.

We make our judgment based on two kinds of evidence. First, the entity column should contain entities of the same concept, and we can derive the confidence of a concept for a given column. Second, the header should contain attributes that describe entities in the entity column, that is, the candidate schema we have derived should match the concept of the entity.

For each $s \in candidate\_schema$ returned by Eq 3, we enumerate every column col and compute its confidence score. Given a column col, we define

- E^{col}: the set of all cells in col, except for the one in the header corresponds to s
- A^{col}: the set of all attributes in s, except for the one in the current column.

We then apply functions $\kappa_A()$ and $\kappa_E()$ on E^{col} and A^{col} to obtain their possible semantics. Specifically, we let SC_A to be the ordered list of (c_i, A_i^{col}, sa_i) returned by $\kappa_A(A^{col})$, and SC_E to be the ordered list of (c_i, E_i^{col}, se_i) returned by $\kappa_A(E^{col})$.

If col is the entity column, we should be able to derive a concept c from E^{col}, and c should also be strongly supported by A^{col}. So the possibility of col being the entity column relies on the confidence of the concepts derived from it.

We join SC_A and SC_E to find common concepts, and we record the corresponding score as a multiplication of sa_i and se_j.

$$h(s, col) = \max\{sa_i \cdot se_j \mid (c_i, A_i^{col}, sa_i) \in SC_A, (c_j, E_j^{col}, se_j) \in SC_E, c_i = c_j\}$$

Finally, the most possible interpretation and entity column are the ones that achieve the maximum score.

$$(final\ schema, entity\ column) = \underset{s, col}{\operatorname{argmax}}\ h(s, col)$$

Using this approach, for the example in Table 2, we will reject the schema of *US presidents*, because although the concept *US presidents* has strong support from the attributes, it has low support from the column that contains the name Arnold Schwarzenegger. Thus, the final schema we find for the table is:

(politicians, {Birthdate, Political Party, Assumed Office}, 0.88)

4 Experimental Results

Since most of the evaluations are related to concepts, we randomly selected 30 concepts with high frequency as our benchmarks, which vary in domains and sizes. For experiments that need to be evaluated manually, we use a consistent scoring criterion for human judges: 1 for correct, 0.5 for partially correct, 0 for incorrect. In the following subsections, we will first present some statistics about the extracted table corpus, and then two applications that can take advantage of the web table understanding. Because of the space limitation, only some statistics of the results will be provided. Readers are referred to our website [4] for complete experimental results.

---

[4] http://research.microsoft.com/en-us/projects/probase/
tablesearch.aspx

4.1 The Web Table Corpus

We implemented the prototype system with a Map-Reduce distributed computing platform. Since all of the following stages are parallelable, Each of them takes at most several hours to complete on our large cluster. We ask human judges to judge the correctness of the results.

1. Header detection: We randomly selected a sample of 200 [5] filtered tables, and ran the header detection algorithm on them. The result is shown below.

Actual no	86.7% correct, 13.3% incorrect
Actual yes	90.7% correct, 9.3% incorrect (2.1% wrong position, 7.1% predict as no header)

Our algorithm correctly identified headers (or lack of which) in 89.5% of the tables. Among the 140 tables with a header, we correctly detected the header in 127 (90.7%) of them.

2. Entity column detection: Because the information in web tables is so diverse, random sampling from web tables may not have a good coverage in both structures and domains. Therefore in this part, we create test set using tables extracted from Wikipedia only. These tables contain structural multiple-domain data in high density. We only focus on precision here as recall is hard to evaluate. We randomly selected 200 tables which our system claims to have an entity column, while ensuring that no two tables are from the same Wikipedia page. Our judges evaluated the correctness of these claims. Results showed that 11 tables actually do not have an entity column, and most of them have two or more main entities, which violate our assumption that there is only one entity column in a table. Among the remaining 189 tables, our algorithm correctly identifies the entity column in 165 tables(87.3%).

4.2 Search Engine

We build a semantic search engine that operates upon table statements, rather than tables themselves. A statement consists of a single entity, its attributes and corresponding values. Usually, a statement corresponds to a row of a table. For example, Table 2 has three statements, each about a politician. A statement represents a basic unit of knowledge of an entity, which is used to answer a query. This is different from previous work [9], which uses tables as the basic unit to answer queries.

Furthermore, instead of performing keyword matching in table search as in previous work, we try to find the semantics of a query, and return a set of statements that match the semantics. We support four semantic components in a query: Concept, Entity, Attribute and Keyword. Different combinations of them lead to different types of query. In this experiment, we focus on one representative pattern of them: Concept + Attribute, because this pattern includes two main factors: the relation between entity and concept, and the relation between attribute and concept. Some examples of pattern include "politicians birthday", "companies industry" and "video games developer". Once we have the set of statements that match a query, we rank them according to two factors:

---

[5] This number is limited since we have to use human judges.

1) match of keywords and attributes between the query and the statement; 2) the quality of the statement.

To avoid indexing all the tables on the web, we implemented the search engine only on Wikipedia's tables. For the 30 concepts, we select three attributes from each of them and generate corresponding queries. Considering we need to have a certain number of results for each query to make the evaluation possible, we ask users to first think of attributes they want to query, and then choose at most three of them with enough results.

To evaluate both the precision and quality of ranking, we ask the users to score each of the at most top 10 results of each query. Let $score_k$ denote the score for the result at position k among n results, then the overall quality of a query result is defined as:

$$\frac{\sum_{k \in [1,n]} score_k/k}{\sum_{k \in [1,n]} 1/k}.$$

In total 82 out of 90 queries return enough results. Among them, 53 queries return all-correct results, and 6 queries have a score lower than 0.6. The average score of all queries is 0.91. Figure 4 shows the average scores of queries for each concept. [6] Among all the 29 concepts which have result, 25 of them received an score higher than 0.8.

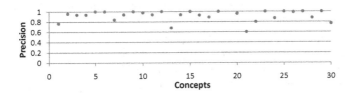

Fig. 4. Average score of queries for each concept

To compare with an existing search engine, we submitted the same queries to Google. For each query, we manually judge the top 10 pages returned to see if they contain the desired information which are forms, lists or categories about the queried class and attribute. Out of the 820 results (10 for each query), 32 results from Wikipedia and 81 results from non-wiki websites are considered useful. But the formats of these pages make it difficult if not impractical to extract the information we need. This experiment showed that traditional keyword search is inadequate in handling such semantic queries.

To better evaluate the quality of entities and values, we compare our results with Google Squared, which also outputs tables as results. Our prototype search engine merges statements with the same schema together as one table. We select at most top 10 schemata ranked by the highest score of their associated statements. For each query, we obtain a list of entities and their attribute values, then compare it with the list returned by Google Squared. We manually checked the correctness of our list and ensure that the entities and values are correct. Figure 5 shows the percentage of our result that is included by, overlapped with or disjoint with the list from Google Squared for all 30

---

[6] some concepts do not have dots or bars in a chart because there aren't enough results for them (similarly hereinafter).

concepts. We argue that our system not only correctly contains attribute-value information that Google Squared currently has for the 30 concepts, but also includes many additional attributes and values that are not found in Google Squared.

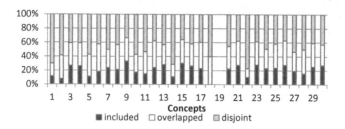

Fig. 5. Comparing with Google Squared

4.3   Taxonomy Expansion

In this experiment, we evaluate how much our prototype system contributes in expanding the Probase taxonomy in terms of the number of entities and attributes.

1. Entity expansion: Expanding entities is straightforward. For example, based on Probase, we conclude that the leftmost column in Table 2 contains names of politicians, even if not every name there is found in Probase. The leftmost column provides an evidence equivalent to a Hearst's pattern "... *politicians such as Barack Obama, Arnold Schwarzenegger and Hillary Clinton* ..." Thus, we can expand Probase by incorporating this new evidence in the same way as we encounter a new Hearst's pattern.

To ensure a high quality, we select top 1000 entities ranked by ambiguity score $a_c(e)$ as seeds for each concept c, then use plausibility score $p_c(e)$ to infer. Here we set δ_p to 0.6. We run our algorithm for one iteration, and discovered 3.4 million existed entities in Probase, but more importantly, 4.6 million new entities for nearly 20,000 concepts. Among the 30 concepts, about 11,000 new entities were discovered from 28 of them, in which 1,410 (12.8%) are also found in Freebase by exact matching. Similar as before, we judge the top 20 entities (if available) manually for each concept, which are sorted by the maximal value of p_c, and use the average score to indicate a concept's quality.

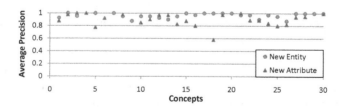

Fig. 6. Precision of new entities and new attributes

The result is shown in Fig 6. The average quality is 0.96, and all concepts have score higher than 0.875. Furthermore, new entities in 12 of the concepts are all correct. In essence, most of the newly discovered entities are correct.

2. *Attribute expansion*: After the entity column is determined, we can expand the attribute list in Probase by adding unknown attributes to representative concepts of the entity column. This process can help us get more attributes in the light of tables, such as "Tel-#" for "phone number". Furthermore, enriching Probase by expanding attribute set and understanding more tables are in positive feedback loop.

We again run our algorithm for one iteration, and discovered 0.15 million new attributes for nearly 14,000 concepts. About 2,700 new attributes were discovered for all the 30 concepts. Like before, we care more about the quality of top-ranked attributes than the general quality. For each concept, new attributes are first filtered though frequency and average $w_c(a)$, and then sorted according to the maximal $w_c(a)$. We still judge the top 20 attributes manually and use the average score to evaluate the concept. Here we set $\delta_a = 0.4$.

The result is shown in Fig 6. We found new attributes for all 25 concepts, and the average quality of them is 0.90. The only outlier is the concept "guitarists", whose quality score is 0.58. That's because "guitarists" are often associated with "songs" in tables. But since Probase's coverage of songs is not good, guitarists are selected as the entity column instead which is not the best choice.

5   Related Work

There has been some early work on information extraction from tables on the Web. Wang *et al.* [10] detects tables by using machine learning methods. It uses features of layout, content type and word usage for training and testing different classifiers. However, the corpus only contained about 10,000 tables collected from 200 web sites. Yoshida *et al.* [11] studied the problem of integrating tables of similar content into a big table. It uses EM method to decide the locations of attributes and values in a table by labeling the table format as a predefined structure, then group tables into several clusters according to their attributes and values. Similar attributes in a cluster are merged together to get a final big table. Another early work [12] focused on mining web tables in a specific domain: travel. It defines similarity between cells, rows, and columns, then uses the similarity to identify the schema of the table.

Cafarella *et al.* [13,9] did some pioneering work in exploring tables on the Web. The goal is to decide if a table contains *relational* data. To achieve this, they tried to identify typed columns in the table, which resulted in schemata for tables that are considered relational. They also studied an interesting application: searching over a corpus of tables. More recently, a related system called Octopus [14] explored the semantic matching among tables and achieved better accuracy and performance. Work in this space relied on correlation statistics inside the tables instead of external knowledge. Furthermore, each table is treated as a single entity (tables are atom units in query results) whereas we consider table as a set of entities and values. Syed *et al.* [15] used *Wikitology*, an ontology which combines some existing manually built knowledge systems such as DBPedia [16] and Freebase [6], to link cells in a table to Wikipedia entities. To resolve the ambiguity issue (a cell may match many Wikipedia entities), it finds the class or the domain of a column (or a row) through majority voting to achieve more accurate matching. Other researchers are interested in labeling cell text, column type and relation between columns(rows) [17,18] using a large corpus such as YAGO, and use the information to assist queries about relation. However, as we described in Section 1, people need knowledge, including common sense and worldly facts, to understand a table, especially when the table is taken out of its context. Without such knowledge, it is often impossible to truly understand the table, or to be able to perform the type of semantic search as described in this paper.

The work presented in this paper focuses on *understanding* tables. To obtain knowledge required for this task, we build it upon an extension of Probase [1], a rich ontology automatically derived from the Web to produce a concept map for the human mind. In our extension, we find attributes for each concept (there are more than 2 million of them in Probase). Some recent work has focused on finding attributes for given classes and concepts. In particular, Pasca *et al.* [19] finds class attributes starting from a few *seed* (entity, attribute) pairs for each class. Furthermore, a ranking mechanism [20] has been developed to score attributes derived from seeds.

Besides tables, some work has also been done to explore other sources of structured data on the Web, for example using data extracted from lists [3] to construct relational data [21] or to expand set of entities [22]. There is also research that transforms text to tables [23,24]. In particular, Pyreddy *et al.* [23] use heuristic rules to tag table components, and Pinto *et al.* [24] use conditional random fields to locate boundaries, headers and rows of text table.

6   Conclusion

This paper presents a framework that attempts to harvest useful knowledge from the rich corpus of relational data on the Web: HTML tables. Through a multi-phase algorithm, and with the help of a universal probabilistic taxonomy called Probase, the framework is capable of understanding the entitles, attributes and values in many tables on the Web. With this knowledge, we built two interesting applications: a semantic table search engine which returns relevant tables from keyword queries, and a tool to further expand and enrich Probase. Our experiments indicate generally high performance in both table search results and taxonomy expansion. This showed that the proposed framework practically benefits knowledge discovery and semantic search.

References

1. Wu, W., Li, H., Wang, H., Zhu, K.Q.: Probase: A probabilistic taxonomy for text understanding. In: SIGMOD (2012)
2. Lee, T., Wang, Z., Wang, H., Hwang, S.: Web scale taxonomy cleansing. In: VLDB (2011)
3. Zhang, Z., Zhu, K.Q., Wang, H.: A system for extracting top-k lists from the web. In: KDD (2012)
4. Liu, X., Song, Y., Liu, S., Wang, H.: Automatic taxonomy construction from keywords. In: KDD (2012)
5. Singh, P., Lin, T., Mueller, E., Lim, G., Perkins, T., Li Zhu, W.: Open Mind Common Sense: Knowledge Acquisition from the General Public. In: Meersman, R., Tari, Z. (eds.) CoopIS 2002, DOA 2002, and ODBASE 2002. LNCS, vol. 2519. Springer, Heidelberg (2002)

6. Bollacker, K., Evans, C., Paritosh, P., Sturge, T., Taylor, J.: Freebase: a collaboratively created graph database for structuring human knowledge. In: SIGMOD (2008)

7. Hearst, M.A.: Automatic acquisition of hyponyms from large text corpora. In: COLING, pp. 539–545 (1992)

8. Song, Y., Wang, H., Wang, Z., Li, H., Chen, W.: Short text conceptualization using a probabilistic knowledgebase. In: IJCAI (2011)

9. Cafarella, M.J., Wu, E., Halevy, A., Zhang, Y., Wang, D.Z.: Webtables: Exploring the power of tables on the web. In: VLDB (2008)

10. Wang, Y., Hu, J.: A machine learning based approach for table detection on the web. In: WWW (2002)

11. Yoshida, M., Torisawa, K., Tsujii, J.: A method to integrate tables of the world wide web. In: International Workshop on Web Document Analysis (2001)

12. Chen, H., Tsai, S., Tsai, J.: Mining tables from large scale html texts. In: ICCL (2000)

13. Cafarella, M.J., Wu, E., Halevy, A., Zhang, Y., Wang, D.Z.: Uncovering the relational web. In: WebDB (2008)

14. Yakout, M., Ganjam, K., Chakrabarti, K., Chaudhuri, S.: Infogather: entity augmentation and attribute discovery by holistic matching with web tables. In: SIGMOD (2012)

15. Syed, Z., Finin, T., Mulwad, V., Joshi, A.: Exploiting a web of semantic data for interpreting tables. In: Proceedings of the Second Web Science Conference (2010)

16. Auer, S., Bizer, C., Kobilarov, G., Lehmann, J., Cyganiak, R., Ives, Z.G.: DBpedia: A Nucleus for a Web of Open Data. In: Aberer, K., Choi, K.-S., Noy, N., Allemang, D., Lee, K.-I., Nixon, L.J.B., Golbeck, J., Mika, P., Maynard, D., Mizoguchi, R., Schreiber, G., Cudré-Mauroux, P. (eds.) ASWC 2007 and ISWC 2007. LNCS, vol. 4825, pp. 722–735. Springer, Heidelberg (2007)

17. Limaye, G., Sarawagi, S., Chakrabarti, S.: Annotating and searching web tables using entities, types and relationships. In: VLDB (2010)

18. Venetis, P., Halevy, A.Y., Madhavan, J., Pasca, M., Shen, W., Wu, F., Miao, G., Wu, C.: Recovering semantics of tables on the web. PVLDB 4 (2011)

19. Pasca, M.: Organizing and searching the world wide web of facts - step two: Harnessing the wisdom of the crowds. In: WWW (2007)

20. Bellare, K., Talukdar, P.P., Kumaran, G., Pereira, F., Liberman, M., McCallum, A., Dredze, M.: Lightly-supervised attribute extraction. In: NIPS (2007)

21. Elmeleegy, H., Madhavan, J., Halevy, A.: Harvesting relational tables from lists on the web. In: VLDB (2009)

22. He, Y., Xin, D.: Seisa: set expansion by iterative similarity aggregation. In: WWW (2011)

23. Pyreddy, P., Croft, W.B.: Tintin: A system for retrieval in text tables. In: ICDL (1997)

24. Pinto, D., McCallum, A., Wei, X., Croft, W.B.: Table extraction using conditional random fields. In: SIGIR (2003)

Bridging the Gaps towards Advanced Data Discovery over Semi-structured Data

Sivan Yogev[1,2] and Haggai Roitman[1]

[1] IBM Research Haifa, Israel
{sivany,haggai}@il.ibm.com
[2] Department of Computer Science, Ben-Gurion University of the Negev, Israel

Abstract. In this work we argue that two main gaps currently hinder the development of new applications requiring sophisticated data discovery capabilities over rich (semi-structured) entity-relationship data. The first gap exists at the conceptual level, and the second at the logical level. Aiming at fulfilling the identified gaps, we propose a novel methodology for developing data discovery applications. We first describe a data discovery extension to the classic ER conceptual model termed *Entity Relationship Data Discovery* (ERD²). We further present a novel logical model termed the *Document Category Sets* (DCS) model, used to represent entities and their relationships within an enhanced document model, and describe how data discovery requirements captured by the ERD² conceptual model can be translated into the DCS logical model. Finally, we propose an efficient data discovery system implementation, and share details of two different data discovery applications that were developed in IBM using the proposed methodology.

Keywords: Conceptual modeling, data discovery, entity relationship.

1 Introduction

Data complexity and its diversity have been rapidly expanding over the last years, spanning from large amounts of unstructured and semi-structured data to semantically rich knowledge. With the increase in data volumes and their diversity, we witness demands for advanced data discovery capabilities over rich (semi-structured) entity-relationship data, beyond existing solutions. Such requirements are being imposed by applications in various domains such as social-media, bioimedical, e-commerce and CRM, cybersecurity and intelligence.

Data Discovery Requirements. Emerging data discovery requirements can be classified into two main types. The first type of requirements includes the support of efficient *keyword search* (textual search) over large scale entity-relationship data, as more and more data nowadays includes large textual portions. Trying to leverage the success of search on the web (e.g., Google search), many applications try to offer their end users much simplified search interfaces (e.g., search box), allowing ordinary users (non-experts) to search data without having to be familiar with the application's data model or format (e.g., relational schema, XML, RDF) and its underlying query language (e.g., SQL, XQuery, SPARQL).

P. Atzeni, D. Cheung, and R. Sudha (Eds.): ER 2012, LNCS 7532, pp. 156–165, 2012.
© Springer-Verlag Berlin Heidelberg 2012

The second type of requirements includes *interactive querying and data exploration* [1] over large scale entity-relationship data. *Faceted search* [2] has become a popular approach for interactive querying on the web (e.g., on e-commerce websites such as Amazon, eBay, etc). Using a faceted search system, users are able to refine their search results based on various categories associated with searchable documents [2]. A faceted search system provides an auxiliary list of facets that report on various categories associated with documents in the current search result, and the number of results classified into each category. This enables users to focus their search by selecting categories that may help guiding the search system towards better coverage of their information needs.

Recently, faceted search also became a very attractive discovery option for application developers who wish to expose end users to similar capabilities over large-scale entity-relationship data [3,4]. In the context of entity-relationship data discovery, potential facets may include entity types, attribute names and values, relationship types, relationship member roles and relationship attributes. *Navigational search* extends faceted search capabilities by providing convenient ways for users to explore large amounts of entity-relationship data. As an example, relationships between entities can be used to traverse the entity-relationship graph [1,5,6]. Such exploration may be well suited for research purposes, where users can explore more facts about entities returned as search results by navigating to other related entities (e.g., related patient's medical entities [5]).

Understanding Data Requirements. Traditionally, from a data engineering perspective, the data requirements of every future application are first analyzed and then formally captured using conceptual modeling tools such as the classic entity-relationship diagram [7], UML class diagram, SysML requirements diagram, etc. Having formally captured the application's requirements, the application's conceptual model is mapped into the logical model used to represent the data within a future data storage. As an example, in this work we focus on the entity-relationship (ER) model. Within this model, entities represent objects that can be uniquely identified and include attributes that describe their properties. Relationships can be used to capture various associations that exist between entities. The ER model is usually coupled with the relational model as the logical data model of choice, where entities and their relationships are translated (mapped) into a relational schema and stored in database tables [7].

Existing Gaps. While it may seem that traditional data engineering methodologies can be used to fulfill the new data discovery requirements, we argue that two major gaps may hinder development of applications that require sophisticated data discovery capabilities over stored rich entity-relationship data.

The first gap exists at the conceptual level, where data requirements should be analyzed and formally captured. Existing conceptual models (and the ER model in particular) were originally designed to answer the following main question: *What data (i.e., entities and relationships) is expected to be stored by the application?* However, such models were not designed to capture new data discovery requirements aiming at answering the following two questions:

1. *What data is keyword (text) searchable?*
2. *What categories exist within the data?*

The first question tries to capture how semi-structured data should be pre-processed to satisfy an application's keyword search requirements, e.g., which data portions should undergo tokenization, lemmatization, stop wording, stemming, etc. In addition, it aims at identifying the data portions, stored by the future application, which may require more sophisticated query answering based on keyword search. The second question aims at identifying the application's categorial data for faceted search. Hence, an extension to traditional conceptual models is required in order to fully capture the new data discovery requirements.

The second gap exists at the logical level in which ER data is "actually" represented in the application's data storage. Traditional data management solutions were usually designed on the assumption that entity-relationship data is mostly structured (i.e., data is stored in relational tables, XML or RDF formats with a well defined schema). Similarly, queries should be formulated using some structured query languages (e.g., SQL, XQuery, SPARQL, etc), where exact answers are expected (i.e., either each result completely satisfies the query or not). However, such solutions support only simplified forms of keyword search such as regex and keyword matching. Compared to that, keyword search may result in imprecise answers and require ranking answers by relevance to the user query. In addition, faceted search is not naturally supported by most existing data management solutions, and similarly to keyword search, it may require nontrivial extensions both in the storage and the query processing levels (e.g., support of ad-hoc indexing schemes [4], top-k ranking, etc).

Our Contributions. Aiming at fulfilling the identified gaps, in this work we propose a novel methodology for developing data discovery applications. According to this methodology, an application's data discovery requirements are first formally captured using an extension to the classic ER model, termed the *Entity-Relationship Data Discovery* model (denoted ERD^2). Next we present a novel logical model termed the *Document Category Sets* (denoted DCS) model, used to represent entities and their relationships within an enhanced document model. We further describe how data discovery requirements captured by the ERD^2 conceptual model should be translated into the DCS logical model. Finally, we share the details of two different data discovery applications that were developed in IBM using the proposed methodology.

2   Related Work

We hereby review related works, classified according to the two identified gaps. At the conceptual level, several new models and extensions to existing ones were proposed in the past for modeling semi-structured data, e.g., WebML [8], UML [9], ERD [10]. However, these proposals were not aimed at modeling data discovery requirements but rather capturing the semi-structured nature of hypertext data on the Web. Bozzon et al. [11] proposed a model driven approach based on WebML for integrating database and search systems for web application development, yet did not address faceted search or any other exploratory aspects over entity-relationship data. Clarkson et al. [12] have studied various types of faceted search user interfaces and proposed to incorporate facets within the ER model. Compared to our work, [12] does not address the requirement for efficient keyword search over entity-relationship data, and by assuming a relational model as the underlying logical model, do not address the logical gap.

At the logical level, Bonino et al. [13] proposed *FaSet* – a formal logical model for faceted search based on set theory. However, the actual implementation of this model is based on a relational data model which makes it unsuitable for data discovery purposes. The basic faceted search data model was designed to handle textual documents associated with categories (facets) taken from some predefined (or auto-generated) taxonomy [2]. State of the art faceted search systems were not designed for handling rich entity-relationship data.

Several other solutions have been proposed in the past for handling data discovery requirements over rich entity-relationship data, including works on dataspace systems [14], keyword and faceted search over databases (e.g., [15,16,3]), and entity oriented search (EOS) (e.g., [17,18]). Trying to face the limitations of traditional database systems in handling large and diverse volumes of ER data, NoSQL systems are proposed lately as promising alternatives [19]. Such systems include among others key-value stores (BigTable), document stores (MongoDB) and graph stores (Neo4j). However, both prior solutions and existing NoSQL systems do not provide formal means for capturing data discovery requirements. In addition, existing solutions only satisfy one of the two data discovery requirements over rich entity-relationship data, but not both.

Finally, in preceding works, we proposed a novel interactive query language for expressive exploratory search over rich ER data based on the DCS logical model [6]. We further demonstrated the effectiveness of this model for implementing novel exploratory search systems [5,20].

3   Modeling Data Discovery Requirements

3.1   Entity Relationship Model - Preliminaries

In this work we borrow several basic notations defined in the classic Entity-Relationship (ER) conceptual model [7]. Within this model, entities (notated by rectangles) represent real or abstract objects that can be uniquely identified. Each entity e has a type $e.type$ (e.g., Car, Person, etc.) and includes one or more attributes $e.a$ (each is notated by an oval attached to the entity) describing its properties (e.g., color, age, etc.). Each attribute a has a name $a.name$ and value $a.value$. The minimal set of attributes that can be used to uniquely identify the entity defines the entity key $e.key$ (and notated by underlining the corresponding attributes). A relationship r (notated by a diamond) can be used to capture various associations that exist among two or more entities (e.g., Car Accident, Car Owner, etc.). Entities that participate in some relationship are termed *relationship members*, and denoted $r.e$. An entity can play various *roles* within the relationship (e.g., Manager vs. Employee in a Management relationship), with the role of $r.e$ denoted as $r.e.role$. A relationship may have attributes $r.a$ which describe its context (e.g., the time and location of a Car Accident relationship between two or more Car entities). A relationship key is formed from the combination of its member keys and relationship key attributes (if exist).

Another basic notation that we borrow from the classic ER model in this work is entity inheritance, notated by triangled "isA" relationships and stating that one entity type (the *super-type*) is a generalization of another type (the *sub-type*). An entity type can have at most one super-type, thus an ER diagram may contain a forest of inheritance trees. For a given entity type t, the *root-type* of t is the root of the inheritance tree in which t resides. For a given entity e with actual type $e.type$, its root type is denoted as $e.root\_type$ and the list containing all types in the path from $e.root\_type$ to $e.type$ is further denoted as $e.all\_types$.

Example ER Model. Through the rest of this paper, examples will be based on a simplified ER model of Wikipedia, including a link relationship describing hyper-links between Wikipedia articles, and another relationship bornIn defined in the YAGO ontology [21]. Figure 1 depicts the Wikipedia ER model.

3.2   Entity Relationship Data Discovery Model

We hereby present the details of the Entity Relationship Data Discovery model (ERD^2). The definition of this discovery model will be done by describing important retrieval tasks, describing the way these tasks are handled using an Information Retrieval (IR) methodology, and adding new constructs to the classic ER model to support the required functionality.

The first type of data discovery requirement that the ERD^2 model should formally capture is keyword (textual) search over attribute values. As an example, a possible requirement for a search service over Wikipedia data may include the ability to answer a query such as *"find articles with the word 'model' in their title"*. Wikipedia contains two articles whose titles are *"Model"* and *"Conceptual model"*. A simplistic interpretation of the above query may be realized by a simple string matching, and return both of the aforementioned Article entities with the same level of importance. However, users of modern search engines have grown to expect ranking of results, with more relevant results appearing higher in the results list. Using the last example, one would expect the article titled *"Model"* to be more relevant, since it exactly satisfies the query, while the other article's title contains additional words.

Within the ERD^2 model, attributes that can be subject of keyword queries are notated with the letter 'S' on the edge connecting the attribute to the entity. It is important to note that non-trivial ranking may require preprocessing of the attribute value, which can include tokenization, lemmatization, stop words removal, stemming, linguistic analysis and other text analysis techniques. Such text preprocessing requirements can be further captured in the ERD^2 model by adding comments below each searchable attribute. As depicted in Figure 1, our example contains three "searchable" attributes: content, title and country. The preprocessing of the content attribute involves word tokenization, word stemming and stop

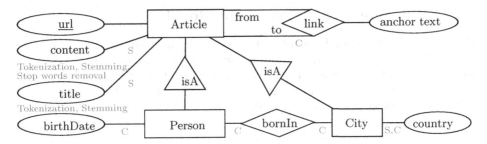

Fig. 1. A simplified ER model of Wikipedia data, with ERD^2 additions to the ER model in grey. Edges marked by 'S' indicate keyword searchable attributes. Edges marked by 'C' indicate attachment of relationship type or attribute as category to the entity. Data discovery related processing instructions are given below attributes.

wording, while preprocessing of the title attribute includes only tokenization and stemming, and the country attribute undergoes no processing.

The example query in the above scenario can be issued only by end-users familiar with the underlying data model, as the user must know that articles have an attribute named 'title'. However, when a user encounters a new dataset such knowledge is usually not available, and for these first steps it is important to support arbitrary keyword queries. As an example, consider the query *"model"*, which can be interpreted as *"find everything containing the word 'model'"*. Such search requirements can be supported by concatenating an entity's preprocessed searchable attributes values into an "unstructured entity content", considering entities as pure textual documents. More details on this issue in Section 5.

How can a novice user move from issuing entity level keyword queries to domain-specific attribute queries? A potential answer may be that relevant attribute names should be exposed along with search results. Turning again to an Information Retrieval methodology, this may be achieved using the *Faceted Search*[4] paradigm. A faceted search system provides an auxiliary list of facets that report on various categories associated with documents in the current search result and the number of result documents that are classified into each category.

In the context of the ER model, facets may include entity types, attribute names and possible values, relationship types, relationship member roles and relationship attributes. In the ERD^2 model, structural aspects that are to be categorized and exposed through faceted search are notated with the letter 'C' on the edge connecting the structural aspect to an entity node. By default, within ERD^2 entity types including their super-types are considered as categories.

Finally, apart from submitting queries that refer to entities, users may wish to discover relationships between various entities using relationship constraints. An example of such requirement is the following query *"people born in Ulm"*. Given the Wikipedia data model, this request can be formulated as *"find all person entities that participate in a 'bornIn' relationships with the city titled 'Ulm'"*. While such a requirement may not require to change the classic ER model, it does require to have a special support within the logical data model with which the ER data will be represented. We shortly discuss this aspect in section 5.

4   The Document Category Sets Model

Having defined the extended conceptual model, we move to the definition of a new logical model termed hereinafter *Document Category Sets Model* (or DCS for short). In the next section we shall present a set of translation rules for mapping ERD^2 data discovery requirements into the DCS model.

The DCS logical model consists of two collections – a collection of categories termed *Taxonomy* (denoted T), and a collection of *Documents* (denoted D). Each category $c \in T$ is defined by a path of nodes $c = v_1/v_2/\ldots/v_l$, with each node containing a non-empty string. A document $d \in D$ is defined by a quartet (id_d, F_d, C_d, CS_d) as follows:

- id_d is a unique document identifier;
- F_d is a collection of fields, with field $f \in F_d$ defined by a pair $(name, value)$;
- C_d is a set of categories with optional "payloads" used to categorize the document, formally $C_d = \{(c, payload) \mid c \in T\}$;
- CS_d is a collection of category-sets, where each category-set $cs \in CS_d$ is a set of categories with optional "payloads", formally $cs = \{(c, payload) \mid c \in T\}$.

Payloads are used to represent additional information related to a certain category in C_d or a category in a category-set $cs \in CS_d$. When a category has no required payload we shall exclude the payload from the category's description.

5   Translating the ERD^2 Model into the DCS Model

We now describe how data discovery requirements defined in the ERD^2 model can be translated into the DCS model. The main idea behind our translation scheme is to have a dual representation of entities captured in the ERD^2 model within the DCS model; once as a document and once as a category. The first document-based representation of entities in the DCS model allows keyword (textual) search over entity data, while the category-based representation in the DCS model allows performing faceted search and relationship-based queries. See `https://www.research.ibm.com/haifa/papers/YogevRoitmanER2012Supp.pdf` for an example illustrating this mapping.

5.1   Category Representation

A fundamental aspect of the translation of the ERD^2 model into the DCS model is the mapping of ERD^2 objects onto the taxonomy T.

An **entity** e is represented by the category c_e whose path in the taxonomy T starts with the literal node "*entity*", followed by a separate node for each type t in $e.all\_types$ (in the order of the inheritance path of $e.type$), and ends with a leaf node for $e.key$. An **attribute** a is represented by the category $c_a = a.type/a.name/a.value$. Some attribute values may be hierarchical, for instance dates can be viewed as a hierarchy of year-month-day. Such hierarchy can be reflected by separating the attribute value into multiple nodes $(year/month/day)$. Finally, each **relationship type** is represented by a category $c_{r.type}$ with two nodes, the prefix literal "relationship" followed by $r.type$.

5.2 Document Representation

Entities. An entity e is represented by a document d_e, with id_{d_e} defined by $entity|e.root\_type|e.key$, where '|' hereinafter serves as a separator in the concatenation of values. F_{d_e} contains the following fields:

1. $f_{key} = (key, e.key)$ for the entity's key;
2. $f_{type} = (type, e.type)$ for the entity's type;
3. $f_a = (e.type|a.name, a.value)$, for each of the entity's searchable attributes $e.a$ (i.e., 'S'-notated in the ERD^2 model), with $a.value$ processed according to the processing instructions (e.g., stemming);
4. $f_{a.name} = (attribute, e.type|a.name)$, for each of the entity's attributes $e.a$, required for supporting efficient retrieval of entities by attribute name;
5. $f_{c_e} = (entity\_cat, c_e)$ provides a mapping from the document representation of e (d_e) to its category representation (c_e), which is essential for combining textual and structural search aspects, as will be explained later on;
6. $f_{content} = (content, text)$, where $text$ contains the concatenation of e's searchable attributes processed values, to efficiently support free-text queries.

C_{d_e} includes the entity's category c_e in order to support faceted navigation through the entity's type and super-types. C_{d_e} further includes the category representation of each attribute marked with 'C' in the ERD^2 model. For each relationship type marked with 'C' in the ERD^2 model, $c_{r.type}$ is added to C_{d_e} with a payload containing the number of relationships of type $r.type$ the entity is a member of. CS_{d_e} is empty for entity documents.

Relationships. A relationship r is represented as a document d_r, whose id is the relationship key. CS_{d_r} contains a single category set cs_r that includes the categories of the relationship type $c_{r.type}$, member entities $c_{r.e}$ and attributes $c_{r.a}$. If a given relationship entity member $r.e$ has a role $r.e.role$, then the category $c_{r.e}$ is added to cs_r with a payload that contains $r.e.role$. This allows filtering entities based on their role within relationships in which they participate. F_{d_r} and C_{d_r} are empty for relationship documents.

6 Example Applications

We now shortly describe data discovery applications in two domains that were already implemented in IBM using our new data discovery methodology. The data discovery requirements of these applications were captured using the ERD^2 model and were then translated into the DCS model and implemented using an inverted index data structure [6].

6.1 Social-Medical Discovery

The "Social-Medical Discovery" system [5] is part of the IBM Patient Empowerment System[1] providing novel social search capabilities for the medical domain.

---

[1] Video: http://www.youtube.com/watch?v=YFRjOB39hvA

The system provides a uniform way for searching social and medical data stored in heterogeneous data repositories. Implemented search scenarios span from simple search services that require to locate relevant information about some patient or medication, to more complex data exploration services that require to query the social-medical "dataspace" to reveal interesting patterns (e.g., relevant patients for some new clinical trial).

6.2   Cultura

Cultivating Understanding and Research Through Adaptivity (Cultura)[2], is a project funded by the 7th Framework Programme of the European Commission. Using our methodology we developed an entity oriented search (EOS) system for data discovery over entity-relationship data extracted from a large collection of historical documents [20]. The implemented system provides new ways in which both historians and researchers can search and explore such historical data.

7   Discussion

In this paper we identified two main gaps within current methodologies for data discovery over rich entity-relationship data imposed by new applications in many domains. The first gap was identified at the conceptual level in which data discovery requirements should be formally analyzed and captured. Trying to address this gap, we proposed an extension to the classic ER model, ERD^2, which is better tailored for the data discovery requirements engineering domain. The second gap was identified at the logical level where application data is "actually" represented. A new logical model, DCS, was also proposed to fulfil this gap. We further described how data discovery requirements captured in the ERD^2 model are to be translated into the logical data representation in the DCS model, providing a complete methodological solution for the data discovery domain. Finally, we shortly described three examples of real applications that were developed in IBM using the new data discovery methodology.

As future work, we would like to further validate the proposed data discovery methodology by applying it in more and more domains. Another interesting question is whether conceptual models other than the ERD^2 model can be augmented with similar data discovery notations and translated into the DCS logical model in a similar manner. Finally, these days we study the problem of entity similarity search in various domains (e.g., patient similarity in the medical domain) and wish to explore extensions to the ERD^2 model that would enable to model application level entity similarity semantics.

References

1. Ruthven, I.: Interactive information retrieval. Annual Rev. Info. Sci & Technol. 42, 43–91 (2008)
2. Tunkelang, D.: Faceted Search. Morgan & Claypool Publishers (2009)

---

[2] http://www.cultura-strep.eu/

3. Basu Roy, S., Wang, H., Das, G., Nambiar, U., Mohania, M.: Minimum-effort driven dynamic faceted search in structured databases. In: Proceedings of CIKM, pp. 13–22. ACM, New York (2008)
4. Ben-Yitzhak, O., Golbandi, N., Har'El, N., Lempel, R., Neumann, A., Ofek-Koifman, S., Sheinwald, D., Shekita, E., Sznajder, B., Yogev, S.: Beyond basic faceted search. In: Proceedings of WSDM, pp. 33–44. ACM (2008)
5. Roitman, H., Yogev, S., Tsimerman, Y., Kim, D.W., Mesika, Y.: Exploratory search over social-medical data. In: Proceedings of CIKM, pp. 2513–2516. ACM, New York (2011)
6. Yogev, S., Roitman, H., Carmel, D., Zwerdling, N.: Towards expressive exploratory search over entity-relationship data. In: Proceedings of WWW (2012)
7. Chen, P.P.-S.: The entity-relationship model-toward a unified view of data. ACM Trans. Database Syst. 1, 9–36 (1976)
8. Ceri, S., Fraternali, P., Bongio, A.: Web modeling language (webml): a modeling language for designing web sites. In: Proceedings of WWW, pp. 137–157. North-Holland Publishing Co., Amsterdam (2000)
9. Baresi, L., Garzotto, F., Paolini, P.: Extending uml for modeling web applications. In: Proceedings of HICSS, p. 3055. IEEE Computer Society, Washington, DC (2001)
10. Hanus, M., Koschnicke, S.: An ER-Based Framework for Declarative Web Programming. In: Carro, M., Peña, R. (eds.) PADL 2010. LNCS, vol. 5937, pp. 201–216. Springer, Heidelberg (2010)
11. Bozzon, A., Iofciu, T., Nejdl, W., Tönnies, S.: Integrating Databases, Search Engines and Web Applications: A Model-Driven Approach. In: Baresi, L., Fraternali, P., Houben, G.-J. (eds.) ICWE 2007. LNCS, vol. 4607, pp. 210–225. Springer, Heidelberg (2007)
12. Clarkson, E.C., Navathe, S.B., Foley, J.D.: Generalized formal models for faceted user interfaces. In: Proceedings of JCDL, pp. 125–134. ACM, New York (2009)
13. Bonino, D., Corno, F., Farinetti, L.: Faset: A set theory model for faceted search. In: Proceedings of WI-IAT, pp. 474–481. IEEE Computer Society, Washington (2009)
14. Halevy, A., Franklin, M., Maier, D.: Principles of dataspace systems. In: Proceedings of PODS, pp. 1–9. ACM (2006)
15. Hristidis, V., Papakonstantinou, Y.: Discover: keyword search in relational databases. In: Proceedings of VLDB. VLDB Endowment, pp. 670–681 (2002)
16. Zhou, Q., Wang, C., Xiong, M., Wang, H., Yu, Y.: SPARK: Adapting Keyword Query to Semantic Search. In: Aberer, K., Choi, K.-S., Noy, N., Allemang, D., Lee, K.-I., Nixon, L.J.B., Golbeck, J., Mika, P., Maynard, D., Mizoguchi, R., Schreiber, G., Cudré-Mauroux, P. (eds.) ASWC 2007 and ISWC 2007. LNCS, vol. 4825, pp. 694–707. Springer, Heidelberg (2007)
17. Lei, Y., Uren, V., Motta, E.: Semsearch: A search engine for the semantic web. Managing Knowledge in a World of Networks, 238–245 (2006)
18. Balog, K., Meij, E., de Rijke, M.: Entity search: building bridges between two worlds. In: Proceedings of SEMSEARCH, pp. 9:1–9:5. ACM (2010)
19. Stonebraker, M.: Sql databases v. nosql databases. Commun. ACM 53, 10–11 (2010)
20. Carmel, D., Zwerdling, N., Yogev, S.: Entity oriented search and exploration for cultural heritage collections. In: Proceedings of WWW (2012)
21. Suchanek, F.M., Kasneci, G., Weikum, G.: Yago: a core of semantic knowledge. In: Proceedings of WWW, pp. 697–706. ACM, New York (2007)

Towards Discovering Conceptual Models behind Web Sites*

Inma Hernández, Carlos R. Rivero, David Ruiz, and Rafael Corchuelo

University of Sevilla, Spain
{inmahernandez,carlosrivero,druiz,corchu}@us.es

Abstract. Deep Web sites expose data from a database, whose conceptual model remains hidden. Having access to that model is mandatory to perform several tasks, such as integrating different web sites; extracting information from the web unsupervisedly; or creating ontologies. In this paper, we propose a technique to discover the conceptual model behind a web site in the Deep Web, using a statistical approach to discover relationships between entities. Our proposal is unsupervised, not requiring the user to have expert knowledge; and it does not focus on a single view on the database, instead it integrates all views containing entities and relationships that are exposed in the web site.

Keywords: URL Patterns, Conceptual Models, Model Discovery.

1 Introduction

The Deep Web comprises a number of web sites that expose data stored in a back-end database, publishing them in a friendly format [8]. Entry points to these web sites are submittable query forms, which return as a response a number of web pages that are generated by filling a template with data [4,11]. The data that fill each template is the result of executing a view over the back-end database [2].

Since query forms are the unique entry points to the Deep Web, the different views that provide the data to fill each template are not accessible. Therefore, the conceptual model of the database, which comprises a number of entities and a number of relationships amongst these entities, remains hidden.

Having access to the conceptual model of a web site is mandatory to perform several tasks, such as integrating different (semantic or non-semantic) web sites [2,14,15], extracting information from the web without supervision [1,7,11], or creating ontologies by means of query forms [16].

As a consequence, there are many proposals in the literature that deal with discovering conceptual models behind web sites [1,2,4,5,6,9,10,11,12,13,16]. Some of these proposals deal with models composed solely of entities, without taking

* Supported by the European Commission (FEDER), the Spanish and the Andalusian R&D&I programmes (grants TIN2007-64119, P07-TIC-2602, P08-TIC-4100, TIN2008-04718-E, TIN2010-21744, TIN2010-09809-E, TIN2010-10811-E, and TIN2010-09988-E).

P. Atzeni, D. Cheung, and R. Sudha (Eds.): ER 2012, LNCS 7532, pp. 166–175, 2012.
© Springer-Verlag Berlin Heidelberg 2012

the relationships between them into account [4,5,6,10,12,13]. Other proposals discover models with entities and relationships [2,16], but they are supervised and require the intervention of the user, providing expert knowledge about each web site. Finally, the rest of the proposals focus on a single template, discovering only one view of the model [1,9,11].

In this paper, we propose a technique to discover the conceptual model behind a web site in the Deep Web. The model our technique is able to discover from each web site does not represent the complete, hidden conceptual model of the back-end database, but the union of the views over that conceptual model, composed of those entities and relationships that are exposed in the web site.

Our technique takes a set of URL patterns as input, each of which represents an entity in a particular web site. It follows a statistical approach to detect relationships between those entities. Our hypothesis is that each relationship is materialised in HTML links that go from pages of one class to pages of another class, so an XPath pattern targeting those links is created to represent each relationship. The URL patterns that support our technique can be either handmade by the user, or automatically built by any of the former proposals [3,6,10,13].

Our proposal presents some advantages: it creates a conceptual model consisting not only of entities, but also of relationships between those entities; it is not supervised, which saves the user a significant amount of time in labelling training sets, and does not require the user to have expert knowledge; and it integrates different views from the different templates in the site. Moreover, our proposal discovers all the possible anonymous relationships in the model, and we leave the user the task of labelling those relationships with an appropriate name and selecting those relationships that are useful for his or her model. Therefore, the set of relationships we automatically discover can be used as a first approach to the model, which can be refined by an expert data modeller, with a significant reduction in time investment [17].

The rest of this article is organised as follows: Section 2 reports on the related work on web site modelling; Section 3 defines our proposal to discover relationships in web sites; Section 4 shows the validation of our technique, using a well-known academical web site; finally, Section 5 lists some of the conclusions drawn from the research and concludes the article.

2   Related Work

There are many proposals related to web site modelling in the literature. Some of these proposals deal with models composed solely of entities [4,5,6,10,12,13], while others deal with more complex models including entities and relationships between those entities [1,2,9,11,16]

Models including only entities are usually discovered by web page clustering proposals, which unsupervisedly classify the pages in the web site. This clustering is based in features either from the page content or its structure [17], which implies that the page must be downloaded beforehand [5,12], or from URL features, which prevents having to download it [4,6,10,13]. In the latter case, the result is a collection of URL patterns representing each class. All the former proposals discover models of web sites that are exclusively composed of entities, but none of them discovers relationships amongst those entities.

Other proposals deal with models composed of both entities and the relationships amongst those entities. These proposals are usually focused on web information extraction, since extractors require such a model, that can be either provided by the user (supervised proposals) [2,16], or automatically inferred after analysing the pages of the web site (unsupervised proposals) [1,9,11].

On one hand, supervised proposals rely on the user to define the model. Tao et al. [16] analysed the problem of learning ontologies from web sites. Their proposal, FOCIH, consists of providing the user with a wizard-like application to design the model, and annotate pages from the web site according to that model. From that annotations, FOCIH infers an ontology, composed of concepts and relationship between the concepts. Atzeni et al. [2] proposed the Araneus Data Model, which defines a user-generated model for each web site that describes the different views of the schema of the web site, including the different entities and relationships. Supervised proposals require the user to have both the expert knowledge about each site to model it, and the expertise in data modelling to create a good model from scratch.

On the other hand, unsupervised proposals infer a model from the analysis of the web site. Kayed et al.[11] proposed FivaTech, a technique to discover the model behind a template, by analysing the DOM tree of a reduced set of web pages generated from that schema. Crescenzi and Mecca[9] proposed RoadRunner, an information extractor which automatically discovers the model behind one template in a web site, and uses this model to extract information. Finally, Arasu and Garcia-Molina [1] proposed EXALG, an information extractor based on grammar inference. The former proposals only discover the model behind one single template in the site, although web sites are usually composed of several templates, one for each type of information it offers. Therefore, each template allows discovering one different view on the back-end database, and all views should be integrated to infer a single conceptual model.

3   Proposal

Our technique takes a set of URL patterns that describe all classes of information offered in a web site as input, and discovers relationships between the classes. In the following subsections, we first introduce a running example, then we define some concepts that support our technique, and finally we describe the technique.

3.1   Running Example: Microsoft Academic Search

Microsoft Academic Search (from now onwards, MsAcademic) is an scholarly web site that offers different classes of information about academic publications (authors, papers, publishing hosts, such as journals or conferences, and research keywords, amongst others). Also, relationships between these classes of information are offered as well, e.g., author pages include a list of papers written by that author, and also a list of papers that cite this author. Furthermore, for each of the former papers, they offer the list of co-authors, the host it was published in, as well as the citations of the paper.

For the sake of simplicity, in this paper we focus on classes *Paper*, *Author*, *Journal*, *Conference* and *Citation*, and the relationships amongst them. An analysis of the MsAcademic site by the pattern building proposal in [10] yields the following URL patterns:

- $p_1 = \langle$ http, academic.research.microsoft.com, Publication, $\star$, $\star\rangle$,
- $p_2 = \langle$ http, academic.research.microsoft.com, Author, $\star$, $\star\rangle$,
- $p_3 = \langle$ http, academic.research.microsoft.com, Journal, $\star$, $\star\rangle$
- $p_4 = \langle$ http, academic.research.microsoft.com, Conference, $\star$, $\star\rangle$
- $p_5 = \langle$ http, academic.research.microsoft.com, Detail, eT, 1, sT, 5, id, $\star\rangle$

For example, pattern p_4 matches all URLs in MsAcademic containing information about conferences (e.g., URL `http://academic.research.microsoft.com/` `Conference/195/er` contains information about the ER conference).

3.2 Preliminaries

Definition 1 (Tokenisation). *Let s be a string, we define $\tau(s) = \langle s_1, s_2, \ldots, s_n\rangle$ as the sequence of tokens that is obtained after tokenising s.*

Note that this definition is applicable to both URLs and XPath locators.

Definition 2 (Pattern). *We define a pattern as a sequence of tokens, such that some of the tokens are literals, whereas others are wildcards. We denote a pattern p as $p = \langle t_1, t_2, \ldots, t_m\rangle$. We distinguish between URL patterns and XPath patterns. The latter are class-dependent, since XPath expressions are calculated in the context of a given page of a certain class c.*

We represent patterns by means of a subset of regular expressions that includes only literals and wildcard expressions. A wildcard is represented with symbol $\star$, and it represents any sequence of characters, excluding token separators defined for a particular tokenisation. Next, we define the problem of finding a match of a given pattern in another string.

Definition 3 (Pattern Matching). *Let p be a pattern $p = \langle p_1, p_2, \ldots, p_l\rangle$, and s be a sequence of tokens $s = \langle s_1, s_2, \ldots, s_l\rangle$, both of length l. We define that s matches p, and we denote it as $s \sim p$ iff each token in p is either a wildcard, or it is equal to the correspondent token in s.*

Note that p can be either a URL or XPath pattern, and that both URLs and XPath expressions are strings, hence we can apply the matching predicate on both URLs and XPaths.

Let $\mathcal{P} = \{p_1, p_2, \ldots, p_n\}$ be a set of URL patterns obtained from website W. Each of those patterns, according to the labels assigned by the user, corresponds to some class, e.g., pattern p_1 corresponds to class *Paper*, pattern p_2 corresponds to class *Author*, and so on.

Definition 4 (Class-Pattern Correspondence). *Let C be the set of classes of information offered by a website W, and P be the collection of URL patterns obtained from W. We define the injective function $\Phi : C \to P$, which assigns to each class in C the different patterns from P that have been labelled as corresponding to that class by the user.*

For example, after obtaining URL patterns p_1, p_2, p_3, p_4 and p_5 in MsAcademic, we assign a label to each pattern, e.g., stating that $\Phi(Paper) = p_1$, $\Phi(Author) = p_2$, $\Phi(Journal) = p_3$, $\Phi(Conference) = p_4$ and $\Phi(Citation) = p_5$.

Definition 5 (Detail Page Set). *Let C be the set of classes of information offered by a website W. We define the set of detail pages of any class $c \in C$, and we denote it by D^c as the set of pages in W containing information of type c.*

Note that the D^c of a class c is composed of those pages whose URLs match the patterns that have been labelled by the user as corresponding to that page.

Definition 6 (Locator). *Let w be a web page, and $u \in w$ be a URL. We define the locator of URL u in w, and we denote it by $XPath(u, w)$ as the XPath expression that points at the position of u in w.*

We use a tree notation to represent tokenisations of XPath expressions based on the PATRICIA trees (XPathTree), which allows representing large collections of strings efficiently and compactly. Every node in an XPathTree is an XPathTreeNode, defined by a label n_i, and it refers to a token t_j. Note that each path from the tree root to a leaf represents a single XPath. An example of a tree containing XPath expressions is presented in Figure 1a. For the sake of readability, each token in each node is preceded by the character that separates it from the previous token.

3.3 Relationships Discovery

We base the discovery of relationships between two classes on the detection of HTML links in pages of one class whose target is a page of another class. We extract the XPath locators of those links, and we apply a statistical-based technique to estimate the variability of each token in each locator. Then, we abstract the tokens with a high variability (again, using a statistical criterion), creating XPath patterns. Finally, each XPath pattern represents a particular relationship between the former classes.

There is an abstract relationship between two classes a and b, if there are links to pages of class b in most pages of class a. However, more than one type of relationship may exist between any given pair of classes a and b. For example, pages of class *Author* in MsAcademic contain both a list of publications, which include coauthors of the publication, and a list of citations, which includes authors that cited this author, as shown in Figure 2. Therefore, there are two different types of relationships in this model between class *Author* and itself: 1) *isCoauthorOf* and 2) *cites*.

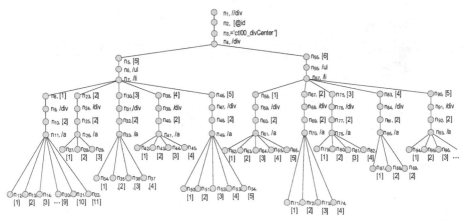

(a) XPathTree containing the XPaths of URLs of class *Author* in pages of class *Author*

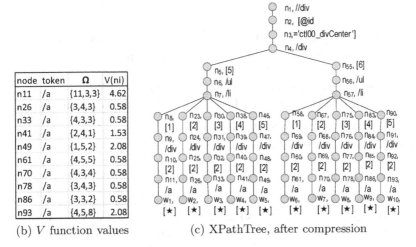

node	token	Ω	V(ni)
n11	/a	{11,3,3}	4.62
n26	/a	{3,4,3}	0.58
n33	/a	{4,3,3}	0.58
n41	/a	{2,4,1}	1.53
n49	/a	{1,5,2}	2.08
n61	/a	{4,5,5}	0.58
n70	/a	{4,3,4}	0.58
n78	/a	{3,4,3}	0.58
n86	/a	{3,3,2}	0.58
n93	/a	{4,5,8}	2.08

(b) *V* function values   (c) XPathTree, after compression

Fig. 1. Representation of the technique, which compresses an XPathTree by abstracting children of nodes with a high variability

Using only URL patterns, we are not able to discern between these different types of relationships, since all URLs match the same pattern, regardless of the type of relationship they represent. Therefore, other features must be extracted from the URLs to classify them according to their role.

We assume that links whose URL matches the same URL patterns may appear in different locations in the page, but all links representing the same relationship appear in similar locations. Therefore, we use the XPath of the different links, which denotes their location in the page. We apply a technique to build patterns for those XPath, which starts by tokenising all XPath locators and inserting all their tokens in order in an XPathTree. Then, we use some criterion to discern tokens that must be abstracted (replaced by a wildcard), based on the concept of variability of a token.

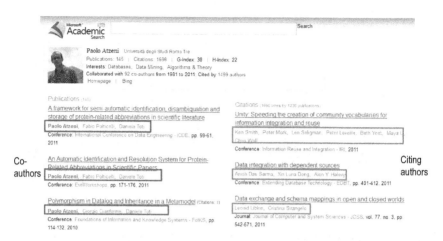

Fig. 2. Detail page of class *Author*

The variability of a token refers to how spread the numbers of tokens that follow that token in different XPath locators in different pages of the same class (i.e., the different numbers of children of the node representing that token in each page) are. Since we do not analyse all pages of a site, but only a representative sample, we estimate the variability by means of the following definition.

Definition 7 (Variability Estimator:). *Let D^c be a set of detail pages of class c, x an XPath expression and n_i be a tree node referring a token t, we define the variability estimator of node n_i, and we denote it as $V(n)$ as the standard deviation of the numbers of children of node n_i in the different pages of D^c.*

Based on these variability estimators, we define a process to generate XPath patterns. For each node n_i in the XPathTree, we check if its variability estimator is significatively high, and in that case, all its children nodes have their token replaced with a wildcard, and the subtrees rooted at them are merged. Contrarily, children of nodes with a low variability are probably part of a pattern, so they are not abstracted, but kept as literals.

Our technique to mine relationships between classes a and b consists of two steps: XPathTree building and XPathTree compressing.

In the first step, we extract all URLs matching pattern $\Phi(b)$ in pages from D^a, and we calculate an XPath locator for each of them. XPath locators are tokenised, and each token is inserted in an XPathTree as a node with a variability estimator. An example of an XPathTree built using this technique is presented in Figure 1a. It contains XPath expressions of URLs matching $\Phi(Author)$ in the running example, extracted from detail pages of class *Author*.

In the second step, we apply a compressing algorithm that performs a depth-first traversal on the XPathTree, and for each visited node, uses its variability estimator to discern nodes with a variability higher than a given parameter $\theta > 0$. Those nodes have their token abstracted into a wildcard ($\star$). As an example, nodes with variability higher than 0.5 are presented in Figure 1b.

After the whole tree has been traversed and processed, each of the resulting tree branches represents a different pattern. Furthermore, each pattern refers to a different type of relationship between classes a and b. As an example, in Figure 1c we show the example tree containing XPath expressions of links between class *Author* and itself, after processing all its nodes. The tree contains ten branches, which correspond to ten XPath patterns.

At the end of this process, for each pair of classes a and b, we have obtained a set of XPath patterns, that represent the different relationships between them. These relationships are anonymous, and it is left to the user the task of labelling them with an appropiate name. Moreover, we have identified all the possible relationships, but some of them might be duplicated (i.e., we discover a relationship between a and b, which is the same as another relationship between b and a). Therefore, the user has the opportunity to select the relationships that are most suitable for his or her model, discarding the rest. Therefore, although we are indeed automatically discovering the relationships between entities, the user still has the complete control over the final model.

As an example, consider the former patterns discovered in Figure 1c. The first five patterns correspond to links to authors that co-author, respectively, the five most recent papers of an author. Meanwhile, patterns sixth to tenth correspond to links to authors of, respectively, the five most recent papers that cite the author. Therefore, the five first patterns correspond to a particular relationship between class *Author* and itself(*isCoauthorOf*), while the five last patterns correspond to a different relationship (*cites*).

4   Validation

We present an experiment to validate our technique. Microsoft Academic Search was analysed to discover the conceptual model behind it, by means of two steps: in the first step, we discovered the entities in the model, using the URL patterns obtained with the technique described in [10]; in the second step we discovered the relationships between these entities, with the former URL patterns as input, and using the technique described in this paper.

We show the relationships discovered for this site in Figure 3a, using a UML class diagram. After the intervention from the user, a possible model obtained from the former relationships is presented in Figure 3b. For example, relationships $r3$, $r4$, $r5$, $r6$ and $r7$, represent respectively the co-authors of the most recent paper of an author, the co-authors of the second most recent paper, and so on. The user analyses these relationships and decides that all these relationships are actually the same, and labels it *isCoauthorOf*.

Using our technique, it is also possible to infer hierarchical relationships between classes, by identifying classes that share a common group of relationship with other classes. For example, in the former example model for MsAcademic, classes *Journal* and *Conference* both share exactly the same types of relationships (*Journal* is related to *Paper* by means of r23 and r24, to *Author* by means of r1 and r2 and to *Citations* by means of r19 and r20. Similarly, *Conference*

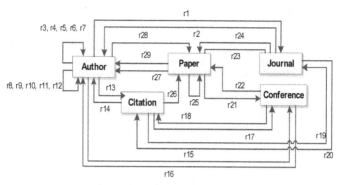

(a) Relationships for MsAcademic

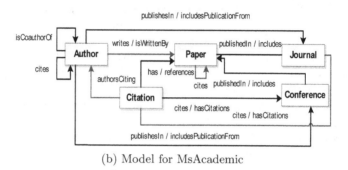

(b) Model for MsAcademic

Fig. 3. Model discovered for the validation site

is related to the same set of classes, with two relationships with each class). Therefore, our technique proposes the user the generalisation of *Journal* and *Conference* into another class, and lets the user name it (e.g., *Host*).

5   Conclusions

In this paper, we present a technique to discover the conceptual model behind a web site in the Deep Web. Using a set of URL patterns as input, we use a statistical approach to discover all the different relationships between those entities. These relationships can be later analysed by the user, who is responsible for labelling them appropriately, and selecting those relationships that are useful for his or her particular model. We validate our proposal using a well-known academical web site, Microsoft Academic Search.

Other proposals have dealt with the problem of discovering the model behind a web site. Some of them discover models composed only of entities, neglecting the discovery of relationships, which we deal with. Others are supervised, which require expert knowledge from the user, while our technique is completely unsupervised. Finally, other proposals discover only one view of the conceptual model, which corresponds to a particular template; contrarily, we discover a model composed of the union of all views over the complete model that include entities and relationships that are exposed in the web site.

References

1. Arasu, A., Garcia-Molina, H.: Extracting structured data from web pages. In: SIGMOD, pp. 337–348 (2003)
2. Atzeni, P., Mecca, G., Merialdo, P.: Managing web-based data: Database models and transformations. IEEE Internet Computing 6(4), 33–37 (2002)
3. Bar-Yossef, Z., Keidar, I., Schonfeld, U.: Do not crawl in the dust: different URLs with similar text. In: WWW, pp. 111–120. ACM (2007)
4. Blanco, L., Bronzi, M., Crescenzi, V., Merialdo, P., Papotti, P.: Automatically building probabilistic databases from the Web. In: WWW, pp. 185–188 (2011)
5. Blanco, L., Crescenzi, V., Merialdo, P.: Structure and semantics of Data-Intensive Web pages: An experimental study on their relationships. J. UCS 14(11), 1877–1892 (2008)
6. Blanco, L., Dalvi, N., Machanavajjhala, A.: Highly efficient algorithms for structural clustering of large websites. In: WWW, pp. 437–446. ACM (2011)
7. Chang, C.-H., Kayed, M., Girgis, M.R., Shaalan, K.F.: A survey of web information extraction systems. IEEE TKDE 18(10), 1411–1428 (2006)
8. Chang, K.C.-C., He, B., Li, C., Patel, M., Zhang, Z.: Structured Databases on the Web: Observations and Implications. SIGMOD Record 33(3), 61–70 (2004)
9. Crescenzi, V., Mecca, G.: Automatic information extraction from large websites. J. ACM 51(5), 731–779 (2004)
10. Hernández, I., Rivero, C.R., Ruiz, D., Corchuelo, R.: A statistical approach to URL-based web page clustering. In: WWW, pp. 525–526 (2012)
11. Kayed, M., Chang, C.-H.: Fivatech: Page-level web data extraction from template pages. IEEE Trans. Knowl. Data Eng. 22(2), 249–263 (2010)
12. Mecca, G., Raunich, S., Pappalardo, A.: A new algorithm for clustering search results. Data Knowl. Eng. 62(3), 504–522 (2007)
13. Deepak, P., Khemani, D.: Unsupervised learning from URL corpora. In: COMAD, pp. 128–139 (2006)
14. Popa, L., Velegrakis, Y., Miller, R.J., Hernández, M.A., Fagin, R.: Translating web data. In: VLDB, pp. 598–609 (2002)
15. Rivero, C.R., Hernández, I., Ruiz, D., Corchuelo, R.: Generating SPARQL Executable Mappings to Integrate Ontologies. In: Jeusfeld, M., Delcambre, L., Ling, T.-W. (eds.) ER 2011. LNCS, vol. 6998, pp. 118–131. Springer, Heidelberg (2011)
16. Tao, C., Embley, D.W., Liddle, S.W.: FOCIH: Form-Based Ontology Creation and Information Harvesting. In: Laender, A.H.F., Castano, S., Dayal, U., Casati, F., de Oliveira, J.P.M. (eds.) ER 2009. LNCS, vol. 5829, pp. 346–359. Springer, Heidelberg (2009)
17. Thonggoom, O., Song, I.-Y., An, Y.: Semi-automatic Conceptual Data Modeling Using Entity and Relationship Instance Repositories. In: Jeusfeld, M., Delcambre, L., Ling, T.-W. (eds.) ER 2011. LNCS, vol. 6998, pp. 219–232. Springer, Heidelberg (2011)

A Distance-Based Spelling Suggestion Method for XML Keyword Search

Sheng Li, Junhu Wang, Kewen Wang, and Jiang Li

School of Information and Communication Technology,
Griffith University, Gold Coast Campus, Australia
{sheng.li,Jiang.li}@griffithuni.edu.au,
{J.Wang,k.wang}@griffith.edu.au

Abstract. We study the spelling suggestion problem for keyword search on XML documents. To address the problems in existing work, we propose a distance-based approach to suggesting meaningful query candidates for an issued query. Our approach uses distance to measure the relationship between keyword matching nodes, and ranks a candidate higher if there are closely-related nodes in the database that match the candidate. We design an efficient algorithm to generate top-k query candidates. Experiments with real datasets verified the effectiveness and efficiency of our approach.

1 Introduction

Keyword search provides a user-friendly mechanism for people to access XML data. When a user types in a keyword query, there are two processes involved: (1) formulating a keyword search query, and (2) typing the query into systems. However, both steps are susceptible to several kinds of errors [6], e.g., step (2) suffers from typographical errors and incorrect keywords. In these cases, a dirty query is formed, which is likely to return empty or low-quality results.

In order to alleviate this problem, *query suggestion (QS)* a.k.a. *query cleaning*, has been proposed, which provides users with alternative queries that better express users' search intention. QS has been studied, and already widely used for text documents. Recently, the study on query suggestion has been explored in the context of relational databases [7] and XML data [6].

It has been observed in [6] that previous methods for QS rely on information such as the query log, but are independent of the database contents. As a result, the quality of suggested queries depends on the quality and quantity of past queries. Besides, these methods are biased towards popular queries, which is unfair to rare keywords that could return meaningful results.

XClean [6] is the first work that focuses on the query suggestion problem for XML Keyword Search. It proposes a probabilistic framework that takes the XML structural characteristics into consideration and improves the result quality of suggested queries. However, XClean still suffers from unfairness in candidate ranking, which we will discuss in Section 2.

P. Atzeni, D. Cheung, and R. Sudha (Eds.): ER 2012, LNCS 7532, pp. 176–189, 2012.
© Springer-Verlag Berlin Heidelberg 2012

In this paper, we propose a novel distance-based XML keyword query suggestion method. The key idea is to evaluate each query candidate by the minimum distance among keyword matching nodes to assess how closely these keywords are related. Our approach successfully suggests query candidates that return high quality results. We design an efficient algorithm to generate top-k suggested queries. Experiments on real datasets demonstrate the better effectiveness of the proposed approach. Our experiments also show that our approach is significantly faster than XClean, which is important as practical query suggestion systems must be fast enough, so that the users can see the suggested queries instantly.

The rest of the paper is organized as follows: We begin with a discussion of the problems in previous work in Section 2. Section 3 elaborates on our distance-based spelling suggestion algorithm. In Section 4, experiments and analysis are presented. Section 5 concludes the paper with a discussion.

2   Previous Work

To the best of our knowledge, XClean is the first and latest work that explores query suggestion on XML data. For self-containment, we briefly recall the method in XClean in this section.

2.1   XClean

Given an input query Q and the XML document T, XClean generates a set of possible candidates C_i. These candidates are ranked by *candidate query probability* $P(C_i|Q,T)$, namely, the probability that C_i is intended when the user issues Q on the XML document T. XClean defines the candidate query probability as:

$$P(C_i|Q,T) = \kappa \cdot P(Q|C_i,T) \cdot P(C_i|T) \tag{1}$$

where κ is a fixed value; $P(Q|C_i,T)$ models the possibility of typing query Q when the intended query is actually C_i over document T. XClean makes the commonly used assumption that $P(Q|C_i,T) = P(Q|C_i)$, which is calculated by the similarity between C_i and Q; $P(C_i|T)$ is the query generation probability given the document T, which is the probability of users issuing the query C_i when the XML document T is presented. The computation of $P(C_i|T)$ is explained below.

XClean adopts the same idea as most works in XML keyword search, which is based on finding subtrees that contain at least one instance for each keyword in an issued query. The XML document can be viewed as a collection of these disjoint subtrees, named *entities*, which are treated as the basic units of information that may interest users [6]. XClean focuses on the *Search-for Node Type* (*SNT*) semantics [1] in XML keyword search, but can also be applied to the *smallest lowest common ancestor* (SLCA) semantics.

After inferring type p for a query candidate C_i, XClean decomposes T into a set of N entities $r_1, \cdots, r_N$ of type p. Each r_j is viewed as a virtual document [6].

$P(C_i|T)$ can then be calculated using equation (2), and equation (3) is the final formula for candidate query probability.

$$P(C_i|T) = \frac{1}{N} \sum_{j=1}^{N} P(C_i|r_j) \tag{2}$$

$$P(C_i|Q,T) = \kappa \cdot P(Q|C_i) \cdot \left(\frac{1}{N} \sum_{j=1}^{N} \prod_{w \in C_i} P(w|r_j) \right) \tag{3}$$

where $P(C_i|r_j) = \prod_{w \in C_i} P(w|r_j)$, $P(w|r_j)$ stands for the probability that keyword w appears in entity r_j, which is calculated by:

$$P(w|r_j) = \frac{count(w, r_j) + \mu p(w|T)}{|r_j| + \mu} \tag{4}$$

where $count(w, r_j)$ is the number of occurrences of w in r_j; $|r_j|$ is the size of entity r_j, treated as a virtual document; $p(w|T)$ is the probability that w appears in T, and μ is a smoothing parameter [6].

2.2   Existing Problems

According to equations (2) and (3), the ranking of a candidate C_i is mainly influenced by $P(w|r_j)$ and the number of subtrees that contain all the keywords. The only structural information that XClean considers is to restrain the keyword matching nodes within a subtree, which may contain many deeper entities. However, nodes distributed in those deeper entities could be loosely-related, which is not considered in XClean.

Consider the XML tree in Fig. 1 and an input keyword *thee*. XClean obtains the variants, denoted as $var_\epsilon(w)$, for the keyword w based on the vocabulary of the dataset. Suppose the edit distance ϵ is set to 1, then the variants of the keyword *thee* is $var_1(w) = \{tree, chee, three\}$. For the example below, both query candidates have the same edit distance to the input query Q. Therefore, we only discuss the query generation probability given the document T, since it proportionally reflects the ranking.

Example 1. Consider the candidates $C_1 = \{volume, 11, search, tree\}$ and $C_2 = \{volume, 11, search, three\}$. The *SNT*s for the candidates are $p = issue$, and there is only *issue* 0.0 whose subtree contains nodes that match all the keywords. Under *issue* 0.0, the frequency of keyword *three* is higher than that of *tree*. According to equation (3), C_2 will be ranked higher than C_1, while C_1 has better quality. It is easily verified that if the SLCA (rather than SNT) semantics is used, the same ranking result will be produced.

As shown in the example above, XClean restrains the nodes matching the keywords in the subtrees of certain *entity* or *SLCA*. However, it does not consider *the relationship between the keyword matching nodes* within each subtree. Besides, it is biased against candidates that have fewer occurrences. To deal with

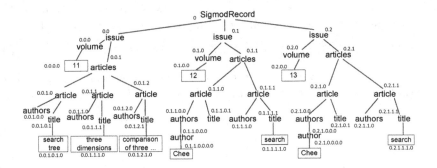

Fig. 1. An example XML tree of SigmodRecord

these weaknesses, we propose a novel approach to generate query suggestions based on the distance of keyword matching nodes.

3 Distance-Based Spelling Suggestion

3.1 Overview of Our Approach

Currently commonly used semantics, such as *SNT*, *LCA*, *SLCA* [10], *ELCA* [4] and so on, are based on the heuristic that far-apart nodes are not as tightly related as nodes that are closer together [9]. Our approach is inspired by the same heuristics. The key idea of our approach is that for each query candidate C_i, we check whether it has nodes physically close to each other that match the keywords in C_i, and use it as one of the major factors in ranking C_i. The ranking of C_i is also affected by several other factors. The details are discussed at the end of this section.

Entities. As observed previously [1,5], an XML document consists of nodes that represent *entities, attributes* and *connection* nodes. XSeek [5] proposed some guidelines to infer those different nodes. In our work, we use the same guidelines and the concept of *entity*, and we refer the label path of the entity as its *entity type*. For example, in Fig. 1 each *article* is an *entity*, and the path *SigmodRecord/issue/articles/article* is its *entity type*. For simplicity, we will use the label of the node to represent the *entity type* when there is no confusion.

Distance between Two Nodes. Given *node u* and *node v*, the distance between them is defined as:

$$dist(u, v) = distB(u, v) + distP(u, v) \tag{5}$$

where $distB(u, v)$ stands for the *basic distance* between nodes u and v, which is the number of edges in the (unique) path linking u and v; $distP(u, v)$ is the *penalty distance* between u and v, which penalizes nodes in different entities to guarantee that a node within an entity of *entity type* E is unable to be closer to nodes in other entities of the same type than to nodes within the same entity.

$distB(u, v)$ can be calculated by finding the *Lowest Common Ancestor* (*LCA*) of nodes u and v, denoted as $LCA(u, v)$. Suppose that $z = LCA(u, v)$, then

$$distB(u, v) = level(u) + level(v) - 2 \times level(z) \tag{6}$$

where $level(z)$ is the level of z in the tree, which can be easily calculated by comparing the Dewey codes of u and v. The *penalty distance* is defined as

$$distP(u, v) = \begin{cases} depth(E) & \text{if } E \text{ exists} \\ 0 & \text{else} \end{cases} \tag{7}$$

where E is the *type* of the top-level *entities* that exist on both *path* $z \to u$ and *path* $z \to v$, where $z = LCA(u, v)$; $depth(E)$, different from $level(u)$, is the depth that the entity type E possesses, which is the maximum number of levels that entities of type E have, e.g., $depth(issue) = 5$ and $depth(article) = 3$.

Consider the tree in Fig. 1, for *issue* 0.1 (denoted as a), it has the same *basic distance* to nodes 0.0.0 (denoted as b) and 0.1.1.1.1 (denoted as c), namely, $distB(a, b) = distB(a, c) = 3$. However, nodes a and b are in different *issue* entities. After adding the *penalty distance*, $dist(a, b) = 3 + depth(issue) = 8$. $dist(a, c) = 3$ is smaller, which reflects the fact the nodes a and c has more meaningful relationship.

Distance between a Node Set and a Node. The distance between a node set M and node v, denoted $dist(M, v)$, is defined as the average distance between v and nodes in M. That is, $dist(M, v) = \frac{\Sigma_{u \in M} dist(u, v)}{|M|}$.

The following definitions are important.

Definition 1 (Nearest w-neighbor). *Let w be a keyword, $U(w)$ be the set of all nodes matching w, M be a set of nodes. A nearest w-neighbor of M is a node in $U(w)$ which has the minimum distance to M among all nodes in $U(w)$. In particular, when M contains a single node q, we call a nearest w-neighbor of M a nearest w-neighbor of q.*

The above definition of nearest w-neighbor extends the definition of nearest w-neighbor in [8] from a single node to a node set. Note that the nearest w-neighbor of a node, or of a set of nodes, is usually not unique.

Definition 2 (Minimum Combination of keyword matching nodes(MC)). *Given a query $Q = \{w_0, w_1, ..., w_l\}$ and a set of nodes $M = \{u_0, ..., u_l\}$, we call M an MC if M satisfies the following conditions:*

1. *node $u_i \in M$ matches keyword w_i;*
2. *$\forall u_i \in M$ ($0 < i \leq l$), u_i is a nearest w_i-neighbor of M_{i-1}, where $M_{i-1} = \{u_0, ..., u_{i-1}\}$.*

We define the distance among the nodes in M as $Dist(M) = \sum_{i=1}^{l} dist(M_{i-1}, u_i)$

Example 2. Consider the tree in Fig. 1 and the query $C_1 = \{volume, 11, search, tree\}$. The set $M_1 = \{0.0.0, 0.0.0.0, 0.0.1.0.1.0, 0.0.1.0.1.0\}$ is an MC for C_1, and its distance is $1 + \frac{5+6}{2} + \frac{5+6+0}{3} = 10.2$.

For a query candidate $C = \{w_0, ..., w_l\}$, there is a set of MCs, denoted as $\mathcal{M}(C)$. Each MC corresponds to a node u_0, and some MCs may share the same u_0 node.

Definition 3 (Minimum/average distance for a query). *For a query candidate $C = \{w_0, w_1, ..., w_l\}$, we define the minimum distance and average distance for query C as $minD(C)$ and $avgD(C)$ respectively:*

$$minD(C) = min_{M \in \mathcal{M}(C)}(Dist(M))$$

$$avgD(C) = average_{M \in \mathcal{M}(C)}(Dist(M))$$

$minD(C)$ and $avgD(C)$ will both be used in the ranking of query candidate C.

Example 3. Consider the tree in Fig. 1, the query candidates $C_1 = \{volume, 11, search, tree\}$, and $C_2 = \{volume, 11, search, three\}$. As seen in Example 2, $minD(C_1) = 10.2$. The set $M_2 = \{0.0.0, 0.0.0.0, 0.0.1.0.1.0, 0.0.1.1.1.0\}$ is an MC for C_2, and its distance is 12.2. It can be easily verified that $minD(C_2) = 12.2$.

Ranking of Query Candidates. To rank query candidates we use the four factors below:

(1) $ed(Q, C_i)$: the edit distance between candidate C_i and the input query Q, which is the sum of the edit distances of the keywords in C_i. The smaller $ed(Q, C_i)$ is, the higher C_i should be ranked.
(2) $minD(C_i)$: the smaller $minD(C_i)$ is, the higher C_i should be ranked;
(3) $avgD(C_i)$: the average distance of the "meaningful" MCs for C_i (an MC is considered *meaningful* if it does not have two nodes such that they are linked through the root or the distance between them is larger than a threshhold), the smaller $avgD(C_i)$ is, the higher C_i should be ranked;
(4) $cf(C_i)$: the number of "meaningful" MCs for C_i, the larger $cf(C_i)$ is, the higher C_i should be ranked;

Among the four factors, $minD(C_i)$ and $ed(Q, C_i)$ are the major factors, while the influences of $avgD(C_i)$ and $cf(C_i)$ are relatively minor. We combine the four factors to rank the candidate as follows:

$$Score(C_i) = \left(1 - \frac{minD(C_i)}{3(l-1)depth(T)}\right) \cdot \exp\left(-\frac{ed(Q, C_i)}{\epsilon \cdot l}\right)$$
$$+ \eta \cdot \left(1 - \frac{avgD(C_i)}{3(l-1)depth(T)}\right)\left(1 - \frac{1}{cf(C_i)+1}\right) \tag{8}$$

where l is the number of the keywords in C_i (We assume $l > 1$. When $l = 1$, we use edit distance and term frequency only in the ranking); $depth(T)$ is the depth of the XML document T; ϵ is the edit distance for generating the variants of each keyword; η is set to adjust the influence of $avgD(C_i)$ and $cf(C_i)$, which is set to 0.01 in our experiments. Note that the distance between any two nodes in T cannot be more than $3depth(T)$, thus $minD(C_i) \leq 3(l-1)depth(T)$, and $avgD(C_i) \leq 3(l-1)depth(T)$. In the above formula, we use the exp function for the edit distance since the quality of query candidates deteriorates more drastically with the increase of edit distance.

3.2 Algorithm

Given a query $Q = \{w_0, ..., w_l\}$, there are $\prod_{i=0}^{l} |var_\epsilon(w_i)|$ possible candidates. For each candidate, we need to find its MCs. The runtime efficiency could be extremely low. We designed an efficient approximate algorithm to find the MCs for each candidate and rank the candidates based on the ranking factors. The main ideas for speeding up the processing are the following:

1. In computing the MCs, we use approximation. Specifically, in computing the nearest w-neighbor of a set $M = \{v_1, \ldots, v_k\}$, we consider the nearest neighbors of $v_1, \ldots, v_k$ only[1], that is, we regard the closest node to M among all $u_1, \ldots, u_l$, where u_i is the nearest w-neighbor of v_i, as the nearest w-neighbor of M. To find the nearest w-neighbor of node v_i, we adopt the TVP index in [8] which makes it cost $O(log(|U(w)|))$ to find the *nearest w-neighbor* of v_i. [8] proposed an efficient method to generate the TVP index for each keyword w. It is not hard to verify that this method is also valid for our distance definition.

2. Like XClean, we make the assumption that nodes connected through the root are unlikely to form an interesting query result. Besides, we use a distance threshold $maxLen$ to restrain the distance between two nodes. Given nodes u and v that match keywords w_i and w_j in a query candidate C, if $dist(u, v) > maxLen$, or u is connected to v through the root, the MC that contains nodes u and v will be considered not meaningful and hence ignored. The default value of $maxLen$ is set to $2depth(E)$, where E is the entity type in T of the maximum depth, but can be configured for different data.

3. Based on the above assumption, given an input query $Q = \{w_0, ..., w_l\}$, we only need to consider keyword match nodes that are within the same subtree rooted at level 1 (for simplicity, let us refer to these subtrees as level 1 subtrees). Therefore, if a level 1 subtree does not contain a match node for all keywords, we will ignore that subtree in the processing. When reading the inverted list of keywords from the database, we generate a list of Dewey codes, called SRList, where each record in the list is the Dewey code of the root of one of the level 1 subtrees. These subtrees are those that cannot be ignored. The details of this process are given in Algorithm 1. Then we only process those keyword-match nodes that are descendants of nodes that match the Dewey codes in SRList, while other keyword match nodes will be ignored.

4. Instead of considering each query candidate separately, we incrementally process the possible candidates by adding the keywords one by one. This is based on the fact that if $\{u_0, \ldots, u_i\}$ is a meaningfulMC for the query $\{w_0, \ldots, w_i\}$, then $\{u_0, \ldots, u_{i-1}\}$ is a meaningfulMC for the query $\{w_0, \ldots, w_{i-1}\}$. First we process keywords in $var_\epsilon(w_0)$ and $var_\epsilon(w_1)$ to generate all the possible candidates from the variants of query $\{w_0, w_1\}$, and record the meaningful MCs matching those query candidates. Then we process $var_\epsilon(w_2)$

---

[1] Although there is a chance to miss the real nearest w-neighbor of M, in practice, such chance is small.

based on the meaningful *MC*s of last step. Incrementally, we can find all the meaningful *MC*s for all variants of Q. This strategy can avoid processing the uninteresting query candidates repeatedly. The details are shown in Algorithm 2.

Algorithm 1. Generate SRList while reading Inverted List from DB

Input: input query $Q = \{w_0, w_1, ..., w_l\}$; edit distance ϵ;
Output: SRList;
1: **for** $i = 0 \to l$ **do**
2:     $var_\epsilon(w_i) \leftarrow makeVariants(w_i)$;
3:     create a new HashMap $SRMap_i$;
4:     **for** each $w_{i_j} \in$ each $var_\epsilon(w_i)$ **do**
5:         **for** each n in Inverted List of w_{i_j} in DB **do**
6:             $g \leftarrow n.getDeweyCodeofAncestorAtLevelOne$;
7:             **if** $i = 0 \vee (i > 0 \wedge SRMap_{i-1}.contains(g))$ **then**
8:                 $SRMap_i.put(g)$; //if g does not exist in $SRMap_i$ already
9:                 store n into the inverted list of w_{i_j} in memory;
10:             **end if**
11:         **end for**
12:     **end for**
13: **end for**
14: SRList = getListFrom($SRMap_l$);

We now explain our algorithms in some detail. When a query $Q = \{w_0, ..., w_l\}$ is issued, like [6], we use a version of the FastSS method [2] to generate variants for each keyword w_i within edit distance ϵ, the set of variants is denoted as $var_\epsilon(w_i)$ in Algorithm 1. When reading a node n from the inverted list of w_{i_j} in DB (Line 5), if n is in the level 1 subtree rooted at g (Line 6), we check whether $i = 0$ or g exists in $SRMap_{i-1}$ (Line 7). If yes, we put g in $SRMap_i$ (Line 8) and store n into the inverted list of w_{i_j} in memory (Line 9). Finally, we can get the SRList which is $SRmap_l$ (Line 14).

Example 4. Consider the query $Q = \{thee, dimensions, volume\}$ for the XML tree in Fig. 1. When the edit distance is 1, the sets of variants for the keywords *thee, dimensions, volume* are {*chee, three, tree*}, {*dimensions*} and {*volume*} respectively. First, we read the inverted list of the variants *thee*, and store the Dewey codes of the roots of level 1 subtrees that contain matching nodes of any of the variants. As a result, 0.0, 0.1, and 0.2 are kept as the keys in $SRMap_0$. When it comes to the variants of *dimensions*, because only the subtree rooted at 0.0 contains the keyword and 0.0 exists in $SRMap_0$, 0.0 is stored in $SRMap_1$. As for keyword *volume*, only 0.0 exists in $SRMap_1$ while 0.1 and 0.2 do not, so $SRMap_2$ contains 0.0 only. Thus SRList={0.0} is returned.

In Algorithm 2, for each w_{0_j} in $var_\epsilon(w_0)$, we retrieve the *inverted list* for it and store it in a hash map $validNodesHM_0$ (Lines 1-4). Here, we only store those nodes whose ancestors' Dewey codes exist in *SRList*. For keywords in $var_\epsilon(w_1)$, ..., and $var_\epsilon(w_l)$, we retrieve their *TVP* information (Lines 5-7) generated during preprocessing.

For each node u_{0_j} from $validNodesHM_0$, we start finding its nearest w_{1_k}-neighbor according to its *TVP*, where $w_{1_k} \in var_\epsilon(w_1)$. If u_{0_j} and its w_{1_k}-neighbor

Algorithm 2. Find MCs for each candidate, and rank candidates

---

Input: Query Q; Inverted Lists; distance threshold $maxLen$; SRList
Output: a list of suggested queries SQ
1: **for** each $w_{0_j} \in var_\epsilon(w_0)$ **do**
2:    iList = getInvertedList(w_{0_j}, SRList); //load nodes in level 1 subtrees whose Dewey codes are in SRList
3:    validNodesHM.put(w_{0_j}, iList);
4: **end for**
5: **for** each $w_{i_j} \in$ each $var_\epsilon(w_i)$ **do**
6:    $TVPMap.put(w_{i_j}, getTVP(w_{i_j}))$; //load TVP index generated during preprocessing
7: **end for**
8: **for** $i = 1 \rightarrow l$ **do**
9:    new $validNodesHM_i$; //create a new hash map
10:    **for** each candidate $C \in validNodesHM.keySet()$ **do**
11:       $\mathcal{M}(C) \leftarrow validNodesHM.get(C)$; //a list of MCs matching C
12:       **for** each $w_{i_j} \in var_\epsilon(w_i)$ **do**
13:          $C' = C \cup \{w_{i_j}\}$; //incrementally generate longer candidates
14:          new $\mathcal{M}(C')$; //create a new empty list of MCs
15:          **for** each combination $M \in \mathcal{M}(C)$ **do**
16:             find w_{i_j}-nearest neighbor of M, denoted as u_j, and its distance $Dist(M, u_j)$;
17:             **if** $minDist(M, u_j) \leq maxLen$ **then**
18:                $M.add(u_j)$; //add u_j to the set M
19:                $M.addDistance(Dist(M, u_j))$; //add the distance from u_j to M
20:                $\mathcal{M}(C').add(M)$; //record M for future computation
21:             **end if**
22:          **end for**
23:          **if** $\mathcal{M}(C')$ is not empty **then**
24:             $validNodesHM_i.put(C', \mathcal{M}(C'))$;
25:          **end if**
26:          $validNodesHM = validNodesHM_i$;
27:       **end for**
28:    **end for**
29: **end for**
30: $SQ = rank(validNodesHM)$;

(denoted w_{1_k}-$NN(u_{0_j})$) are only connected through the root node or do not meet the threshold $maxLen$, we ignore the combination of u_{0_j} and w_{1_k}-$NN(u_{0_j})$ for further calculation (Line 17). Otherwise, we add this combination along with its distance information into a *hash map*, denoted as $validNodesHM_1$, of which the key is the string containing the keywords and the data is a list of combinations of keyword matching nodes (Line 24). After the algorithm finishes probing through the $var_\epsilon(w_i)$, the hash map $validNodesHM_i$ containing meaningful combinations will be used as the search space for the next iteration (Line 26), searching for meaningful connection with nodes carrying keywords in $var_\epsilon(w_{i+1})$. Then new hash map $validNodesHM_{i+1}$ will be generated, replacing $validNodesHM_i$ for the next iteration. This process continues until it finishes probing $var_\epsilon(w_l)$. Finally, all the possible query candidates are stored in $validNodesHM_l$ as the key set. For each query candidate C_i, there is a list of MCs that match all the keywords in C_i and each MC contains the distance information. By enumerating all the MCs in the list for C_i, we are able to retrieve information that could assist in ranking (Line 30).

Example 5. Consider the query Q in Example 4. As seen before, SRList=$\{0.0\}$. Suppose $maxLen = 7$. Initially, $validNodesHM$ contains two records:

(1) "three" $\rightarrow$ {{0.0.1.1.1.0}, {0.0.1.2.1.0}}; (2) "tree" $\rightarrow$ {{0.0.1.0.1.0}}. Then based on these two records, Lines (10-28) find the meaningful MCs for the candidates {three dimensions} and {tree dimensions}, and store the result in $validNodesHM_1$. Since only {0.0.1.1.1.0, 0.0.1.1.1.0} satisfy the maxLen threshold, $validNodesHM_1$ contains one record: "three dimensions" $\rightarrow$ {{0.0.1.1.1.0, 0.0.1.1.1.0}}. Then this record will be used to find the MCs when the keyword $volume$ is added to the query, and we will find $validNodesHM_2$ which contains only one record "three dimensions volume" $\rightarrow$ {{0.0.1.1.1.0, 0.0.1.1.1.0, 0.0.0}}.

Complexity Analysis. In our algorithm, we need to find the nearest w_i neighbor of $M_{i-1} = \{u_0, ..., u_{i-1}\}$. To do so, we need to find the nearest w_i-neighbor of u_j for every node $u_j \in M_{i-1}$, and this takes $O(i \log |U(w_i)|)$ using the TVP index. Then we need to find the distance between every pair of nodes u_j, v_k, where u_j is a node in M_{i-1} and v_k is the nearest w_i-neighbor of some node in M_{i-1}. Using the Dewey codes, the distance between u_j and v_k can be found in constant time, thus in the worst case, it requires $O(i \log |U(w_{i+1})| + i^2)$ to find the nearest w_i-neighbor of M_{i-1}.

Given a query $Q = \{w_0, ..., w_l\}$, let X_i be the total number of MCs in $validNodesHM_i$. For each MC in $validNodesHM_{i-1}$, it costs $O(i \log |U(w_{i_k})| + i^2)$ to find its nearest w_{i_k}-neighbor. Let $Y_i = \sum_{w_{i_k} \in var_e(w_i)} (i \log |U(w_{i_k})| + i^2)$, it costs $O(X_{i-1} \cdot Y_i)$ to generate $validNodesHM_i$ in each iteration. As a result, the total runtime is $O\left(\sum_{i=1}^{l} X_{i-1} \cdot Y_i\right)$ in the worst case.

4   Experiments

In this section, we present and analyze the experimental results.

4.1   Experimental Setup

All our experiments were carried out on a PC with Intel Core i5 3.20GHz CPU and 4GB RAM. The operating system was Windows 7, and we used the Berkeley DB Java Edition. The algorithms were implemented using Eclipse with JDK 1.6. We used two real world datasets in the experiments: (1) **DBLP** is a snapshot of the DBLP database. Its size is 400MB, max depth is 6, and average depth is 5.58; (2) **SigmodRecord** is an XML file which contains SIGMOD Records in February 2007. Its size is 467KB, max depth is 6, and average depth is 5.14.

For query cleaning, finding the *ground truth* is difficult [3,6]. We first designed a set of initial clean queries from keywords that occur in the entity nodes, such as

Table 1. Sample Queries

Dataset	Clean	Dirty
DBLP	rose architecture fpga	rase architecture fpga
	harry role processing	garry rale procesing
Sigmod	logical modeling arie	logichal modeling ariel
	compiler david john	compile davis joan

title, *author*, and *volume*. We then used random edit operations ($\epsilon = 1$) to obtain dirty queries from them. We generated 50 clean queries and 50 dirty queries on each dataset, some query samples are given in Table 1. All the queries can be found at http://www.ict.griffith.edu.au/~jw/ER12/query.xls. Like XClean, we employed human assessors to manually identify the *ground truth*.

4.2   Algorithms and Measures

Because XClean is currently the only work on XML keyword query cleaning, we use it as the baseline of comparison. To measure the effectiveness of algorithms, we use the same measurements that XClean uses:

MRR (Mean Reciprocal Rank) is defined as $\text{MRR} = \frac{1}{|\mathcal{Q}|} \cdot \sum_{Q \in \mathcal{Q}} \frac{1}{rank(Q_g)}$, where $\mathcal{Q}$ is a set of queries, and $rank(Q_g)$ is the rank of the ground truth Q_g [6]. The larger the value of MRR is, the better the quality is.

Precision@N is defined as $precision@N = \frac{A}{|\mathcal{Q}|}$, where A is the number of queries whose top-N suggestions contain the ground truth, which indicates what percentage of given queries that users will be satisfied if they are presented with at most N suggestions for each query [6]. The higher Precision@N is, the better the quality.

The MRR and precision@N for XClean and our algorithm are shown in Fig. 2, Fig. 3, and Fig. 4.

4.3   Analysis

Effectiveness. Our algorithm works perfectly with clean queries on both datasets. For XClean, depending on the query semantics used, it may return low quality suggestions. For instance, when the *SNT* semantics is used on DBLP, XClean ranks some clean queries lower than they should be. For example, consider the candidate $C = \{harry, role, processing\}$. There is an *inproceedings* node matching all the keywords. However, type *article* is returned as the *SNT* (when r in Formula (6) of [1] is set to 0.75). Since there is no *article* that matches all the keywords, C is ranked very low by XClean. By adopting *SLCA*, this problem can be avoided. However, problems such as that in Example 1 still exist for XClean.

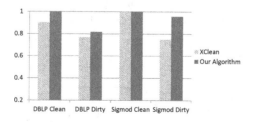

Fig. 2. MRR

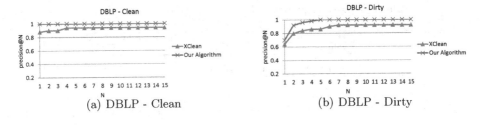

Fig. 3. Precision@N on DBLP with clean and dirty queries

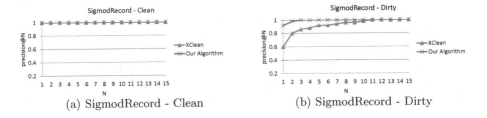

Fig. 4. Precision@N on SigmodRecord with clean and dirty queries

Neither algorithm achieves a close-to-1 MRR value for dirty queries on the two datasets. Other reasonable candidates could be ranked higher than the clean queries. For example, for the clean query[2] and Q_C ={$roome, computer, measurement$}, when its dirty query Q_D ={$roame, compute, measureament$} is issued to DBLP, candidate C_1 ={$route, computer, measurement$} is ranked first, followed by Q_C, since there is an article which has the keywords $route, computer,$ and $measurement$ in its title.

As shown in Fig. 3 and Fig. 4, our algorithm outperforms XClean in effectiveness, especially on the SigmodRecord dataset which possesses a more complicated structure with different entity types in different levels. In our experiments, all top ranked candidates by our approach return meaningful results. By adjusting the distance threshold $maxLen$, we can filter out candidates of poor quality that could result in empty or uninteresting results. For example, consider the dirty query Q_D ={$volume, 28, reconciliation, smeth$}. The suggested queries are given in Table 2. In the whole Sigmod document, the frequencies for keywords $sheth, seth$ and $smith$ are 13, 6 and 3, respectively. Within the issue whose volume is 28, the frequencies for the three keywords are 2, 1 and 1, respectively. XClean ranks candidates C_1, and C_2 higher than candidate C_3. For C_1 and C_2, the closest relationship between $reconciliation$ and $Sheth/Seth$ is within issue 28, where there is one article whose title contains $reconciliation$ and another article which has one author named $Sheth/Seth$.

In our algorithm, C_3 is ranked first. For C_3, the minimum distance between $reconciliation$ and $smith$ within a same issue is 5. In issue 28, there is

---

[2] There is an article with author $roome$ and title containing $computer$ $measurement$.

Table 2. Suggested queries of {*volume, 28, reconciliation, smeth*}

Rank by XClean	Suggested Query	$minD$	Rank by Our Algorithm
1	C_1: volume 28 reconciliation sheth	14.2	2
2	C_2: volume 28 reconciliation seth	14.2	3
3	C_3: volume 28 reconciliation smith	12.5	1

an article whose title contains *reconciliation* and which has an author named *Smith*. Compared with XClean, our algorithm measures the relationship among keyword matching nodes to make sure meaningful candidates are ranked higher.

Efficiency. With the edit distance ϵ set to 1, the average runtime of the 100 queries for each of the two datasets is given in Table 3. Our algorithm significantly outperforms XClean[3], especially on the DBLP dataset. We ran both algorithms on DBLP datasets of different sizes, the result is shown in Fig. 5. Both algorithms appear scalable with the data size. In our experiments, we found that the runtime varies dramatically with the queries. The runtime of our algorithm not only depends on the number of nodes matching $var_\epsilon(w_i)$ for each w_i in query Q, but also depends on how fast the search scope shrinks. In our experiments, we generated the variants of each keyword with the same edit distance ϵ. However, in practical application, it is more reasonable to set different ϵ for keywords of different lengths.

Table 3. Average runtime in milliseconds over 100 queries, $\epsilon = 1$

Dataset	XClean	Our Algorithm
DBLP 400MB	450.12	7.96
Sigmod 467KB	0.76	0.33

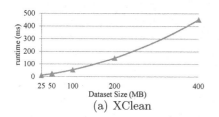

(a) XClean

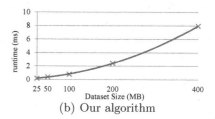

(b) Our algorithm

Fig. 5. Average runtime over DBLP of different size, $\epsilon = 1$

---

[3] In [6] the authors assume that keywords shorter than 4 are all clean, so no variants will be generated, which partly explains why the time reported in [6] is much shorter than what we found in our experiments.

5   Conclusion and Discussion

In this paper, we proposed a distance-based spelling suggestion method to return a list of suggested queries for an issued keyword query for XML data. The key idea is to calculate the distance among keyword matching nodes for a query to measure the relationship among those nodes. The closer the relationship is, the more meaningful the query is. Experiments have shown that our approach not only provides higher-quality suggested queries, but is also much faster than XClean.

One might wonder why we have not adopted the probability-based approach of XClean and simply improve the ranking by taking the distance between keyword matching nodes into consideration. There are two reasons for this. First, computing the minimum distances takes time, and therefore, although it may improve the ranking quality, it would make it slower than XClean. By adopting a different approach, we are able to achieve better efficiency as well as better effectiveness. Second, XClean returns different suggested queries for different keyword search semantics (e.g., SNT, VLCA). The quality of suggested queries is thus dependant on, for instance, whether the correct SNT can be inferred. If an incorrect SNT is inferred, then all suggested queries will be of poor quality, and this problem cannot be fixed by taking distance into consideration in the probability-based approach.

Acknowledgement. This work is supported by the Australian Research Council Discovery Grant DP1093404.

References

1. Bao, Z., Ling, T.W., Chen, B., Lu, J.: Effective XML keyword search with relevance oriented ranking. In: ICDE, pp. 517–528 (2009)
2. Bocek, T., Hunt, E., Stiller, B.: Fast Similarity Search in Large Dictionaries. Technical Report ifi-2007.02, Department of Informatics. University of Zurich (April 2007)
3. Cucerzan, S., Brill, E.: Spelling correction as an iterative process that exploits the collective knowledge of web users. In: EMNLP, pp. 293–300 (2004)
4. Guo, L., Shao, F., Botev, C., Shanmugasundaram, J.: XRANK: Ranked keyword search over XML documents. In: SIGMOD Conference, pp. 16–27 (2003)
5. Liu, Z., Walker, J., Chen, Y.: A semantic XML search engine using keywords. In: VLDB, pp. 1330–1333 (2007)
6. Lu, Y., Wang, W., Li, J., Liu, C.: XClean: Providing valid spelling suggestions for XML keyword queries. In: ICDE, pp. 661–672 (2011)
7. Pu, K.Q., Yu, X.: Keyword query cleaning. PVLDB 1(1), 909–920 (2008)
8. Tao, Y., Papadopoulos, S., Sheng, C., Stefanidis, K.: Nearest keyword search in XML documents. In: SIGMOD Conference, pp. 589–600 (2011)
9. Termehchy, A., Winslett, M.: Using structural information in XML keyword search effectively. ACM Trans. Database Syst. 36(1), 4 (2011)
10. Xu, Y., Papakonstantinou, Y.: Efficient keyword search for smallest LCAs in XML databases. In: SIGMOD Conference, pp. 537–538 (2005)

Cross-Language Hybrid Keyword and Semantic Search

David W. Embley[1], Stephen W. Liddle[2], Deryle W. Lonsdale[3],
Joseph S. Park[1], Byung-Joo Shin[4], and Andrew J. Zitzelberger[1]

[1] Department of Computer Science,
[2] Information Systems Department,
[3] Department of Linguistics and English Language,
Brigham Young University, Provo, Utah 84602, U.S.A.
[4] Department of Computer Science & Engineering
Kyungnam University, Kyungnam, Korea

Abstract. The growth of multilingual web content and increasing internationalization portends the need for cross-language information retrieval. As a solution to this problem for narrow-domain, data-rich web content, we offer ML-HyKSS: **M**ulti**L**ingual **Hy**brid **K**eyword and **Se**mantic **S**earch. The primary component of ML-HyKSS is a collection of linguistically grounded conceptual-model instances called extraction ontologies. Extraction ontologies can recognize keywords and applicable semantics; when coupled with cross-language mappings at the conceptual level, they enable cross-language information retrieval and query processing. Our experimental results are promising, yielding good results for cross-language information retrieval with contrasting languages, application content, and cultures.

Keywords: Cross-language information retrieval, hybrid keyword and semantic search, multilingual web content, extraction ontologies.

1 Introduction

With the growth of multilingual web content, it is becoming increasingly important to enable users whose native language is A to find useful information in web pages written in a language B, which they do not know. As an example, consider a person in the USA who wishes to send flowers for a friend's funeral in France and thus wishes to enter the search-engine query "address for funeral of Mrs. Gabrielle DUPIEREUX of Braine-le-Comte" or who wishes to send flowers in place of making a condolence call to a family in Korea and thus wants to search for "mortuary location and interment date for Minhaegyeong's father".

For these queries and many others like them, answers are on the web, but in a language the user does not know. Furthermore, keyword search, though helpful, is insufficient. The keywords "funeral", "Mrs. Gabrielle DUPIEREUX", and "Braine-le-Conte" might be sufficient to return a page with the address—although, of course, with translations "funérailles" for "funeral" and "Madame" for "Mrs."

P. Atzeni, D. Cheung, and R. Sudha (Eds.): ER 2012, LNCS 7532, pp. 190–203, 2012.

Ideally, however, the search engine should return the address highlighted within the identified source page. Here, "address" is a meta-word rather than a keyword, and should therefore be recognized semantically. In the Korean example, none of the words in the query is likely to be helpful for keyword search—even when translated. Many of them, however, when translated would semantically relate to a Korean obituary, which would contain information about the hospital for the condolence call, the day of the burial, and the relatives.

These queries call for CLIR (Cross-Language Information Retrieval) [1,2]. The typical approach to CLIR consists of query translation followed by monolingual retrieval, where systems perform query translation with machine-readable, bilingual dictionaries and machine translation. Our approach to CLIR differs significantly, in that we translate queries only after having semantically conceptualized them and after having separated semantic and keyword components. Semantic conceptualization requires a linguistically grounded conceptual-model instance as its key component, which limits the approach to applications that are data-rich and narrow in scope.

We call our CLIR engine ML-HyKSS (**M**ulti-Lingual **Hy**brid **K**eyword and **S**emantic **S**earch, pronounced "M-L-hikes"). Like search engines, ML-HyKSS assumes the existence of an indexed document collection. Indexes for ML-HyKSS, however, are not just for keywords, but also for recognized semantic concepts. In our prototype implementation, we use Lucene for keyword indexing and extraction ontologies [3,4], which are linguistically grounded conceptual models, for semantic indexing. In Section 2 we describe extraction ontologies and explain how we semantically index a document collection. ML-HyKSS processes a query for a single language, as we also explain in Section 2, by applying extraction ontologies to the query itself to extract semantic constraints and to isolate non-semantic keywords resulting in a formal conceptualized query. ML-HyKSS then locates relevant documents by matching constraints and keywords of the formal conceptualized query with the semantic and keyword indexes. For cross-language query processing ML-HyKSS transforms a formal conceptualized query Q_A in language A to a formal conceptualized query Q_B in language B. ML-HyKSS can then apply Q_B to the already indexed language-B document collection. Answer transformation is an inverse mapping. In Section 3 we describe the mappings, and thus both query and answer translation. We present experimental results in Section 4 and conclude in Section 5 by summarizing our contributions:

1. Development of cross-language query translation at the conceptual level as an effective alternative to the more traditional language-level translation.
2. Implementation of a prototype system showing the viability of cross-language hybrid keyword and semantic search over diverse languages and applications.

Our earlier work on multilingual ontologies focused on architecture [5]. The work here includes hybrid keyword and semantic search queries—a significant step beyond the types of queries proposed earlier. The work also includes an implementation that establishes the viability of hybrid query processing and conceptual-level query translation.

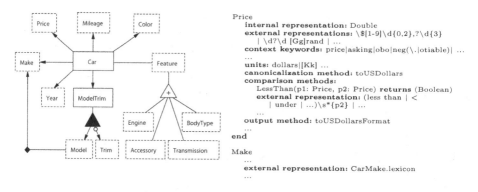

Fig. 1. Car Ad Model Instance   **Fig. 2.** Car Ad Data Frames

2 Query-Processing with Extraction Ontologies

An *extraction ontology* [3,4] is a conceptual model augmented linguistically to enable information extraction. (To make the paper self-contained we briefly introduce extraction ontologies in Section 2.1.) Extracted information constitutes a database structured with respect to the schema induced by a specified conceptual-model instance. When coupled with links into the document collection for each fact extracted, the database constitutes a semantic index. Then with the semantic index in place as well as a standard keyword index, ML-HyKSS can process hybrid semantic/keyword free-form queries[1] (see Section 2.2).

2.1 Extraction Ontologies

The primary components of an extraction ontology are object sets, relationships sets, constraints, and linguistic recognizers. Figure 1 shows a conceptual-model instance for a car-ad application with its object and relationship sets and its constraints, and Figure 2 shows part of two linguistic recognizers—one for *Price* and one for *Make*. Together they constitute an extraction ontology.

The conceptual foundation for an extraction ontology is a restricted fragment of first-order logic. Each object set (denoted graphically in Figure 1 by a rectangle) is a one-place predicate. Each predicate has a *lexical* or a *non-lexical* designation: lexical predicates (denoted by dashed-border rectangles) restrict domain-value substitutions to be literals in domain sets, and non-lexical predicates (denoted by solid-border rectangles) restrict substitutions to be object identifiers that represent real-world objects. Each n-ary relationship set (denoted graphically by a line with n object-set connections) is an n-place predicate. Black diamonds on relationship sets denote prepopulated, fixed relationship sets.

---

[1] The hybrid search systems we are aware of require structured queries rather than free-form queries (e.g., [6,7]), require a keyword-based structured query language (e.g., [8]), or require queries that are an extension of formal conjunctive queries (e.g., [9,10]). None is multilingual.

Black triangles denote aggregation groupings of relationship sets in an is-part-of hierarchy. Constraints are of three types: (1) referential integrity including optional/mandatory participation (with optional participation denoted by an "o" on the optional side) (2) functional (denoted by an edge with an arrowhead) and (3) generalization/specialization constraints (denoted by a white triangle).

Like Buitelaar et al. [11], we linguistically ground ontologies, turning a conceptual specification into an extraction ontology. Each object set has a *data frame* [4], which is an abstract data type augmented with linguistic recognizers that specify textual patterns for recognizing instance values, applicable operators, and operator parameters. Figure 2 shows part of the data frames for the object sets *Price* and *Make* in Figure 1. Although any kind of textual pattern recognizer is possible, our implementation supports regular expressions as exemplified in *Price* and lexicons as exemplified in *Make*, or combinations of regular expressions and lexicons. Each relationship set may also have a data-frame recognizer. Relationship-set recognizers reference and depend on data-frame recognizers for each connected object set. In addition, relationship sets may be prepopulated with with a fixed set of relationships that can provide additional context to aid in linguistic grounding. Thus, in Figure 1, the *Make* "Honda" would be additional context information for *Model* "Accord" in the *Car Ad* ontology.

In a data frame, the **internal representation** clause indicates how the system stores extracted values internally, and the **external representation** clause specifies the instance recognizers. The textual distance of matches from *context keywords* helps determine which match to choose for ambiguous concepts within an ontology. The string "25K", for example, could be a *Mileage* or a *Price* but would be interpreted as a *Price* in close context to words such as *asking* or *negotiable*. A **units** clause expresses units of measure or value qualifications that help quantify extracted values. A **canonicalization method** converts an extracted value and units to a unified internal representation. Once in this representation, **comparison methods** can compare values extracted from different documents despite being represented in different ways. These methods can correctly confirm, for example, that "$9,500" is less than "12 grand". The **external representation** clause within a method declaration recognizes text indicating method applicability. The recognizer for the *LessThan* method in Figure 2, for example, identifies comparison phrases in queries like "under 12 grand". The *p2* within curly braces denotes the position in the regular expression of the expected appearance of a *Price* for parameter *p2*. The **output method** is responsible for displaying internally-stored values to the user in a readable format.

2.2 Extraction-Ontology-Based Query Processing

Before query processing begins, ML-HyKSS preprocesses a document collection and creates a keyword index (with Lucene) and a semantic index (with extraction ontologies). ML-HyKSS applies extraction ontologies to text documents to find instance values in the documents with respect to the object and relationship sets in the ontology. The extraction process uses the linguistic recognizers in data frames and the constraints of the conceptual-model structure along with several heuristics to extract instance values. Past work shows that extraction ontologies perform well in terms of precision and recall for the extraction task when documents are rich in recognizable constants and narrow in ontological breadth [3,4]. ML-HyKSS returns its semantic index as RDF triples that contain, in addition to the extracted data values, internal values converted to appropriate internal representations by data-frame canonicalization methods, standardized external string representations obtained by data-frame output methods, and information identifying the document and location within the document of extracted text.

To explain and illustrate how ML-HyKSS processes queries, we consider as a running example, the query "Hondas in 'excellent condition' under 12 grand". When ML-HyKSS applies the extraction ontology in Figures 1 and 2 to this query, the data frames recognize "Hondas" and "under 12 grand" as constraints on *Make* and *Price* respectively. Given the nodes *Make* and *Price* in the conceptual-model graph in Figure 1, ML-HyKSS generates a formal SPARQL query in a straightforward way: it generates joins over the edges *Car-Make* and *Car-Price* and also appropriate selection conditions in *FILTER* statements, allowing constraint satisfaction to be *OPTIONAL*, as an open-world assumption requires.

Query generation for expected queries over expected ontologies is straightforward. Like keyword-only queries in which users query for pages that satisfy all, or as many of the keywords as possible, we expect that for hybrid queries users wish to maximize the semantic constraints that are satisfied. Thus, we generate conjunctive queries and always allow constraint satisfaction to be optional. Further, since we expect most ontologies for ML-HyKSS applications to have a simple acyclic structure (or acyclic after excluding prepopulated fixed relationship sets like *Model-Make* in Figure 1), ML-HyKSS can generate queries like it does for the running example in a straightforward way: join over ontology edges that connect identified nodes, and filter conjunctively on identified conditions. For cycles that lead to multiple paths between the same two nodes, such as if the conceptual-model instance in Figure 1 had a second edge between *Car* and *Feature* denoting features that could be added (as opposed to ones the car already has), we expect that relationship-set data-frame recognizers would be able to identify clues in queries that would select one path or another, or in the absence of identified clues, interact with the user to disambiguate the query. Also, like typical search engines, we provide advanced search capabilities for users who wish to pose queries that involve disjunctions and negations.

To add keyword search, making query processing hybrid, ML-HyKSS removes, in addition to stopwords, all meta-words denoting application concepts and all phrases denoting semantic constraints, except equality constraints. For our running example, ML-HyKSS thus removes "in" as a stopword and "under 12 grand" as an inequality constraint, but not "Hondas", which denotes the equality comparison constraint *Make = Honda*. It would also have removed "price" or "cost" as meta-words if they had appeared. Note that it is the plural, "Hondas", which is a keyword, even though the *Make* data-frame recognizer converts the plural to a singular for its use in semantic equality comparisons. The keywords for our running example are thus "Hondas" and the phrase "excellent condition".

I understood: show me Vehicle.Make and Vehicle.Price where Vehicle.Make = Honda and Vehicle.Price < $12,000
Keywords: Hondas "excellent condition"   Advanced Search

Rank	Document	Keywords	Make	Price
1	2002 Honda Accord LX Sedan(highlighted)	excellent condition(1)	Honda	$4,995
2	2002 Honda Accord LX Sedan(highlighted)	excellent condition(1)	Honda	$4,995
3	1997 Honda Accord EX(highlighted)	hondas(1)	Honda	$3,200
4	1993 Honda Accord White(highlighted)	excellent condition(1)	Honda	$3,100
5	2002 Honda Accord!!! 106k 5,995(highlighted)		Honda	$5,995

Fig. 3. Query Result

2002 Honda Accord LX Sedan - $4995 (Orem)

                                                                          miscategorized

                                                                          prohibited

Date: ****-**-**, **.****** MST   spam/overpost
Reply to: see below
                                                                          best of craigslist

Excellent condition! 158k miles with 2.3L Vtec engine, automatic transmission, new inspections and timing belt, power windows, locks, and mirrors, tilt wheel and cruise control, air conditioning, nice Kenwood CD stereo system. Restored title.

Fig. 4. Highlighted Result

ML-HyKSS uses its keyword and semantic indexes to identify documents that satisfy its semantic constraints and contain its keywords. It then ranks returned documents by the simple linear interpolation formula: $keywordScore \times keywordWeight + semanticScore \times semanticWeight$ where the scores measure the degree of match with a document and the weights measure the relative contribution of keywords and semantic constraints in the query. Figure 3 shows the top-five ranked results when ML-HyKSS processes the running query over a collection of car ads from craigslist.com. Figure 4 shows the result of clicking highlighted for the top-ranked document.

3   Cross-Language Query Processing

Similar to the work of Dorr et al. [12], we adopt a star or pivot or interlingua-based architecture for cross-language query processing with a central language-agnostic ontology A. Each application has a single central ontology, together with, for each language/locale[2] L an extraction ontology and an L-to-A/A-to-L mapping specification. Omitting its linguistic grounding, the conceptual-model instance in Figures 1 and 2 is a language-agnostic central ontology.[3] Language-agnostic ontologies are not extraction ontologies because we do not ground them

---

[2] For any language, different locales may use different terminology and thus warrant different extraction ontologies—e.g., British English vs. American English.

[3] Although we could use any real or artificial language to represent a language-agnostic ontology, similar to [13], we use English.

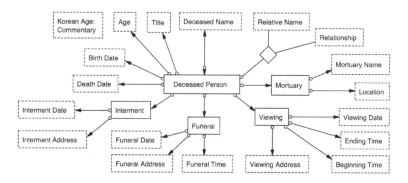

Fig. 5. Obituary Ontology

linguistically. They do, however, have full conceptual-model instances with all of an application's object and relationship sets and constraints, and with **internal representation** declarations, **units** declarations, and **method** signatures. The obituaries ontology in Figure 5 is a second example, which we use as we continue to explain how cross-language hybrid query processing works.

For narrow-domain, data-rich applications, we expect extraction ontologies for different languages/locales to be similar, but not identical. In French obituaries, for example, titles such as *Madame* and *Monsieur* regularly appear but are rare in English. As we added French, we therefore added the object set *Title*, which was not part of the original English ontology. In Korea, friends and family make condolence calls usually at the hospital where the deceased body lies in wait between death and interment. Thus, as we added Korean, we also added the information for condolence calls—*Mortuary* and its *Name* and *Location*. Since users may wish to query about all concepts in all languages/locales, we propagate additions to all existing extraction ontologies as well. We thus keep all ontologies structurally synchronized—structurally identical.

Cross-language mappings are compositional through the central ontology. For extraction-ontology languages L_1 and L_2 and central language-agnostic ontology A, the L_1-to-L_2 mapping is the composition of the L_1-to-A and A-to-L_2 mappings. Thus, adding a new language to the system is a linear pay-as-you-go endeavor. Further, because data-rich, narrow-domain ontologies are largely alike in all languages and cultures, and because cross-language lexical resources have been developed over the years and are readily available on the web, the cost of adding a new language/locale to the system is much less than the cost of developing extraction ontologies and mappings from scratch.

Mappings are of several types, and each has its pecularities and its linguistic resources to aid in rapid development:

– *Lexicons*. Lexicon mappings substitute one word by another, or one word by a small number of others. For common concepts such as colors (e.g., for car ads in Figure 1), corresponding translations are available in cross-language dictionaries. Interestingly, these mappings are not always one-to-one as many might expect (e.g., "blue" in Korean is 파랑색 and 파란색 and 청색). Concept words not usually in dictionaries such as makes and models of cars are often readily available in lists on the web (e.g., pull-down menus in `http://paruvendu.fr` contain all French make/model combinations, and tabs in `http://www.encar.com` lead to Korean makes and models). Further, identity translations to English are common for many of these words.

— *Units and Measures.* ISO standard conversion formulas for units and measures are commonly available and coding them is straightforward. In our prototype implementation for car ads and obituaries, we use kilometers for mileage, integers for car years and person ages, julian-calendar specifications for dates, and a 24-hour clock for time.

— *Currency.* Because services exist that directly convert amounts in one currency to amounts any other currency, mappings for currency conversions should be an exception to the star-architecture composition rule. However, since the service[4] we were able to conveniently use for our prototype implementation only has conversion rates for US Dollars, we also implement currency conversion as a composition.

— *Transliteration.* Like direct conversion among currencies, transliteration mappings are likely to be best if they map directly from one language to another. We, however, are unable to find a general transliteration service. In our current application, we only need transliteration to and from Korean for English and French and thus use a Hangul/Latin-Language transliterator.[5]

— *Keywords.* Since keywords can be any word or quoted phrase, we use a general translation service, the Bing translator[6] for our implementation. As it should be, the translation is direct, not indirect through the central ontology.

— *Commentary.* Ontologies may contain free-form commentary to explain unfamiliar concepts, such as Korean age. In our implementation, we use the Bing translation service, which translates the Korean explanation for *Age*:

> 한국의 나이 계산법은 갓 태어난 아이에게 한 살을 부여한다. 그 이후로는, 나이가 새해에 변경된다. 이 방법으로는 12월 31일에 태어난 아이는 다음날에 나이가 두 살로 변경된다. 이 나이 계산법은 공식적으로(법적으로) 사용되지 않으며, 일상적으로는 한국에서 널리 통용되고 있다.

to:

> Korean age reckoning is a newborn child was one year old. Since then, the age is changed in the new year. In this way, children born on December 31, the next day is changed to two years of age. In this age of reckoning is not used officially (and legally) on a daily basis, and widely accepted in Korea.

Translations are usually understandable but not ideal. When important, human translators can provide better translations in "pay-as-you-go" fashion.

By making use of structural correspondences and defined mappings, query transformation from language L_1 to language L_2 can occur at the "deep" conceptual level, as opposed to the more common textual "surface" level. When an extraction ontology interprets a query, it generates an internal representation of the query before generating a formal SPARQL query. The internal representation consists of (1) an acyclic join path in the conceptual-model graph connecting all nodes

---

[4] `http://raw.github.com/currencybot/open-exchange-rates`

[5] `http://sori.org/hangul/conv2kr.cgi`

[6] `http://api.microsofttranslator.com/V2/Http.svc/Translate`

relevant to the query, (2) an identification of lexical object sets on the path, (3) for each constraint, its Boolean method along with canonicalized values for identified parameters, and (4) keywords and keyword phrases. For our running example—"Hondas in 'excellent condition' under 12 grand"—the join path from Figure 1 is { Car-Price, Car-Make}, the lexical object sets are Price and Make, the constraints are LessThan(x, 12000) and Equal('Honda'), and the keywords are Hondas and the phrase 'excellent condition'. Since all extraction ontologies for an application are structurally identical, the transformed join paths, identified lexical object sets, and methods are immediate. For values and keywords, the mappings provide the transformation. Thus the internal representation for the running query in French consists of referenced lexical object sets Marque and Prix, constraints Marque="Honda" and Prix < 9148 €, and keywords Hondas and 'excellent état'. For Korean the internal representation consists of object sets 가격 and 제조사, constraints 제조사 = 혼다 and 가격 < 1340만원, and keywords 혼다 and '좋은 상태가'. When value mappings are one-many, we generate disjunctions; thus, if an English query asks for a blue car, the Korean constraint becomes the disjunction (색상 = 파랑색 ∨ 색상 = 파란색 ∨ 색상 = 청색). Since all internal representations map directly to hybrid SPARQL and keyword queries in the same way, ML-HyKSS immediately obtains and executes the transformed query.

For answer values returned, we use the mappings to transform values and keywords back into the original language. For the query in the introduction, "mortuary location and interment date for Minhaegyeong's father", the answer returned is "Soonchunhyang, Hospital, Hannam, Seoul" on "November 2nd". The date is a units compositional mapping from 11월 2일 to the julian date 306 (the year 2011 assumed) and then to "November 2nd", written as a standard date without the year in English as specified in the **output method** of the English extraction ontology. Mortuary name and location are direct Hangul-to-Latin transliterations (not quite correct, as "Hannam" should be "Hannamdong"). For the query in the introduction, "address for funeral of Mrs. Gabrielle DUPIEREUX of Braine-le-Comte", the answer returned is "l'église Saint-Géry de Braine-le-Comte". Note that neither names nor addresses require any translation for French-to-English.

Advanced queries are also possible in ML-HyKSS. For example, the filled in form in Figure 6 lets a user augment the running query to find either a Honda or a Toyota, but not an Accord, that is red or yellow and still under 12 grand and in 'excellent condition'. HyKSS generates the form from the structure of the extraction ontology along with its declared methods. (Thus, to enable advanced-form capabilities, nothing need be done beyond declaring an extraction ontology.) ML-HyKSS initializes the form with values obtained from a user query and writes them in the entry blanks of the form according to the ontology's **output methods**—thus "Honda" appears as the Make while "Hondas" appears as a keyword, and "$12,000" appears for the "<" Price operator (not "12 grand"). After initialization, a user can click on ‹OR› for Make and enter "Toyota" and on ‹OR› for Color and enter "red" in one field and "yellow" in the other. Clicking on the NOT checkbox for Model and entering "Accord" adds the exclusion. Interestingly, users can enter data values without concern for format because ML-HyKSS invokes its data-frame recognizers to interpret user entries.

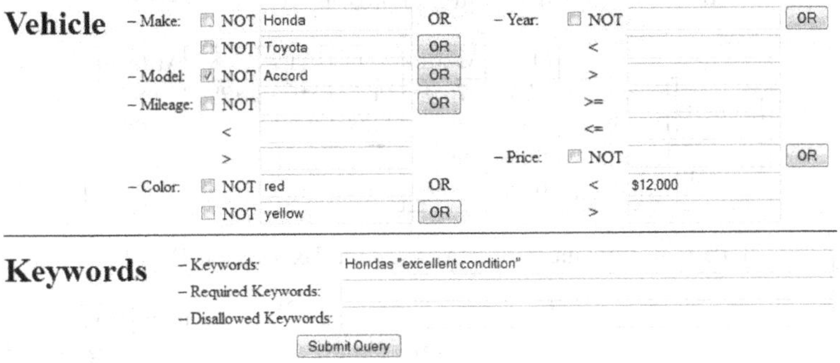

Fig. 6. Advanced Form Query. (Generated from the *Vehicle* extraction ontology we used in our experimental work—not from the extraction ontology in Figures 1 and 2.)

Like mappings, extraction ontologies also have linguistic resources to aid in rapid development. For example, to quickly obtain a dictionary for French given names, we took the list from a book of suggested French baby names.[7] Further, since extraction ontologies are often similar, many regular-expression recognizers can be readily adopted or adapted. For **context keywords**, WordNet synsets are useful. For example, instead of "interment date" a user might say "burial date". The WordNet synset for "interment" is {"burial", "entombment", "inhumation", "interment", "sepulture"}, which includes "burial" and thus, ML-HyKSS will recognize the query as asking for the interment date in the extraction ontology in Figure 5. In general, we can use techniques for automatic query enhancement [14] to automatically populate **context keywords** clauses.

4   Experimental Results

Cross-language query-processing accuracy depends on (1) extraction accuracy in all languages when initially indexing the semantics in a document collection with respect to an application ontology and (2) cross-language query transformation so that nothing is lost or spuriously added in the transformation.

4.1   Extraction Accuracy

To check extraction accuracy for French, we gathered 500 car ads from on-line sites in France and Canada. The car ads in www.craigslist.fr appear as free-form French, but the rest are mostly semi-structured. For obituaries, we gathered 1500 obituaries from several different online sites in France, Belgium, Canada, and Switzerland—all free-form, but largely conforming to informal conventions. Similarly, for Korean, we gathered 430 car ads from 13 different sites (all semi-structured), and 502 obituaries from 36 different sites (all free-form). We randomly selected about 100 of each of the four combinations to constitute

---

[7] http://www.journaldesfemmes.com

Table 1. Car Ad within-language Extraction Results

		Make	Model	Year	Price	Color	Mileage
French	Recall	87%	76%	96%	89%	82%	98%
	Precision	65%	67%	90%	95%	47%	92%
Korean	Recall	99%	99%	100%	100%	100%	95%
	Precision	99%	99%	100%	100%	100%	95%

Table 2. Obituary within-language Extraction Results

		Title	Name	Death Date	Funeral			Mortuary Name	Relative Name & Relation
					Date	Time	Place		
French	Recall	76%	42%	80%	69%	43%	38%	N/A	—
	Precision	99%	63%	88%	70%	30%	83%		
Korean	Recall	N/A	97%	97%	50%	50%	100%	99%	97%
	Precision		97%	97%	100%	100%	67%	94%	94%

validation and blind test sets (respectively 20 and 80 of the 100) and used the rest for training—training in the sense that we looked at many of them as we linguistically grounded our ontologies.

Tables 1 and 2 show the results. The car ads domain is ontologically narrow, and accordingly our extraction ontologies perform quite well on this domain. Precision and recall for Korean car ads are high because these ads mostly have a regular structure, allowing our Korean expert to quickly tune the extraction ontology. Results for French car ads are lower both because French ads are more free-form and because we were not able to spend enough time with our French expert tuning the extraction ontology (as one consequence, we did not finish relative-name extraction—hence the empty cell in Table 2). Overall, the numbers are in line with what we have come to expect in this domain [4]. The obituaries domain is much broader and extraction is more challenging—particularly for names and places. Even so, our Korean expert was able to quickly tune the extraction ontology, and performance for most concepts was remarkably high. French extraction was hampered by greater variability and complex sentence structures. For example, there are only 187 names in our Korean surname lexicon, compared with 228,429 in our French surname lexicon, which partially explains the relatively high performance for Korean name extraction. Most Korean obituaries do not mention a funeral, and in our test set there were only two examples of funerals. Our extractor did well with one of the two declared funerals, so the 50% recall we report for funeral date and time represents only one missed concept for the entire corpus. Another cultural difference is that the mortuary name is important in a Korean obituary because that is where friends and family make condolence calls. In contrast, French obituaries refer to viewings, funerals, and interments, much like typical English obituaries. Since mortuary name does not appear in the French obituary ontology, and title does not appear in the Korean ontology, we mark their cells as "N/A". Performance for concepts not listed in Table 2 (e.g. viewing, interment, birth date) is similar to the performance for concepts that are listed.

Table 3. Cross-Language Query Transformation Results

Car Ad Queries	Recall			Precision		
	σ	π	κ	σ	π	κ
French-to-English	77%	86%	100%	81%	90%	74%
Korean-to-English	98%	100%	100%	93%	99%	52%

4.2 Cross-Language Query Accuracy

Obtaining a query set for hybrid keyword/semantic search is not as straightforward as it might seem. Search-engine users learn quickly to adapt their queries to the capabilities of a search engine. Users quickly learn not to ask queries like our running example when they see that a top returned result is something far removed from what they want, like grand pianos in 'excellent condition'. Nevertheless, it is possible to explain, and in our case even show how they work, and ask subjects to imagine queries that should work.

To obtain query sets, we asked the students in two senior-level database classes to generate two car-ad queries they felt an earlier demo version of a free-form query processor (not ML-HyKSS) interpreted correctly and two queries they felt the the demo version misinterpreted, but should have interpreted correctly. The students generated 137 syntactically unique queries, of which 113 were suitable for testing ML-HyKSS. To obtain Korean and French queries, we faithfully translated 50 of these 113 into each language.

Table 3 shows the results of interpreting the queries in their respective languages and transforming the internal representation of each query, as understood, into the internal representation of the query in English. In the table the σ columns are for generated database selection operations (Boolean conditions such as *Price* < \$12,000 and *Make* = "Honda" for our running query), the π columns are for generated database projection operations (object sets referenced from which results are returned such as *Price* and *Make* for our running query), and the κ columns are for keywords and keyword phrases ("Hondas" and "'excellent condition'" for our running query). We remark, as explained earlier, that σ and π translations are always correct because ML-HyKSS translates them at the conceptual level by matching methods and object sets respectively, which are necessarily in a one-to-one correspondence. Thus, the less than perfect σ and π results all come from inaccurate within-language query interpretation. It is significant that no σ and π translation errors can occur since these errors are common in cross-language information retrieval systems that translate at the textual language level and then apply information retrieval techniques.

The lower recall and precision for French conditionals (σ) points to a need for better recognizers. For example, we missed recognizing "une 1990 ou moins récente" as the conditional *Year* $\leq$ 1990. Better recognizers, along with more complete synonym sets for ontological concepts, would also increase the recall for requested French results (π). Conditional recognition failures also account for some of the lower keyword (κ) precision, especially for French, as words in missed semantic conditionals remain as possible keywords. Expanded stopword lists in French would remove spurious keywords like "list" and "want". Stopwords in Korean make little sense because most of the standard English-like stopwords are prefixes and suffixes and become part of characters themselves. An attempt to remove them after translation often fails because translations themselves are often poor; e.g., 인, which in our query should translate as "which is"—both English stopwords—instead was translated as "inn" (or "hotel").

5   Concluding Remarks

We have demonstrated the viability of ML-HyKSS—a conceptual-model-based, cross-language, hybrid keyword and semantic search engine. ML-HyKSS indexes both the semantics and the keywords of a data-rich document collection for an application written in language L_1 and allows users to search the collection with free-form queries in language L_2. While not exhaustive in coverage, our prototype demonstrates its ability to work with a diversity of languages and applications. How well ML-HyKSS performs depends on how well it identifies and interprets an application's semantics in a document collection and in user free-form queries. Experimental results show that ML-HyKSS is able to identify and index the semantics in data-rich, narrow-domain French and Korean documents with an average F-measure of about 90% for semi-structured documents (car ads) and about 75% for unstructured documents (obituaries). Interpretation of French and Korean queries have average F-measures of about 94% for identifying semantic constraints, 87% for identifying referenced concepts of interest, and 77% for identifying keywords. Non-semantic keyword translations are often less than ideal, but since ML-HyKSS translates queries at the conceptual level across language-diverse but structurally identical ontologies, its semantic translations are necessarily correct. Hence, results returned are surprisingly accurate.

Although we have accomplished much (built an ML-HyKSS prototype system, enabled advanced-form hybrid search for non-conjunctive queries, struggled through and resolved issues with international encodings, dealt with diverse languages and locales, and learned how to leverage language resources and international standards to mitigate the construction of extraction ontologies), more can still be done. As future work, we plan to enrich semantic-constraint recognition for superlative queries (e.g., "cheapest car") and aggregate queries (e.g., "average age at death") by including additional types of recognition phrases in data-frame methods. We also intend to pursue ideas for automating the construction of interlingual mappings and linguistic-grounding components of extraction ontologies from available language resources.

Acknowledgements. We are grateful for the work of Tae Woo Kim and Rebecca Brinck in annotating Korean and French document sets.

References

1. Olive, J., Christianson, C., McCary, J. (eds.): Handbook of Natural Language Processing and Machine Translation: DARPA Global Autonomous Language Exploitation. Springer (2011)

2. Peters, C., Braschler, M., Clough, P.: Multilingual Information Retrieval: From Research to Practice. Springer (2012)
3. Embley, D.W., Campbell, D.M., Jiang, Y.S., Liddle, S.W., Ng, Y.-K., Quass, D.W., Smith, R.D.: A Conceptual-Modeling Approach to Extracting Data from the Web. In: Ling, T.-W., Ram, S., Li Lee, M. (eds.) ER 1998. LNCS, vol. 1507, pp. 78–91. Springer, Heidelberg (1998)
4. Embley, D.W., Liddle, S.W., Lonsdale, D.W.: Conceptual Modeling Foundations for a Web of Knowledge. In: Embley, D.W., Thalheim, B. (eds.) Handbook of Conceptual Modeling: Theory, Practice, and Research Challenges, ch. 15, pp. 477–516. Springer, Heidelberg (2011)
5. Embley, D.W., Liddle, S.W., Lonsdale, D.W., Tijerino, Y.: Multilingual Ontologies for Cross-Language Information Extraction and Semantic Search. In: Jeusfeld, M., Delcambre, L., Ling, T.-W. (eds.) ER 2011. LNCS, vol. 6998, pp. 147–160. Springer, Heidelberg (2011)
6. Castells, P., Fernandez, M., Vallet, D.: An adaptation of the vector-space model for ontology-based information retrieval. IEEE Transactions on Knowedge and Data Engineering 19(2), 261–272 (2007)
7. Bhagdev, R., Chapman, S., Ciravegna, F., Lanfranchi, V., Petrelli, D.: Hybrid Search: Effectively Combining Keywords and Semantic Searches. In: Bechhofer, S., Hauswirth, M., Hoffmann, J., Koubarakis, M. (eds.) ESWC 2008. LNCS, vol. 5021, pp. 554–568. Springer, Heidelberg (2008)
8. Pound, J., Ilyas, I.F., Weddell, G.: Expressive and flexible access to web-extracted data: A keyword-based structured query language. In: Proceedings of the 2010 International Conference on Management of Data (SIGMOD 2010), Indianapolis, Indiana, pp. 423–434 (June 2010)
9. Wang, H., Tran, T., Liu, C.: CE^2: Towards a large scale hybrid search engine with integrated ranking support. In: Proceedings of the 17th ACM Conference on Information and Knowledge Management (CIKM 2008), Napa Valley, California, pp. 1323–1324 (October 2008)
10. Zhang, L., Liu, Q., Zhang, J., Wang, H., Pan, Y., Yu, Y.: Semplore: An IR Approach to Scalable Hybrid Query of Semantic Web Data. In: Aberer, K., Choi, K.-S., Noy, N., Allemang, D., Lee, K.-I., Nixon, L.J.B., Golbeck, J., Mika, P., Maynard, D., Mizoguchi, R., Schreiber, G., Cudré-Mauroux, P. (eds.) ASWC 2007 and ISWC 2007. LNCS, vol. 4825, pp. 652–665. Springer, Heidelberg (2007)
11. Buitelaar, P., Cimiano, P., Haase, P., Sintek, M.: Towards Linguistically Grounded Ontologies. In: Aroyo, L., Traverso, P., Ciravegna, F., Cimiano, P., Heath, T., Hyvönen, E., Mizoguchi, R., Oren, E., Sabou, M., Simperl, E. (eds.) ESWC 2009. LNCS, vol. 5554, pp. 111–125. Springer, Heidelberg (2009)
12. Dorr, B.J., Hovy, E., Levin, L.: Machine translation: Interlingual methods. In: Brown, K. (ed.) Encyclopedia of Language and Linguistics, 2nd edn. Elsevier (2004)
13. Lonsdale, D.W., Franz, A.M., Leavitt, J.R.R.: Large-scale machine translation: An interlingua approach. In: Seventh International Conference on Industrial and Engineering Applications of Artificial Intelligence and Expert Systems (IEA/AIE 1994), Austin, Texas, pp. 525–530 (May/June 1994)
14. Carpineto, C., Romano, G.: A survey of automatic query expansion in information retrieval. ACM Computing Surveys 44(1) (January 2012)

Mapping Semantic Widgets to Web-Based, Domain-Specific Collections

Scott Britell and Lois M.L. Delcambre

Department of Computer Science, Portland State University
Portland, OR 97207 USA
{britell,lmd}@cs.pdx.edu

Abstract. Domain-specific, online collections and digital libraries are often composed of richly structured content and modern content management systems make it easy to define multiple heterogeneous structures with similar semantics. But developers typically write code for each structure, a repetitive perhaps difficult process. In this paper, we present canonical structures, small schema fragments, consisting of unnamed entities, relationships, and attributes that can be instantiated into domain structures by providing names for all structural elements. Domain structures represent commonly occurring patterns within a given application domain that can be mapped repeatedly, as appropriate, to local structures. We formally define canonical and domain structures, mappings to local structures, and functionality. We have implemented several, generic semantic widgets using our approach in an operational educational web repository.

1 Introduction

The benefits and the challenges of schema integration have been known and studied for decades. Whenever data in various databases share some sort of common semantics, schema integration seeks to provide mappings or correspondences that highlight the shared semantics and provide new capabilities such as an integrated query facility, over the participant databases. Our work is motivated by the emergence of structured web sites—typically implemented using a content management system where the web site developer can typically define any number of content types (analogous to entity types in an Entity-Relationship diagram (ERD)) related by any number of field types that reference other content types (roughly analogous to relationship types in an ERD). This means that even within a single web site hosting content within a single application domain, there can be multiple schemas (e.g., content types and fields that relate content types) to describe similar content.

As one motivating example, we have developed a web-based repository to host robotics educational materials for use in middle and high schools in Oregon and Washington in the US. Since teaching styles can be different, e.g., from a lecture-based approach to a more active-learning-based approach, the structure of the various curricula is different. Given this natural—and beneficial—heterogeneity,

P. Atzeni, D. Cheung, and R. Sudha (Eds.): ER 2012, LNCS 7532, pp. 204–213, 2012.

we are working to provide powerful new capabilities in such a site with these goals: (1) embrace a very broad range of structures (within a single application domain), (2) allow the local content and local structure to dominate (to make the new capabilities easy to use), and (3) specify and implement new capabilities generically—so that they can easily work with future structures that may appear within a site.

Our approach is to define a set of data model fragments called *canonical structures* comprised of unnamed entities, relationships, and attributes. While unnamed, these fragments carry certain semantics such as relationship type, e.g., aggregation. Canonical structures can be instantiated as many times as desired within an application domain as *domain structures.*

Domain structures represent notable, commonly occurring structures within a domain. Example domain structures include lessons with associated assessments, contact information containing phone numbers, or the schema fragment necessary to represent the scheduling of a financed purchase. Domain structures are mapped to local schemas by local content creators providing simple correspondences.

Once mapped, domain structures support tasks ranging from traditional information integration to the construction of generic widgets for the repository. In this paper we focus on the definition and use of generic widgets based on canonical and domain structures. We call these widgets *semantic widgets*. The power of our approach is that semantic widgets can be implemented generically—based on the canonical and domain structures but can provide immediate benefit to each local site. The key aspect of our approach is that the mappings allow the local schema names and local structure to dominate.

We provide a motivating example in Section 2. We define canonical and domain structures in Section 3, mappings to local schemas in Section 4, and semantic widgets in Section 5. Section 6 contains a discussion, Section 7 contains related work and Section 8 concludes the paper.

2   Overview

We present an example of a semantic widget for the STEMRobotics[1] educational repository in order to demonstrate the concepts described in this paper. STEMRobotics is a digital library of robotics curricula for use in middle and high school with diverse structure.

Figure 1 shows a sample of the course structures that appear in the repository[2]. The left structure represents a traditional course where the Course is made up of one or more Units and each Unit consists of one or more Lessons. A Lesson may have associated instructional materials (STEM IM) or assessments (STEM Assess). An instructional material may be used in a Lesson as the primary material (Pri) or as extended (Ext) or supplemental (Supp) materials. Any given instructional material (STEM IM) can be used as the primary material

---

[1] http://stemrobotics.cs.pdx.edu
[2] For the sake of clarity, the structures in STEMRobotics have been simplified here.

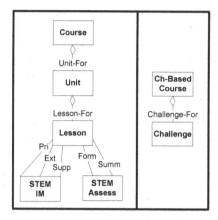

Fig. 1. The variety of different course structures present in STEMRobotics

in one lesson and as an extended or supplemental material in another lesson. Assessments may also be related to a lesson in several ways, where a formative assessment (Form) is used to help a teacher decide how to teach material to a student while a summative assessment (Summ) is used to evaluate a student's understanding of the material. Like instructional materials, an assessment may be formative in one lesson and summative in another.

The structure in the right column of Figure 1 shows a challenge-based course (Ch-Based Course) consisting only of Challenges that direct the student to build a robot that can perform a given task. The challenge may have associated instructional materials on how to perform the given task. Note that the challenge serves as an educational material (i.e., it directs the activities of the student) and also serves as an assessment of the student's knowledge of the material; there are no other assessments in the challenge-based course.

Our goal is to define and build generic widgets that can work with the diversity in structure, type, and names in this site. This then requires some means of resolving the heterogeneity while still preserving and using the names, types, and relationships present in the local schemas. We also want to allow a content creator to add additional structure in a navigation menu to group the content, for example, adding "Differentiated Instructional Material" for some of the relationships between a Lesson and an Instructional Material. Creators may also choose to have aggregations of their materials by type, for example "Formative Assessments" in a Lesson. While these aggregations introduce structure in the navigation widget, the names arise from the entity and relationship types in the local schema.

3   Canonical and Domain Structures

This section presents a formal definition of canonical structures and their instantiation in a domain as domain structures. Figure 2 shows a set of canonical

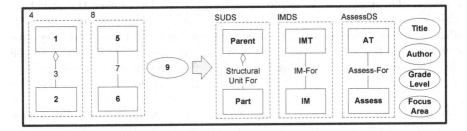

Fig. 2. The canonical structures (left) and their instantiations in the educational domain (right)

structures on the left side and their instantiation as domain structures on the right side for our educational repository. We first define *canonical entities* and *canonical relationships*.

Definition 1. *A canonical entity with a globally unique identifier ceID is* $CE(ceID)$. *Let* $\mathfrak{C}_E$ *be the set of all canonical entity identifiers.*

Definition 2. *A canonical relationship with a globally unique identifier crID, a relationship type, and a list of canonical entity identifiers* $[ceID]$ *where* $ceID \in \mathfrak{C}_E$ *is* $CR(crID, type, [ceID])$. *Let* $\mathfrak{C}_R$ *be the set of all canonical relationship identifiers.*

Canonical entities are the boxes on the left side of Figure 2. Individual entities and relationships are labeled with unique identifiers. The relationship with identifier "3" is represented as $CR(3, Part\text{-}Of, [1, 2])$, where the order of the entity identifiers determines the position of the entities in the relationship. The relationship type is *Part-Of*, with the semantics of a many-to-many aggregation relationship.

Definition 3. *A canonical structure with a globally unique identifier csID, a set of canonical entity identifiers* $E \subseteq \mathfrak{C}_E$, *and a set of canonical relationship identifiers* $R \subseteq \mathfrak{C}_R$ *is* $CS(csID, E, R)$.

Canonical structures are defined by the entities and relationships that they consist of. While we could have inferred the entities in relationships from the list of entity identifiers in each relationship, we maintain a complete set of entities for a canonical structure because the structure may include entities that do not participate in any relationships.

A *canonical attribute* is a place holder for an attribute with a globally unique identifier caID, $CA(caID)$, shown as the oval on the left side of Figure 2.

A canonical structure is instantiated in a domain as a *domain structure* by naming each part of the canonical structure. A canonical structure may be instantiated multiple times within a domain. The right side of Figure 2 shows three domain structures within the educational domain. The first structure ("SUDS") represents materials in a structural unit hierarchy, for example the units within a course. Here canonical structure 4 is instantiated by naming the structure ("SUDS"), its relationship ("Structural Unit For"), and its entities ("Parent", "Part"). Canonical structure "8" is instantiated twice: the first represents instructional materials and things that they may be related to ("IMDS") and the second represents assessments and the things that they assess ("AssessDS"). A domain structure and its instantiation are defined as follows.

Definition 4. *A canonical structure, $CS(csID, CE, CR)$, is instantiated as a domain structure, $DS(dsID, dsName, csID, DE, DR)$, where*
$DE = \{(deID, eName, ceID)|ceID \in CE\}$ *and*
$DR = \{(drID, rName, crID)|crID \in CR\}$ *where eName and rName are the names provided for entities and relationships. Let $\mathfrak{D}_S$ be the set of all domain structure identifiers, $\mathfrak{D}_E$ be the set of all domain entity identifiers, and $\mathfrak{D}_R$ be the set of all domain relationship identifiers.*

Similarly, a canonical attribute is instantiated in a domain by naming the attribute.

Definition 5. *A canonical attribute, $CA(caID)$, is instantiated as a domain attribute, $DA(daID, daName, caID)$, where daName is the name of the attribute. Let $\mathfrak{D}_A$ be the set of all domain attribute identifiers.*

We then define an *application*, $\mathfrak{A}$, as the set of all domain entity, relationship, structure, and attribute identifiers within a given domain, $\mathfrak{A} = \mathfrak{D}_E \cup \mathfrak{D}_R \cup \mathfrak{D}_A \cup \mathfrak{D}_S{}^3$.

4   Mapping to Local Structure

Once domain structures and attributes have been defined they must be mapped to local schemas in order to be used by semantic widgets. Figure 3 shows the mapping of the domain structures shown in Figure 2 to the four local structures from Figure 1. A mapping of a domain structure to a local schema consists of a set of correspondences between the elements of a domain structure or domain attribute and elements of the local schema where a domain structure or attribute element corresponds to either a local entity, relationship, or attribute id. We leave the description of the local schema largely uninterpreted; we simply rely on a set of named structural elements and named attributes—designated as a *schema set*, $\mathbb{S}_x$ for a given schema x.

Definition 6. *For an application $\mathfrak{A}$ and a schema set $\mathbb{S}_x$, a domain-local correspondence with a globally unique identifier corrID is given by*
$Corr(corrID, dID, lID)$ *where $dID \in \mathfrak{A}$ and $lID \in \mathbb{S}_x$.*
Let $Corr_{DS}$ be the set of correspondences between a domain structure and a local structure, let $Corr_{\mathbb{S}_x}$ be the set of all correspondences from an application $\mathfrak{A}$ to a local schema set $\mathbb{S}_x$, and let $Corr_{\mathfrak{A}}$ be the set of all correspondences for an application.

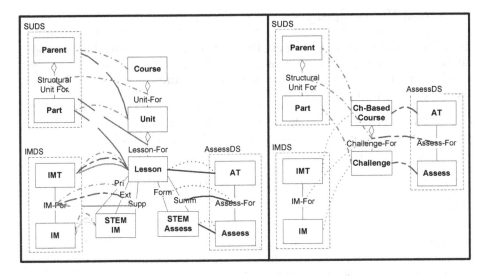

Fig. 3. Correspondences from the domain structures SUDS, IMDS, and AssessDS are shown to each of the local structures shown in Figure 1

In the left box of Figure 3 we see that the "SUDS" domain structure is mapped twice to the local schema of the traditional course: (1) "Units" as parts of a "Course" (shown with a blue/dot-dash line) and (2) "Units" as a parent made up of "Lessons" as parts (shown with red/long-dash lines). The "IMDS" structure is mapped three times, once for each relationship between "Lesson" and "STEM IM". "AssessDS" is also mapped twice, once for each relationship between a "Lesson" and a "STEM Assess". Each different color/pattern of mapping lines in Figure 3 indicates a set of correspondences between a domain structure and the local schema. The elements of the challenge-based course in the right box are mapped to all three domain structures since a "Challenge" acts as a part of the challenge-based course as well as an instructional material and an assessment. Each of these mappings (i.e., one set of correspondences of the same color/pattern) is a $Corr_{DS}$ and each set of $Corr_{DS}$ for a single course is a $Corr_{S_x}$. While all the local entities and relationships in this example have domain structures mapped to them, there is no requirement that there be a mapping to every element in the schema set.

Attributes from the local schema are mapped in a similar fashion. A correspondence documents the local entity or relationship and the local attribute to which the domain attribute is attached in the IID of the correspondence. As a result, there is no need to attach attributes to entities in the domain structures; the composition of domain-local correspondences and domain attribute correspondences will determine the attributes available to a domain structure for a given mapped local structure. Note: a domain attribute can be associated with any domain structure.

---

[3] For the purposes of this paper, we assume that the identifiers for domain entities, relationships, structures and attributes are disjoint.

5 Semantic Widgets

Given the domain structures and attributes that have been defined for an application and the associated correspondences to local schemas, a semantic widget is defined by the set of domain structures and attributes that it is based on plus additional widget-specific modifiers.

Definition 7. *A semantic widget with a globally unique name wname, given an application $\mathfrak{A}$, and the correspondence set $Corr_{\mathfrak{A}}$, is defined as*
$\psi(wname, \delta, m_{\epsilon}, m_{corr})$ *where* $\delta \subseteq \mathfrak{A}$
and $m_{\epsilon} = \{(mod, dID)\}$ *where* $dID \in \delta$
and $m_{corr} = \{(mod_{corr}, corrID)\}$ *where* $corrID \in Corr_{\mathfrak{A}}$
where mod and mod_{corr} are user-supplied modifier names for domain elements and correspondences.

As mentioned in Section 2, when mapping domain structures to local schemas, a content creator may wish to determine how their content appears in a given widget or even add content or structure that doesn't exist in the local schema. This information is supplied via the correspondence modifier names in the semantic widget definition. Notice that a widget creator does not specify the composition of the domain structures or attributes in the specification of the semantic widget. The composition is instead determined from the domain-local correspondences.

The effect of a semantic widget is then defined as the set of all local parts that correspond to the domain structures and attributes used in the widget.

Definition 8. *The* effect *of a semantic widget ψ, given the correspondence set of an application, $Corr_{\mathfrak{A}}$, is defined as*
$\Sigma(\psi(wname, \delta, m_{\epsilon}, m_{corr}), Corr_{\mathfrak{A}}) =$
  $\{lID | \exists Corr(corrID, dID, lID) \wedge dID \in \delta \wedge corrID \in Corr_{\mathfrak{A}}\}$

6 Discussion

We have instantiated our software to host two repositories of educational material (STEMRobotics mentioned earlier and a site for middle school science curricula[4].

In addition to the navigation widget presented in this paper, we have implemented a generic cloning widget. The clone widget is designed to allow teachers to clone a course in STEMRobotics to tailor it to their specific needs by rearranging, dropping, or adding lessons or units. The cloning process creates a copy of any entity mapped to the SUDS domain structure while linking the copies to the original instructional materials and assessments, e.g. anything mapped to the IMDS or AssessDS domain structures. Thus, instructional materials and assessments are shared by the clone and the original.

If we had used classical data integration techniques [10] with a generic widget built against a global schema, an instructional material, a tutorial, and a

---

[4] http://msscience.cs.pdx.edu

challenge would likely be mapped to the same entity type (e.g., instructional structural unit or possibly course) and a semantic widget such as our navigation widget would be unable to display the local schema names. Our use of schema fragments allows content creators to map domain structures to the local structures numerous times potentially in contradictory ways. For example, an author can map the structural unit domain structure onto their content with instructional materials as an aggregation of units where it is used—in contrast to the way the SUDS hierarchy is mapped above.

A strength of our approach is the way in which the domain structures are fairly under-specified. Attributes need not be attached to entities in the domain structures; they can simply be mapped to local attributes. Through the composition of our mappings, an entity in our domain structure, in effect, has an attribute because the structure was mapped to a local part that has an attribute mapped to a domain attribute. In a similar way, our choice to leave the description of the local structure largely uninterpreted suggests that we will be able to map domain structures and thus bring the power of semantic widgets and other features to a broad range of local sites.

7 Related Work

Canonical structures and domain structures can be viewed as design patterns similar to those proposed in data modeling [3] and ontology creation [13]. Blaha [3] defines domain-independent *patterns* that compose generic data model constructs and domain-dependent *seed models* that can be used as a starting point for a schema. Our canonical structures are roughly at the level of Blaha's patterns and our domain structures are at the seed model level. Similarly, ontology design patterns [13] represent domain-independent models that may be elaborated for a domain. In both of these approaches, the patterns are used to build new systems; existing systems are not mapped to these patterns. We approach pattern use in an opposite manner: we overlay patterns on existing systems in order to extend their functionality.

A number of model-driven web development approaches such as WebML [4], Araneus [11], HDM [8], RMM [9] and AutoWeb [7] allow web information systems to be specified at the conceptual level and allow navigation structures to be defined for various content types—similar to the semantic widget described in this paper. Our approach differs in that it can be used to extend the functionality of systems already implemented in various ways as well as enabling the addition of other semantic widgets besides navigation.

Our approach of using simple correspondences borrows from schema mapping tools like Clio [6], where users can draw lines between source and target schemas to generate schema mappings.

In the semantic web [2], Tim Berners-Lee proposed a "web of data"[5] that would allow the interoperability of web systems. Technologies such as RDF[6] and

---

[5] http://linkeddata.org

[6] http://www.w3.org/RDF/

the Web Ontology Language[7] have allowed users to document web content in semantically meaningful ways. But allowing users to arbitrarily create names and relationships for their content has led to heterogeneity problems well known in the database community [1]. Traditional integration approaches have been adopted such as using foundational ontologies [14] and ontology matching and alignment [5] to tackle design and integration issues, but these approaches still rely on an ontology as a global schema and do not present a local-dominant view of data.

More recent semantic web technologies such as RDFa[8] and Microformats[9] allow content creators to embed RDF relationships or schema fragments directly within a webpage. By embedding known schema elements from sources such as Schema.org[10], content creators can quickly add semantics to their content to take advantage of search engine and other functionality that can make use of these schema elements—much like our correspondences and semantic widgets. We see our work as complementary to this effort. We intend to allow developers using our approach to automatically post their semantics (based on the domain structures and the correspondences) to the "web of data".

Paggr [12], makes use of the collection of RDF data known on the web to build reusable and extensible web page widgets written using SPARQL[11] as a common interface to the web of data. By publishing RDF triples, content creators can take immediate benefit of these widgets. But the Paggr widgets only show the RDF data present, not the local content whereas, our correspondences allow us to access the known semantic information, in the domain structures, while still having access to the local content.

8   Conclusion

In this paper we have presented a formal foundation for canonical and domain structures and how they map to local schemas. We also have shown how canonical and domain structures and their mappings can be leveraged to build generic semantic widgets that can address global (i.e., our domain) structures while still accessing and displaying local schema information. Our implementation within a web-based content management system shows the feasibility of our approach. As further evidence, we have also implemented and formalized the use of the Open Archives Initiative (OAI) Protocol for Metadata Harvesting (PMH)[12] as well as the Object Reuse and Exchange (ORE)[13] protocol to serialize and transfer the structures that are defined using our approach (e.g., in our STEMRobotics site) to a web-based portal[14]. More than that, we ported our navigational widget to

---

[7] http://www.w3.org/2004/OWL/

[8] http://www.w3.org/TR/rdfa-syntax/

[9] http://microformats.org/wiki/Main_Page

[10] http://schema.org

[11] http://www.w3.org/TR/rdf-sparql-query/

[12] http://www.openarchives.org/pmh/

[13] http://www.openarchives.org/ore/

[14] http://computingportal.org

the portal and thus provided a generic navigational tool to all richly structured content mapped to our domain structures.

Acknowledgments. This work was supported in part by the National Science Foundation (0840668). We would like to thank David Maier for his knowledge and insights and we would also like to thank our collaborators and content authors Randy Steele, Don Domes, and Devin Hunter.

References

1. Anderson, J.Q., Rainie, L.: The Fate of the Semantic Web (2010), http://www.pewinternet.org/Reports/2010/Semantic-Web.aspx
2. Berners-Lee, T., Hendler, J., Lassila, O.: The Semantic Web. Scientific American 284(5), 34–43 (2001)
3. Blaha, M.: Patterns of Data Modeling. CRC Press (June 2010)
4. Ceri, S., Fraternali, P., Bongio, A.: Web Modeling Language (WebML): a modeling language for designing Web sites. Computer Networks 33(1-6), 137–157 (2000)
5. Euzenat, J., Shvaiko, P.: Ontology matching. Springer Publishing Company, Incorporated (2010)
6. Fagin, R., Haas, L.M., Hernández, M., Miller, R.J., Popa, L., Velegrakis, Y.: Clio: Schema Mapping Creation and Data Exchange. In: Borgida, A.T., Chaudhri, V.K., Giorgini, P., Yu, E.S. (eds.) Mylopoulos Festschrift. LNCS, vol. 5600, pp. 198–236. Springer, Heidelberg (2009)
7. Fraternali, P., Paolini, P.: Model-driven development of web applications: the autoweb system. ACM Trans. Inf. Syst. 18(4), 323–382 (2000)
8. Garzotto, F., Paolini, P., Schwabe, D.: Hdm a model-based approach to hypertext application design. ACM Trans. Inf. Syst. 11(1), 1–26 (1993)
9. Isakowitz, T., Stohr, E.A., Balasubramanian, P.: Rmm: a methodology for structured hypermedia design. Commun. ACM 38(8), 34–44 (1995)
10. Lenzerini, M.: Data integration: a theoretical perspective. In: PODS 2002, pp. 233–246. ACM, New York (2002)
11. Merialdo, P., Atzeni, P., Mecca, G.: Design and development of data-intensive web sites: The Araneus approach. ACM Transactions on Internet Technology 3(1), 49–92 (2003)
12. Nowack, B.: Paggr: Linked data widgets and dashboards. Web Semantics 7(4), 272–277 (2009)
13. Presutti, V., Gangemi, A.: Content Ontology Design Patterns as Practical Building Blocks for Web Ontologies. In: Li, Q., Spaccapietra, S., Yu, E., Olivé, A. (eds.) ER 2008. LNCS, vol. 5231, pp. 128–141. Springer, Heidelberg (2008)
14. Staab, S., Studer, R.: Handbook on Ontologies, 2nd edn. Springer (2009)

Evaluation Measures for Similarity Search Results in Process Model Repositories

Markus Guentert, Matthias Kunze, and Mathias Weske

Hasso Plattner Institute at the University of Potsdam
Prof.-Dr.-Helmert-Straße 2-3, 14482 Potsdam
markus.guentert@student.hpi.uni-potsdam.de,
{matthias.kunze,mathias.weske}@hpi.uni-potsdam.de

Abstract. With the increasing uptake of business process management efforts in companies, similarity search in large process model repositories has gained significance, as it forms a cornerstone of effective process model management and reuse. Similarity search uses a process model as query and retrieves all models, which resemble the query, in a ranked order. So far, the quality of the ranking has not been investigated.

In this paper, we propose quality measures for similarity search results in order to address this problem, providing information on how good and how differentiated the results are. Our measures assess result statistics, which are derived from the similarity to the query model, and the agreement of different rankings, produced by diverse similarity measures. We apply our findings to a reference process model collection and comprehensively evaluate their prediction towards human assessment of process similarity.

Keywords: Similarity search, evaluation measures, search result quality, model repository, business process management.

1 Introduction

In the field of business process management (BPM), we see organizations maintain collections of hundreds or thousands of business process models [19,1]. Effective management of this knowledge asset requires powerful means to identify similar models and search efficiently within the model repository. For instance, identifying similar behavior among a set of process models might indicate that several variants or even duplicates of the same model exist, which should be unified into one model, or refactored in order to improve readability or propagate changes [22]. Existing process models in such a collection can be used before and during process design to identify related processes and reuse them or fragments thereof. Hence, effective use of process model repositories requires meaningful capabilities to search within them.

However, in practice, we generally observe keyword text search and folder navigation as means to retrieve models from a collection, which generally are of limited value, if one searches for specific fragments or behavioral aspects. Researchers in

P. Atzeni, D. Cheung, and R. Sudha (Eds.): ER 2012, LNCS 7532, pp. 214–227, 2012.
© Springer-Verlag Berlin Heidelberg 2012

the field of BPM have addressed this by means of similarity search [14], i.e., users express their search intention in a process model, and models that resemble this query in certain aspects, e.g., structural or behavioral, are proposed as a search result. By now, a large body of work exists that addresses similarity notions and evaluates the effectiveness of similarity measures towards approximating human assessment of similarity, as will be discussed in Section 2. However, we observed barely any work that approached the quality of a search result, answering questions such as: How well can search results be distinguished from each other? Do models exist that stand out significantly from the rest of the models? Is the ranking provided by the similarity measure any meaningful to the searcher?

In this paper, we propose five evaluation measures that address these very questions. As we cannot judge on the relevance of a search result in absence of a human assessment for each possible query, these measures examine distribution aspects within a similarity search result to evaluate confidence, discrimination, and the ranking provided by similarity search. These measures are supposed to assist a searcher in disseminating a similarity search result, before exhaustively inspecting returned models. In a comprehensive evaluation, we examine the applicability and expressiveness by an experiment that takes three different similarity measures into account and compares the evaluation measures against human judgement. While these measures are applicable to similarity search results in general, we restrict examples and evaluation of this paper to the field of BPM for illustrative purposes.

The remainder of this paper is structured as follows: In Section 2, we briefly discuss literature and findings that are related to our research, before we introduce essential foundations and discuss the problem approached in this paper, in Section 3. Subsequently, we propose five measures to evaluate the quality and assess the ranking of a similarity search result in Section 4, before we analyze these measures in Section 5. Section 6 concludes the paper and points at future work.

2   Related Work

In business process management, the application of similarity to retrieve process models from large repositories has gained increasing interest within the last five years. Consequently, a large number of similarity measures to business process models has been proposed. Dumas et al. [14] present an overview of the basics of process model similarity approaches.

The similarity of business processes is generally established on a mapping of corresponding concepts between a pair of processes, e.g., their activities [14,23]. Once corresponding activities have been identified, one can identify relations between activities in one model and compare them to relations between the corresponding activities in another model. These relations are generally based on three complementary aspects, element labels, the graph structure of a process model, or its behavioral aspects. The former regularly compute similarity by the share of common activities. Structural approaches typically leverage common subgraph isomorphisms [6] or cost based algorithms, e.g., the graph edit distance [5]. Behavioral approaches address the behavior of a process model in terms of its reachable states, execution traces it can produce, or abstractions thereof.

An overview of various similarity measures, both behavioral and structural, can be found in [16] along with a qualitative comparison of the respective similarity notions. A quantitive analysis and comparison of similarity measures is presented in [3]. The majority of the approaches recapitulated there does not address efficiency, and only few address effectiveness, i.e., how well a similarity measure simulates human assessment of similarity, cf. [13,16].

Deciding whether a result is relevant to a given query is subjective, as it depends on the intention of the users formulating the query and their assumptions about the result [9]. Hence, additional aspects to assess the search result are required. [4] discuss the degradation of meaningfulness of similarity search results in high dimensional spaces, from which we have been inspired to research evaluation measures presented below. [21] define different aspects of data quality: intrinsic, contextual, representational, and accessibility data quality. For this work, we especially refer to the intrinsic and contextual data quality.

By intrinsic data quality we understand process model quality features, such as readability, coherence, modularity, and the absence of errors [20,17]. One of the main quality features in web search applications is the amount of incoming links to a particular page or source of the search result. Projected to process models, such quality features could be the creator of the model, the number of process instances, the number of model revisions, and community ratings, cf. [2]. However, in many process model repositories, such information is not available, which encouraged this work to assess the quality of a search result purely on its own.

Contextual data quality is defined as "contextually appropriate for the task". This matches with our intuition of relevance. [21] argue that quality attributes are ideally collected from data consumers and, therefore, cannot be purely defined on a theoretical foundation. This motivates our contribution to evaluate how a human assessment of process similarity compares against theoretically assessed data by means of similarity metrics, cf. Section 5.

3   Problem Statement

In large process model collections, data to be searched is complex and widely unstructured. Hence, exact match search is not meaningful, and traditional indexing not possible, as no expressive ordering of this data exists [24]. Similarity search, therefore, assumes nothing but a distance that can be computed from a pair of data points.

Given a set S of data points of domain $\mathcal{P}$ and a query q of the same domain $\mathcal{P}$, similarity search obtains a search result R_q containing all data points p from the data set S that are within a given proximity $r(q)$ to the query. Figuratively, similarity search can be visualized as a sphere of radius $r(q)$ around a query model; all data points within this sphere are contained in the search result, cf. Fig. 1. Similarity of pairs of data points is defined in terms of a distance function $d : \mathcal{P} \times \mathcal{P} \rightarrow \mathbb{R}$, and thus $\forall p \in S \setminus R_q : d(q,p) > r(q)$. The search result $R_q = \langle p_1, p_2, ...p_k \rangle$ is a sequence comprised of $k = |R_q|$ distinct data points, sorted in ascending order of their distance to the search query, i.e., $\forall i, k \in \mathbb{N}, 1 \leq i < k : d(q, p_i) \leq d(q, p_{i+1})$.

Naively, the restriction of similarity search to pairwise distances requires exhaustively comparing each data point p in the data set with the query q to obtain all points within $r(q)$. However, if the similarity is a metric, i.e., its distance function features in particular the triangle inequality, indexes can be built that partition the data set and allow pruning partitions during search [24]. Various data structures and algorithms for similarity search have been developed; surveys and discussions of them can be found in [7,15]. Efficient search in metric spaces can be generally distinguished into range search and k-nearest-neighbor search [24]. In the former category, a search radius $r(q)$ is explicitly given, and all data points that are within this distance to q are obtained. This approach has the drawback that the number of results cannot be known in advance, and different similarity metrics can hardly be compared as they may use different scales. An extension to range search is k-nearest-neighbor search, which searches for the k most similar data points to the query; $r(q)$ is then the distance from q to the furthest data point in R_q.

Recalling Section 2, none of the related work towards business process similarity addressed the presentation of search results. A similarity search would return all results up to a certain threshold of similarity, $r(q)$, or the k top results. A ranking is provided that arranges a data points of a search result by increasing distance to the given query, which we refer to as *naive ranking*, hereafter.

Fig. 1 illustrates three different search results, which all produce the same naive ranking, $\langle A, B, C, D, E \rangle$. However, their quality and meaning significantly differs. The first result, cf. Fig. 1a, shows a uniform distribution of distance differences between the result data points. As the distances increase linearly, we are able to derive a non-ambiguous ranking for the search results. In the second result, cf. Fig. 1b, one result is notably better than the other results. In the third search result, cf. Fig. 1c, all results are equally far away from the query model, i.e., the results are not differentiated in distance. No result is any close to the query. Hence, the naive ranking may be ambiguous.

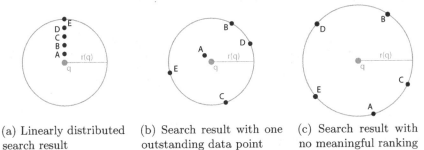

(a) Linearly distributed search result

(b) Search result with one outstanding data point

(c) Search result with no meaningful ranking

Fig. 1. Illustration of different similarity search results with equal naive ranking

These examples show that it is inappropriate to determine an absolute distance value, up to which models should be considered as good results and be presented to the user. The search radius threshold is usually determined dynamically according to the number of results to be retrieved. The number of results to be retrieved is arbitrary and does not reveal the relevance of any of its results. Finally, a naive ranking does not consider, how well the result models are differentiated and whether there are any good results at all.

4 Search Result Evaluation Measures

As discussed previously, we cannot reliably assess the quality, i.e., the relevance, of a search result, as this is subjective to the user and their task. In business process model collections we usually do not have additional information present, which could possibly influence the ranking and act as an indicator for the quality of a model. Thus, we provide assistance in evaluating the search result from a quantitative perspective. In this section, we propose a set of evaluation measures to assess the structure and arrangement of a search result. Therefore, we rely only on the search result R_q, i.e., we have no knowledge about the query, nor can we compare the result to data points not included in the search result, which could contain results more relevant to the user.

For our measures, we rely on three characteristic distance functions of the result sequence, $f : R_q \to \mathbb{R}$. Fig. 3 illustrates these values.

- $min(R_q) = d(q, p_1)$ is the minimum distance, i.e., the distance of the closest data point to the query in the result set.
- $med(R_q) = d(q, p_i), i = \lceil \frac{k+1}{2} \rceil$ is the median distance to the query, which intuitively resembles the maximum distance of the "better half" of the results. We refer to the better half of results hereafter as *most results*.
- $max(R_q) = d(q, p_k), k = |R_q|$ is the maximum distance, i.e., the distance of the farthest data point to the query in the result sequence. In case of k-nearest-neighbor search, this is equal to $r(q)$.

All evaluation measures presented hereafter express search result statistics and produce values $x \in \mathbb{R} \wedge x \in [0, 1]$, where the best case is expressed by a value of 1 and the worst or sufficiently unsatisfying cases by a value of 0.

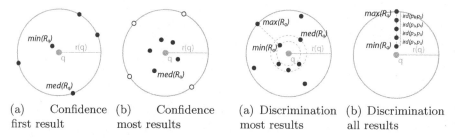

(a)   Confidence   (b)   Confidence   (a) Discrimination   (b) Discrimination
first result   most results   most results   all results

Fig. 2. Illustration of result confidence   **Fig. 3.** Illustration of result discrimination

4.1   Result Confidence

Confidence expresses the belief that the best results are outstanding from the perspective of the rest of the results. This is expressed by its distance to a reference value that determines the maximum distance of the results to be compared. In order to assess the confidence of the best result, i.e., the closest data point, we compare its distance to the query model with the median distance of the result sequence. The median is more stable and not as strongly affected by the number of results retrieved as the arithmetic mean or maximum distance in the search result because the median is generally robust against outliers in a data set.

Definition 1 (Confidence first result). *Given a search result sequence R_q, the* confidence of the first result *is a function $C_1 : R_q \to [0,1]$ such that*

$$C_1(R_q) = 1 - \frac{min(R_q)}{med(R_q)}$$

In the best case, the minimum distance to the query in the results equals 0, which implies that according to the applied similarity measure there is at least one model in the search result, which is equal to the query model. In the worst case the minimum equals the median, which implies that the best result is not any better than the majority of the results. Fig. 2a illustrates a result sequence in which the first result is clearly better than the median of the search result. The confidence of the first result is therefore high.

As a confidence measure for most results we compare the sequence of superior models, denoted by $\overline{R_q}$, of the results with the inferior models, $\underline{R_q}$, by dividing the results into two groups.

Definition 2 (Confidence most results). *Given a search result sequence $R_q = \langle p_1, p_2, ..., p_k \rangle$, we define two subsequences $\overline{R_q} = \langle p_1, ..., p_i \rangle$ and $\underline{R_q} = \langle p_{i+1}, ..., p_k \rangle$ with $i = \lceil \frac{k+1}{2} \rceil$ and $k = |R_q|$, which partition the result sequence in two sequences.*
The confidence of most results *is a function $C_{most} : R_q \to [0,1]$ such that*

$$C_{most}(R_q) = 1 - \frac{med(\overline{R_q})}{med(\underline{R_q})}$$

A large difference in values $med(\overline{R_q})$ and $med(\underline{R_q})$ indicates that the majority of the superior models is significantly better than the inferior models. Fig. 2b illustrates such a setting, determined by $med(\overline{R_q})$ (white data points) and $med(\underline{R_q})$ (black data points). In contrast, if both values are close to equal, one cannot reasonably argue that the superior models are truly better, as they become rather indistinguishable.

4.2   Result Discrimination

Discrimination describes the distribution of the distances between the query and the search results, since well differentiated results have a higher probability to show an unambiguous ranking compared to a search result sequence where most of the data points do not differ in their distance to the query model.

Beyer et al. [4] discussed the meaningfulness of similarity search in high-dimensional vector spaces. They proposed a stability measure based on the capacity of a query to obtain results that can be distinguished in an unambiguous fashion. Consequently, this provides a measure how well search results can be discriminated. Whereas the proposed measure provided only a binary value, i.e., a query either produced a stable result or not, we adapted this notion and extended it, such that it provides a continuous measure as explained above.

Consequently, the stability of a search result is defined by the interval between the distance of the best result and the median of all results, normalized by the range of distances to the query in the search result, or put differently, we compare the range of the superior half of results against the whole range of results.

Definition 3 (Discrimination most results). *Given a search result sequence R_q, the discrimination of most results is a function $\mathcal{D}_{most} : R_q \to [0, 1]$ such that*

$$\mathcal{D}_{most}(R_q) = \frac{med(R_q) - min(R_q)}{max(R_q) - min(R_q)}$$

In the worst case, the interval in which most results lie is 0. Most results are therefore not differentiated in distance to the query model, and a ranking of these results cannot be unambiguous. Fig. 3a depicts a rather moderate interval illustrated by the width of the ring denoted by the two dashed circles. It further shows the range of search results, which is the difference in distances from the closest and most distant data point to q (rather than the search radius $r(q)$). In the best case, the width of the ring spreads the whole range, which indicates that the superior majority of search results are well distinguishable.

As $\mathcal{D}_{most}$ does not address the ranking of the inferior search results, we also propose a measure that evaluates the distribution of all data points within the search result. Here, we consider a linear increase in distances to the query as the ideal case, i.e., the distance differences between consecutive data points are equal, which is depicted in Fig. 3b.

Definition 4 (Discrimination all results). *Given a search result sequence R_q, the inter-result difference function* ird $: (\mathcal{P} \times \mathcal{P}) \to \mathbb{R}$ *defined as*

$$ird(p_i, p_j) = |d(q, p_i) - d(q, p_j)|$$

the discrimination of all results is a function $\mathcal{D}_{all} : R_q \to [0, 1]$ such that

$$\mathcal{D}_{all}(R_q) = \begin{cases} 1 - \frac{\sum\limits_{1 \leq i < k} \left| ird(p_i, p_{i+1}) - \frac{max(R_q) - min(R_q)}{k-1} \right|}{2 \cdot (max(R_q) - min(R_q))} & \text{if } max(R_q) \neq min(R_q) \\ 0 & \text{else} \end{cases}$$

where $k = |R_q|$.

Here, we compare the inter-result difference of each neighboring pair in a sequence, $ird(p_i, p_{i+1})$ against the ideal difference provided by the range of the result sequence and the number of inter-result distances, which is one less than the cardinality of the search result, $|R_q|$. In the worst case, none of the results can be distinguished by its distance, which we denote with $\mathcal{D}_{all}(R_q) = 0$.

4.3   Ranking Agreement

As briefly discussed in Section 2, a vast number of process similarity measures exists, addressing linguistic, structural, and behavioral features of process models. Hence, it is possible to assess the similarity of two data points by various measures; each assessing different aspects. Every similarity measure may produce different distance values for the data points of a search result to the query and thus, different rankings.

Comparing the distance values between the query and a data point among different similarity measures turned out to be inappropriate, because different similarity measures use different scales, e.g., the simplest form of graph edit distance of a process model pair may return a natural number, i.e., the number of edit operations, whereas distance measures that leverage the Jaccard coefficient, generally produce a real value in the range $[0, 1]$. Even if the distance values are fairly different, the measures could still agree on the ranking. We therefore rely on the ranking as reasonable comparison instance. With the ranking agreement we assess to which extent different similarity measures agree on the ranking of the results.

Given a search result sequence R_q, we denote a ranking i of the same search results produced by some means as R_q^i. Hence, all rankings over R_q contain the same data points, but they may be ordered differently. A ranking produced by the same similarity function as the one to obtain R_q is called a *naive ranking*; other rankings could be created by other similarity functions, or, for example, through human assisted reordering.

The function $rnk : (\mathcal{P} \times R_q^i) \to \mathbb{N}$ returns the position of a result data point $p \in R_q$ in a ranking R_q^i. Consider, for example, the following rankings for a search result R_q: $R_q^1 = \langle A, B, C, D, E \rangle$, $R_q^2 = \langle C, B, A, D, E \rangle$, and $R_q^3 = \langle A, C, B, E, D \rangle$. Here, $rnk(C, R_q^1) = 3$, $rnk(C, R_q^2) = 1$ and $rnk(C, R_q^3) = 2$.

Definition 5 (Ranking agreement). *Let R_q be a result sequence with $|R_q| = k$ and a set of rankings $\mathcal{R}_q = \{R_q^1, ..., R_q^m\}$, $|\mathcal{R}_q| = m$, over R_q, that contain the same data points, i.e., $\forall 1 \leq j \leq m : p \in R_q \Leftrightarrow p \in R_q^j$. We define the ranking difference function, diff $: \mathcal{R}_q \to [0, 1]$, as follows:*

$$diff(\mathcal{R}_q) = \sum_{1 \leq s < m} \left(\sum_{s < t \leq m} \left(\sum_{p \in R_q} |rnk(p, R_q^s) - rnk(p, R_q^t)| \right) \right)$$

The ranking agreement *is a function* $\mathcal{A} : \mathcal{R} \to [0, 1]$ *defined as*

$$\mathcal{A}(\mathcal{R}_q) = 1 - \frac{diff(\mathcal{R}_q)}{\binom{m}{2} \cdot \frac{1}{4}(2 \cdot k^2 + (-1)^k - 1)}$$

For every combination of metrics (s, t) and for every data point p in the search result, we compare the positions of p in the respective pair of rankings and sum their absolute ranking difference. In order to normalize the overall ranking difference we divide by the worst possible ranking difference, determined by the number of metric combinations, $\binom{m}{2}$, where m is the number of metrics incorporated, and the worst possible pairwise ranking difference, which is produced by $\frac{1}{4}\left(2 \cdot k^2 + (-1)^k - 1\right)$. If all rankings agree in the ordering of the data points in R_q, their ranking difference $diff(\mathcal{R}_q)$ will be 0, leading to the best case value of $\mathcal{A}(\mathcal{R}_q) = 1$.

In contrast to existing ranking correlation coefficients, e.g., Kendall's τ [8], our ranking agreement is sensitive to the ranking distance of a data point in two rankings, rather than considering concordance or non-concordance only.

Above example shows three different rankings of data points A, B, C, D, E. If we compare the rankings of R_q^1 and R_q^2, we get $diff(\{R_q^1, R_q^2\}) = |1 - 3| + |2 - 2| + |3 - 1| + |4 - 4| + |5 - 5| = 4$. The ranking difference of all rankings $\mathcal{R}_q = \{R_q^1, R_q^2, R_q^3\}$ yields 14, which leads to a ranking agreement $\mathcal{A}(\mathcal{R}_q) = 1 - 14/36 \approx 0.61$.

5   Evaluation

We introduced evaluation measures to assess the quality of a search result in terms of confidence, i.e., how much the best search results stand out, discrimination, i.e., whether search results can be well distinguished and therefore be ranked unambiguously, and ranking, i.e., whether the ranking provides a potentially helpful assessment of relevance to the user.

In order to judge on the expressiveness and applicability of our measures, we conducted a set of experiments in the field of business processes. We obtained similarity search results in the SAP reference model collection [10], which comprises process models in the form of event-driven process chains [18], and applied our evaluation measures to them.

5.1   Material and Method

As a basis for our evaluation, we chose three complementary process model similarity measures to obtain results sets and rankings. The *activity mapping distance* (AMD) determines the ratio of shared activities to the union of all activities in two models by means of the well know Jaccard distance; it does not address graph structure or behavior of a process model and similarity is only affected by the number of similarly labeled activities. The *graph edit distance* (GED), introduced in [12], considers process model structure computing the minimal cost to transform one graph into another by atomic graph operations: insert, delete, and substitute nodes and edges. Finally, we considered a behavioral similarity measure, the *behavioral profile distance* (BPD), presented in [16], that evaluates the amount of shared behavior in terms of execution order of activities.

To compare our measures with human assessment, we leveraged data that has been assembled in an experiment addressing similarity of process models conducted by Dijkman et al. [12,13]. In the experiment, the authors randomly selected 100 document models from the SAP reference model [10] and from those ten query models, of which eight underwent slight changes. The authors and 20 process experts were then given the task to assess the similarity of every pair of query and document model on a likert scale from 1 to 7. Pairs with a score of 5 or higher indicated that a document model was relevant for the query [13]. We have applied our evaluation to the same sets of process models. However, 15 out of 100 document models and one query model had to be sorted out, because of ambiguous instantiation semantics [11,16]. Thus, our setup comprises 85 result models and nine query models.

With each of the nine query models and above similarity measure, we conducted k-nearest-neighbor searches for $k \in [3, 50]$, applied our evaluation measures to them, and compared this against the human judgement of similarity. Therefore, we mapped the average human assessed likert score for every pair of document and query models to distance values $1 - \frac{\text{likert score}}{8}$ in order to produce an equal distribution of distance values without the values 0 and 1. Our results are presented in Fig. 4 and 5.

5.2 Discussion of Results

Confidence. The confidence evaluation measure indicates whether the best results are significantly better than the rest of the results. In Fig. 4a, the confidence of the first result, C_1, is increasing as more models are obtained. This behavior of the measure is reasonable, because the more "bad" results we retrieve, the worse our reference value, the median, gets. Accordingly, the ratio between the minimum and the median is shifting towards the benefit of the best result. However, the measure converges and does not produce any values near 1, which further supports the intuition that the median is a stable reference value.

Confidence over the most results, C_{most}, cf. Fig. 4b, increases similarly, yet it reveals a peak around a result size $|R_q| = 15$ and then slowly decreases. This is due to the fact that the measure of interest, $med(R_q)$, is shifted by the size of the search result. When the peak is overstepped by adding inferior models, the majority of search results are not significantly better than all results anymore. From this fact we conclude that in our model collection—on average— the top 8 models perform comparatively best against the remaining models in the search result. This characteristic can effectively be used to assess the number of meaningful models in response to the query just by the search result.

For both confidence measures, we observe that the human assessment (solid grey curve) is significantly better than what similarity measures provide, which is not surprising. Nevertheless, we conclude that a high value in confidence indicates a good assessment of relevance of the provided search results.

Discrimination. This measure evaluates whether the ranking of the result sequence is unambiguous, where discrimination of most results, D_{most}, in particular considers the superior half of the search results in comparison to all results. Fig. 4c shows that this increases, which indicates that the majority of superior models can be ranked unambiguously. However, a high value of this measure at a large size of results points to the fact, that the set of inferior models remains within a rather small distance interval and may not be ranked unambiguously.

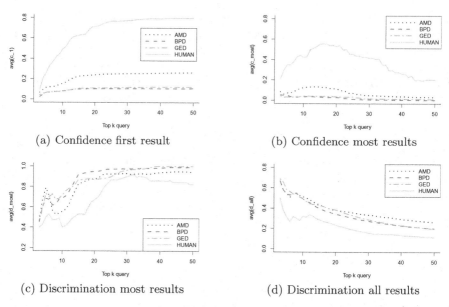

(a) Confidence first result

(b) Confidence most results

(c) Discrimination most results

(d) Discrimination all results

Fig. 4. Application of evaluation measures to similarity search results, compared to human assessment

This assumption is supported by the discrimination of all results, cf. Fig. 4d. With increasing cardinality of the search result, the majority of results deviate largely from a uniform inter-result distance distribution, that is, the ranking of models after position 20 may not be meaningful anymore and users may need to be supported with additional data to distinguish the models and assess relevance for their intention.

Both diagrams, Fig. 4c and 4d, show that the human assessment of discrimination is considerably lower than what similarity measures provide. This is due to the fact, that the assessment of relevance is subjective to the users and their task. Consequently, we observed different users to rate the similarity of the same pairs of query and document models differently, which leads to a lower agreement among them, and thus a higher ambiguity of the identified ranking. As the human judgement was based on a seven point likert scale, there were furthermore only seven values to distinguish the similarity of a process model pair.

Ranking Agreement. Fig. 5 illustrates the average agreement of pairs of rankings provided by the similarity measures introduced above, cf. Fig. 5a, and compared to humans, cf. Fig. 5b. We notice that AMD and BPD strongly agree on the ranking, which is a peculiarity of the model collection, which serves as a reference model collection, that is, pairs of activities that appear in two different process models are very likely to be executed in the same order in both models. The agreement between AMD and the GED therefore also correlates with the agreement between BPD and GED. In absolute terms, the agreement of {AMD,GED} as well as {BPD,GED} is mostly diverse.

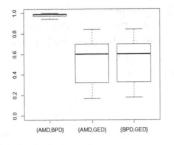
(a) Pairwise ranking agreement of AMD, BPD, GED

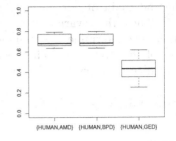
(b) Ranking agreement of AMD, BPD, GED with human assessment

Fig. 5. Ranking agreement of search results

Fig. 5b presents the ranking agreement between each metric and the human assessed similarity, i.e., expresses, how good a similarity measure simulates human assessment of similarity. We observe that AMD and BPD perform considerably better compared to GED—at least for the tested model setup. The agreement may be even higher if the human assessed data would be more differentiated. As the data only uses differentiated values from a likert scale, the ranking for 85 result models can never be non-ambiguous. The presented results have been derived from one possible ranking of the human assessed data.

For the data gathered from the human assessment of process similarity, we notice that the tendencies of every measure resemble those obtained by means of similarity measures. Whereas [13] has demonstrated that the application of similarity measures provides good approximation of human classification of relevant and non-relevant results, we hereby indicate that also the quality and differentiation of the results produced by similarity measures matches with the human perception.

The results support our intuition that our measures add to the naive ranking produced by the similarity metrics, in a way, that they give additional hints on the number of meaningful models, and the quality of the provided ranking. Hence, we improve the prediction of quality of the search results for a user querying models in a repository.

5.3   Correlation within the Measures

In order to assure that each of our proposed metrics provides expressiveness on its own, we determined the correlation within the measures for confidence and discrimination. A strong correlation between these measures would indicate that at least one of them is obsolete. Therefore we assessed the covariance and Pearson correlation coefficient of all pairs of evaluation measures produced in above experiments, listed in Table 1.

Table 1. Covariance and Pearson correlation coefficient for evaluation measures

Measures		Covar.	Pearson
$\mathcal{C}_1$	$\mathcal{C}_{most}$	0.0024	0.364
$\mathcal{C}_1$	$\mathcal{D}_{most}$	0.0068	0.214
$\mathcal{C}_1$	$\mathcal{D}_{all}$	-0.0023	-0.113
$\mathcal{C}_{most}$	$\mathcal{D}_{most}$	-0.0028	-0.258
$\mathcal{C}_{most}$	$\mathcal{D}_{all}$	0.003	0.438
$\mathcal{D}_{most}$	$\mathcal{D}_{all}$	-0.0107	-0.312

As the codomain of our measures is $[0, 1]$, we expect small values for the covariance. Thus, the covariance as such does not yet prove linear independence. Additionally we determine the Pearson correlation coefficient which normalizes the values of the covariance. The Pearson correlation values lie between -0.31 and 0.44, which provides no indication for linear dependency of the measures.

6   Conclusion

Business process similarity search has become a research topic of considerable interest. Different similarity criteria that assess different aspects of a process model lead to a variety of distance measures. These measures are typically used to obtain and naively rank a search result. However, questions towards the quality of a search result are generally neglected. As business process model collections usually do not provide meta information which may act as an indicator for the quality of a model, we evaluate search results from a quantitative perspective.

In this paper, we presented measures to evaluate the quality of similarity search results, examining structural aspects of the search result. We propose measures that assess the confidence that the best search results stand out compared to inferior search results, evaluate the ambiguity of a provided ranking, and examine the ranking of a search result with regards to other similarity measures. Our measures have been applied to a business process model collection, for which a human assessment of process similarity is available. The evaluation of our measures reflects the tendency of the human assessment. We therefore argue that our measures give good indication about the quality of search results and thus, assist the user in assessing the meaningfulness of a provided search result.

Although we resort to business process model repositories and use a process model collection and experiment data set for our evaluation in the scope of this paper, our results apply to all use cases that exhibit a large collection of contemporary data to be searched within by similarity, but lacks a large user basis from which quality features and usage statistics can be drawn, e.g., scientific workflow repositories, or audio and video documentation databases.

As future work we intend to apply our measures to other domains which make use of similarity search. Furthermore we plan to evaluate the distance distribution in the entire repository to act as a further indicator towards the quality of a search result. Another stream of work shall address the incorporation of these measures into the presentation of search results in a way that it effectively assists users in distinguishing provided search results.

References

1. Akkiraju, R., Ivan, A.: Discovering Business Process Similarities: An Empirical Study with SAP Best Practice Business Processes. In: Maglio, P.P., Weske, M., Yang, J., Fantinato, M. (eds.) ICSOC 2010. LNCS, vol. 6470, pp. 515–526. Springer, Heidelberg (2010)
2. Awad, A., Sakr, S., Kunze, M., Weske, M.: Design by Selection: A Reuse-Based Approach for Business Process Modeling. In: Jeusfeld, M., Delcambre, L., Ling, T.-W. (eds.) ER 2011. LNCS, vol. 6998, pp. 332–345. Springer, Heidelberg (2011)

3. Becker, M., Laue, R.: Analysing Differences between Business Process Similarity Measures. In: Daniel, F., Barkaoui, K., Dustdar, S. (eds.) BPM Workshops 2011, Part II. LNBIP, vol. 100, pp. 39–49. Springer, Heidelberg (2012)
4. Shaft, U., Ramakrishnan, R.: When Is Nearest Neighbors Indexable? In: Eiter, T., Libkin, L. (eds.) ICDT 2005. LNCS, vol. 3363, pp. 158–172. Springer, Heidelberg (2005)
5. Bunke, H., Allermann, G.: Inexact Graph Matching for Structural Pattern Recognition. Pattern Recognition Letters 1(4), 245–253 (1983)
6. Bunke, H.: A Graph Distance Metric Based on the Maximal Common Subgraph. Pattern Recognition Letters 19(3-4), 255–259 (1998)
7. Chávez, E., Navarro, G., Baeza-Yates, R., Marroquín, J.: Searching in Metric Spaces. ACM Comput. Surv. 33(3), 273–321 (2001)
8. Conover, W.J.: Practical Non-Parametric Statistics, 2nd edn. John Wiley and Sons, New York (1980)
9. Croft, W.B., Metzler, D., Strohman, T.: Search Engines: Information Retrieval in Practice. Addison-Wesley (2010)
10. Curran, T., Keller, G., Ladd, A.: SAP R/3 Business Blueprint: Understanding the Business Process Reference Model. Prentice-Hall (1997)
11. Decker, G., Mendling, J.: Process Instantiation. Data Knowl. Eng. 68, 777–792 (2009)
12. Dijkman, R., Dumas, M., García-Bañuelos, L.: Graph Matching Algorithms for Business Process Model Similarity Search. In: Dayal, U., Eder, J., Koehler, J., Reijers, H.A. (eds.) BPM 2009. LNCS, vol. 5701, pp. 48–63. Springer, Heidelberg (2009)
13. Dijkman, R., Dumas, M., van Dongen, B., Käärik, R., Mendling, J.: Similarity of Business Process Models: Metrics and Evaluation. Inf.Sys. 36(2), 498–516 (2011)
14. Dumas, M., García-Bañuelos, L., Dijkman, R.: Similarity Search of Business Process Models. IEEE Data Eng. Bull. 32(3), 23–28 (2009)
15. Hjaltason, G.R., Samet, H.: Index-driven Similarity Search in Metric Spaces. ACM Trans. Database Syst. 28(4), 517–580 (2003)
16. Kunze, M., Weidlich, M., Weske, M.: Behavioral Similarity – A Proper Metric. In: Rinderle-Ma, S., Toumani, F., Wolf, K. (eds.) BPM 2011. LNCS, vol. 6896, pp. 166–181. Springer, Heidelberg (2011)
17. Mendling, J.: Metrics for Process Models: Empirical Foundations of Verification, Error Prediction, and Guidelines for Correctness. Springer (2008)
18. Nüttgens, M., Rump, F.J.: Syntax und Semantik Ereignisgesteuerter Prozessketten (EPC). In: Promise, pp. 64–77 (2002)
19. Rosemann, M.: Potential Pitfalls of Process Modeling: Part B. Business Process Management Journal 12(3), 377–384 (2006)
20. Vanderfeesten, I., Cardoso, J., Reijers, H., Van Der Aalst, W.: Quality Metrics for Business Process Models. In: BPM and Workflow Handbook, pp. 1–12 (2006)
21. Wang, R.Y., Strong, D.M.: Beyond Accuracy: What Data Quality Means to Data Consumers. Journal of Management Information Systems 12(4), 5–33 (1996)
22. Weber, B., Reichert, M.: Refactoring Process Models in Large Process Repositories. In: Bellahsène, Z., Léonard, M. (eds.) CAiSE 2008. LNCS, vol. 5074, pp. 124–139. Springer, Heidelberg (2008)
23. Weidlich, M., Dijkman, R., Mendling, J.: The ICoP Framework: Identification of Correspondences between Process Models. In: Pernici, B. (ed.) CAiSE 2010. LNCS, vol. 6051, pp. 483–498. Springer, Heidelberg (2010)
24. Zezula, P., Amato, G., Dohnal, V., Batko, M.: Similarity Search: The Metric Space Approach. Springer (2005)

The MOSKitt4ME Approach: Providing Process Support in a Method Engineering Context*

Mario Cervera, Manoli Albert, Victoria Torres, and Vicente Pelechano

Centro de Investigación en Métodos de Producción de Software
Universidad Politécnica de Valencia
Camino de Vera s/n, 46022 Valencia, Spain
{mcervera,malbert,vtorres,pele}@pros.upv.es

Abstract. It is commonly agreed that software developments methods must be defined (or adapted) in-house in order to meet the particular needs of the organizations where they are to be applied. To help meet this challenge, Method Engineering (ME) research aims to provide solutions to efficiently deal with the definition and adaptation of methods, and the construction of the supporting software tools. However, while the product part of methods is fully considered by most ME approaches, the specification and enactment of the process part is less well-supported. To fill this gap, this work presents a methodological ME approach and a Computer-Aided Method Engineering (CAME) environment (MOSKitt4ME) that support the design and implementation of the process part of methods in the context of Model-Driven Engineering. The proposal is illustrated by means of a real case study from the Valencian Regional Ministry of Infrastructure, Territory and Environment.

Keywords: Method Engineering, CAME Environment, Process Support, Model-Driven Engineering.

1 Introduction

The definition of a software development method suitable for all situations is now considered unfeasible [10,13]. For this reason, software organizations need to define (or adapt) their methods in-house in order to meet their specific needs. To help meet this challenge, Method Engineering (ME) research aims to provide solutions [5,15,20,22] to efficiently deal with the definition and adaptation of methods and also with the construction of the supporting software tools.

Similarly to software engineering, which is concerned with all aspects of software and its development, ME is concerned with all aspects of methods and their definition. Thus, most ME approaches define precise engineering solutions that address the definition of the two interrelated aspects that generally comprise methods: product and process [5,20,22]. However, while it is commonly

---

* This work has been developed with the support of MICINN under the project EVERYWARE TIN2010-18011.

P. Atzeni, D. Cheung, and R. Sudha (Eds.): ER 2012, LNCS 7532, pp. 228–241, 2012.

agreed that the product aspect of methods represents the artifacts to be produced during the method execution, the process aspect is usually understood in two slightly different ways. Some consider the process aspect as the overall development process of the method, which encompasses all the activity-related issues needed for software development [17]. By contrast, most ME approaches (e.g., [5,16,21,22]) use the term process at a smaller scale, considering a process as the description of how a single method product must be built.

In this work, we consider processes at the greater scale (hence, we denote hereafter the overall process of methods simply as the method *process part*[1]). We argue that supporting the specification and the enactment of the process part of methods brings important benefits. A precise, complete, and well-structured process specification may be useful to facilitate the understanding of how software development is performed within an organization. Furthermore, the enactment of this process specification in a software environment may be useful to guide software engineers throughout the actual development process and also to automate it as far as possible. However, to the best of our knowledge there is no ME approach that supports the specification and the enactment of the process part of methods, and also provides complete software support to these issues.

In order to fill this gap, this paper describes a methodological approach and a supporting software environment that support the specification of the process part of methods and also the construction of the software tools required to enact this process part. The proof of concept is performed in the context of a ME approach and a Computer-Aided Method Engineering (CAME) environment, called MOSKitt4ME[2], that have been presented by the authors in [6,7]. This ME approach is based on Model-Driven Engineering (MDE) techniques and thereby it proposes defining methods as method models based on the SPEM 2.0 standard [18] (*method design* phase) and semi-automatically building the supporting CASE environments by means of model transformations (*method implementation* phase). This paper focuses on extending both the ME proposal and the MOSKitt4ME tool to enable process specification (during method design) and process enactment (during method implementation). Specifically, the proposal and the CAME environment have been enhanced with an executable process modeling language (BPMN 2.0 [19]) to properly support the specification of the process part of methods. Moreover, the proposal and the CAME environment have also been extended to support the generation of CASE environments that incorporate process enactment through the use of a process engine.

The proposal and the CAME environment presented in this paper are being used at the Valencian Regional Ministry of Infrastructure, Territory and Environment, also known as CIT. Specifically, the gvMetrica method has been specified using MOSKitt4ME and a CASE environment has been generated from this specification. Moreover, gvMetrica is currently being executed in real projects using this CASE environment. The application of the proposal in a real context is allowing us to take initial feedback that can be used to improve it.

---

[1] Likewise, we denote the product aspect of methods as the method *product part*.

[2] MOSKitt4ME is an extension of MOSKitt (http://www.moskitt.org/) for ME.

The paper is structured as follows. First, section 2 presents a brief state-of-the-art review that highlights some of the process support limitations that present current ME approaches. Then, section 3 provides an overview of our proposal. Section 4 describes how the process specification is performed during the method design. Section 5 describes how the process enactment is supported in the method implementation. Finally, section 6 draws some conclusions.

2   State of the Art

The term Method Engineering was introduced in the mid-eighties by Bergstra *et al.* in [3]. Thereafter, many research efforts have attempted to provide solutions to the challenges that ME entails. Some of the most relevant contributions are those by Brinkkemper [4,5], Prakash [20,21], Ralyté [22,23] and Karlsson [15,16]. These works are based on a modular view of methods, whereby methods are built by assembling different kinds of methods modules, namely method fragments, method blocks, method chunks and method components respectively. With regard to the process support, these modules focus on the processes necessary to develop specific method products. The method process part is defined when these modules are assembled, generally by means of precedence relationships that establish their execution order. Thereby, the resulting process is quite limited in terms of control flow, since complex behavior (such as the expressed by branching conditions, events, synchronizations, etc.) cannot be defined.

Another important limitation of these approaches refers to process enactment. The proposals by Ralyté [22,23] and Karlsson [15,16] do not consider process enactment. Prakash [20,21] defines an enactment algorithm that is oriented towards the construction of specific method products and, therefore, does not align with process enactment as we intend to support in our ME approach. The support provided for process enactment in Brinkkemper's proposal [4,5] is more akin to our work. However, since only two types of relationships (i.e., precedence and conditional precedence) can be established between process fragments, the resulting processes, and consequently their enactment, may be too limited for software engineers working on real development projects.

Other important contribution to the ME field is the OPEN Process Framework (OPF) [12]. The OPF provides a repository of method modules (in OPF called method components) that are defined in terms of a meta-model. This meta-model has recently been upgraded to fit the ISO/IEC 24744 standard [14]. While this standard does support the specification of the process part of methods, it provides limited support for the definition of processes with a complex control flow. In ISO/IEC 24744 process elements are defined as *work units*, which are allocated in *stages*. The control flow of the process is established by the stages, which can only be associated via precedence relationships. Aharoni *et al.* note this problem in [2] and suggest enriching processes by means of an extension of the stage concept that allows work units to be combined within stages in more meaningful ways than simple inclusion (e.g., concurrently or iteratively). Another important limitation refers to the lack of formalization of process execution semantics, which hinders process enactment via a process engine.

Another recent standard initiative is represented by the SPEM 2.0 standard [18]. SPEM 2.0 defines a meta-model for development methods that also presents the problems stated above regarding process (control flow) specification and process enactment. However, SPEM 2.0 presents an important advantage that helps overcome these problems, making it a suitable language to be used in our proposal. Specifically, SPEM 2.0 provides powerful mechanisms for enhancing process definitions via behavioral modeling formalisms such as BPMN 2.0 [19]. This allows method engineers to enhance the process definition in terms of control flow specification and process executability.

To sum up, we consider that the main process support limitations of ME approaches are in terms process control flow specification and enactment. To fill this gap, this paper extends the work presented by the authors in [6,7]. Our main intent is to meet ME needs regarding support to the process part of methods.

3   Overview of the MOSKitt4ME Approach

Figure 1 shows an overview of our ME approach. The proposal is situated within the context of MDE. Following MDE principles, methods are first defined as models (*method design* phase) and these models are then used by model transformations to generate the supporting CASE environments (*method implementation* phase). During the method design, the method engineer defines both the product and process parts of the method. This is carried out by assembling method fragments that are available in a method base repository [7]. Once the method design is finished, a CASE environment that provides support to both the product and process parts of the method is obtained during the method implementation. This is done by means of a Model-To-Text (M2T) transformation.

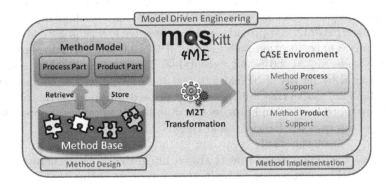

Fig. 1. Overview of MOSKitt4ME

This paper focuses on the process support provided in both phases of our ME approach. The main goal of the work is to allow method engineers not only to properly specify the process part of methods (during method design) but also to bring this specification to execution (during method implementation). In order to achieve this goal, our proposal has been defined based on a set of needs that must be met. We consider these needs as a first step towards suitable process support in ME and CAME technology.

With respect to method design, we need a language to properly specify the process part of the method. This language must:

- be expressive enough to enable the representation of complete, understand-able, unambiguous, and well-structured processes.
- fully formalize process execution semantics so that process enactment support can be provided in the CASE environment supporting the method.

With respect to method implementation, we need to enhance CASE environments with mechanisms that provide support to:

- the execution of the process specification. Process engines provide a set of enactment facilities (such as task orchestration, task automation, constraint enforcement, etc.) that allow CASE environments to provide guidance to software engineers throughout the actual development process and also to partially automate its performance.
- the management of the method products consumed or produced during the process execution. Process engines must be able to invoke the software tools that enable the creation and manipulation of the method products. There-fore, these tools (editors, generators, etc.) must be integrated in the CASE environment supporting the method.

In order to meet the needs regarding method design, our proposal combines the use of SPEM 2.0 and BPMN 2.0 for various reasons. On the one hand, SPEM 2.0 provides better support than BPMN 2.0 for method modeling. On the other hand, BPMN 2.0 provides more expressiveness than SPEM 2.0 with respect to process specification. In addition, BPMN 2.0 fully formalizes execution semantics, while SPEM 2.0 is not executable. Thereby, the method is defined in our proposal by means of SPEM 2.0, and the process part is complemented with a BPMN 2.0 model that enhances the process definition in terms of process (control flow) specification and process executability. We provide a more detailed comparative analysis of SPEM 2.0 and BPMN 2.0 in [8].

In order to meet the needs regarding method implementation, we have de-veloped a software component that is always integrated in the generated CASE environments. This component is built upon a process engine that enables the execution of the process definition by orchestrating method tasks and invoking tools for the development of the method products. In order for the CASE envi-ronments to contain the required tools, we make use of reusable assets. These assets are stored in a repository and are associated to method elements dur-ing the method design. Then, the M2T transformation takes these assets and integrates them in the generated CASE environment.

4   Process Specification during Method Design

Figure 2 shows our proposal for method design. It is composed of three main steps: *method definition, method configuration,* and *executable process definition.* These steps are described below, focusing on how the process specification is performed. Then, subsection 4.1 illustrates these steps with an example.

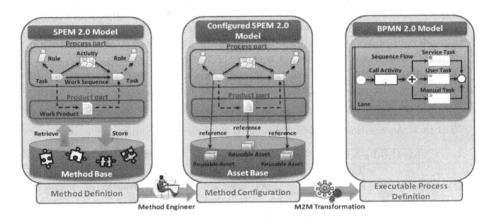

Fig. 2. Method design in MOSKitt4ME

Method Definition. In this step, the method engineer builds the method model by means of SPEM 2.0. As figure 2 illustrates, the main elements that the method engineer must define to build the method process part are *Tasks, Activities, Work Sequences,* and *Roles.* Tasks represent basic units of work. Activities group tasks and other activities, forming breakdown structures. The root activity of these breakdown structures is named *Delivery Process.* Work sequences represent precedence relationships between tasks and activities. Roles are performers of method tasks. On the other hand, the method product part is composed of *Work Products,* which are inputs and outputs of method tasks.

The construction of this model can be performed by assembling reusable method fragments retrieved from a method base repository. Further details about how these fragments are managed in our ME approach can be found in [6,7].

Method Configuration. In this step[3], the method engineer defines the tools that will allow software engineers to perform the method tasks during the process enactment, and the guides that will assist them during the tasks performance. To do so, the method engineer links tasks (and work products) with *reusable assets* that are retrieved from an asset base repository. These assets can be *tool assets,* which contain tools (editors, model transformations, etc.) that enable the creation of products, and *guidance assets,* which contain guidelines (textual descriptions, process models, etc.) about the performance of tasks.

---

[3] Note that this phase differs from the ME approach of method configuration [15].

The association of reusable assets with tasks and products is performed based on the following observations: tasks can be associated with tool assets containing model transformations. These transformations will be executed when the tasks are invoked during the process enactment. These tasks are considered automatic. Manual tasks are not associated with tool assets. Tasks can also be associated with guidance assets. The guidelines contained in these assets will be used as guidance for the user during the process enactment. On the other hand, products must be associated with tool assets defining the notation that will be used during the process enactment for the creation of the products. These assets can contain meta-models or editors. A meta-model defines the abstract syntax of a notation. An editor defines both the abstract and concrete syntax.

To summarize, table 1 gathers the associations allowed between method elements and reusable assets. Further details can be found in [6,7].

Table 1. Associations allowed between method elements and reusable assets

Method Element Type	Reusable Asset Type
Task	Guidance Asset
	Tool Asset (Model Transformation)
Work Product	Tool Asset (Meta-model)
	Tool Asset (Editor)

Executable Process Definition. In this step, the method engineer defines an executable representation of the method process part. We have automated this step by means of a Model-to-Model (M2M) transformation that takes the configured SPEM 2.0 model as input and automatically generates a set of BPMN 2.0 processes. Then, these processes can be manually modified to complete the process specification[4]. This is often needed because BPMN 2.0 provides more expresiveness than SPEM 2.0 with respect to process elicitation.

In order to provide more insights on how the transformation obtains the BPMN 2.0 processes, we summarize in table 2 the mappings between the SPEM 2.0 and BPMN 2.0 concepts. In particular, table 2 contains mappings for all SPEM 2.0 concepts that belong to the process part of a method (see figure 2). Logical elements such as decision nodes are not included since they cannot be represented in SPEM 2.0. The rationale of the mappings is provided below.

1. A SPEM 2.0 *DeliveryProcess* is mapped into a BPMN 2.0 *Process*.
2. A SPEM 2.0 *Activity* is mapped into a BPMN 2.0 *CallActivity* and a BPMN 2.0 *Process*. The *CallActivity* invokes the *Process* when it is executed.
3. A SPEM 2.0 *Task* is mapped into a BPMN 2.0 *ServiceTask* if and only if a reusable asset containing a model transformation is associated to the *Task*.

---

[4] Ideally, the M2M transformation will be implemented as a bidirectional transformation. Thus, changes in the BPMN 2.0 model affecting the SPEM 2.0 model will be automatically transmitted (and vice versa).

Table 2. Mappings between SPEM 2.0 and BPMN 2.0

	SPEM 2.0	BPMN 2.0
1	DeliveryProcess (root Activity)	Process
2	Activity (nested)	Process and CallActivity
3	Task (with ReusableAsset)	ServiceTask
4	Task (with ReusableAsset associated to output WorkProduct)	UserTask
5	Task (without ReusableAsset)	ManualTask
6	WorkSequence	SequenceFlow
7	Role	Lane

4. A SPEM 2.0 *Task* is mapped into a BPMN 2.0 *UserTask* if and only if the *Task* is not automatic (no reusable asset containing a model transformation is associated to it) and a reusable asset is associated to an output *WorkProduct* of the *Task*.

5. A SPEM 2.0 *Task* is mapped into a BPMN 2.0 *ManualTask* if and only if no reusable asset is associated to the *Task* and its output *WorkProducts*.

6. A SPEM 2.0 *WorkSequence* is mapped into a BPMN 2.0 *SequenceFlow*. The source of the *SequenceFlow* is set to the BPMN 2.0 element generated from the *predecessor* of the *WorkSequence*. The target is set to the BPMN 2.0 element generated from the *successor* of the *WorkSequence*.

7. A SPEM 2.0 *Role* is mapped into a BPMN 2.0 *Lane* if and only if the *Lane* has not been previously generated.

4.1 An Example of Process Specification in MOSKitt4ME

In order to illustrate our proposal for process specification, we present a practical example carried out in MOSKitt4ME. The example process has been modeled in real settings at the CIT. Specifically, it corresponds to an excerpt of gvMetrica that deals with the design of information systems.

Method Definition. This step is performed via the EPF Composer [11], an Eclipse-based editor that has been integrated in MOSKitt4ME to enable the creation of SPEM 2.0 models. Figure 3 shows the example process after being defined by means of this editor. As the figure shows, the first steps of the process are to define the system architecture, the user interface, and the business logic. Then, a model defining the system database schema is obtained from the models specifying the business logic. The database model is then used to automatically obtain the database code. Once all the system artifacts have been created, the test cases can be defined. Finally, the design validation is carried out to validate all the work performed during the process.

As depicted in figure 3, the process is represented in SPEM 2.0 as a breakdown structure that is mainly composed of *Activities* (e.g., *Data Persistence Design*) and *Tasks* (e.g., *Database Model Generation*), which reference performing *Roles* (e.g., *Analyst*) as well as input and output *Work Products* (e.g., *UML*

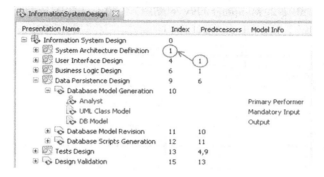

Fig. 3. Example process in SPEM 2.0

Class Model). Moreover, there are *Work Sequences* that are established between method elements (e.g., *System Architecture Definition* is a predecessor of *User Interface Design*).

Method Configuration. This step is performed via a repository client that allows method engineers to retrieve reusable assets from the asset base [7]. As an example, let us consider the task *Database Model Generation*. The execution of this task obtains a database model from a UML class model. To specify this behavior, the method engineer can associate the task with an asset containing the Eclipse plug-ins that implement the UML2DB transformation provided by the MOSKitt tool. On the other hand, the work products of the task can be associated with assets containing the UML meta-model and the MOSKitt SQLSchema meta-model respectively.

Executable Process Definition. This step is performed via the M2M transformation described above, which has been implemented in MOSKitt4ME as an extension of the EPF Composer. Figure 4 shows the BPMN 2.0 processes resulting from applying this transformation to the example process. These processes are represented in terms of the Activiti Designer [1], an Eclipse-based graphical editor that has been integrated in MOSKitt4ME to support BPMN 2.0.

To illustrate how the processes shown in figure 4 have been generated, we present below some examples of the application of the mappings of table 2.

1. The SPEM 2.0 *Delivery Process "Information System Design"* is mapped into a BPMN 2.0 *Process* (diagram "*A*" in figure 4).
2. The SPEM 2.0 *Activity "Data Persistence Design"* is mapped into a *Call Activity* and a BPMN 2.0 *Process* (diagram "*B*" in figure 4).
3. The SPEM 2.0 *Task "Database Model Generation"* is mapped into a *Service Task* since it has a M2M transformation associated to it as a reusable asset.
4. The SPEM 2.0 *Task "Database Model Revision"* is mapped into a *User Task* since it is not automatic but has an output product with a reusable asset associated to it (not shown in the example).

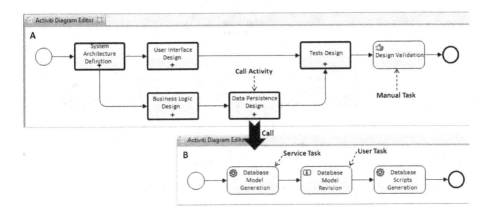

Fig. 4. Generated BPMN 2.0 processes

5. The SPEM 2.0 *Task "Design Validation"* is mapped into a *Manual Task* since it does not have any reusable asset associated to it.
6. The SPEM 2.0 *Work Sequences* are mapped into the *Sequence Flows* that connect the BPMN 2.0 elements in both diagrams.

After the generation of the BPMN 2.0 processes, the method engineer can manually modify them to specify more complex control flows. For instance, the method engineer can add a *Gateway* to specify that the *Call Activity "Tests Design"* must not be executable until all its predecessors are finished.

5   Process Enactment during Method Implementation

The method implementation phase is in charge of the construction of the CASE tool support for the method defined during the method design. In MOSKitt4ME, CASE environments are semi-automatically built by means of a M2T transformation. This transformation takes a method specification as input and obtains a software environment that integrates the tools contained in the reusable assets (as well as a set of static components). We detail this transformation in [9].

In this section we focus on how CASE environments are structured to support the process enactment. As figure 5 shows, CASE environments are divided into three parts: components providing *method product support*, components providing *method process support*, and the *Project Manager Component* (PMC). These parts are introduced below. Subsection 5.1 illustrates them with an example.

Method Product Support. The software components that provide product support must enable the creation of the method products. These components correspond to the reusable assets associated to the method elements during the method configuration. Specifically, *Tool assets* allow software engineers to create the method products by means of model transformations or software

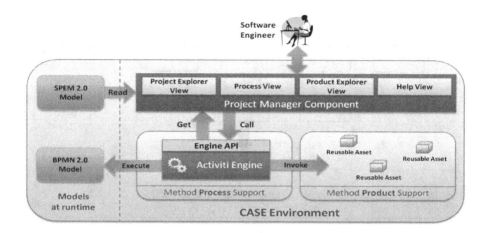

Fig. 5. Method implementation in MOSKitt4ME

applications such as graphical or textual editors. *Guidance assets* do not allow software engineers to create the method products, but they do provide guidance on how the method tasks must be performed to properly develop these products.

Method Process Support. The software components that provide process support must enable the execution of the BPMN 2.0 model. This functionality is provided by the Activiti Engine [1]. The behavior of the engine according to the different task types is the following:

- *Service tasks*: when a service task becomes active, it is automatically executed. The execution of a service task invokes the model transformation that is associated to the task as a reusable asset.
- *User tasks*: when a user task becomes active, the engine invokes the software tools that enable the creation of the output products of the task. These tools are associated to the products as reusable assets.
- *Manual tasks*: when a manual task becomes active, the engine does not perform any action.
- *Call activities*: when a call activity becomes active, the engine automatically starts a new instance of the BPMN 2.0 process referenced by the call activity.

Project Manager Component. The PMC provides a graphical user interface for the CASE environment and assists software engineers during the process enactment. To achieve this goal, it makes use of the Activiti engine to execute BPMN 2.0 process instances and also makes use of the SPEM 2.0 model to extract information about the method that is not represented in the BPMN 2.0 model. The PMC is divided into the following Eclipse views:

- *Project Explorer*: This view is provided as part of the Eclipse platform and has been integrated in the PMC to show the projects that have been created and the resources they contain (files, folders, etc.). From this view, the software engineer can create new projects, delete existing projects, etc.

- *Process*: This view shows the current state of the process instance associated to the project that is selected in the Project Explorer view. From this view, the software engineer can invoke the execution of the tasks that are executable. Once a task is finished, the PMC invokes the engine API to set the task as executed and proceed to the next state of the process. The Process view also enables task filtering based on the role of the users.
- *Product Explorer*: This view shows a hierarchical picture of the artifacts that have been produced during the course of the project that is selected in the Project Explorer view. This hierarchy is based on domains, subdomains, and work product elements, which are read from the SPEM 2.0 model. The Product Explorer view also enables product filtering based on the user role.
- *Help*: This view is provided as part of Eclipse and has been integrated in the PMC to provide guidance to software engineers during the performance of the method tasks. This view is dynamically updated based on the task that is selected in the Process view. The guides to show are known by the PMC because they were associated to the selected task as a reusable asset.

5.1 An Example of Process Enactment in MOSKitt4ME

In order to illustrate how processes are enacted in MOSKitt4ME, we continue with the example introduced in subsection 4.1. Let us consider that a new project has been created by means of the Project Explorer view. When this project is selected, the Process and Product Explorer views are updated accordingly. At this point, the Process view shows the initial state of the process and the Product Explorer view is empty. Now, let us consider that the first three activities have been executed, and the next executable task is *Database Model Generation*. Since this task is automatic, the PMC automatically invokes the associated transformation. Once the transformation is finished, the task is set as executed and the process proceeds to the next state. This state is illustrated in figure 6.

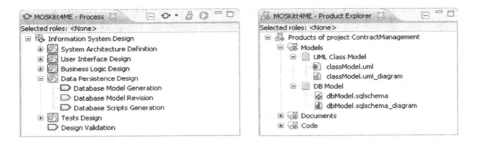

Fig. 6. Process and Product Explorer views

The Process view (left)[5] shows the tasks and activities in different colors depending on whether they have already been executed (blue), are executable

---

[5] Additional screenshots of the tool can be found in our previous work [6,7,9] and also in https://users.dsic.upv.es/~vtorres/moskitt4me/

(green), or are not executable (red). Note that, even though the Process view shows the process in terms of SPEM 2.0, the process instance corresponds to an instance of a BPMN 2.0 process. This is possible because there is a one-to-one correspondence between SPEM 2.0 tasks and BPMN 2.0 tasks.

The Product Explorer view (right) depicts some artifacts that have been produced during the process enactment. In this case, it is showing the input and output products of the last executed task (i.e., *Database Model Generation*).

Now, let us consider that the user wants to proceed with the process execution. To do this, the user selects the task *Database Model Revision* in the Process view. This action has a twofold effect. The Help view is updated to show textual guidance about the task and, since the task is a user task, the PMC opens the software tool that allows the software engineer to carry out the task.

Once all the tasks have been executed, the process engine deletes the process instance and, therefore, the project can be considered as concluded.

6   Conclusions

In this paper we present an extension of our ME approach and CAME environment (MOSKitt4ME) [6,7] so as to provide adequate support to process specification during method design and process enactment during method implementation. This extension builds on the idea of combining the use of SPEM 2.0 and BPMN 2.0. This is based on the fact that BPMN 2.0 can resolve SPEM 2.0 limitations with respect to process elicitation and process executability.

MOSKitt4ME is currently being used in real settings at the CIT. Since practitioners at the CIT are familiar with the techniques, languages, and tools used in our work (i.e., MDE, SPEM 2.0, BPMN 2.0, Eclipse, etc.), MOSKitt4ME has had a low learning curve. This brings important benefits, such as shorter periods of training, and lower overhead. The use of the tool in a real context is also providing us initial feedback that allows us to identify limitations of our work. For instance, MOSKitt4ME does not yet properly deal with the dynamic nature of projects. Therefore, as future work we aim to provide support for variability and evolution of methods (and CASE tools) at runtime.

References

1. Activiti, http://www.activiti.org/
2. Aharoni, A., Reinhartz-Berger, I.: A Domain Engineering Approach for Situational Method Engineering. In: Li, Q., Spaccapietra, S., Yu, E., Olivé, A. (eds.) ER 2008. LNCS, vol. 5231, pp. 455–468. Springer, Heidelberg (2008)
3. Bergstra, J., Jonkers, H., Obbink, J.: A software development model for method engineering. In: Esprit 1984: Status Report of Ongoing Work (1985)
4. Brinkkemper, S.: Method engineering: Engineering of information systems development methods and tools. Information and Software Technology 38, 275–280 (1996)
5. Brinkkemper, S., Saeki, M., Harmsen, F.: Meta-modelling based assembly techniques for situational method engineering. Inf. Syst. 24, 209–228 (1999)

6. Cervera, M., Albert, M., Torres, V., Pelechano, V.: A Methodological Framework and Software Infrastructure for the Construction of Software Production Methods. In: Münch, J., Yang, Y., Schäfer, W. (eds.) ICSP 2010. LNCS, vol. 6195, pp. 112–125. Springer, Heidelberg (2010)

7. Cervera, M., Albert, M., Torres, V., Pelechano, V.: Turning Method Engineering Support into Reality. In: Ralyté, J., Mirbel, I., Deneckère, R. (eds.) ME 2011. IFIP AICT, vol. 351, pp. 138–152. Springer, Heidelberg (2011)

8. Cervera, M., Albert, M., Torres, V., Pelechano, V.: A comparative analysis of SPEM 2.0 and BPMN 2.0. Tech. rep., Universidad Politécnica de Valencia (2012), http://www.pros.upv.es/index.php/en/technical-reports/

9. Cervera, M., Albert, M., Torres, V., Pelechano, V., Bonet, B., Cano, J.: A technological framework to support model driven method engineering. In: Actas De Los Talleres De Las JISBD, pp. 47–56 (2010)

10. Cockburn, A.: Selecting a project's methodology. IEEE Software 17, 64–71 (2000)

11. Eclipse Process Framework, http://www.eclipse.org/epf/

12. Firesmith, D., Henderson-Sellers, B.: The OPEN Process Framework: An Introduction. Addison-Wesley (2002)

13. Henderson-Sellers, B., Ralyté, J.: Situational method engineering: State-of-the-art review. J. UCS. 16, 424–478 (2010)

14. ISO/IEC: Software Engineering: Metamodel for Development Methodologies. ISO/IEC 24744 (2007)

15. Karlsson, F., Ågerfalk, P.J.: Method configuration: adapting to situational characteristics while creating reusable assets. Information and Software Technology 46, 619–633 (2004)

16. Karlsson, F., Ågerfalk, P.J.: Towards structured flexibility in information systems development: Devising a method for method configuration. J. Database Manag. 20, 51–75 (2009)

17. Niknafs, A., Asadi, M.: Towards a process modeling language for method engineering support. In: 2009 WRI World Congress on Computer Science and Information Engineering, vol. 07, pp. 674–681. IEEE Computer Society Press (2009)

18. OMG: Software & Systems Process Engineering Metamodel, v2.0 (2007)

19. OMG: Business Process Model and Notation, v2.0 (2011)

20. Prakash, N.: Towards a formal definition of methods. Requir. Eng. 2, 23–50 (1997)

21. Prakash, N.: On method statics and dynamics. Inf. Syst. 24, 613–637 (1999)

22. Ralyté, J., Rolland, C.: An Approach for Method Reengineering. In: Kunii, H.S., Jajodia, S., Sølvberg, A. (eds.) ER 2001. LNCS, vol. 2224, pp. 471–484. Springer, Heidelberg (2001)

23. Ralyté, J., Rolland, C.: An Assembly Process Model for Method Engineering. In: Dittrich, K.R., Geppert, A., Norrie, M. (eds.) CAiSE 2001. LNCS, vol. 2068, pp. 267–283. Springer, Heidelberg (2001)

Extending Conceptual Data Model for Dynamic Environment

Nicolas Lumineau[1,2], Frédérique Laforest[1,4],
Yann Gripay[1,3], and Jean-Marc Petit[1,3]

[1] Université de Lyon, CNRS
[2] Université Lyon 1, LIRIS, UMR5205
[3] INSA-Lyon, LIRIS, UMR5205
{Nicolas.Lumineau,Yann.Gripay,Jean-Marc.Petit}@liris.cnrs.fr
[4] Université Jean Monnet, LT2C
{Frederique.Laforest}@telecom-st-etienne.fr

Abstract. The design of data-centric pervasive applications in dynamic environments is raising more and more interests in many application domains. To design pervasive applications, we take advantage of recent advances in database management systems dealing with streams and services: new systems exist that simplify pervasive application deployment. We believe that the field is now mature and that conceptual data models like Entity-Relationship models could be revisited at the light of the pervasive application requirements. In this paper, we propose to extend an Entity-Relationship model to a new conceptual model, the so-called XD-ER equipped with some key notions: *dynamic datasource types* to model both streams and services and *dynamic relationship types* to link dynamic datasource types to classical entity types. Based on the SoCQ data model, we point out how to transform a conceptual XD-ER schema into XD-Relations straightly implementable in the SoCQ engine. The use of our model is shown through a running example [1].

Keywords: Conceptual model, pervasive application, data streams.

1 Introduction

Pervasive applications [18] gather distributed and heterogeneous components like sensors, off the shelf web services spread over the Internet or on local area networks providing new kinds of data. The development of pervasive applications is quite difficult and requires the ability to integrate distributed data providers, encompassing databases, data streams and services. This integration requires efforts at different levels. At the implementation level, the concept of service and its associated standards have allowed a big step forward. At the logical level, efforts have been made for stream data management [19] and for the integration of data, streams and services [3,14,20]. At the conceptual level, few works have

---

[1] This work was partially supported by the ANR (French National Research Agency) project Optimacs (ANR-08-SEGI-014, 2008-2012).

tackled the problem of representing all data available for one application in an integrated data view, whatever the data sources.

Data from streams have a particular lifecycle: they are transient data that should be caught by the running queries when they arrive [2]. Similarly, data from services appear only after service calls, that happen at query time [20]. Such characteristics cannot be presented in existing data modeling conceptual models like the Entity-Relationship model [10] (ER) or its extensions [17,15,6]. Only static data are easily caught with ER model. Rather surprisingly, conceptual data models dealing with dynamic data like streams and services have not received much attention from the research community. For example, UML diagrams are not satisfactory because many diagrams are necessary. In opposition to this static point of view, many approches focus on data exchange modeling and take care of streams and/or services. We identify two main types of approaches: dataflow modeling and event modeling. Data flow modeling [5,7] ease the design of data exchange and of processes performed on data. Event modeling [11] allows specifying the data producer, data consumer and a communication channel through which they exchange data. Both approaches are not data schema-oriented. To sum up, the boundaries between static and dynamic data have moved over the last decade. Existing conceptual models do not include the concepts required to describe data coming from the different kinds of data providers. Recent advances in database management systems deal with streams and services: new systems exist nowadays with primitives that simplify the design of pervasive applications, e.g [14]. It points out that among the three levels of the ANSI-SPARC architecture, solutions for physical and logical levels now exist. Nevertheless, to the best of our knowledge, no conceptual model has ever been proposed to design data, streams and services in an integrated manner.

The main contribution of this work is a conceptual model that allows expressing the specificity of data, stream and services. We propose to extend ER to the eXtended Dynamic Entity-Relationship model (XD-ER) where a difference is made between classical data stored in databases–that build the so-called static data universe–and data sources like streams and services–that build the so-called dynamic data universe of the pervasive environment of the application–. We provide a specification of the XD-ER model that exhibits some key notions: 1) dynamic datasource type to model both methods and streams and 2) dynamic relationship type to link datasources to static data. Based on the SoCQ data model [1,14,13], we point out how to transform a conceptual XD-ER schema into the logical XD-Relational model implementable in the SoCQ engine [1]. The complete process is illustrated through a concrete scenario related to the transportation of medical containers.

The rest of the paper is organized as follows. The next section provides a motivating scenario that illustrates our proposal. The specification of our conceptual model called XD-ER is given in Section 3. In Section 4 we define the rules allowing tranforming a XD-ER schema into a XD-Relational model and we conclude in section 5.

2   Motivating Scenario: Medical Container Tracking

Our motivating scenario is about the transportation of fragile biological matter in sensor-enhanced containers. Quality criteria for biological matter transportation are defined by legislation. For example, living matters have to be quickly handled by technical biologists before some given deadline; they cannot tolerate too high or too low temperatures. The transported matter status is continuously checked by a temperature sensor to verify temperature variation, a timer to control the deadline beyond which the transportation is unnecessary and a GPS to know the container position at anytime. Moreover, each container contains memory to store event logs.

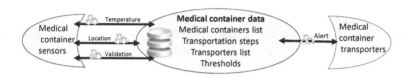

Fig. 1. Interactions in the medical container scenario

The transportation is divided into succeeding steps, each step being ensured by a different transporter. At each step, the corresponding transporter has to monitor quality criteria. Sensors on the container can emit alerts that are forwarded to the transporter. For each step, a supervisor determines thresholds for the different quality criteria the container must follow. When a threshold is exceeded, the container sends a text message to the transporter. A validation is done each time a new transporter takes the responsibility of the container. To validate the state of a container, the event log is read to detect any failure.

In this scenario,only little information is static and can be stored in classical databases: the medical containers descriptions, the transportation steps, the transporters and the thresholds. All other data in this environment are dynamically produced by distributed services and accessed through method invocations (e.g. get current location) or stream subscriptions (e.g. temperature notifications). Moreover, services can provide additional functionality like sending some messages (e.g. by SMS) when an alert is triggered.

3   XD-ER for Dynamic Environments

In this section, we present the concepts of the eXtended Dynamic-ER model that takes its foundations in the ER model [8]. Our objective is to propose a model gathering the data requirements of the user, whatever the data sources. This is done through some extensions based on the concepts of entity types, relationship types and attributes. We propose to include the specificities of dynamic environments composed of databases, services and streams. In the rest of the paper, the term *functionality* encompasses all kinds of dynamic data sources, *i.e.* both streams and methods provided by services in the environment. We illustrate this specification with our motivating scenario about the tracking of medical containers.

3.1 Specification

As XD-ER is an extension of the ER model, the specifications of entity types and relationship types are not modified. As defined in [10], "an entity is a 'thing' in the real world with independent existence [...] an entity may be an object with a physical existence, [...] or it may be an object with a conceptual existence". Moreover, a relationship among entities "is a logical link between entities". Using these definitions for dynamic datasources would not allow to make correct transformations into a logical model. We specify in the following the new notions required to describe pervasive applications with XD-ER.

Dynamic Datasource Type. A *dynamic datasource type* is a concept regarding data with changing availability. It can be seen as a "set of distributed functionalities" in the real world with independent and autonomous existence and a changing availability. A *dynamic datasource type* has a particular status outside the database storage: it takes part of the dynamic universe of the pervasive environment. A dynamic datasource type may represent a service gathering a set of methods and/or streams of data produced by a sensor. Properties of dynamic datasource types represent configuration items of the datasource (for example the precision required from a sensor). The difference between methods and streams is expressed in dynamic relationship types (see below).

In a XD-ER diagram, a *dynamic datasource type* is graphically represented by a triangle as illustrated in figure 2. Under its name, properties are listed with their type.

Example 1. Entity types *RecipientMobile* and *MedicalContainer* represent the recipient mobile of the transporters and static data on the medical containers, (see Figure 3). Dynamic datasource types are used to represent the services able to supply temperature, location, container validation and messaging with the dynamic datasource types *TemperatureService*, *LocationService*, *ValidationService* and *AlertService*.

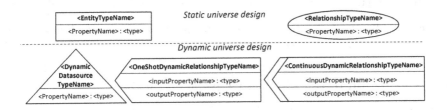

Fig. 2. Elements of the conceptual model XD-ER

Dynamic Relationship Type. A *dynamic relationship type* builds the bridge between the static universe and the dynamic universe. A *dynamic relationship type* is a link between a classical entity type (resp. relationship type) and a dynamic datasource type that indicates that the entity type (resp. relationship type) may consume data provided by the dynamic datasource type. A dynamic relationship type refers to one functionality of the dynamic datasource type. Note that no dynamic relationship type can be created between dynamic datasource types. We define two kinds of dynamic relationship types according to the related functionality of the dynamic datasource type:

- A *Continuous dynamic relationship type* is a link between an entity type and a functionality of a dynamic datasource type that provides a stream.
- A *One-shot dynamic relationship type* is a link between an entity type and a functionality of a dynamic datasource type that provides its data on demand.

In a XD-ER diagram, dynamic relationship types are graphically represented by flags. As illustrated in Figure 2, we consider a simple flag for a one-shot dynamic relationship types and a flag with an arrow for a continuous dynamic relationship types.

Example 2. A relationship type is defined between entity types *MedicalContainer* and *RecipientMobile*. As a medical container consumes data from a temperature service either in a one-shot manner or in a continuous manner, two dynamic relationship types are defined between the *TemperatureService* dynamic datasource type and the *MedicalContainer* entity type. The continuous dynamic relationship type named *temperatureNotification* represents temperature streams subscription and the one-shot dynamic relationship type named *getTemperature* represents temperature data retrieved through a method call on the service.

Attributes of the XD-ER Model. Attributes are the properties of entity types, relationships types [10] or dynamic datasource types. A dynamic relationship type has properties corresponding to input and output attributes defined by the corresponding functionality. Input attributes represent data used to call a method implementing the corresponding functionality. Output attributes represent data that the dynamic datasource is able to supply. Those attributes are only valued on the fly at query evaluation time when entities are available.

Example 3. The dynamic relationship type *temperatureNotification* has two output properties : *temperatureValue* and *temperatureDate*.

3.2   Application to the Medical Container Tracking

Figure 3 provides an XD-ER schema of the Medical Container Tracking. In the static universe, transporters mobile phones are represented by an entity type *RecipientMobile* and medical containers are defined by an entity type *MedicalContainer*. We assume that a mobile and a container have an unique identifier that we use as keys. The *RecipientMobile* entity type also has a property *phone* that represents the phone number of the recipient mobile. As recipient mobiles supervise containers, a relationship type named *supervise* links *RecipientMobile* and *MedicalContainer*. A transporter may supervise many medical containers; one container is supervised by many transporters: one per transportation step. So the *supervise* relationship type has many-to-many cardinality. Supervision is based on temperature thresholds that are defined for each container and each transportation step. It is set as a property of *supervise*. This relationship also contains date and place of transportation steps.

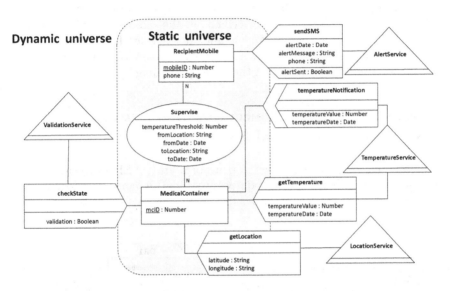

Fig. 3. A XD-ER schema used for the tracking of medical containers. Dynamic relationships form a bridge between the static universe and the dynamic one.

The dynamic universe is composed of four services: *temperatureService* provides information about the temperature observed in the container, *validationService* provides a validation of the container status, *locationService* gives the position of the container, and *alertService* sends SMS. Each service is modeled by a dynamic datasource type of the same name.

The temperature services propose two functionalities: a stream of captured temperatures called *temperatureNotification* and a method *getTemperature* that can be called to get the last temperature capture. To specify that our medical container has the possibility to continuously query the functionality *temperatureNotification* belonging to *temperatureService*, we define the continuous dynamic relationship type *temperatureNotification* between *MedicalContainer* and *temperatureService*. *TemperatureNotification* has no input, i.e. no data is necessary to use its functionality *temperatureNotification*. Its outputs are couples with two elements: *temperatureValue* and *temperatureDate*.

The location of a medical container can be obtained through the method *getLocation* of the *LocationService* service. To specify that our medical container has the possibility to query on demand the functionalities *getTemperature*, *checkState* and *getLocation* respectively belonging to *TemperatureService*, *ValidationService* and *LocationService*, we define three one-shot dynamic relationship type between *MedicalContainer* and those dynamic datasource types. These dynamic relationship types have no input and they provide the following ouputs: *temperatureValue* and *temperatureDate* for the functionality *getTemperature*, *latitude*, *longitude* for the functionality *getLocation*, and *validation* for the functionality *checkState*. In the same way, the validation of a medical container state can be obtained through the method *checkState* of the *ValidationService* service.

The functionality *sendSMS* is specified by the one-shot dynamic relationship *sendSMS* between *RecipientMobile* and *AlertService*. In this case, *sendSMS* needs three inputs: *alertDate*, *alertMessage* and *phone*. We consider an output *alertSent* which allows committing the message sending.

4   From the XD-ER Conceptual Model to the XD-Relational Logical Model

In this section, we point out how to transform an XD-ER diagram composed of entities, relationships, dynamic datasources and dynamic relationships, into SoCQ DDL [1]. The reader may refer to [1,14] for details about SoCQ.

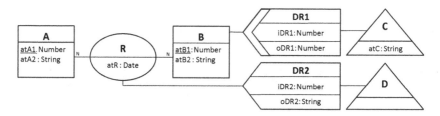

Fig. 4. Two entity types (A,B), one relationship type (R), two dynamic datasource types (C,D) and two dynamic relationship types (DR1, DR2)

To achieve the transformation of a XD-ER schema into XD-Relational Model, our methodology requires first to transform entity types and relationship types as usual. The complex issue of attribute naming w.r.t. the URSA (Universal Relation Schema Assumption) is not tackled in this paper neither are the algorithmic issues of such a mapping [16]. In a second time, we transform dynamic datasource types and dynamic relationship types using new rules we introduce.

The transformation rules for entity types and relationship types are the same as the well-known rules to transform an ER schema into the relational model [12]. For instance in Figure 4, the entities *A* and *B* are transformed into two relations respectively *XRA* and *XRB*. The formal definition of these relations

- type(XRA) = 2,
- realSchema(XRA) = {$atA1,atA2$},
- virtualSchema(XRA) = $\emptyset$,
- key(XRA) = {$atA1$},
- BP(XRA) = $\emptyset$,
- infinite(XRA) returns false.

(a)

- type(XRB) = 2,
- realSchema(XRB) = {$atB1,atB2$},
- virtualSchema(XRB) = $\emptyset$,
- key(XRB) = {$atB1$},
- BP(XRB) = $\emptyset$,
- infinite(XRB) returns false.

(b)

- type(XRR) = 3,
- realSchema(XRR) = {$atA1,atB1,atR$},
- virtualSchema(XRR) = $\emptyset$,
- key(XRB) = {$atA1,atB1$},
- BP(XRR) = $\emptyset$,
- infinite(XRR) returns false.

(c)

Fig. 5. Definition of XD-Relation : (a) XRA, (b) XRB and (c) XRR

are respectively depicted in Figure 5.a and Figure 5.b. Moreover, the many-to-many relationship R is transformed into relation XRR (see Figure 5.c).

After this first transformation step, we consider the dynamic datasource types and the dynamic relationship types. These transformations modify the relation schemas previously created. In the following, we say a entity/relationship type is related to a dynamic datasource type when a dynamic relationship type exists between them.

Rule 1 (Dynamic datasource type transformation). *Transforming a dynamic datasource type implies to update the relation schema of the related entities/relationships. The name of each dynamic datasource type becomes an attribute of type 'SERVICE' in the relation schema of related entity/relationship types.*

For instance in Figure 4, we consider the dynamic datasource types C and D respectively related to entity B by dynamic relationship type DR1 and to relationship R by dynamic relationship type DR2. This rule implies to update the definition of XRB and XRR, as follows:

- type(XRB) $\leftarrow$ type(XRB) + 1,
- realSchema(XRB) $\leftarrow$ realSchema(XRB) $\cup$ {s_C} ,
- type(XRR) $\leftarrow$ type(XRR) + 1,
- realSchema(XRR) $\leftarrow$ realSchema(XRR) $\cup$ {s_D}.

Note that s_C (resp. s_D) matchs with the datasource type name C (resp. D). Moreover, the attributes of a dynamic datasource type (e.g. atC in Figure 4) are parameters of service implementation that we do not consider in the XD-relational model.

Rule 2 (Dynamic relationships transformation). *Transforming a dynamic relationship type implies to update the relation schema of related entities/relationship types as follows : A* Binding Pattern *coupling a service declaration and a prototype definition is added to this relation schema. A prototype is defined according to the name of the dynamic relationship type. The input schema of this prototype is built from the input attributes of the dynamic relationship type. The output schema of this prototype is built from the output attributes of the dynamic relationship type. Lastly, all output attributes of the dynamic relationship type are added to the virtual schema of the relation schema.*

In Figure 4, the dynamic relationship type $DR1$ links entity B with dynamic datasource type C. The definition of XRB is updated as follows :

- schema($Input_{XRB}$) ← schema($Input_{XRB}$) ∪ $\{I_{DR1}\}$
- schema($Output_{XRB}$) ← schema($Output_{XRB}$) ∪ $\{O_{DR1}\}$
- virtualSchema(XRB) ← $virtualSchema$(XRB) ∪ $\{O_{DR1}\}$
- BP(XRB) ← BP(XRB) ∪⟨$prototype_{DR1}, service_C$⟩, where:
    - $prototype_{DR1} = p_{DR1}$,
    - $service_C = s_C$.

Thus, p_{DR1} matchs with the dynamic relationship type name $DR1$. Similar updates are achieved on XRR to express the link between R and D *via DR2*.

Rule 3 (Transformation of continuous dynamic relationship types). *A continuous dynamic relationship type modifies the state of the XD-Relation which is considered as infinite.*

In Figure 4, the continuous dynamic relationship type $DR1$ links entity B with dynamic datasource type C. Lastly, infinite(XRB) returns true.

Note that the ColisTrack scenario described in this paper has been implemented as a testbed in [13] and more details are available in [1].

5   Conclusion

In this paper, we describe an extension of the Entity-Relationship conceptual model to deal with streams and services, two fundamental notions of pervasive applications. The rules we propose allow to easily build the logical model which is then implementable with the SoCQ system. We show the complete process from conceptual design to implementation through a scenario about the supervision of medical containers transportation. Other pervasive applications can be easily designed with XD-ER like the famous Linear Road scenario [4], even if it does not include services.

The objective of XD-ER is not to be twined with the SoCQ model but has been thought to be generic enough to be implemented with other systems, like the Hypatia model [9] or the service-oriented model proposed in [20].

References

1. SoCQ Project, http://liris.cnrs.fr/socq/Main/HomePage
2. Abadi, D.J., et al.: The Design of the Borealis Stream Processing Engine. In: Proceedings of CIDR 2005, pp. 277–289 (2005)
3. Abiteboul, S., Manolescu, I., Taropa, E.: A Framework for Distributed XML Data Management. In: Ioannidis, Y., Scholl, M.H., Schmidt, J.W., Matthes, F., Hatzopoulos, M., Böhm, K., Kemper, A., Grust, T., Böhm, C. (eds.) EDBT 2006. LNCS, vol. 3896, pp. 1049–1058. Springer, Heidelberg (2006)
4. Arasu, A., Cherniack, M., Galvez, E., Maier, D., Maskey, A.S., Ryvkina, E., Stonebraker, M., Tibbetts, R.: Linear road: a stream data management benchmark. In: Proceedings of VLDB 2004, pp. 480–491 (2004)
5. Bhattacharyya, S.S., Deprettere, E.F., Keinert, J.: Dynamic and multidimensional dataflow graphs. In: Handbook of Signal Processing Systems, pp. 899–930. Springer US (2010)
6. Blaha, M.R., Premerlani, W.J., Rumbaugh, J.E.: Relational database design using an object-oriented methodology. Commun. ACM 31, 414–427 (1988)
7. Bruza, P.D., van der Weide, T.P.: The semantics of data flow diagrams. In: Proc. of the International Conference on Management of Data, pp. 66–78 (1993)
8. Chen, P.P.-S.: The entity-relationship model toward a unified view of data. ACM Trans. Database Syst. 1, 9–36 (1976)
9. Cuevas-Vicenttín, V., Vargas-Solar, G., Collet, C., Ibrahim, N., Bobineau, C.: Coordinating Services for Accessing and Processing Data in Dynamic Environments. In: Meersman, R., Dillon, T.S., Herrero, P. (eds.) OTM 2010. LNCS, vol. 6426, pp. 309–325. Springer, Heidelberg (2010)
10. Elmasri, R., Navathe, S.B.: Fundamentals of Database Systems, 2nd edn. Benjamin/Cummings (1994)
11. Etzion, O., Niblett, P.: Event Processing in Action, 1st edn. Manning Publications Co., Greenwich (2010)
12. Garcia-Molina, H., Ullman, J.D., Widom, J.: Database systems - the complete book, 2nd edn. Pearson Education (2009)
13. Gripay, Y., Laforest, F., Lesueur, F., Lumineau, N., Petit, J.-M., Scuturici, V.-M., Sebahi, S., Surdu, S.: ColisTrack: Testbed for Pervasive Environment Management System. In: Proc. of EDBT 2012 (2012)
14. Gripay, Y., Laforest, F., Petit, J.-M.: A simple (yet powerful) algebra for pervasive environments. In: Proc. of EDBT 2010, pp. 359–370 (2010)
15. Kroenke, D.M.: Database Processing Conceptual Database Design, 5th edn. Prentice Hall, Inc. (1995)
16. Markowitz, V.M., Shoshani, A.: On the correctness of representing extended entity-relationship structures in the relational model. In: Proceedings of SIGMOD 1989, pp. 430–439 (1989)
17. Teorey, T.J., Yang, D., Fry, J.P.: A logical design methodology for relational databases using the extended entity-relationship model. ACM Comput. Surv. 18, 197–222 (1986)
18. Weiser, M.: The computer for the 21st century. Scientific American 265(3), 94–104 (1991)
19. Widom, J., et al.: STREAM: The Stanford Stream Data Manager. IEEE Data Engineering Bulletin 26(1), 19–26 (2003)
20. Xue, W., Luo, Q.: Action-Oriented Query Processing for Pervasive Computing. In: Proceedings of CIDR 2005, pp. 305–316 (2005)

Contracts + Goals = Roles?[*]

Lam-Son Lê and Aditya Ghose

School of Computer Science and Software Engineering
Faculty of Informatics, University of Wollongong
New South Wales 2522, Australia
{lle,aditya}@uow.edu.au

Abstract. The concept of role has been investigated in various fields of computer science as well as social sciences. While there is no clear consensus on how roles should be represented, a survey of the literature suggests that we should address both responsibilities and rights in the modeling of roles [1]. Based on this, we argue that the responsibilities and rights of roles can be captured by leveraging the notions of business contract and goal (in the sense of the goals of an actor being constrained by the rights associated with the role played by the actor) in the realm of requirements engineering. We leverage existing work on the formalization of business contracts [2] and the formulation of goals in the i^* modeling framework [3]. We devise formal techniques for reasoning about the composition and substitutability of roles and illustrate them through a running example.

Keywords: i^*, Business Contracts, Role Modeling, Formal Methods.

1 Introduction

Role modeling is a mechanism to separate concerns in the early phases of systems development (e.g. analysis) when developers have not actually built the system of interest. At this stage, model elements (e.g. components, agents, actors) that represent components of the system to be built do not yet exist. A role captures a coherent piece of behavior in the system. Roles may interact with one another in a process (in which case they are called *process roles*). A role could be regarded as an interface through which one might access the system or its subsystems (in which case it is called an *interface role*). In subsequent phases of the development process, designers assign roles to components. The representation of roles in the analysis and design phases thus provides critical input to the subsequent phases of the development lifecycle. This rationale also applies to social settings such as institutional design, where roles capture the expected behavior of positions that need to be filled by specific individuals or business entities. Systems (or institutions) can be designed/developed from scratch or can be obtained via the composition of pre-existing components (or via the composition of socio-technical

[*] Funding of this research was provided by the Smart Services CRC Initiative
http://www.smartservicescrc.com.au/

P. Atzeni, D. Cheung, and R. Sudha (Eds.): ER 2012, LNCS 7532, pp. 252–266, 2012.
© Springer-Verlag Berlin Heidelberg 2012

systems containing both system components and people, obtained through a combination of outsourcing, crowd-sourcing and recruitment). Either way, the components (or entities/people) must be capable of playing the roles that were assigned to them.

The so-called principle of separation of concerns matters in the requirements phase too. Unfortunately, the notion of role has received relatively little attention in the literature on requirements engineering (RE). For example, in the literature on the i^* framework, roles are simply regarded as abstract actors [4]. We take as a starting point the approach of Zhu et al [1] to role modeling where roles are associated with both responsibilities and rights. Given that systems often involve collaboration between roles, we need to understand the semantics of putting their responsibilities and rights together. Specifically, we need to answer the following two questions well before the system (or organization) of interest is built: (a) do the component roles of a system conjointly deliver what the system is expected to deliver? (b) how do we know if a role can safely substitute for another role without affecting the requirements of the system it belongs to?

To address these questions, we propose a formal framework for modeling and reasoning about process roles. We are inspired by the notion of socially-enhanced actors in the early-phase requirements that was made popular by the i^* framework [3], as well as the contract-based enterprise specification defined in the Reference Model of Open Distributed Processing (RM-ODP) standard [5]. Modeling roles in such a context involves representing (i) a contract a role is committed to (i.e. role's responsibility); (ii) the rights associated with a role that constrains the space of goals that an actor assuming that role can pursue. We leverage our previous work on semantic business processes, the work of Governatori et al [2] on business contracts and the well-established field of goal-oriented modeling in RE.

Paper Structure. Section 2 presents the background of our work. We discuss the representation of roles in RE and provides a running example in Section 3. Section 4 formally describes the framework. Section 5 surveys related work. Section 6 concludes the paper and outlines our future work.

2   Background

We base our work on semantic business processes, formalization of business contracts and the notion of role in a broad sense such as information systems.

2.1   Semantically Annotated Business Processes

An semantically annotated business process model is a process in which every task has been annotated with immediate effects. To determine the functionality delivered up to a given point of time during the occurrence of an annotated process, we reason about the cumulative effect. We suppose that analysts can associate context-independent effect to each step represented in the process. There exists a technique that contextualizes these effects, i.e., to compute cumulative effects. The technique, called ProcessSEER, involves doing two stages of computation [6]. In the first stage, we derive a set of possible *scenario label*(s) for the given point in the process view. Each scenario label is a precise lits of steps that define a path leading from the start point to a the point being considered. In the second stage, the contiguous sequence of steps in each scenario label is taken into account to accumulate effects annotated to steps along this scenario in a pair-wise fashion.

A functionality annotation states the effect of having functionality delivered at a specified task. The effect can be textual. Alternatively, it could be written in first-order logic (FOL) or some computer-interpretable form. The total functionality delivered up to a certain task is the accumulation of all effects of the precedent tasks. We assume that the delivery annotations have been represented in conjunctive normal form (CNF) where each clause is also a prime implicate (this provides a non-redundant canonical form) [7]. The cumulative effect of tasks can inductively be defined as follows. The cumulative effect of the very first task is equal to its delivery annotation. Let $\langle Tk_i, Tk_j \rangle$ be an ordered pair of consecutive tasks such that Tk_i precedes Tk_j; let e_i be an effect scenario associated with Tk_i and e_j be the delivery annotation associated with Tk_j. Without loss of generality, we assume that e_i and e_j are sets of clauses. The resulting cumulative effect, denoted by $acc(e_i, e_j)$ is defined as follows.

- $acc(e_i, e_j) = e_i \cup e_j$ if $e_i \cup e_j$ is logically consistent
- Otherwise $acc(e_i, e_j) = e'_i \cup e_j$ whereby $e'_i \subseteq e_i$ such that $e'_i \cup e_j$ is consistent and we do not have any $e''_i \subseteq e'_i \subseteq e_i$ such that $e''_i \cup e_j$ is consistent

The task of accumulating functionality annotations is non-trivial since there might be various paths that can be traversed during the occurrence of the process up to the point of time being considered. We call the path leading to a certain point of time a *scenario label*. A scenario label can either be a sequence, denoted by the $\langle \rangle$ delimiters, or a set denoted by the $\{\}$ delimiters or combinations of both. The set delimiters are used to deal with parallel splits, and distinct elements in a set can be performed in any order [6]. Elements in a scenario label could be tasks (which have delivery annotations) or control elements (e.g. mutually exclusive split, parallel split). In addition to pair-wise effect accumulation across scenario labels, we need to make special provision for the following: (i) accumulation across AND-joins, and (ii) accumulation of effects over input/output flows.

2.2   Business Contract Modeling

Business contracts specify obligations, permissions and prohibitions as mutual agreements between business parties [8], as well as actions to be taken when a contract is violated. Governatori et al [2] have proposed such a contract modeling language which includes a non-boolean connective, $\otimes$, to represent contrary-to-duty obligations (i.e., what should be done if the terms of a contract are violated). Deontic operators capture the contractual modality (i.e. obligations, permissions and prohibitions) [9]. Governatori et al represent a contractual rule as $r : A_1, A_2 \ldots A_n \vdash C$ where each A_i is an antecdent of the rule and C is the consequent. Each A_i and C may contain deontic operators but connectives can only appear in C.

As an example, $r : \neg p, q \vdash O_{seller}\alpha \otimes O_{seller}\beta$ is a contractual rule (identified by r) stating that if antecedents $\neg p$ and q hold, then a seller is obliged to make sure that α is brought about. Failure to do so results in a violation, for which a reparation can be made by bringing about β (the connective $\otimes$ can therefore be informally read as "failing which").

Definition 1. *[2] Contractual rules r and r' can be merged into rule r'' as follows where X denotes either an obligation or a permission.*

$$\frac{r : \Gamma \vdash O_s A \otimes (\bigotimes_{i=1}^{n} O_s B_i) \otimes O_s C \quad r' : \Delta, \neg B_1, \neg B_2, \ldots, \neg B_n \vdash X_s D}{r'' : \Gamma, \Delta \vdash O_s A \otimes (\bigotimes_{i=1}^{n} O_s B_i) \otimes X_s D}$$

The $\otimes$ operator is associative but not commutative. This property matters when reasoning about the subsumption and merging of contractual rules. Definition 1 defines how contract rules might be merged. Governatori et al also devise a machinery for determining if one contractual rule subsumes another as presented in Definition 2.

Definition 2. *[2] Let's consider two rules $r_1 : \Gamma \vdash A \otimes B \otimes C$ and $r_2 : \Delta \vdash D$ where $A = \bigotimes_{i=1}^{m} A_i$, $B = \bigotimes_{i=1}^{n} B_i$ and $C = \bigotimes_{i=1}^{p} C_i$. Then r_1 subsumes r_2 (i.e. r_2 can safely be discarded if we have r_1) iff*

1. *$\Gamma = \Delta$ and $D = A$; or*
2. *$\Gamma \cup \{\neg A_1, \ldots, \neg A_m\} = \Delta$ and $D = B$; or*
3. *$\Gamma \cup \{\neg B_1, \ldots, \neg B_n\} = \Delta$ and $D = A \otimes \bigotimes_{i=0}^{k \leq p} C_i$*

2.3 Roles in Information Systems

Zhu et al. [1] provide a comprehensive survey of role modeling in various fields including object-oriented modeling, multi-agent systems, role-based access control, computer-supported cooperative work, social psychology and management. They suggest that role modeling should address both the responsibilities and rights of a role. They also define two categories of roles: interface roles and the process roles. The former is used in systems analysis and design while the latter in systems implementation.

Steimann [10] has studied role modeling mainly from the perspective of object-oriented modeling and identified 15 features of roles in relation to object types and relationships. He has also suggested merging the concepts of interface and role since they both refer to externally-visible behavior of an object that plays a certain role by realizing an interface [11]. In revisiting Bachman's role concept in data modeling [12], Steimann [13] observes that role expectations capture not only behavior (e.g. a scientific reviewer is expected to comment on and evaluate papers she was assigned to) but also "role qualities" (e.g. a scientific reviewer should conduct her reviews thoroughly, constructively and objectively).

These views suggest that we should not be confined by the bounds of action-oriented behavior modeling when it comes to role modeling.

3   Roles in the Requirements Phase

We make our standpoint based on the two modeling frameworks that recently became international standards - the i^* and the RM-ODP.

i^* is a popular notation for capturing the rationale and intention in the early-phase requirements engineering. It defines the concepts of actor, agent and role. Being socially-motivated and inspired of multi-agent systems, i^* adopts the notion of autonomy and intentionality when it comes to the representation of actors in systems analysis and design [3]. The authors of this framework argue that, to capture the social aspects in RE, actors and roles should be represented as being intentional and autonomous, as opposed to being programmable and mechanistic[1]. This framework defines strategic rationale business models where we model a role in an RE phase as something that has internal objectives and activities.

RM-ODP, a standardization effort that defines essential concepts for modeling distributed enterprise systems, positions the concept of role in relation with contract, objective and policy of an enterprise specification, which can be thought of as the requirements for an ODP system [5]. In RM-ODP, we can make the coarse-grained organizational representation of an enterprise system using the enterprise concept of community. A community object is supposed to have (component) enterprise objects. These component objects are said to play roles described in the enterprise specification, in order to fulfill objectives of the community in question. The contract specified for a community object states how to assign the component enterprise objects to roles.

The analysis we make on these two standards would suggest that, in RE, roles might better be described in terms of objectives and contracts instead of plain action-oriented behavior. Contract-oriented behavior accounts for role's commitments (or role's qualities by Steimann [13]). Goal-oriented behavior addresses the intentionality and the autonomy of roles. This vision also coincides with those that have been around in the realms of Information Systems and Conceptual Modeling presented in Subsection 2.3. To reason about role's goal, we leverage the well-established field of goal-oriented modeling. To reason about role's contract, we need to consider all processes in which the role in question participates in order to fulfill the contracts it commits to.

---

[1] Many object-oriented analysis & design methodologies share a vision whereby objects' behavior is defined (by the designer) at design-time. At run-time, objects are instantiated and their operations are invoked on a presumably single thread of control in ways they were designed [14].

3.1   Running Example

Let us consider the business model of a car rental company (as a service). It has roles that are expected to provide the following services: identity check & deposit, cars pickup & return and cars maintenance. We call them Receptionist (or rl_{11} for short to be referred to formally later on), Fleet Manager (or rl_{12}) and Mechanics (or rl_{13}) respectively. Note that the rental car company as a whole also plays a role.

Figure 1 gives an overview of this business model using the i^* notation. In this diagram, i^* actors whose text is in bold represent roles that are considered in this example. Each role have their own goals and contracts. Their goals are represented under the i^* notation of hard goal inside their boundaries. Their

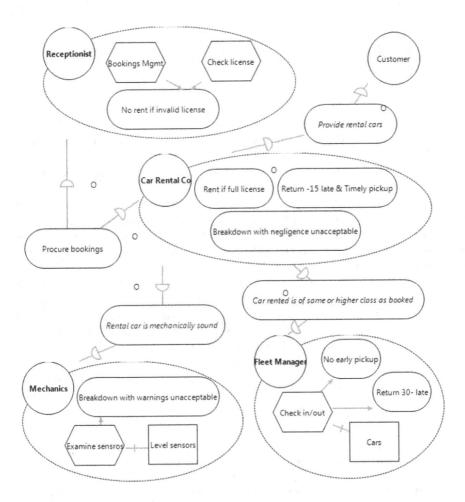

Fig. 1. Representation of roles in the car rental using the i^* notation

Table 1. Informal description of roles participating in the car rental business

Role	Role's Goal	Role's Contract
Role played by the car rental company	To reserve the right not to serve customers who don't have a full driver license; to avoid accepting car return of 15+ minutes late and early pickup; to hold customers responsible for breakdowns caused by their negligence in checking oil/coolant.	The customer expects to be handed in a car of which class was specified in her reservation. If the rental car company fails to do so, they are obliged to provide her with an alternative car of a comparable class. If this option is not available either, the company will offer her some discount on her next rent.
Receptionist (rl_1)	To cancel bookings made by customers who don't have a valid driver license.	We assume that Receptionist always delivers its functionality as expected. No contractual rules are specified for this role.
Fleet Manager (rl_2)	To allow customers to return their rental cars 30- minutes late; to avoid accepting early pickup.	If Fleet Manager can't provide the customer with a car of the class she requested in her reservation, an alternative car of a comparable class shall be provided. If this obligation is violated, Fleet Manager will be charged an amount of money.
Mechanics (rl_3)	To avoid taking responsibility for breakdowns caused by customers who ignored dashboard warnings about the engine oil and coolant during their prolonged renting.	If the car requested by the customer has not been properly serviced, and it is impossible to provide an alternative car of a comparable class, Mechanics is obliged to do a quick service to the selected car. Failure to do so will result in Mechanics being charged.

contracts are also represented under the i^* goal notation but have italicized text[2]. Each contract features a dependency between a pair of roles that engage in it. Table 1 informally describes these goals and contracts.

To formulate the contracts, let us consider a process that starts when a customer shows up to pick up a rental car she has previously booked and ends when she returns her rented car. At a stage of this process, each of these roles is expected to deliver certain functionality. Figure 2 is a diagram (presented in side view to be fit in a single page) that represents this process using the Business Process Model and Notation (BPMN)[3]. Receptionist processes the registration for customers who have booked in advance and checks the customer's driver license. Fleet Manager deals with pick-up & return procedures (and checking for any damages the customers might do to their rental cars). Mechanics takes care of the rental cars and makes sure that any they are in sound condition before the customers pick them up. In this diagram, each task[4] has a name that starts with a prefix telling which role conducts the task.

Receptionist, Fleet Manager and Mechanics each enter in a contract with the role played by the car rental company. They are expected to deliver the aforementioned functionality. In case they fail to deliver it, they are supposed to deliver alternative functionality as a reparation for their violation.

---

[2] i^* does not support the modeling of business contracts. We use this visual trick to make contracts and goals look differently in the diagram of Figure 1.

[3] OMG Business Process Management Initiative http://www.bpmn.org/

[4] We also annotate each task with functionality drawn under a rectangular callout, which is formally represented and useful for reasoning about contracts in the next section.

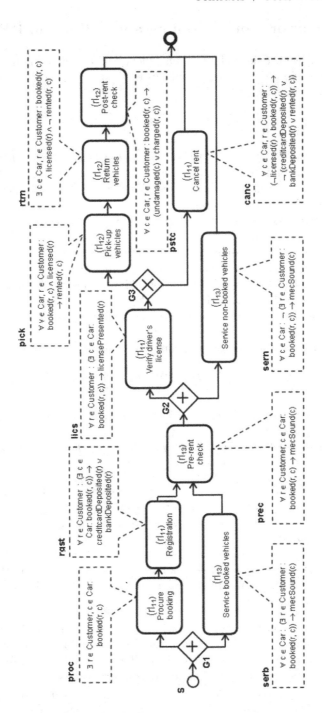

Fig. 2. Roles `Receptionist`, `Fleet Manager` and `Mechanics` in the process of renting a car

4   A Framework for Reasoning about Roles in RE

In this section, we provide a formal machinery for reasoning about the composition and extension of roles, which addresses the research statement presented in the previous section.

4.1   Role Composition

Let $SysBiz$ denote a system or a business that performs a function to deliver value to external or internal actors. Let RL_{sb} be the role played by $SysBiz$ as a whole. This role interacts with roles played by the external actors (hence, external roles). We formally represent RL_{sb} as a tuple $\langle R_{sb}, G_{sb} \rangle$ where R_{sb} denotes a contractual rule that RL_{sb} and the external roles have agreed upon; G_{sb} denotes RL_{sb}'s goal.

In the following, we will use $r_1 \sqsubseteq r_2$ to denote that r_2 subsumes r_1, for contractual rules r_1 and r_2. Let $rl_1, rl_2 \ldots rl_n$ be the component roles that collaborate in order for $SysBiz$ to deliver its function. These roles are assumed by actors that are internal to $SysBiz$, and RL_{sb} can be viewed as a role obtained via the composition of the component roles. Each role rl_i is a tuple $\langle r_i, g_i \rangle$ where r_i denotes a contractual rule agreed upon by rl_i and RL_{sb}; g_i denotes rl_i's goal. Definition 3) formalizes role composition.

Definition 3. *Role RL_{sb} is the composition of roles $rl_1, rl_2 \ldots rl_n$ (and roles $rl_1, rl_2 \ldots rl_n$ constitute role RL_{sb}) if and only if the following hold.*

- *$R_{sb} \sqsubseteq q$ (i.e. rule q subsumes rule R_{sb}) where q is the result of merging $r_1, r_2, \ldots r_n$ according to Definition 1.*
- *RL_{sb}'s goal entails all goals of roles $rl_1, rl_2 \ldots rl_n$. Formally, we have $G_{sb} \models g_1 \wedge g_2 \wedge \ldots \wedge g_n$ and $g_1 \wedge g_2 \wedge \ldots \wedge g_n \not\models \bot$.*

The intuition of Definition 3 is that we relax goals while strengthening rules when we compose roles. Example 1 illustrates this definition. To reason about the contractual behavior of roles, we annotate tasks in Figure 2 with semantic effect written in FOL. Each task has a delivery annotation that is drawn under a rectangular callout. The delivery annotations are written in FOL. We will refer to these annotations by 4-character nick names that appears next to their callout pictograms. The start event and gateways are enumerated as S, G1, G2 and G3 as can be seen in the figure.

Example 1. Let us consider the moment in the rental process (see Figure 2) when a customer is about to pick up a rental car. Roles rl_1, rl_2 and rl_3 are expected to deliver lics, pick and prec respectively. The contractual rule for delivering pick of role rl_2 (see Table 1) can be formally expressed as

$$\boxed{r_2 : \neg pick \vdash O_2 AltCar \otimes O_2 Charged}$$ where *pick* is as depicted in Figure 2 and $AltCar \equiv \forall c \in Car, r \in Customer : booked(r, c) \wedge licensed(r) \wedge \neg rented(r, c) \rightarrow (\exists ac \in Car : comparableClass(ac, c) \wedge mecSound(ac) \wedge rented(r, ac))$.

The formal representation of the contractual rule for role rl_3 is as follows.

$$r_3 : \neg prec, \neg AltCar \vdash O_2 QuickServ \otimes O_3 Charged$$

Given that role rl_1 does not have a contractual role, merging r_2, r_3 will yield $\boxed{r_{23} : \neg pick, \neg prec \vdash O_2 AltCar \otimes O_3 QuickServ \otimes O_3 Charged}$ according to Definition 1.

Now, let us formally represent the contractual rule for the role played by the rental car company in the following.

$$r_{co} : \neg delivery \vdash O_{co} AltCar \otimes O_{co} DiscountNextRent$$ where $delivery$ denotes the cumulative effect of the rental process at the aforementioned moment.

According to the definition of function acc in Subsection 2.1, $\{\neg prec, \neg pick\} \cup \neg QuickServ = \neg delivery$ (see Appendix B for a proof). If the amount of money charged to rl_3 exceeds the mount of discount offered to the customer then according to criterion 3 of Definition 2, r_{23} subsumes r_{co}. To reason about the entailment of goals, we formalize them in FOL. The goals of rl_1, rl_2, rl_3 and the role played the rental car company are formalized as follows.

$g_1 \equiv \forall r \in Customer : \neg hasValidLicense(r) \rightarrow serviceDenied(r)$

$g_2 \equiv \forall c \in Car, r \in Customer : booked(r, c) \rightarrow$
$(timelyPickup(r, c) \wedge returnWithin30(r, c))$

$g_3 \equiv \forall c \in Car, r \in Customer : breakdownWithWarnings(r, c) \rightarrow heldResponsible(r)$

$g_{co} \equiv \forall c \in Car, r \in Customer : (\neg hasFullLicense(r) \rightarrow serviceDenied(r)) \wedge (booked(r, c) \rightarrow (timelyPickup(r, c) \wedge returnWithin15(r, c)) \wedge (breakdownWithoutCheck(r, c) \rightarrow heldResponsible(r))$

Using theorem proving techniques, we can conclude that goal g_{co} entails g_1, g_2 and g_3. A proof for this can be found in Appendix A.

4.2 Role Subtyping

Subtyping is a binary relation on roles. A role is a subtype of another role if the former has weaker goal but a stronger contractual rule than the latter. We call the former the subtyping role. Definition 4 formally captures this point.

Definition 4. Let $rl_a = \langle r_a, g_a \rangle$ and $rl_b = \langle r_b, g_b \rangle$ be two roles. Role rl_a is a subtype of role rl_b if $r_b \sqsubseteq r_a$ (i.e. rule r_a subsumes rule r_b) and $g_b \models g_a$ (i.e. goal g_b entails goal g_a).

Example 2. Let us consider an alternative fleet manager (denoted as rl_{2a}) for Fleet Manager that would allow customers to return their rental cars within 30 minutes after the end of their rent and accept slightly early pickup (e.g. within 10 minutes before the start of their rent). In case neither the car selected nor an on-premise alternative car could be provided, this fleet manager may offer an alternative car sourced from other car fleets under its management. If this obligation is violated, it will be charged the same amount of money as specified in the contract of the original fleet manager.

We formalize the contract of role $rl_{2a} = \langle r_{2a}, g_{2a} \rangle$ as follows.

$\boxed{r_{2a} : \neg pick \vdash O_{2a}AltCar \otimes O_{2a}AltCarFromNetwork \otimes O_{2a}Charged}$ where
$pick$ and $AltCar$ are the same as in Example 1.

$g_{2a} = \forall c \in Car, r \in Customer : booked(r,c) \rightarrow (timelyPickup(r,c) \vee earlyPickup(r,c)) \wedge returnWithin30(r,c))$

According to criterion 3 of Definition 2, g_{2a} subsumes g_2 (in a similar way to Example 1). In addition, we can deduce that $g_2 \models g_{2a}$ using the distribution property of $\wedge$ over $\vee$. Thus, according to Definition 4, role rl_{2a} is a subtype of role rl_2.

Role subtyping is essential for reasoning about the substitutability of roles in a role composition. The intuition is that a subtyping role can substitute for another role (i.e. the former can safely replace the latter in processes where the latter is expected) if it expresses more commitments but reserves less rights than the other does.

Theorem 1. *Let $R = \{rl_1, rl_2 \ldots rl_n\}$ be the component roles that constitute role RL_{sb}. If role rl_s is a subtype of role $rl_k \in R$, then roles $rl_1 \ldots rl_{k-1}, rl_s, rl_{k+1} \ldots rl_n$ constitute a composite role that is a subtype of RL_{sb}.*

Proof. Let $rl_i = \langle r_i, g_i \rangle, RL_{sb} = \langle r_{sb}, g_{sb} \rangle$. Let p and q be contractual rules such that p is the result of merging $r_1, r_2, \ldots r_n$ and q is is the result of merging $r_1 \ldots r_{k-1}, r_s, r_{k+1} \ldots r_n$ according to Definition 1

According to Definition 4, $r_k \sqsubseteq r_s$ and $g_k \models g_s$. Applying these to what is stated in Definition 3, we have $R_{sb} \sqsubseteq p \sqsubseteq q$ and $G_{sb} \models g_1 \wedge \ldots \wedge g_k \wedge \ldots \wedge g_n \models g_1 \wedge \ldots \wedge g_{k-1} \wedge g_s \ldots \wedge g_{k+1} \ldots \wedge g_n$. According to Definition 4, $rl_1 \ldots rl_{k-1}, rl_s, rl_{k+1} \ldots rl_n$ constitute a subtype role of RL_{sb}.

Theorem 1 is useful for checking whether a composite role can substitute for another composite role from the perspective of roles that are external to both. Example 3 illustrates this point.

Example 3. Let us consider another rental company that deals with an alternative fleet manager (rl_{2a}) discussed in Example 2. It also deals with the very receptionist and mechanics described in our running example (Subsection 3.1). As such, this rental company is able to offer early pickup (10 minutes maximum), late return (20 minutes maximum) and a wide range of alternative rental cars. It is a subtype role of the one described in the running example according to Definition 4. In other words, we have $\langle rl_1, rl_{2a}, rl_3 \rangle$ is a subtype of $\langle rl_1, rl_2, rl_3 \rangle$ because rl_{2a} is a subtype of rl_2.

5   Related Work

As discussed in Section 3, the RM-ODP standard defines the enterprise specification, which could be regarded as requirements for the system to be built. Roles are specified in this requirements specification. To reach international consensus for becoming a standard, RM-ODP is sometimes exceedingly generic and avoids providing patterns. Definitions of role in this standard range from an identifier of behavior, a subset of the total behavior (of an object) to an abstraction of the behavior that belongs to collaborative behavior [15]. In our work, we come up with a definition of roles that is separated from concrete components, agents, entities, etc. who may come into existence in the subsequent phases of development. This is in line with existing work on the semantics of roles and role-related methodological issues [16,10,14]. However, we do not relate the subtyping hierarchy of roles to that of entities like Steimann's work [10].

With respect to relationships between roles, our work shares the relaxing/strengthening principle with the notion of behavioral subtyping [17] whereby preconditions are eased and postconditions are strengthened in a subtype. Our notion of role substitutability in Theorem 1 is nevertheless not in line with the concept of object aggregation in contemporary object-oriented modeling where substitutability between component objects does not imply substitutability between composite objects.

Regarding the i^* standard, we provide an alternative conceptualization of roles in RE by introducing the concept of contractual rules [2] to the representation of roles. A role not only has goals (i.e. rights) but also features contracts (i.e. responsibility). We explicitly represent the responsibilities and rights of roles who are considered intentional and autonomous in i^*.

6   Conclusion

Are business contracts plus goals equal to roles in RE? In this paper, we propose a formal approach for modeling intentional and autonomous roles in the realm of RE. We base our work on a unified view in role modeling that suggests we explicitly represent both the responsibilities and the rights of a role [1]. We argue that the role's responsibility and rights can be captured by contractual rules and goals, respectively. We leverage existing work on the formalization of business contracts [2] and the formulation of hard goals in the i^* modeling framework [3] in order to devise techniques for reasoning about the composition and the substitutability of roles. Technically, goals are relaxed by means of entailment whereas contracts are strengthened via subsumption in a subtype role or a role composition.

Discussions. The contribution of our work is twofold. First, we propose to conceptually model RE roles in terms of goals and contracts. Second, we deal with the composition and substitutability of roles by leveraging existing work on business contracts and goal-oriented behavior. However, our framework may not cope with scenarios where role's contractual rules cannot effectively be merged and/or entailment between role's goals cannot be formulated. We have not aligned our work to contemporary enterprise modeling frameworks such as ArchiMate[5].

Future Investigations. Further work includes investigating the non-functional properties of roles in RE, which can be done by leveraging the $i*$ concepts of soft goal and belief while making logic-based contractual rules more quantifiable. Another direction of future work is to detect conflicts between constituent roles when composing them (e.g. their goals contradict with one another). Yet another direction for further work would be feeding our role model to a design phase. The research question here is on the semantics and methodological issues of designing entities (or agents) so that they can play roles specified in the requirements phase.

References

1. Zhu, H., Zhou, M.: Roles in Information Systems: A Survey. IEEE Transactions on Systems, Man, and Cybernetics, Part C: Applications and Reviews 38(3), 377–396 (2008)
2. Governatori, G., Milosevic, Z.: A Formal Analysis of a Business Contract Language. International Journal of Cooperative Information Systems 15(4), 659–685 (2006)
3. Yu, E., Giorgini, P., Maiden, N., Mylopoulos, J.: Social Modeling for Requirements Engineering. The MIT Press (January 2011)
4. Liu, L., Yu, E., Mylopoulos, J.: Security and Privacy Requirements Analysis within a Social Setting. In: Proceedings of 11th IEEE International Requirements Engineering Conference, pp. 151–161 (September 2003)
5. ISO/IEC: ITU-T X.903 | ISO/IEC 10746-3 Information Technology - Open Distributed Processing - Reference Model - Architecture. International Standard, SC 7 and ITU (2010)
6. Hinge, K., Ghose, A., Koliadis, G.: Process SEER: a Tool for Semantic Effect Annotation of Business Process Models. In: Proceedings of the 13th IEEE International Conference on Enterprise Distributed Object Computing, pp. 49–58. IEEE Computer Society, Auckland (2009)
7. Raut, M., Singh, A.: Prime Implicates of First Order Formulas. International Journal of Computer Science and Applications 1(1), 1–11 (2004)
8. Linington, P., Milosevic, Z., Cole, J., Gibson, S., Kulkarni, S., Neal, S.: A Unified Behavioural Model and a Contract Language for Extended Enterprise. Data & Knowledge Engineering 51(1), 5–29 (2004)
9. Gabbay, D.M., Woods, J.: Logic and the Modalities in the Twentieth Century. Handbook of the History of Logic, vol. 7. North-Holland (July 2006)
10. Steimann, F.: On the Representation of Roles in Object-Oriented and Conceptual Modelling. Journal of Data & Knowledge Engineering 35(1), 83–106 (2000)
11. Steimann, F.: Role = Interface: A merger of concepts. Journal of Object Oriented Programming 14(4), 23–32 (2001)

---

[5] Archimate Homepage http://www.archimate.nl/

12. Bachman, C.W.: The Role Data Model Approach to Data Structures. In: Proceedings of International Conference on Databases, pp. 1–18. University of Aberdeen: Heyden & Son (1980)
13. Steimann, F.: The Role Data Model Revisited. Applied Ontology Journal 2(2), 89–103 (2007)
14. Zambonelli, F., Jennings, N.R., Wooldridge, M.: Developing Multiagent Systems: the Gaia Methodology. ACM Transaction on Software Engineering Methodology 12, 317–370 (2003)
15. Genilloud, G., Wegmann, A.: A Foundation for the Concept of Role in Object Modelling. In: Proceedings of 4th International Enterprise Distributed Object Computing Conference, pp. 76–85 (September 2000)
16. Guarino, N.: Concepts, Attributes and Arbitrary Relations: Some Linguistic and Ontological Criteria for Structuring Knowledge Bases. Journal of Data & Knowledge Engineering 8(3), 249–261 (1992)
17. Liskov, B.H., Wing, J.M.: A Behavioral Notion of Subtyping. ACM Transactions on Programming Languages and Systems 16(6), 1811–1841 (1994)

A   Goals Entailment

We have $hasFullLicense(r) \equiv hasValidLicense(r) \wedge probationPassed(r)$ (i.e. full driver's license is a valid license that has successfully undergone a probation), $breakdownWithoutCheck(r, c) \equiv breakdownWithWarnings(r, c) \vee levelsDropped(c)$ (i.e. breakdown was caused by levels dropped significantly, of which the customer might be warned if the rental car was equipped with sensors) and $returnWithin15(r, c) \equiv returnWithin30(r, c) \wedge within15$. To make the proof easier to follow, we denote predicates that appear in the formalization of these goals as follows. $hasFullLicense(r) \equiv F$, $hasValidLicense(r) \equiv V$, $probationPassed(r) \equiv P$, $breakdownWithoutCheck(r, c) \equiv C$, $breakdownWithWarnings(r, c) \equiv W$, $serviceDenied(r) \equiv D$, $timelyPickup(r, c) \equiv T$, $booked(r, c) \equiv B$, $returnWithin30(r, c) \equiv R$, $returnWithin15(r, c) \equiv Q$, $heldResponsible(r) \equiv H$, $within15 \equiv X$, $levelsDropped(c) \equiv L$.

We can safely add quantifier $\forall c \in Car$ to g_1 because g_1 does not matter on rental cars. This is to unify goals g_1, g_2, g_3 and g_{co} in terms of quantifiers. Syntactically, they now all start with quantifiers $\forall c \in Car, r \in Customer$. For the sake of simplicity, we ignore these quantifiers in the proof. Given the aforementioned denotation, we have $g_{co} \equiv (\neg F \to D) \wedge (B \to (T \wedge Q)) \wedge (C \to H)$, $g_1 \equiv \neg V \to D$, $g_2 \equiv B \to (T \wedge R)$ and $g_3 \equiv W \to H$.

Note that $\neg F \to D$ equals to $F \vee \neg D equiv (V \wedge P) \vee \neg D equiv (V \vee \neg D) \wedge (P \vee \neg D) equiv (\neg V \to D) \wedge (\neg P \to D)$, which obviously entails g_1. In addition, since $Q \equiv R \wedge X$, we have $B \to (T \wedge Q) \equiv \neg B \vee (T \wedge R \wedge X) \equiv (\neg B \vee (T \wedge R)) \wedge (\neg B \vee X) \equiv (B \to (T \wedge R)) \wedge (B \to X)$, which entails g_2. Finally, $C \to H = \neg C \vee H = \neg(W \vee L) \vee H equiv (\neg W \wedge \neg L) \vee H = (\neg W \vee H) \wedge (\neg L \vee H) \equiv (W \to H) \wedge (L \to H)$, which entails g_3. Thus $g_{co} \models g_1 \wedge g_2 \wedge g_3$.

B   Accumulating Effects

Negating the semantic annotations given in Figure 2 will yield $\{\neg prec, \neg pick\} \equiv \{\exists r \in Customer, c \in Car : booked(r, c) \wedge \neg mecSound(c); \exists r \in Customer, c \in Car : booked(r, c) \wedge licensed(r) \wedge \neg rented(r, c)\}$. In addition, we have $QuickServ \equiv \forall r \in Customer, c \in Car : booked(r, c) \wedge \neg mecSound(c) \rightarrow serviced(c)$ (i.e. all cars, booked by the customers, that are not mechanically sound shall be serviced quickly). Thus, $\neg QuickServ \equiv \exists r \in Customer, c \in Car : booked(r, c) \wedge \neg mecSound(c) \wedge \neg serviced(c)$.

The scenario label for the three constituent roles at the moment when task Pick-up vehicles has been executed is $\langle S, G1, \{\langle proc, rgst, prec \rangle, servb\}, prec, G2, lics, G3, pick \rangle$. We proceed in cumulating semantic annotations along this label scenario (see Subsection 2.1). Since $proc$, $rgst$, $serb$, $prec$, $lics$ and $pick$ do not contain contradictory clauses (actually $serb$ and $prec$ have the same clause), we can simply collect their clauses. Negating these clauses would yield $\exists r \in Customer, c \in Car : booked(r, c) \wedge \neg mecSound(c); \exists r \in Customer, c \in Car : booked(r, c) \wedge \neg mecSound(c); \exists r \in Customer, c \in Car : booked(r, c) \wedge licensed(r) \wedge \neg rented(r, c)$.

So, we have $\{\neg prec, \neg pick\} \cup \neg QuickServ = \neg delivery$.

Specialization in *i** Strategic Rationale Diagrams[*]

Lidia López, Xavier Franch, and Jordi Marco

Software Engineering for Information Systems Research Group (GESSI)
Universitat Politècnica de Catalunya (UPC)
Barcelona, Spain
{llopez,jmarco}@lsi.upc.edu, franch@essi.upc.edu

Abstract. The specialization relationship is offered by the *i** modeling language through the `is-a` construct defined over actors (a subactor `is-a` superactor). Although the overall meaning of this construct is highly intuitive, its semantics when it comes to the fine-grained level of strategic rationale (SR) diagrams is not defined, hampering seriously its appropriate use. In this paper we provide a formal definition of the specialization relationship at the level of *i** SR diagrams. We root our proposal over existing work in conceptual modeling in general, and object-orientation in particular. Also, we use the results of a survey conducted in the *i** community that provides some hints about what *i** modelers expect from specialization. As a consequence of this twofold analysis, we identify, define and specify two specialization operations, extension and refinement, that can be applied over SR diagrams. Correctness conditions for them are also clearly stated. The result of our work is a formal proposal of specialization for *i** that allows its use in a well-defined manner.

Keywords: *i** framework, i-star, goal-oriented modeling, specialization, generalization, subtyping, inheritance.

1   Introduction

The *i** (pronounced *eye-star*) framework [1] is currently one of the most widespread goal- and agent-oriented modeling and reasoning frameworks. It has been applied for modeling organizations, business processes and system requirements, among others.

In the heart of the framework lies a conceptual modeling language, that we will name "the *i** language" throughout the paper. It is characterised by a core whose constructs, although subject of discussion in some details [2], are quite agreed by the community. A rough classification of the core distinguishes six main concepts: actors, intentional elements (IE), dependencies, boundaries, IE links and actor association links [3]. They can be used to build two types of diagrams: Strategic Dependency (SD) diagrams, composed by actors, dependencies and actor association links among them; and Strategic Rationale (SR) diagrams, that introduce IEs, with their respective links, inside actors' boundaries, and reallocate the dependencies from actors to IEs.

Among actor association links, we may find a typical conceptual modeling construct: *specialization*, represented by the `is-a` language construct. The *i** Guide

---

[*] This work has been supported by the Spanish project TIN2010-19130-C02-01.

P. Atzeni, D. Cheung, and R. Sudha (Eds.): ER 2012, LNCS 7532, pp. 267–281, 2012.

[4] defines this construct as follows: "The is-a association represents a generalization, with an actor being a specialized case of another actor". In other words, this construct is defined at the SD level as: an actor *a* (*subactor*) may be declared as a specialization of an actor *b* (*superactor*) using is-a. No more details are given and in particular, the effects that a specialization link may have on SR diagrams is not stated.

Despite the widespread use of specialization in *i** models, a systematic analysis of the literature reveals that none of these works has defined formally the effects of the is-a link beyond the sketchy definition we have presented above, or proposed methodological guidelines for its usage. In particular, and this is the focus of our work, given the relationship *a* is-a *b*, the consequences at the SR diagram involving *a* are not clear. Therefore, several questions have not a well-defined answer. For instance, consider the model at Fig. 1: how are IEs belonging to *Customer* inherited in *Family*?, what modifications are valid over these inherited elements?, do dependencies as *Easily Bought* also apply to *Family*?, may *Buy Travel* have additional subtasks in *Family*?, etc. This uncertainness makes the modeller hesitant about the use of specialization and then about the correctness of the *i** models that use this construct.

Fig. 1. Fragment of *i** SR model with two actors linked with is-a

The work presented here addresses these questions and specifically tries to answer the following research question divided into subquestions:
- RQ. Given an actor specialization relationship declared at the SD level, what modeling operations can be defined at the SR level?

SQR1.   What is the relevant background to make this decision?
SQR2.   What are the effects of these operations?
SQR3.   What are the correctness conditions to be fulfilled for their application?

The rest of the paper is structured as follows. Section 2 presents the background for our work from which we identify two specialization operations, extension and refinement, defined formally in sections 4 and 5 upon the algebraic specification of *i** and the correctness notion given in Section 3. Section 6 provides the conclusions and future work. Basic knowledge of *i** is assumed, see [1] and [4] for details.

2   Background and Specialization Operations in *i**

The idea of organizing concepts into is-a hierarchies emerged very early in Information Systems and Software Engineering. The main concepts that appear around taxonomies are *specialization* or how to make something generic more concrete; its counterpart *generalization*; and *inheritance* as the mechanism that determines how the characteristics from the most generic concept are transferred to the most concrete one.

2.1   The Concept of Specialization and Its Use in Conceptual Modeling

In this subsection we focus on specialization in three areas of interest: knowledge representation, software development and conceptual modeling.

Knowledge Representation. Quillian introduced inheritance as part of his definition of semantic networks [5]. Brachman and Levesque distinguished two kinds of inheritance semantics [6]. In *strict inheritance*, a concept inherits all the attributes of its ancestors on the `is-a` hierarchy and can add its own attributes. In *defeasible inheritance*, it is allowed cancelling attributes from the ancestors. Although cancellation can help to represent knowledge, it poses some problems to infer information [7].

Object-Oriented (OO) Programming Languages. Simula 67 [8] was the first programming language proposing the notions of class and inheritance. It adopted a strict inheritance strategy. Later on, languages as Smalltalk-80, Delphi, C++, C# and Java aligned with defeasible inheritance allowing modifying the implementation of a method (overriding). Visual Basic for .NET allowed in addition cancelling properties (shadowing). As a kind of compromise between the strict and defeasible approaches, Eiffel introduced the concept of design by contract [9] to delimit the changes included in an overridden method and facilitating the declaration of class invariants.

Conceptual Modeling. First works on conceptual modeling focused on semantic data models for database logical design. Smith and Smith introduced the notion of generalization in database modeling according to the concept of strict inheritance [10]. Afterwards, conceptual modeling languages and methodologies for specification and design in the OO paradigm started to proliferate. For instance, Borgida et al. proposed a software specification methodology based on generalization/specialization that uses the concept of strict inheritance adding the refinement of attributes [11]. Concerning languages, the UML became the dominant proposal [12]. Inheritance is used in class diagrams in the same way it was used the semantic data models.

Table 1. Summary of specialization behaviour in different areas

Area	Approach	New feature	Add Invariant	Redeclare feature
Knowledge Representation	Strict	New Attributes	No	No
	Defeasible			Attribute Cancellation
OO Languages	Simula 67	New Properties & Methods	Simulation accessing properties via methods	No
	Smalltalk-80, Delphi, C++, C#, Java			Overrides for methods Simulation for properties accessing via methods
	Visual Basic			Overrides and Shadows for properties and methods
	Eiffel		Adding invariants	Renaming and Redefinition for routines and procedures using contracts
Conceptual Modeling	Semantic data models	New Attributes & Methods	No	No
	UML		No	
	Borgida & Mylopoulos		For attributes	

Table 1 classifies these approaches using Meyer's Taxomania rule [9]: *"Every heir must introduce a feature, redeclare an inherited feature, or add an invariant clause".*

2.2   Specialization in the *i** Framework: Antecedents

Inheritance appeared in *i** from the very beginning. Yu used the is-a relationship as actor specialization in his thesis [1]. This link is only used in SD models between actors but it is not formally defined; the only observable effect in the examples is the addition of new incoming dependencies to the subactor. No examples are given of SR diagrams for subactors so the precise effects of is-a at this level remain unknown.

The is-a construct has been used in several works with the same meaning than Yu's. A non-exhaustive list is: [13][14] as a regular modeling construct; [15] for model-driven generation; [16] for modeling actor states, and [17] for deriving feature models. In all of these works the level of detail given is as insufficient as in [1].

2.3   A Community Perception on Specialization from *i** Researchers

In order to complete our preliminary analysis, we conducted a survey to know *i** modelers' concept of specialization. It was conducted from June to September 2010. Most of the answers come from attendees to the 4th Intl' iStar Workshop, where the survey was first presented. It was responded anonymously. We finally got 21 valid answers. Even if it seems a low number, it has to be considered that the core community of researchers is not too big. As an indicator, we explored the literature review presented in [18] and counted 196 authors contributing to the 146 papers found; thus the survey's population was about the 10% of this core community of authors.

The questions were very basic and are listed in Table 2; the full text, including the proposed answers, is available in [19].

Table 2. Questions appearing in the survey on *i** specialization

Q1. How often do you use is-a links in the *i** models that you develop?
Q2. If you use is-a links, do you have any doubts about their usage?
Q3. If A is-a B, what is the consequence regarding dependencies at SD model level?
Q4. If A is-a B, what is the consequence regarding the SR model level?

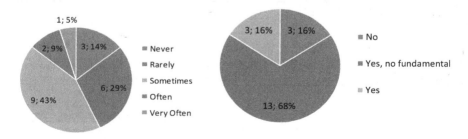

Fig. 2. Survey on *i** specialization: results of Q1 (left) and Q2 (right)

Fig. 2 shows the results for the first two questions, which are of exploratory nature and admitted just one answer. According to these results, the construct is frequently used (57% answered *sometimes* or more in Q1) but mostly with some concerns about its usage (84% answered *yes* in Q2). This contradiction is explained because in fact 68% answered Q2 as: *yes, but these doubts are not fundamental for my models.*

Fig. 3 shows the results of the last two questions, which are of interpretative nature and admitted more than one answer. According to these results, when actor *a* is-a actor *b*, new elements can be added in the actor *a* (86% for dependencies (Q3); 90% for intentional elements (Q4)). There is less agreement about modification (38% and 14% respectively). Finally, almost none of the respondents supported the option of removing elements (5% and 10% respectively).

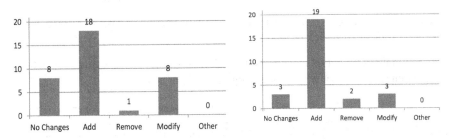

Fig. 3. Survey on *i** specialization: results of Q3 (left) and Q4 (right)

2.4   Conclusion

Considering the review presented in this section, it can be concluded that specialization consists on adding new and modifying the inherited information. Meyer summaryzes these operations in his Taxomania Rule, which can be applied to *i** as:

— **Extension** (from Taxomania rule: "introducing a feature"). A new IE or dependency, related somehow to inherited elements, is added to the subactor.
— **Redefinition** ("redeclaring an inherited feature"). An IE or dependency that exists in the superactor is changed in the subactor.
— **Refinement** ("adding an invariant clause"). The semantics of an inherited IE or dependency is made more specific.

Our goal is to align *i** specialization with the general concept of specialization (Section 2.1), considering the uses made by *i** researchers (Section 2.2) and their reported preferences (Section 2.3). For this reason, we do not consider redefinition in this work, since it is not used in main conceptual modeling proposals and clearly rejected by the *i** community ("Remove" in the survey), whilst we adopt extension ("Add"), since the introduction of new features is the essence of specialization. As for refinement ("Modify"), where the most diversity exist, we include it due to the highly strategic nature of *i**, which demands a richer conceptual modeling language. The questions that arise are then:

— What extension and refinement operations do exist?
— Which is their formal definition?
— Which are the correctness conditions?

We answer these questions in the next sections. First, we need to formalize the definition of $i*$ SR models to be able to write definitions and correctness proofs.

3   Notion of Correctness

In this section we introduce the notion of satisfaction required to reason about specialization correctness. For the purposes of this work, we make some simplifications over the language:

— actors are restricted to general actors (without roles, positions and agents);
— actors links are restricted to actor specialization (`is-part-of` is not considered);
— an IE cannot be decomposed using more than one IE link type simultaneously;

To avoid the need of distinguishing continuously special cases, and since we are interested in SR models, we assume that:

— the rationale of all actors is declared (i.e., at least one IE exists inside each actor);
— dependency ends are always connecting IEs and not actors.

Table 3. Formal definition of the $i*$ language as used in this paper

$i*$ concept	Definition	Components
$i*$ (SR) model	$M = (A, DL, DP, AL)$	A: set of actors; DL: set of dependencies DP: set of dependums; AL: set of actor specialization links
Actor	$a = (n, IE, IEL)$	n: name; IE: set of IEs; IEL: set of IE links
IE	$ie = (n, t)$	n: name; t: type of IE, $t \in$ {goal, softgoal, task, resource}
IE link	$l = (p, q, t, v)$	p, q: IEs (source and target). actor(p)=actor(q) (intra-actor links) t: type of IE link, $t \in$ {means-end, task-decomp., contribution} target(means-end) $\neq$ softgoal, target(task-decomposition) = task, target(contribution) = softgoal v: contribution value, $v \in CT^+ \cup CT^- \cup$ {Unknown} $CT^+ = $ {Make, Some+, Help}, $CT^- =$ {Break, Some-, Hurt}
Dependency	$d = ((dr,s_r),$ $(de,s_e),$ $dm)$	dr, de, dm: IEs (depender, dependee and dependum, resp.) s_r, s_e: strengths, $s_r, s_e \in$ {open, committed, critical} actor(dr) $\neq$ actor(de) (an actor cannot depend on itself)
Dependum	$dm = (n, t)$	dm: IE
Actor specialization link	$l = (a, b)$	a, b: actors (subactor and superactor). No cycles allowed
Derived concepts	Definition	
Main IEs of an actor	mainIEs(a) = { $ie \in IE$	ancestors(IEL, ie) = $\varnothing$ }, mainIEs(a) $\neq \varnothing$	
Decomposition link	decompositionLink(l) $\Leftrightarrow$ type(l) $\in$ {means-end, task-decomposition} $\vee$ (type(l)=contribution $\wedge$ value(l) $\in$ {And, Or})	

Table 3 summarizes the formal definition of the resulting $i*$ language under these simplifications and assumptions. An $i*$ (SR) model contains actors, dependencies, dependums and actor specialization links. Actors contain IEs connected by IE links of different types. Dependencies connect IEs and have a dependum (that is also an IE). Throughout the paper, we can use auxiliary predicates and functions to obtain components of a model element (e.g., in fourth and fifth rows, we use the function *actor* that returns the actor that contains a given IE). We introduce a couple of auxiliary derived concepts that are used when defining the specialization operations.

We can now address the notion of specialization correctness, in other words, what conditions have to be fulfilled in order to consider this specialization correct. We consider the notion of satisfaction as the baseline to define correctness: subactor's satisfaction must imply superactors' satisfaction. This property ensures that the subactor a may be used in those contexts where the superactor is expected.

Definition 1. Actor specialization satisfaction.
Given an *i** model $M = (A, DL, DP, AL)$ and two actors $a, b \in A$ such that $(a, b) \in AL$, we define actor specialization satisfaction as: $sat(a, M) \Rightarrow sat(b, M)$

Given our simplifications and assumptions, each actor contains at least one main intentional element, hence we reduce the actor satisfaction to IE satisfaction.

Definition 2. Actor satisfaction.
Given an *i** model $M = (A, DL, DP, AL)$ and an actor $a \in A$, $a = (n, IE, IEL)$ with $IE \neq \varnothing$, we define a's satisfaction, sat(a, M), as the satisfaction of all its main IEs:
$$\forall ie \in mainIEs(a): sat(ie, M).$$

Satisfaction of an IE depends on the IE links that reach that IE. If there are no links, satisfaction is up to the modeler. If there are decomposition links or dependencies, a logical implication may be established. In the case of contributions to softgoals, we adopt Horkoff and Yu's [20] proposal.

Definition 3. IE satisfaction.
Given a model $M = (A, DL, DP, AL)$ and an IE $ie \in IE$, we define ie's satisfaction, $sat(ie, M)$, according to the cases below (note that the second and third cases can happen simultaneously with the fourth, then both conditions apply):
– ie is neither decomposed nor has outgoing dependencies: satisfaction has to be explicitly provided by the analyst/modeler.
– ie is decomposed by decomposition links: satisfaction depends on the link type:
  • task-decomposition: according to the *i** definition (an incomplete AND-decomposition), the sources are AND-ed:
    $$\forall ie_{and}: (ie_{and}, ie, task\text{-}decomposition, \perp) \in IEL: sat(ie, M) \Rightarrow sat(ie_{and}, M)$$
  • means-end: according to the *i** definition, the sources are OR-ed:
    $$\forall ie_{or}: (ie_{or}, ie, means\text{-}end, \perp) \in IEL: sat(ie_{or}, M) \Rightarrow sat(ie, M)$$
– ie is softgoal with contribution links: satisfaction is defined as in [20].
– ie has outgoing dependencies: satisfaction depends on dependum's:
  • $\forall ((ie, s_{ie}), (de, s_{de}), dm) \in DL: sat(ie, M) \Rightarrow sat(dm, M)$
Note that the implication cannot be an equivalence because the ie can be decomposed and then its satisfaction would depend on its decomposition.

At this point, we have completely defined the notion of specialization satisfaction and may therefore proceed to define extension and refinement operations.

4   Extension Operations

Extension means adding a new model element to the subactor. There are two types of elements to consider:

- IEs. An IE can be added extending an inherited IE or as a main IE:
  - IE extension. In the subactor, some IE is added as a decomposition of an inherited IE.
  - New main IE. Some IE is added as a main IE due to the subactor has a new intentionality that is not covered by the superactor's main IEs.
- Dependencies. A dependency can be added to an IE *ie* in two different directions:
  - Outgoing dependencies. This case is not allowed. The reason is that if a super-actor is able to satisfy *ie* by itself, its subactors must be able to do so as well.
  - Incoming dependencies. Adding a new incoming dependency does not affect *ie*'s satisfaction, but the satisfaction of the IE that acts as depender. This means that this dependency needs not to be considered in the analysis of *ie*.

As a conclusion, we need two extension operations for IEs, but none for dependencies. We present in the rest of the section these two operations.

CASE 1. *IE Extension.* An IE inherited from a superactor can be extended in a subactor by adding a new decomposition link:

— Task-decomposition link: Since task-decompositions are not necessarily complete, it is always possible to add a new IE that provides more detail in the way in which a task is performed. By defining a task-decomposition link, the linked element is considered AND-ed with the elements that decompose the task in the superactor.
— Means-end link: An element may be considered as a new means to achieve an end. By defining a means-end link, the linked element is considered OR-ed with the means that appear in the superactor.

Fig. 4 presents two examples of extension. In the diagrams, inherited elements in the subactor are shown in dotted lines. The subactor UTA shows the extension of a superactor TA's non-decomposed task (Name a price). The FTA adds a third means to an inherited end (Travels Contracted Increase) that was already decomposed in TA; this new IE, playing the role of means, has just sense in the case of the subactor. In both cases, the IE that is being subject of the operation is further decomposed; additionally, in FTA, some IEs contribute to two softgoals inherited from the superactor, shown also in dotted lines to indicate that they are same as in the superactor.

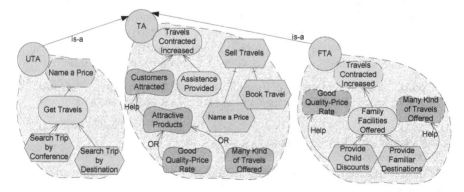

Fig. 4. Specialization operations: adding task-decomposition (UTA) & means-end (FTA) links

Extension Operation 1.   Intentional element extension with a decomposition link.
Declaration. extendIEWithDecompositionLink(M, a, ie_t, ie_s, t), being:

- $M = (A, DL, DP, AL)$, an *i** model
- $a = (n_a, IE_a, IEL_a)$, $a \in A$, the subactor where the IE extension takes place
- $ie_t \in IE_a$, the inherited IE to be extended (the target)
- ie_s, the new IE to be linked to ie_t (the source)
- t, the type of decomposition link, either means-end or task-decomposition.

Preconditions. 1) ie_s is semantically correct with respect to ie_t given the type of link:
- means-end: $sat(ie_t, M) \Rightarrow sat(ie_s, M)$
- task-decomposition: $sat(ie_s, M) \Rightarrow sat(ie_t, M)$

2) ie_s is not main element in the superactor: $ie_s \notin$ mainIEs(superactor(a))

Effect. *extendIEWithDecompositionLink(M, a, ie_t, ie_s, t)* yields a model *M'* defined as:
$M' = substituteActor(M, a, a')$, being *substituteActor* a function that replaces every occurrence of a in M by a', $a' = (n_a, IE_a \cup \{ie_s\}, IEL_a \cup \{(ie_s, ie_t, t, v)\})$.

Theorem. The operation *extendIEWithDecompositionLink(M, a, ie_t, ie_s, t)* is correct.
Proof. We demonstrate by induction that this operation keeps actor specialization correctness, i.e. $sat(a', M') \Rightarrow sat(b, M')$ (see Definition 1).

Induction Base Case (IBC). In the IBC, this operation is the first specialization operation applied over the subactor a, i.e. $IE(a) = IE(b) \wedge IEL(a) = IEL(b)$ [P1]

[1] $sat(a', M') \equiv \forall ie \in mainIEs(a'): sat(ie, M)$, applying Definition 2 over a'
[2] $\equiv \forall ie \in mainIEs(a): sat(ie, M)$, since main elements do not change:
            $(ie_s, ie_t, t, v) \in IELs(a') \wedge$ precondition $2 \Rightarrow mainIEs(a') = mainIEs(a)$
[3] $\equiv \forall ie \in mainIEs(b): sat(ie, M)$, since [P1] $\Rightarrow mainIEs(b) = mainIEs(a)$
[4] $\equiv \forall ie \in mainIEs(b): sat(ie, M')$, since b is the same in M and M'
[5] $\equiv sat(b, M')$, applying Definition 2 over b

Induction Hypothesis (IH). We assume a state in which after several specialization operations applied, still the correctness condition holds:
$sat(a, M) \Rightarrow sat(b, M)$

Induction Step (IS). If this operation is applied over a subactor a that satisfies the correctness condition, the resulting subactor a' satisfies it too:
$sat(a', M') \Rightarrow sat(b, M')$

[1] $sat(a', M') \equiv \forall ie \in mainIEs(a'): sat(ie, M')$, applying Definition 2 over a'
[2] $\equiv \forall ie \in mainIEs(a): sat(ie, M)$, since ie_s is not added as main IE
[3] $\equiv sat(a, M)$, applying Definition 2 over a
[4] $\Rightarrow sat(b, M)$, applying the IH
[5] $\equiv \forall ie \in mainIEs(b): sat(ie, M)$, applying Definition 2 over b
[6] $\equiv \forall ie \in mainIEs(b): sat(ie, M')$, since b is the same in M and M'
[7] $\equiv sat(b, M')$, applying Definition 2 over b

CASE 2. *Main IEs Addition.* The subactor has an intentionality that is not covered by the superactor's main IEs. Therefore, a new main IE needs to be added. Fig. 5

presents an example of adding a new main IE in the subactor. Again, this new element is further decomposed and its decomposition includes an inherited element (drawn in dotted lines) at the second level of decomposition.

Extension Operation 2.   Actor extension with a main intentional element
Declaration. extendActorWithMainIE(M, a, ie$_{new}$), being:
- $M = (A, DL, DP, AL)$, an $i*$ model
- $a = (n_a, IE_a, IEL_a)$, $a \in A$, the subactor where the new IE is added
- ie$_{new}$, the new IE to be added as main IE

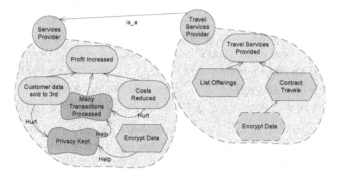

Fig. 5. Actor specialization operations: adding main IEs

Precondition. ie$_{new}$ is really enlarging subactor's intentionality:
- $\neg (sat(ie_{new}, M) \Rightarrow \forall ie \in mainIEs(a): sat(ie, M))$

Effect. extendActorWithMainIE(M, a, ie$_{new}$) yields $M' = substituteActor(M, a, a')$, where $a' = (n_a, IE_a \cup \{ie_{new}\}, IEL_a)$ and *substituteActor* defined as above.

Theorem. The operation *extendActorWithMainIE(M, a, ie$_{new}$)* is correct.
Proof. By induction, very similar to the former proof. The only notable difference is that since the new IE is added as main element, some equivalence needs to be converted into implication. For instance, in the IBC, step [2] changes into:

[2a] $\equiv \forall ie \in mainIEs(a): sat(ie, M') \wedge sat(ie_{new}, M')$, since ie$_{new}$ is added as main IE
[2b] $\Rightarrow \forall ie \in mainIEs(a): sat(ie, M')$, since $X \wedge Y \Rightarrow X$

5   Refinement Operations

Refinement means replacing an existing model element by another that somehow constraints the inherited behaviour. There are three types of elements to consider:

— IEs: any IE in the model can be refined.
— Contribution links: the value of a contribution link can be enforced in the subactor.
— Dependencies: an inherited dependency can be refined either by enforcing the IE placed as dependum or by making stronger any of the two strengths.

As a conclusion, we need three refinement operations, presented next. Their correctness is demonstrated at [19] (proofs are very similar to CASE 1 above).

CASE 3. *IE Refinement.* A subactor *a* can refine an IE *ie* inherited from its superactor *b* with the following meaning depending on its type:

— Goal, softgoal: the set of states attained by *ie* in *a* is a subset of those attained in *b*.
— Task: the procedure to be undertaken when executing *ie* in *a* is more prescriptive (i.e. has less freedom) than the procedure to be undertaken when executing *ie* in *b*.
— Resource: the entity represented by *ie* in *a* entails more information than the entity represented by *ie* in *b*.

Fig. 6 presents two examples of IE refinement. On one hand it shows the refinement of a non-decomposed resource (Travel Information) in which information related to families (e.g., number and age of children) is included in the subactor. On the other hand, it refines a decomposed task (Charge Travel), with the particularity that what is needed is not an IE but an additional dependency that expresses the dependence on some other actor for undertaking the task in the subactor. As usual, IEs in dotted lines represented IEs inherited from the superactor and not changed in the subactor.

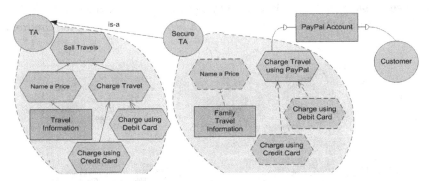

Fig. 6. Specialization operations: refining a resource top and a decomposed task bottom

Refinement Operation 1. Intentional element refinement.
Declaration. refineIE(M, a, ie$_s$, n$_{ref}$), being:

- $M = (A, DL, DP, AL)$, an *i** model
- $a = (n_a, IE_a, IEL_a)$, $a \in A$, the subactor where the IE refinement takes place
- $ie_s = (n, t) \in IE_a$, the inherited IE to be refined
- n_{ref}, the name to be given to the refined IE

Precondition. the new IE is enforcing the inherited one: $sat((n_{ref}, t), M) \Rightarrow sat((n, t), M)$
Effect. refineIE(M, a, ie$_s$, n$_{ref}$) yields a model $M' = substituteIE(M, a, ie_s, ie_{ref})$, being $ie_{ref} = (n_{ref}, t)$ and *substituteIE* a function that replaces *ie$_s$* of *a* in *M* by *ie$_{ref}$* in *M'*.

CASE 4. Contribution Link Refinement. Contribution link refinement means changing the value of a contribution link going from an IE to a softgoal, both of them appearing in the superactor. Of course, not all the changes must be allowed, since it is necessary to guarantee that the satisfaction of the refined link's value implies the link under refinement's value. This is done by using the typical order relation among contribution link values [20]: Unknown > Some+ > Help > Make, and Unknown > Some- > Break > Hurt. Note that we keep positive and negative values separated, meaning that we do not allow changing the "sign" of the contribution.

Fig. 7 presents two examples of contribution link refinement. The left figure shows a refinement where the involved IEs are the same in both actors, just the contribution value changes. In the right figure, the source IE has been also refined, meaning that the subactor is the result of two refinement operations.

Refinement Operation 2. Contribution link refinement.
Declaration. refineContributionLink(M, a, iel, v), being:

- $M = (A, DL, DP, AL)$, an i^* model
- $a = (n_a, IE_a, IEL_a)$, $a \in A$, the subactor where the IE link refinement takes place
- $iel=(ie_s, ie_t, contrib, v_l) \in IEL_a$, the inherited contribution link to be refined
- v, the value to be given to the refined contribution link

Precondition. The new contribution value is enforcing the inherited one: $v < v_l$.
Effect. refineContributionLink(M, a, iel, v) yields a model *M'* defined as:
$$M' = substitute(M, a, a'), \text{ where } a'=(n_a, IE_a, (IEL_a\backslash\{iel\}) \cup \{(ie_s, ie_s, contrib, v)\})$$

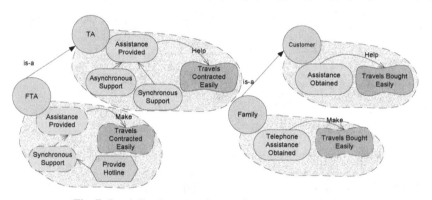

Fig. 7. Specialization operations: refining contribution links

CASE 5. *Dependency Refinement.* A dependency can be refined only if at least one of the actors involved in the refined dependency is a subactor. Both the dependum and the strengths may be refined. In the case of the dependum, since it is an IE, the rules are the same to those introduced in CASE 3, although technically there is a difference: in CASE 3 the refined IEs were IE appearing inside an actor, whilst here the refined IE appears in dependencies that are external to actors. In other words, given an i^* model $M = (A, DL, DP, AL)$, CASE 3 is defined over A whilst CASE 5 is defined over DP. Concerning strengths, it is similar to CASE 4 (refinement of a value) with the relationship Open > Committed > Critical (being Committed the default case).

Fig. 8 presents two examples of dependency refinement. In the bottom dependency (Customer Info), just the dependum is refined, it also presents the particularity that both dependency ends correspond to subactors. In the top dependency (Travel Offerings), besides the dependum, the dependee's strength is refined too.

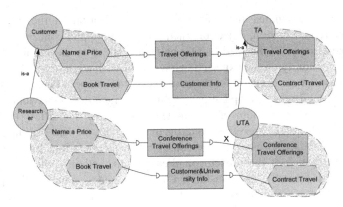

Fig. 8. Specialization operations: refining dependums

Refinement Operation 3. Dependency refinement

Declaration. refineDependency(M, d, dm_{ref}, sdr_{ref}, sde_{ref}), being:

- $M = (A, DL, DP, AL)$, an *i** model
- $d = ((dr, s_{dr}), (de, s_{de}), dm)$, $d \in DL$, the inherited dependency under refinement
- dm_{ref}, sdr_{ref} and sde_{ref} the dependum and strengths for the refined dependency

Note that *d* is the inherited dependency, where at least one of the depender or depen-dee is a subactor, not to confound with the original dependency that will not change.
Precondition.

- The new dependum is enforcing the inherited one: $sat(dm_{ref}, M) \Rightarrow sat(dm, M)$.
- The new strengths are less or equal than the older: $sdr_{ref} \leq s_{dr} \wedge sde_{ref} \leq s_{de}$
- At least one component changes: $dm_{ref} \neq dm \vee sdr_{ref} \neq s_{dr} \vee sde_{ref} \neq s_{de}$

Effect. refineDependency(M, d, dm_{ref}, sdr_{ref}, sde_{ref}) yields a model *M'* defined as:

$$M' = (A, DL \setminus \{d\} \cup \{((dr, sdr_{ref}), (de, sde_{ref}), dm_{ref})\}, DP \cup \{dm_{ref}\}, AL)$$

Note that *d* is removed since it is substituted by the new dependency. On the contrary, *d*'s dependum, *dm*, is not removed since the specialized dependency (the one being inherited) still makes use of it.

6   Conclusions and Future Work

In this paper we have presented a proposal for defining *i** specialization in a formal manner at the level of SR diagrams. According to the main research question, the aim has been to study the consequences of a specialization relationship declared at the SD level. We have identified two main specialization operations, extension and refinement, and for them, we have identified two and three concrete operations, respectively. Concerning the three derived subresearch questions stated at the introduction:

- SQR1: we have studied the literature on specialization in the disciplines of knowledge representation, object-oriented programming and conceptual modeling, and we have compiled the works so far on *i** specialization as well as ran a survey in the *i** community on the expected behaviour of such a construct. This study has been the basis of our decision for the two specialization operations.
- SQR2: for each of the five operations, we have defined their behaviour in terms of the algebraic specification of *i** models. We have identified the required preconditions for these operations in terms of properties on their parameters.
- SQR3: we have also proven the correctness of these operations by demonstrating that the satisfaction of the subactor implies the satisfaction of the superactor. We have defined formally the satisfaction concept and conducted the proofs by induction. The paper includes one of the proofs with all details, whilst the others are in a separated document due to space reasons.

These operations can be combined in any arbitrary order during the modeling process: our proofs show that satisfaction is kept provided that the original model was correct.

The work presented here has assumed a few simplifications on the *i** language. Most of them are really not important although some may require further attention, specifically the exclusion of the is-part-of construct of our analysis (see below).

Future work spreads along several directions. First, the Taxomania rule considers a third type of specialization operation, *redefinition*, which we have not included in the present work. We plan to analyse in detail under which conditions this operation could be applied and then define it in a similar way than extension and refinement. Second, we aim at providing an ontological-based semantics to *i** specialization. At this respect, we have recently started to apply the UFO foundational ontology over *i** [21][22], and we plan to include specialization in this work. Third, the problem of loose definition of the specialization relationship is not the only point of ambiguity of the *i** language. A similar situation can be found for the rest of actor links: is-part-of, plays, occupies and covers. Therefore, we plan to address this problem following the same method as with specialization and as a further step, to explore the relationships of all of these actor association links with is-a.

References

[1] Yu, E.: Modelling Strategic Relationships for Process Reengineering. PhD. Computer Science University of Toronto, Toronto (1995)

[2] López, L., Franch, X., Marco, J.: Making Explicit Some Implicit *i** Language Decisions. In: Jeusfeld, M., Delcambre, L., Ling, T.-W. (eds.) ER 2011. LNCS, vol. 6998, pp. 62–77. Springer, Heidelberg (2011)

[3] Cares, C., Franch, X., Perini, A., Susi, A.: Towards Interoperability of *i** Models using iStarML. CSI 33(1), 69–79 (2011)

[4] The *i** Wiki, http://istar.rwth-aachen.de (last accessed March 2012)

[5] Quillian, M.: Semantic Memory. In: Minsky, M. (ed.) Semantic Information Processing. The MIT Press (1968)

[6] Brachman, R.J., Levesque, H.J.: Knowledge Representation and Reasoning. Elsevier Inc. (2004)

[7] Brachman, R.J.: I Lied About the Trees, or Defaults and Definitions in Knowledge Representation. AI Magazine 6(3), 80–93 (1985)

[8] Dahl, O.: SIMULA 67 Common Base Language. Norwegian Computing Center (1988)

[9] Meyer, B.: Object-Oriented Software Construction, 2nd edn. Prentice-Hall (1997)

[10] Smith, J.M., Smith, D.C.P.: Database Abstractions: Aggregation and Generalization. Journal ACM Transactions on Database Systems 2(2), 105–133 (1977)

[11] Borgida, A., Mylopoulos, J., Wong, H.K.T.: Generalization/Specialization as a Basis for Software Specification. In: On Conceptual Modelling, Intervale, pp. 87–117 (1982)

[12] Unified Modeling Language (UML) site, http://www.uml.org/

[13] Franch, X.: On the Lightweight Use of Goal-Oriented Models for Software Package Selection. In: Pastor, Ó., Falcão e Cunha, J. (eds.) CAiSE 2005. LNCS, vol. 3520, pp. 551–566. Springer, Heidelberg (2005)

[14] Mouratidis, H., Jürjens, J., Fox, J.: Towards a Comprehensive Framework for Secure Systems Development. In: Martinez, F.H., Pohl, K. (eds.) CAiSE 2006. LNCS, vol. 4001, pp. 48–62. Springer, Heidelberg (2006)

[15] Castro, J., Lucena, M., Silva, C., Alencar, F., Santos, E., Pimentel, J.: Changing Attitudes Towards the Generation of Architectural Models. JSS 85(3), 463–479 (2012)

[16] Goldsby, H.J., Sawyer, P., Bencomo, N., Cheng, B.H.C., Hughes, D.: Goal-Based Modeling of Dynamically Adaptive System Requirements. In: ECBS, pp. 36–45 (2008)

[17] Clotet, R., et al.: Dealing with Changes in Service-Oriented Computing Through Integrated Goal and Variability Modelling. ICB Research Report 2008 (22), 43–52 (2008)

[18] Cares, C., Franch, X.: A Metamodelling Approach for *i** Model Translations. In: Mouratidis, H., Rolland, C. (eds.) CAiSE 2011. LNCS, vol. 6741, pp. 337–351. Springer, Heidelberg (2011)

[19] Lopez, L., Franch, X., Marco, J.: Specialization in *i** Strategic Rationale Diagrams. Research Report ESSI-TR-12-4. Universitat Politècnica de Catalunya (2012)

[20] Horkoff, J., Yu, E.: Finding Solutions in Goal Models: An Interactive Backward Reasoning Approach. In: Parsons, J., Saeki, M., Shoval, P., Woo, C., Wand, Y. (eds.) ER 2010. LNCS, vol. 6412, pp. 59–75. Springer, Heidelberg (2010)

[21] Franch, X., Guizzardi, R., Guizzardi, G., López, L.: Ontological Analysis of Means-End Links. In: CEUR Workshop proceedings, vol. 766, pp. 37–42 (2011)

[22] Guizzardi, R., Franch, X., Guizzardi, G.: Applying a Foundational Ontology to Analyze the *i** Framework, pp. 1–11 (2012)

Conceptualizing and Specifying Key Performance Indicators in Business Strategy Models

Alejandro Maté[1], Juan Trujillo[1], and John Mylopoulos[2]

[1] Lucentia Research Group, Department of Software and Computing Systems,
University of Alicante, Spain
{amate,jtrujillo}@dlsi.ua.es
[2] Department of Computer Science , University of Toronto, Toronto, ON, Canada
jm@cs.toronto.edu

Abstract. Key Performance Indicators (KPI) measure the performance of an organization relative to its objectives. To monitor organizational performance relative to KPIs, such KPIs need to be manually implemented in the form of data warehouse queries, to be used in dashboards or scorecards. Moreover, dashboards include little if any information about business strategy and offer a scattered view of KPIs and what do they mean relative to business concerns. In this paper, we propose an integrated view of strategic business models and conceptual data warehouse models. The main benefit of our proposal is that it links strategic business models to the data through which objectives can be monitored and assessed. In our proposal, KPIs are defined in Structured English and are implemented in a semi-automatic way, allowing for quick modifications. This enables real-time monitoring and what-if analysis, thereby helping analysts compare expectations with reported results.

Keywords: Business Intelligence, Conceptual Data Warehouse models, KPI, SBVR, OLAP.

1 Introduction

Key Performance Indicators (KPI) are used by organizations to monitor the performance of their processes and business strategies [10]. KPIs are traditionally defined with respect to a business strategy and objectives by using a Balanced Scorecard [7], to indicate what is to be monitored in different areas of the organization thereby providing a global view of the organization's status. These KPIs are then included in different dashboards, providing a detailed view of each specific area of an organization [3]. However, this approach entails that KPIs are (i) created in isolation, without describing inter-relationships between each other, and (ii) manually implemented by IT specialists. Unfortunately, this approach leads to several problems. First, it does not provide the decision maker with any information about the goal being monitored by a KPI and its effect on the rest of the business strategy. Second, even if a complex dashboard is developed, the

P. Atzeni, D. Cheung, and R. Sudha (Eds.): ER 2012, LNCS 7532, pp. 282–291, 2012.

decision maker is unable to validate if the KPI is correctly measuring its intended goal. Third, it is unable to verify that business strategy and the implemented KPIs are consistent with each other.

In order to address some of these issues, researchers have proposed to apply (semi-)formal techniques for strategy modeling. In [11] an i* profile for Data warehouses (DW) is used, focusing on building the underlying DW. On the other hand, the Business Intelligence Model (BIM) [1,6] is proposed to support the analysis step, once the DW is already built. BIM allows us to model the business strategy including processes related to each goal, indicators, and potential situations which may affect goals, thus supporting SWOT (Strengths, Weaknesses, Opportunities, Threats) analysis [5]. In this way, BIM provides a comprehensive view of the business strategy along with KPIs and their relationships.

In our previous work [9,11], we defined a hybrid DW development approach in the context of the Model Driven Architecture (MDA) framework [12], in order to support the decision making process. In this paper, we complement our previous approach by proposing a semi-automatic process that obtains the value of each KPI, thus providing a comprehensive view of the organization's status. A summary of the process described throughout the paper is depicted in figure 1, and each step is further described on the corresponding section.

The remainder of this paper is structured as follows, Section 2 presents the basic concepts of BIM and introduces running examples for the rest of the paper. Section 3 describes how KPIs can be defined by using the Semantics of Business Vocabulary and Business Rules (SBVR) proposal [13]. Section 4 presents how to validate the KPIs defined against the DW schema. Section 5 describes how the proposed model can be used as a strategic dashboard which can also be navigated. Section 6 presents related work in this area. Finally, Section 7 summarizes conclusions and sketches future works.

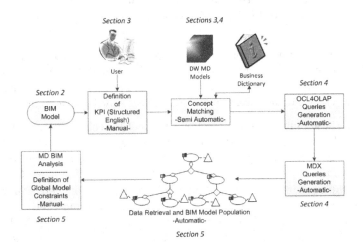

Fig. 1. Overview of the steps included in our approach

2   Basic Concepts and an Illustrating Example

In order to describe the basic concepts that underlie strategies and KPIs, which we will use throughout the paper, we present a running example, modeled after a fictitious vehicle manufacturer case study, the Steel Wheels company. The Steel Wheels company desires to improve its monitoring and decision making process. In order to improve this process, the company starts by modeling the business strategy, described textually in a business plan. The process of modeling the business strategy is performed by extracting four key types of elements from the business plan: goals, business processes, situations, and indicators, included in the BIM metamodel [6]. The definition of these key types is as follows:

Goals (a). Goals capture the objectives of the organization being modelled and depict a situation that an actor wishes to achieve [4,11,16]. For example, the main objective of Steel Wheels is achieving the "Revenue Increased" goal. In order to achieve it, the strategy can be decomposed into two alternative paths: "Costs Cut" and "Fancy Designs Created". As goals in BIM are related to their definition in the business plan, they also include their business perspective (Financial, Customer, Processes and Learning) from the Balanced Scorecard [7].

Processes (b). Business processes are responsible for the realization of the lowest level goals. In this way, "Innovative vehicle design" and "Design quality evaluation" processes realize the goal "Attractive vehicles designed", while "Sales zone planification" realizes the goal "Dealership distribution optimized".

Situations (c). Situations enable SWOT analysis over the business strategy and influence goals either positively or negatively depending on the relationship between them. For example, having "Positive Customer Reviews" is a strength (internal, positive) of the "Fancy Designs Created" strategy, which helps the goal "Customer satisfaction". On the other hand, the situation "Economic Crisis" is a threat (external, negative) for the business, and hurts the goal "Sales Increased".

KPIs (d). KPIs act as monitoring elements, measuring values related to goals or situations. Each indicator presents a target value (value to be achieved), a threshold (margin between good and bad performance), a current value and worst value. According to these values, the KPI is normalized in the range [-1,1], describing how good or how bad the measured element is performing.

By modeling these four types of elements, the Steel Wheels company BIM model is obtained. The result is depicted in figure 2. The Steel Wheels business strategy has one main goal: increase its revenue. In order to achieve this goal, two alternative ways to achieve this goal (courses of action) may be followed. On the one hand, the company can decrease the manufacturation costs of their vehicles, thus making them affordable for most customers but also lowering their quality. On the other hand, the currently chosen course of action, is to create high-quality designs, which are more expensive but also more attractive to customers. In turn, this course of action improves the image of the company, hopefully increasing its profit.

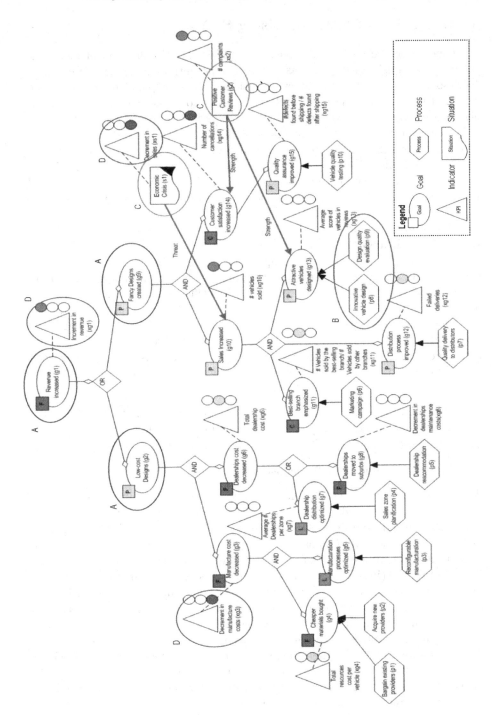

Fig. 2. Steel Wheels business strategy loaded with data

3   Definition of KPIs by Using SBVR

After modeling the business strategy, we can identify the different elements involved in the business strategy, as well as the potential courses of action. However, so far this model only allows decision makers to plan and estimate the values of the different KPIs. Therefore, in order to gather feedback from business processes, we have to define how business KPIs are calculated.

We can consider two different kinds of KPIs which can be defined over the models, according to how they are calculated: atomic and composite. On the one hand, atomic KPIs are those whose value is obtained from the DW. For example, the KPI "Number of vehicles sold" which retrieves the total amount of vehicles sold from the Sales table would be an atomic KPI. On the other hand, the KPI "Increment in revenue" could be created as a composite KPI, obtaining its value from the difference of "Gross profit" minus "Manufacture Costs" KPIs. For further information on composite KPIs see [1].

Atomic KPIs are defined in our approach by using Structured English [13], based on the SBVR language proposal from the Object Management Group. Decision makers can use a special font which identifies business concepts involved in the definition of a KPI, as well as the relationship between them. In this way, decision makers may use the keyword[1] font, term font, Name font, and verb font to provide semantic definitions of KPIs. Each of these fonts provides a specific semantic meaning, which allows us to match business concepts with multidimensional elements without constraining their definition. Some examples of KPIs involved in our strategy are:

- "Increment in revenue", may be defined as a combination of two KPIs "Revenue in the current year" minus "Revenue in the previous year":
  - Total benefit in This year
  - Total benefit in Previous year
- "Number of vehicles sold", defined as: Total sales in This year.

As shown in the examples, and according to the SBVR specification [13], keyword font can be related to unary operations over measures, which allow us to aggregate the data obtained. On the other hand, term font can be mapped to metadata from the multidimensional schema, such as fact attributes and dimension attributes (properties), dimensions, and levels (concepts). Next, Name font is used to refer to individuals and exact values. As such, it can refer to instances of levels or exact numeric values. Finally, verb font can be used to mark verbs or prepositions. All these mappings are formalized in Section 4, in the form of a grammar which recognizes the language used to define indicators.

The benefits of following this approach are that (i) indicators can be defined in a user-friendly, controlled language, and (ii) they can be included into a Business Dictionary (BD), thus they can be referenced and queried by other applications, or used to generate documentation. This helps other decision makers into defining their own indicators as well as re-using existing ones.

---

[1] Font colors have been changed in order to improve the readability in grayscale color. SBVR original colors are: keyword, term, Name, and verb respectively.

4   Validation Using the Multidimensional Schema

Once we have defined a series of KPIs by using Structured English, they must be validated in order to guarantee that (i) the necessary data is stored in the DW, and (ii) the indicators are correctly defined according to the DW structure.

In order to perform the validation of each indicator, first, the multidimensional representation of the DW must be obtained (figure 3 and figure 4). The multidimensional information required are existing facts (center of analysis), fact attributes (measures, related to the performance of the business process), dimensions (context of analysis), levels and attributes of the dimension levels.

The first schema represents the information about budget asignation and actual costs ("QuadrantAnalysis" fact). The fact attributes included in this schema are the planned "Budget" cost for each entry, the real "Actual" cost, and the difference between them, "Variance". As context of analysis, we know information about each "Region", "Department" and "Position", allowing us to browse the assignations in the budget as we need.

The second schema represents the information available regarding the "Steel-WheelsSales" process (fact). The fact attributes included in this process are the "Quantity" of each product sold, as well as the total amount of the sale, "Sales". Regarding the context of analysis, we have information about a "Product", such as its "Vendor" and the product "Line". Additionally, we also have information regarding "Customers", such as their name and address, the "Markets" where the sale was performed, the month of the year when the product was sold ("Time"), and the current status of the corresponding order (Cancelled, Delivered, On Hold, etc.).

These multidimensional schemata are used to (i) analyze if the concepts used by the decision maker to define KPIs do exist in the multidimensional models of

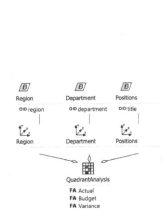

Fig. 3. (a) Multidimensional model for analyzing costs

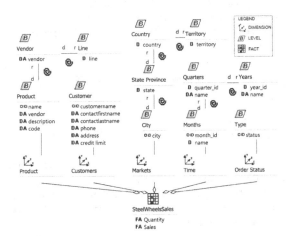

Fig. 4. (b) Multidimensional model for analyzing sales

the DW and to (ii) support the mapping from business concepts to DW elements. If a business concept has not been stored in the BD yet, the concept is matched against the multidimensiona schemata, asking for disambiguation to the the user if the mapping is not found.

After having identified the different concepts used by the decision maker in his definition, we proceed to translate the SBVR definition of the KPI into an OCL4OLAP representation [14]. OCL4OLAP is an extension of the OCL formal language, allowing us to query models which present a multidimensional structure (i.e. facts and dimensions). The translation of SBVR to OCL is considered to be a challenging transformation [2], since OCL does not consider concepts such as business rules. Since we focus on the definition of indicators, we restrain the possible transformations to definitions of indicators specified by means of formulas over multidimensional conceptual models. First, the definition of the specified indicator is recognized by the grammar described in figure 5, then, the indicator is translated to OCL4OLAP through the following process:

1. Values identified correspond with sets of cells in the cube specified by the multidimensional schema. By specifying a term corresponding to a fact attribute, the decision maker is implying that he is interested in operating with the value, thus a *dimensionalProject(cube::factattribute)* OCL4OLAP operation is performed. This operation extracts the relevant set of cells from the cube, allowing further operations to be performed.

```
Indicator -> Value ( Predicate )
Value -> ( Value BinaryOP ) Value1
Value1 ->  term
Value1 -> ( UnaryOP ) Value
Predicate -> ( Predicate AND ) Predicate1
Predicate1 ->  Dimension | Instance
Dimension -> Keyword1 Dimension1
Dimension1 -> term FactType1 term ( Condition )
Dimension1 -> term ( Condition )
Instance -> FactType2 Instance1
Instance1 -> ( Name OR ) Name term
Keyword1 -> by | of
FactType1 -> of
FactType2 -> in | of
Condition -> Keyword2 ( UnaryOP ) term ( CompOp Name )
Keyword2 -> with
CompOp -> equal to | higher than | lower than
CompOp -> equal or higher than | equal or lower than
UnaryOP -> sum | maximum | minimum | count | average
BinaryOP -> plus | minus | divided by | times Number
Number -> [0-9]*
```

Fig. 5. Grammar for recognizing indicators described using SBVR font

2. Whenever an unary operation is performed over a given value, the corresponding OCL operation over the set of values previously projected is applied. These operations may be *sum()*, *count()*, or other unary operators.
3. Whenever a binary operation is performed over a pair of values, the result is obtained by iterating over the set of cells corresponding to each value. Therefore, a binary operation is translated to an *iterate(value1,value2,result=0 | result = value1 operator value2)* operation over the cube.
4. Predicates specify sets of conditions over certain dimensions, levels, and values. First, required dimensions are added to the query by means of *addDimension(Dimension,additivity)*. Then, the level of detail is adjusted though *rollUp(Dimension,level,additivity)*. Finally, conditions are translated by means of *sliceDice(cell | condition)*.
5. Once all the necessary operations have been performed, the initial aggregation function specified is applied (typically *sum()*), in order to obtain the value of the indicator. If no aggregation function has been specified, then, the default additivity function of the cube is applied.

After we have obtained the OCL4OLAP representation, we validate the correctness of the OCL constraint against the multidimensional schema. Since OCL4OLAP is an extension of OCL without the addition of new constructs, an OCL compiler can be used to validate the constraint. Finally, it is translated into a MultiDimensional eXpression (MDX[2]) query, as specified in [14].

5   Data Extraction

Once we have obtained the MDX representation for each KPI, their value is retrieved, loaded into the atomic KPIs, and normalized according to the values specified [6] for each of them. Afterwards, composite indicators are calculated, resulting in a comprehensive view of the business strategy including the performance associated to each element. The result can be seen in figure 2.

According to the results obtained, the Steel Wheels company is meeting its main goal (green light), increasing its revenue. As expected, since the company is focusing on the "Fancy Designs created" course of action, the KPIs point out that the "Low-Cost Designs" approach is performing overall between average and bad, presenting two indicators (*xg4* and *xg10*) with yellow light and one indicator (*xg3*) with red light. On the other hand, the "Fancy Designs created" approach is obtaining average results. This approach is exceeding the target amount of sales (*xg10*), but although the "# of complaints" is low (*s1* is active), the "Number of cancellations" is anormally high (*xg14* presents a red light), thus customer satisfaction may decrease, hurting the image of the company.

Our approach allows the decision maker to analyze the business strategy by using real-data, as opposed to estimations only. This allows the decision maker to identify potential problems in the business processes, e.g. there may be a potential problem in the distribution and delivery processes, as well as in the

---

[2] http://msdn.microsoft.com/en-us/library/ms145595.aspx

business strategy, e.g. despite average results in the delivery process we are meeting our goals in sales and revenue increase, have the problems in our delivery process not impacted our revenue yet?.

6   Related Work

We briefly present related work in the areas of KPIs and business models. In [15], the authors specify a series of Awareness Requirements over a requirements elicitation model of the DW, in order to model constraints which should be monitored, but pay little attention to analysis capabilities of the indicators and the process of providing such information to the user. Thus, as the focus of their work is on DW design, it can be considered as a complementary approach to ours. On the other hand, in [7], the Balanced Scorecard is proposed. The Balanced Scorecard has been one of the cornerstones in decision making for a long time. Its great advantage is that it maintains a global vision of the business strategy along with KPIs. However, this vision is not modeled, thus the relationships between strategies, goals and indicators are unknown. Furthermore, it does not provide analysis capabilities, such as those provided by dashboards. Dashboards [3] are proposed as means to provide a detailed view of certain KPIs. While dashboards provide detailed information about a subset of KPIs they are focused on, they lack a global view of business strategy, and must be manually implemented, introducing an overhead in the process and potentially introducing errors. Finally, Strategy Maps [8] describe how the organization creates value combining the different perspectives present in the Balanced Scorecard. However, are built in an informal way, and do not provide any mechanism to assess the effectiveness of the strategy modeled.

7   Conclusions and Future Work

We have presented a novel approach to relate KPIs defined in the business plan and the Balanced Scorecard [7] with business strategies and goals. Our proposal presents several advantages. First, all indicators are related to their respective goals, thus the decision maker can precisely identify which goals are having problems. Second, our approach not only allows the decision maker to model the business goals and indicators, but also allows him to analyze the business strategy using all the information in the underlying Data Warehouse, thus transforming the business strategy into a powerful dashboard. Third, KPIs are defined by using Structured English, allowing the decision maker to perform quick modifications, without requiring knowledge of how is the Data Warehouse structured at logical level. Finally, our approach supports a combination of real data and what-if analysis, allowing analysts to compare expectations with reported results, thereby helping them identifying existing problems.

Finally, since our approach involves decision makers in the process, we plan to test the approach by applying it to a real case study and evaluate the results obtained. We will focus on the interaction between decision makers and the system to analyze the effectiveness of using Structured English to define KPIs.

Acknowledgments. This work has been partially supported by the MESO-LAP (TIN2010-14860) and SERENIDAD (PEII-11-0327-7035) projects from the Spanish Ministry of Education and the Junta de Comunidades de Castilla La Mancha respectively, and by the European Research Council (ERC) through advanced grant 267856, titled Lucretius: Foundations for Software Evolution (04/201103/2016) http://www.lucretius.eu. Alejandro Maté is funded by the Generalitat Valenciana (ACIF/2010/298).

References

1. Barone, D., Jiang, L., Amyot, D., Mylopoulos, J.: Composite Indicators for Business Intelligence. In: Jeusfeld, M., Delcambre, L., Ling, T.-W. (eds.) ER 2011. LNCS, vol. 6998, pp. 448–458. Springer, Heidelberg (2011)
2. Cabot, J., Pau, R., Raventós, R.: From uml/ocl to sbvr specifications: A challenging transformation. Information Systems 35(4), 417–440 (2010)
3. Eckerson, W.: Performance dashboards: measuring, monitoring, and managing your business. Wiley (2010)
4. Giorgini, P., Rizzi, S., Garzetti, M.: Goal-oriented requirement analysis for data warehouse design. In: DOLAP 2005, pp. 47–56 (2005)
5. Hill, T., Westbrook, R.: Swot analysis: it's time for a product recall. Long Range Planning 30(1), 46–52 (1997)
6. Jiang, L., Barone, D., Amyot, D., Mylopoulos, J.: Strategic Models for Business Intelligence. In: Jeusfeld, M., Delcambre, L., Ling, T.-W. (eds.) ER 2011. LNCS, vol. 6998, pp. 429–439. Springer, Heidelberg (2011)
7. Kaplan, R., Norton, D.: The balanced scorecard–measures that drive performance. Harvard Business Review 70(1) (1992)
8. Kaplan, R., Norton, D.: Strategy maps: Converting intangible assets into tangible outcomes. Harvard Business Press (2004)
9. Luján-Mora, S., Trujillo, J., Song, I.: A UML profile for multidimensional modeling in data warehouses. Data & Knowledge Engineering 59(3), 725–769 (2006)
10. Marr, B., Schiuma, G., Neely, A.: Intellectual capital–defining key performance indicators for organizational knowledge assets. Business Process Management Journal 10(5), 551–569 (2004)
11. Maté, A., Trujillo, J.: A Trace Metamodel Proposal Based on the Model Driven Architecture Framework for the Traceability of User Requirements in Data Warehouses. In: Mouratidis, H., Rolland, C. (eds.) CAiSE 2011. LNCS, vol. 6741, pp. 123–137. Springer, Heidelberg (2011)
12. Object Management Group: A Proposal for an MDA Foundation Model (2005)
13. Object Management Group: SBVR specification (2008)
14. Pardillo, J., Mazón, J.N., Trujillo, J.: Extending OCL for OLAP querying on conceptual multidimensional models of data warehouses. Information Sciences 180(5), 584–601 (2010)
15. Silva Souza, V.E., Mazón, J.-N., Garrigós, I., Trujillo, J., Mylopoulos, J.: Monitoring strategic goals in data warehouses with awareness requirements. In: Proceedings of the 27th Annual ACM Symposium on Applied Computing, pp. 1075–1082. ACM (2012)
16. Yu, E.: Modeling strategic relationships for process reengineering. Social Modeling for Requirements Engineering (2011)

Data Support in Process Model Abstraction

Andreas Meyer and Mathias Weske

Hasso Plattner Institute at the University of Potsdam
Prof.-Dr.-Helmert-Str. 2–3, D-14482 Potsdam, Germany
{andreas.meyer,mathias.weske}@hpi.uni-potsdam.de

Abstract. Process model abstraction is an effective approach to reduce the complexity and increase the understandability of process models. Several techniques provide process model abstraction capabilities, but none of them includes data in the abstraction procedure. To overcome this gap, we propose data abstraction capabilities for process model abstraction. The approach is based on use cases found in literature as well as encountered in practice. Altogether, we introduce a framework for data abstraction in process models and provide algorithmic guidance to apply it in practice. The approach is evaluated by an implementation and a scenario of a workshop organization, that contains process models on different levels of abstraction.

Keywords: Business Process Management, Process Model Abstraction, Data, Data Abstraction Framework.

1 Introduction

Business process management is a an important approach to manage work in organizations [1], with process models being the key artifacts. Process models contain activities and their execution ordering as well as data the process creates or modifies. Each process model serves a specific goal. Goals range from alignment of processes with strategic goals to finding operational deficiencies and process implementation using IT systems. Traditionally, several process models have been maintained to capture these different aspects. For avoiding inconsistencies through several independent models for a given process, business process model abstraction has recently been established [2,3,4]. Unfortunately, these techniques deal with control flow abstraction only, failing to support further modeling artifacts, such as data. This paper extends business process model abstraction with data abstraction techniques. As a result, not only activities in a business process can be considered at different levels of abstraction, but also data.

As starting point, we consider a business process model that is structurally sound, i.e., it contains a single source node and a single sink node. Process models consist of activities, events, gateways, and data objects, associated to activities. The approach utilizes data models, that relate data objects with each other. Abstraction decisions are based on the occurrences of data objects in the process model and their relations in the data model. The approach supports abstraction

P. Atzeni, D. Cheung, and R. Sudha (Eds.): ER 2012, LNCS 7532, pp. 292–306, 2012.

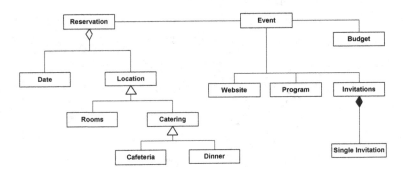

Fig. 1. Data model for the workshop organization example

use cases, derived from [5,4] as well as from discussions with experts. Additionally, analyses of process models encountered in practice, teaching, and scientific publications, e.g., [6], contributed to the set of use cases that are covered. Applicability of the approach is shown by an implementation and an evaluation of the abstraction results.

The paper proceeds as follows: The next section defines fundamental concepts for data abstraction. In Section 3, we present the data abstraction approach and provide algorithmic guidance for its application. Then, we discuss the evaluation in Section 4. Section 5 is devoted to related works, before concluding the paper.

2   Fundamental Concepts

In this section, we introduce fundamental concepts for data abstraction in business processes. A business process consists of two parts: (i) a data model, which presents the data dependencies between all data objects, and (ii) a set of process models, which describe the real world process from different perspectives. We start the discussion with the definition of the first part – a data model.

Definition 1 (Data Model)
A *data model DM* is represented as subset of an UML class diagram [7] utilizing the concepts of classes, data relations, and generalization sets. Classes represent data object classes DC. Data relations R describe the dependencies between data object classes. They are either undirected associations $R_{Assoc} \subseteq DC \times DC$ or directed parental data relations $R_P \subseteq DC \times DC$. Parental data relations are of types composition (R_{Comp}), aggregation (R_{Aggr}), or generalization (R_{Gen}) such that $R_P = R_{Comp} \cup R_{Aggr} \cup R_{Gen}$. The UML concept of generalization sets is extended to relation clusters RC such that clustering is available for any type of the introduced parental data relations R_P.   ◇

We use subscripts, e.g., DC_{DM}, $R_{P,DM}$, and RC_{DM} to denote the relations of the sets to data model DM and omit subscripts where the context is clear. Note that we restrict the number of relations between two data object classes to one. We refer to a data model containing data object classes, parental data relations, and relation clusters only as directed graph. Figure 1 shows a data model with the

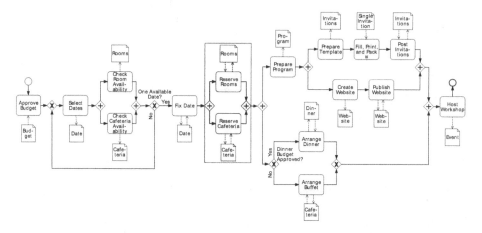

Fig. 2. Process model for the workshop organization example

boxes representing the data object classes and the edges representing different relations as used in UML. Next, we proceed with the definition of a data object.

Definition 2 (Data Object)
A *data object* $o \in DO$ in a business process is associated with a data object class in the data model such that $\varphi(o) \in DC$, where $\varphi : DO \to DC$. The collection of all data objects is denoted as DO.   ◇

Each set of data objects is associated to a business process and therefore, each data object is mapped to exactly one data object class of the appropriate data model. Subscripts are used to explicitly state these dependencies and are omitted where the context is clear. Generally, a data object is any piece of information utilized in the business process. It may represent, but is not limited to, documents, forms, database fields, database tables, variables, and products. But a data object does not represent full IT systems or databases. The second part of a business process is the set of process models, which we define next.

Definition 3 (Process Model)
A *process model* is a tuple $PM = (A, E, G, D, C, F, \delta_C, \delta_O)$, where A is a finite set of activities, E is a finite set of events, G is a finite set of gateways, D is a finite set of data nodes (A,E,G,D are pairwise disjoint), $C \subseteq (A \cup E \cup G) \times (A \cup E \cup G)$ is the control flow relation, $F \subseteq (A \times D) \cup (D \times A)$ is the data flow relation, $\delta_C : D \to DC$ maps each data node $d \in D$ to a data object class $dc \in DC$ such that $\delta_C(d) = dc$, and $\delta_O : D \to DO$ maps each data node $d \in D$ to a data object $do \in DO$ such that $\delta_O(d) = do$.   ◇

We use subscripts, e.g., A_{PM}, D_{PM}, and $\delta_{O,PM}$ to denote the relations of the sets and functions to process model PM and omit subscripts where the context is clear. We refer to the set $N = (A \cup E \cup G)$ as flow nodes of the process model. Please note that we define δ_C for convenience reasons only to increase understandability of the approach formalization following in Section 3. $\varphi(\delta_O(d)) = dc$ holds for $d \in D$. A process model can be arbitrarily structured, but needs to be structurally sound, i.e., it contains a single source node and a single sink node and each node is on a path from the source to the sink node. Figure 2 shows a structural sound process model visualizing one perspective of a *workshop organization* business process, which is adapted from an event organization process at IBM Research Zurich provided in [6]. A process model can also be partitioned into numerous process fragments, which may intersect. Each process fragment has a single entry and a single exit point. The highlighted part of the given process model is an example.

Definition 4 (Process Fragment)
A *process fragment pf* is a connected subgraph of a process model such that $N_{pf} \subset N$, $D_{pf} \subset D$, $C_{pf} \subset C$, and $F_{pf} \subset F$.   ◇

Finally, a business process is defined as follows.

Definition 5 (Business Process)
A *business process* is a tuple $BP = (\mathcal{PM}, DM)$, where $\mathcal{PM}$ is a finite set of process models PM representing perspectives of the same business process and DM is the corresponding data model describing the data structure.   ◇

Process model abstraction is generally defined in Definitions 1 to 4 in [5]. The work focuses on abstraction scenarios affecting the control flow of a process model. However, these foundations are defined on modeling artifacts in general instead of flow nodes so that they are also applicable to data abstraction.

Basing thereon, the proposed approach extends existing techniques for process model abstraction, which allow to aggregate an identified process fragment into a single activity, e.g., [2,3]. Thereby, the process model gets abstracted stepwise. In one abstraction step, the process fragments may only intersect in their entry and exit points, i.e., in their first and last flow node. To this, we refer as process fragment based control flow abstraction. Figure 3 visualizes one abstraction step for a subset of the process model introduced above. The left box shows a part of the initial process model, whereas the right box shows the corresponding part of the abstracted one. Abstraction of process fragments is highlighted by arrows in the figure. Both sequences, highlighted there in the left box, do have the same *and* gateways as entry and exit points.

Definition 6 (Control Flow Abstraction)
Given process fragment $pf \in PF_{PM}$ with PF_{PM} representing the set of all process fragments in the original process model PM, the *control flow abstraction function* $\mu : PF_{PM} \to A_{PM'}$ maps the process fragment to a single activity $a \in A_{PM'}$ such that $\mu(pf) = a$ and a is part of the abstracted process model PM'.   ◇

3   Data Abstraction

This section presents the actual approach to abstract data in process models. For each of the abstracted process fragments, all according data objects are evaluated and handled with respect to a set of rules, which will be introduced below.

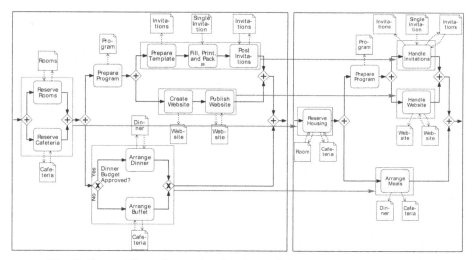

Fig. 3. Control flow abstraction for the workshop organization example

Thereby, data abstraction results are directly influenced by the given control flow abstraction as the choice of the abstracted process fragments determines the scope of data abstraction, i.e., the set of data objects considered for abstraction. Vice versa, made data abstractions do not influence further control flow abstractions applied to the process model. Consequently, control flow and data abstraction do not intermingle in the approach presented in this paper. Figure 4 shows the result of data abstraction (see right box) if the control flow abstraction from Figure 3 (see left box) is given. Information about combining data objects is taken from the corresponding data model (see Figure 1). For instance, data objects *Rooms* and *Cafeteria* are combined to data object *Location* as visualized by the corresponding arrow. Formally, we define data abstraction as follows.

Definition 7 (Data Abstraction)
Given two data objects $o, o' \in DO$, the *data abstraction function* $\lambda : DO \to DO \cup \varnothing$ maps the original data object o to the corresponding abstracted one o' such that $\lambda(o) = o'$ or the empty set if the data object is removed from the process model. Analogously, the data nodes abstraction function $\lambda_D : D \to D \cup \varnothing$ maps an original data node to the corresponding abstracted one or none.   ◇

Note that both abstraction functions eventually lead to the same result as $\lambda(o) = \lambda(\delta_O(d)) = \delta_O(\lambda_D(d)) = \delta_O(d') = o'$ with $d \in D_{PM}$, $d' \in D_{PM'}$, and $o, o' \in DO$. For convenience reasons, we define both functions.

The rule set, the data abstraction function λ is operating on, is naturally identified by reasoning processes considering use cases cases adapted from process model abstraction use cases presented in [5,2,8,4] as well as from discussions with experts. For reasoning, we chose a set of process models containing data nodes, which have been presented in scientific publications, e.g., [6,9] and which have been created in our research group during teaching and in projects. The process models were harmonized by transformation into BPMN process models and integrating subprocess modeling. Complexity of process models is mostly similar to

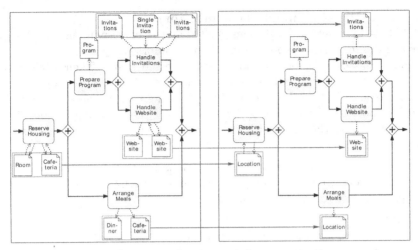

Fig. 4. Data abstraction for the workshop organization example

the one presented in Figure 2 or slightly smaller. We also considered one process model containing twice the number of flow nodes and 5times the number of data associations to incorporate insights taken from comparably large process models into the data abstraction process. However, the size of the process model did not influence the process of data abstraction. The data models have been created manually for the given process models.

During analyses, we utilized the process model abstraction approach from [3] and aligned the data nodes with respect to our perception for each abstraction step. Alignment of data nodes includes the *removal* of a data node from the abstracted process model, *keeping* a data node as is, and *combination* of several data nodes to one data node. Based on understanding our data abstraction decisions, we discovered the set of rules introduced and formalized below. Afterwards, we also provide algorithmic support for rule set application.

3.1 Preliminaries for Rule Set Formalization

Prior to proceeding with definition of the data abstraction rules, we define several notions to be used during rule set formalization. We start with the notion of the transitive closure over parental data relations, which comprises all data object classes being a predecessor of the specified data object class in a directed graph.

Definition 8 (Transitive Closure over Parental Data Relations)
Let dm be a connected graph derived from the data model. Then $R^+(dc)$ denotes the *transitive closure over parental data relations* of data object class $dc \in dm$. $R^+(dc)$ represents all data object classes, which are a predecessor of dc.   ◇

Second is the notion of a least common denominator of a set of data nodes, which is defined as follows.

Definition 9 (Least Common Denominator)
Let $DN \subseteq D$ be a set of data nodes such that the corresponding data object classes $\delta_C(dn)$ for all data nodes $dn \in DN$ are comprised by the set OC and all data object classes $dc_i \in OC$ are nodes in a directed graph. Then, $CD(OC) = \bigcap (R^+(dc_i) \cup dc_i)$ represents the set of common denominators of all nodes of set OC. Taken any data object class $dc \in OC$, the *least common denominator* $LCD(DN)$ is either data node $\delta_C^{-1}(dc)$, if $dc \in CD(OC)$, or data node $\delta_C^{-1}(n)$ with $n \in OC$ being the node with the smallest distance to dc in the directed graph.   ◇

Finally, we introduce the notions of associations between a data node and an activity or a process fragment respectively. These associations are important as they represent data utilization within the process model.

Definition 10 (Data Node–Activity–Association)
A set of data nodes associated to a particular activity $a \in A$ is denoted as $DOA(a)$ with $DOA(a) = DOA_i(a) \cup DOA_o(a)$, where $DOA_i(a)$ represents the set of data nodes being read and $DOA_o(a)$ represents the set of data nodes being written by activity a.   ◇

Definition 11 (Data Node–Process Fragment–Association)
A set of data nodes associated to a particular process fragment pf is denoted as $DOF(pf)$.   ◇

3.2   Abstraction Rules

In the following paragraphs, we formally define all rules we propose to abstract data in process models. For some rules, we do provide explanatory examples. These utilize the data relations presented in Figure 1 with one extension: The data object class *Proposal* is added by connecting it to the data object class *Event* via an association relation. Rules R1 and R2 deal with process quality during data abstraction. Rule R3 identifies opportunities to combine several data objects to a single one. The decision about *keeping* or *deleting* a data node is determined based on rules R4, R5, and R6.

For achieving high process quality in the abstracted process models, consistency in terms of level of abstraction of the data objects must be ensured within a process model. Therefore, all data node representations of a data object must be equally abstracted, if at least one of the data nodes gets abstracted. Rule R1 ensures abstraction consistency by enforcing identical data abstraction of all data nodes representing the same data object.

R1 Abstraction Consistency
Let $d \in D$ be a data node, which gets abstracted to $\lambda_D(d)$, then $\forall d_i \in D$ with $\delta_O(d_i) = \delta_O(d) : d_i$ gets also abstracted such that $\lambda_D(d_i) = \lambda_D(d)$.   ◇

A second indicator of high process quality are aligned data and control flow representations of a process model. Assuming such alignment exists before process model abstraction, such alignment must also exist afterwards. The alignment is achieved by avoiding abstraction of data nodes being associated to activities, which have not been selected for control flow abstraction. The only exception is data abstraction enforced by rule R1.

R2 Preserving Granularity Alignment

Let PM be a process model and let PM' be the corresponding abstracted process model. Further, let $a \in A$ be an activity in the original process model such that $a \in PM$. Then, if $a \in PM' : \forall d_i \in DOA(a) : \lambda_D(d_i) = d_i$ holds.   ◇

Generally, data abstraction shall reduce the number of data nodes where applicable. Within a process fragment, selected for control flow abstraction, several data object classes of corresponding data nodes might have common predecessors in a directed graph representation of the data model (see Definition 1). The same holds true for data object classes corresponding to data nodes, which are associated to a single activity, e.g., after reduction of a process fragment to an activity. In these cases, the appropriate data nodes can be abstracted to one data node – the one, which corresponds to the least common denominator in the directed graph. Figure 5 visualizes one abstraction alternative for rule R3 by combining data nodes *Date* and *Rooms* to data node *Reservation* because of data object class *Rooms* being a specialization of data object class *Location* and the aggregation relation for data object class *Reservation* given in the data model in Fig. 1.

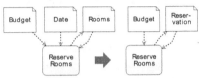

Fig. 5. R3 data object combination

R3 Data Object Combination

Let PM be a process model and let PM' be the corresponding abstracted process model. Let $pf \in PM$ be a process fragment and $a \in PM'$ be an activity in the abstracted process model such that $\mu(pf) = a$. Further, let Z be a set of data nodes such that $Z \subseteq DOF(pf)$ and $LCD(Z) \neq \varnothing$. Or let b be an activity in a process model and let Z be a set of data nodes such that $Z \subseteq DOA(b)$ and $LCD(Z) \neq \varnothing$. Then, $\forall d_i \in Z : \lambda_D(d_i) = LCD(Z)$.   ◇

After data abstraction, an activity might not be associated to data nodes any more. But, data highly influences process execution. Therefore, the main dependencies to enable an activity shall be preserved on all levels of abstraction in the process model. Rule R4 introduces the *lower bound of data nodes* in a process model – a configurable number stating the minimum data associations to exist after data abstraction. Assuming, the lower bound has been set to one, this means, that no activity is allowed without an association to a data node after data abstraction, if there was at least one data node present in the process fragment which got abstracted. Following, deletion of any data node is omitted in one abstraction step, if that step would remove all remaining data nodes from the process fragment.

R4 Lower Bound of Data Nodes

Let a be an activity, which represents process fragment pf of the initial process model after abstraction such that $\mu(pf) = a$. Then $|DOA(a)| \geq x$ if $|DOF(pf)| \geq x$ with $x \in \mathbb{N}^*$ being the specified lower bound.   ◇

A data object can be utilized in a process fragment and by at least one other activity, which is not part of that process fragment. Data objects spread over different areas of a process model have an increased importance with respect to correct execution as the degree of spread determines a data object's impact to the process model. Due to the increased importance, a data object must not be deleted during data abstraction, if it is utilized in the process fragment selected for control flow abstraction and associated to at least one more activity. Fig. 6 visualizes data abstraction based on rule R5. After abstraction, data node *Rooms* is preserved as it is utilized in the fragment selected for abstraction and it is associated to one more activity, in this case *Check Room Availability*.

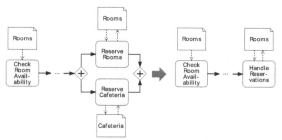

Fig. 6. R5 reoccurring data object

R5 Reoccurring Data Object

Let PM be a process model and let PM' be the corresponding abstracted process model. Further, let a_1, a_2 be two activities and let pf be a process fragment such that $pf \in PM$, $a_1 \in PM$, $a_2 \in PM'$, and $\mu(pf) = a_2$. Also let $d_1 \in D_{PM}$ be a data node such that $d_1 \in DOF(pf)$. Then, $\lambda_D(d_1) \in DOA(a_2)$, if $\exists d_2 \in D_{PM}$: $d_2 \in DOA(a_1)$ such that $\delta_O(d_1) = \delta_O(d_2)$.   ◊

Data objects only utilized in a process fragment can be of importance for the complete process model. For instance, a data object can be gathered from an external source and be input to a process fragment, where it gets manipulated. Therefore, we require a data object to be associated to each activity of one path through the process fragment. Additionally, the data object must also be read by the first or written by the last activity of that path to ensure that the data object is either input or output to the process fragment. Figure 7 visualizes data abstraction based on rule R6. The data nodes *Location* an *Proposal* are the only ones being associated to each activity on one path. Thereby, data node *Proposal* is neither read by the first nor written by the last activity of the corresponding path. In contrast, data node *Location* is read by the first activity in the corresponding path. Therefore, data node *Location* is kept while data node *Proposal* is deleted during the abstraction step.

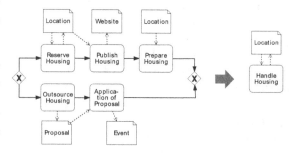

Fig. 7. R6 full data object activity association

R6 Full Data Object Activity Association
Let PM be a process model and let PM' be the corresponding abstracted process model. Let $pf \in PM$ be a process fragment and $a \in PM'$ be an activity in the abstracted process model such that $\mu(pf) = a$. Let T be a set of paths through pf. Further, let B be a set of activities such that $\forall a_i \in B : a_i \in pf$ and there are no other activities in pf. Also, let $Z \subseteq \delta_O^{-1}(o)$ be a set of data nodes representing data object $o \in DO$ such that $\forall d_j \in Z : d_j \in DOF(pf)$. If $\exists t \in T$ such that $\forall a_i \in t : \exists d_j \in Z : d_j \in DOA(a_i)$, then $d \in DOA_i(a)$, if $\exists d_j \in Z : d_j \in DOA_i(a_1)$, and $d \in DOA_o(a)$, if $\exists d_j \in Z : d_j \in DOA_o(a_2)$, with a_1 being the first and a_2 being the last activity of path t in terms of order of execution. d is a data node such that $\delta_O(d) = \lambda(o)$.   ◇

3.3   Algorithm

After defining the rules, which describe the abstraction of data in process models, we proceed with algorithmic guidance on how to apply these rules in the context of process model abstraction fully automatically. The algorithm proposed below describes one way towards process model abstraction including data. But further ways based on the given rule set are possible – especially by utilizing a different control flow abstraction technique. We require that each business process meets several assumptions for process model abstraction: (i) the data model is explicitly defined and provides the truth about data relations, i.e., it is used as reference, (ii) the approach of triconnected abstraction [3] is used for control flow abstraction, (iii) best results are achieved with structured process models although unstructured process models can be covered as well by abstracting them to a single activity while handling data with respect to the rule set, (iv) labeling of activities, being the result of control flow abstraction, is done either manually or by appending the labels, (v) labeling is consistent throughout the complete business process such that same modeling elements get assigned the same name, and (vi) the mappings between data object classes, data objects, and data nodes are done by label matching, i.e., data object classes, data objects, and a data nodes with the same name can be related to the same data object used in the business process.

The algorithm below provides the main steps to abstract a process model once. For further iteration steps, this algorithm has to be applied iteratively. The algorithm works for all structural sound process models. However, block structured process models increase the quality of process model abstraction by allowing more diverse abstracted process models. A process model is block structured, if and only if each control flow node with multiple outgoing edges has a corresponding control flow node with multiple incoming edges and vice versa. Otherwise, it is unstructured. Unstructured process models may get structured by applying structuring techniques as described in [10], which can be used as preparation step to increase the expected quality of the abstracted process models. However, structuring is out of scope for the proposed algorithm and process models are abstracted as given.

Algorithm 1: Process model abstraction including data

$PF \leftarrow identifyProcessFragments(PM); //$Abstraction fragments selection
for all pf **in** PF **do**
  $DN \leftarrow getDataNodes(pf); //$R2
  $DN \leftarrow dataObjectCombination(DN); //$R3
  if pf is a polygon or rigid **then**
    $K \leftarrow fullDataObjectActivityAssociation(DN); //$R6
    $K \leftarrow reoccurringDataObject(DN); //$R5
    $K \leftarrow lowerBoundOfDataNodes(DN,K); //$R4
  else $\{pf$ is a bond$\}$
    $K \leftarrow DN$
  end if
  $a \leftarrow triconnectedAbstraction(pf); //$control flow abstraction
  $a \leftarrow addDataFlow(K);$
end for
$PM \leftarrow abstractionConsistency(); //$R1

First, the given process model PM is analyzed and the RPST representation is created. RPST refers to the tree representing the structure of the process model [11]. Each process fragment gets typed and related in that tree – the process structure tree. Each control flow edge is typed as *trivial*. Sequences are typed as *polygon*. Blocks are typed as *bonds*. All unstructured and not yet typed process fragments are typed as *rigids*. Trivials denote the leafs in the process structure tree and are of no interest for the proposed algorithm. Each non trivial process fragment, which is deepest on one path of the process structure tree is considered for abstraction and therefore added to the set PF of process fragments.

Afterwards, for each element of this set, the following steps are undertaken. First, data nodes DN of a particular process fragment are extracted. Then, rule R3 is applied to this set of data nodes to combine them where applicable. Next, decisions are taken whether a data object is kept in or removed from the abstracted process model. If the particular process fragment is of type *polygon* or *rigid*, rules R6, R5, and R4 are applied in this order. Each of these rules selects the data nodes from DN the set, which are considered important enough to keep with respect to the content of the rule. If the particular process fragment is of type *bond*, all data nodes are kept. This procedure works in this algorithm as process fragments typed as bonds in the process structure tree and selected for abstraction contain none or exactly one activity on each path between the *and* gateways – the entry and exit points of the process fragment. Next, the particular process fragment is reduced to a single activity a and the control flow is added accordingly before the data nodes, added to set K, are associated to activity a. Thereby, the type of association, i.e., read and write access, is derived from process fragment before abstraction and from rule R6. Basically, a data node first read by any activity in the process fragment is also read by the abstracted activity. A data node written by any activity before abstraction is also written by the abstracted activity regardless whether and when it is read. In contrast,

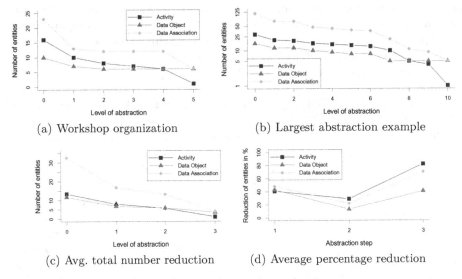

(a) Workshop organization

(b) Largest abstraction example

(c) Avg. total number reduction

(d) Average percentage reduction

Fig. 8. Data abstraction evaluation

a data node read after it has been written by any activity is not read by the abstracted activity. Only rule R6 may interfere with this typing of associations.

After performing these steps to all identified process fragments, rule R1 is applied to the abstracted process model to ensure consistent data abstraction. Rule R2 – preserve granularity alignment – is implicitly fulfilled by considering only data nodes being part of a process fragment, which is selected for abstraction.

4   Evaluation

After introducing the data abstraction framework, we will proceed with an evaluation based on our implementation. It comprises the algorithm as stated above and is based on jbpt[1] – a Java-based library containing techniques for managing and analyzing business processes; e.g., a common representation of process models, into which process models from different modeling notations like BPMN and EPC can be transformed. We use this representation as input to the algorithm.

Figure 8 presents results from applying the implementation to the set of process models we analyzed for rule set discovery. The set comprises ten process models with varying numbers of flow nodes and data associations from 7 to 52 and 21 to 114 alternatively. Figures 8(a) and 8(b) present the total number of activities, data objects, and data associations between them existing in the process models at different levels of abstraction, which are marked by the x-axis. 0 refers to the original model. Figure 8(c) presents the average reduction of model artifacts of all ten process models. As the smallest ones only contained three levels of abstraction, the levels of abstraction of all process models have been reduced by merging abstraction steps in an equally distributed manner. All

---

[1] http://code.google.com/p/jbpt/

three figures show that the number of data objects as well as data associations are continuously reduced – in terms of tendency, the reduction is comparable to the activity reduction, i.e., the proportion stays roughly the same. On average, the first and the last step of abstraction reduce a process model most (see Figure 8(d)) as the steps in-between usually affect a small fraction of the process model only. There may even arise situations when reduction is very limited as, for instance, in the workshop organization example in the abstraction step from level two and three. Therefore, merging succeeding abstraction steps may help to achieve larger distinctions between different levels of abstraction. However, reduction of data objects stops when all parental relations are resolved and the remaining data objects drive the entire process. The last abstraction step reduces the number of activities to one and, in case there exists a data object being the main driver of the process and the others depend on it, the number of data objects may be reduced again – most likely to one. One example for this behavior was seen in an order process with order being the main data object and, for instance, invoice and customer data objects depending on the order data object. Comparable reduction of activities and data associations allows still readable process models after abstraction with respect to visualized data information. An overwhelming of stakeholders is avoided.

Generally, we conclude that the rule set technically works and introduces data support to process model abstraction.

5   Related Work

There do exist many approaches and techniques to deal with process model abstraction. [2] provides an extensive overview about abstraction scenarios and the application of rules for the actual abstraction step. Thereby, abstraction is based on identification and abstraction of process fragments. Similarly, the approach of triconnected abstraction utilizes the refined process structure tree [11,12] and abstracts process fragments identified by the tree into single activities. In the field of process modularity [13], techniques are utilized to aggregate common activities into one subprocess. In terms of process abstraction, this approach is similar to process fragment based abstraction [4].

In [14], the authors provide an abstraction approach based on semantic information of process models. Further approaches for process abstraction are activity clustering utilizing information provided by process mining techniques [8] and abstraction via graph reduction rules [4]. Thereby, each reduction rule specifies a structural fragment of the process which is abstracted. Iterative application of all rules leads to an abstracted process model. [15] introduces Petri net transformations to simplify, i.e. reduce, a process model. The approach of model transformation is also utilized in [16]. The authors transform a graph into a view, which meets a stakeholder's requirements using activity aggregation operations. [17] provides an approach presenting methods for abstraction of specific elements of BPMN 1.2 process models. Thereby, the focus was set to activities. The considered model artifacts also exist in BPMN 2, so that the approach can be applied there as well.

Additionally, [2,8,4] discuss abstraction criteria, which provide on the one hand use cases for applying abstraction techniques and on the other hand insights about which information shall be kept during the process of abstraction. In [5], the authors present a set of use cases for process model abstraction in general. These use cases were elicited from BPM experts. Moreover, the authors introduce a generalization for abstraction operations.

All works mentioned above do have one aspect in common: They focus on control flow constructs. The challenge of handling data in process model abstraction has not been tackled in the field of business process management, although some of the works provide a glimpse of data support, e.g., [5]. The abstraction operation generalization is provided the way, that data abstraction is also covered. However, the focus of the paper is still control flow abstraction as the use cases following the generalization clearly show.

6 Conclusion

We presented an approach to handle data in process model abstraction. Therefore, the contributions are i) a framework to abstract data and ii) algorithmic guidance to apply it to business processes. The framework comprises the fundamental concepts to incorporate data into process model abstraction and the rule set. The rule set works with arbitrarily structured process models, although best results are achieved with block structured ones. Additionally, the rule set assumes the application of a control flow abstraction technique, which identifies single entry single exit process fragments and aggregates each to a single activity. Following, we support a class of process model abstraction techniques rather than a specific one only. Further requirements are not required to apply the rule set as long as the fundamental concepts are not contradicted. Consequently, the framework can be applied to a variety of process modeling notations. Rule identification bases on process model analyses considering use cases with respect to data, which in turn are based on literature study as well as process analysis with experts.

Directly ongoing work will deal with extending the proposed data abstraction approach with coverage of data states as these represent the manipulation of data objects during process execution. Besides, one of our next steps is to conduct a field experiment to further validate the results of the rule set including the extension with data states by humans. We will also target data abstraction based on the type of relation between data object classes by differentiating the actual actions with respect to the type. Further, abstraction of data objects being associated to activities in a *xor* block arises the need to handle mandatory and optional data access in the abstracted process model. Types of data objects influence the usefulness of data object combinations. Combination of two forms into a single form might be useful, whereas combining two products, containing the same formal properties as the forms, to one product might lead to an inappropriate result. We seek to find solutions for this challenge.

References

1. Weske, M.: Business Process Management: Concepts, Languages, Architectures, 2nd edn. Springer (2012)
2. Polyvyanyy, A., Smirnov, S., Weske, M.: Business Process Model Abstraction. In: Handbook on Business Process Management 1, pp. 149–166. Springer (2010)
3. Polyvyanyy, A., Smirnov, S., Weske, M.: The Triconnected Abstraction of Process Models. In: Dayal, U., Eder, J., Koehler, J., Reijers, H.A. (eds.) BPM 2009. LNCS, vol. 5701, pp. 229–244. Springer, Heidelberg (2009)
4. Eshuis, R., Grefen, P.: Constructing Customized Process Views. Data & Knowledge Engineering 64(2), 419–438 (2008)
5. Smirnov, S., Reijers, H., Nugteren, T., Weske, M.: Business Process Model Abstraction: Theory and Practice. Technical Report 35, Hasso Plattner Inst. (2010)
6. Wahler, K., Küster, J.M.: Predicting Coupling of Object-Centric Business Process Implementations. In: Dumas, M., Reichert, M., Shan, M.-C. (eds.) BPM 2008. LNCS, vol. 5240, pp. 148–163. Springer, Heidelberg (2008)
7. OMG: Unified Modeling Language (UML), Version 2.2 (February 2009), http://www.omg.org/spec/UML/2.2/ (accessed June 11, 2012)
8. Günther, C.W., van der Aalst, W.M.P.: Fuzzy Mining – Adaptive Process Simplification Based on Multi-perspective Metrics. In: Alonso, G., Dadam, P., Rosemann, M. (eds.) BPM 2007. LNCS, vol. 4714, pp. 328–343. Springer, Heidelberg (2007)
9. Wang, J., Kumar, A.: A Framework for Document-Driven Workflow Systems. In: van der Aalst, W.M.P., Benatallah, B., Casati, F., Curbera, F. (eds.) BPM 2005. LNCS, vol. 3649, pp. 285–301. Springer, Heidelberg (2005)
10. Polyvyanyy, A., García-Bañuelos, L., Dumas, M.: Structuring Acyclic Process Models. Information Systems (2011)
11. Vanhatalo, J., Völzer, H., Koehler, J.: The Refined Process Structure Tree. Data & Knowledge Engineering 68(9), 793–818 (2009)
12. Polyvyanyy, A., Vanhatalo, J., Völzer, H.: Simplified Computation and Generalization of the Refined Process Structure Tree. In: Bravetti, M. (ed.) WS-FM 2010. LNCS, vol. 6551, pp. 25–41. Springer, Heidelberg (2011)
13. Reijers, H.A., Mendling, J.: Modularity in Process Models: Review and Effects. In: Dumas, M., Reichert, M., Shan, M.-C. (eds.) BPM 2008. LNCS, vol. 5240, pp. 20–35. Springer, Heidelberg (2008)
14. Smirnov, S., Reijers, H.A., Weske, M.: A Semantic Approach for Business Process Model Abstraction. In: Mouratidis, H., Rolland, C. (eds.) CAiSE 2011. LNCS, vol. 6741, pp. 497–511. Springer, Heidelberg (2011)
15. Berthelot, G.: Transformations and Decompositions of Nets. Petri Nets: Central Models and Their Properties, 359–376 (1987)
16. Bobrik, R., Reichert, M., Bauer, T.: View-Based Process Visualization. In: Alonso, G., Dadam, P., Rosemann, M. (eds.) BPM 2007. LNCS, vol. 4714, pp. 88–95. Springer, Heidelberg (2007)
17. Smirnov, S.: Structural Aspects of Business Process Diagram Abstraction. In: Commerce and Enterprise Computing, pp. 375–382. IEEE (2009)

Synthesizing Object Life Cycles from Business Process Models

Rik Eshuis and Pieter Van Gorp

Eindhoven University of Technology
P.O. Box 513, NL-5600 MB, Eindhoven, The Netherlands
{h.eshuis,p.m.e.v.gorp}@tue.nl

Abstract. Business process models expressed in UML activity diagrams can specify the flow of stateful business objects among activities. Such business process models implicitly specify the life cycles of those objects. To check the consistency of a business process model with an existing object life cycle or to generate or configure software supporting the business process, these implicit life cycles need to be discovered. This paper presents an approach for synthesizing an object life cycle from a business process model in which the object occurs in different states. The synthesized object life cycles are expressed as hierarchical statecharts. The approach makes implicit life cycles contained inside business process models explicit. The synthesis approach has been implemented using a graph transformation tool and has been applied to case studies.

Keywords: statecharts, UML, model transformation.

1 Introduction

Requirements on software systems that support operational business processes are typically captured in business process models. The Unified Modeling Language (UML) [15] has a notation, activity diagrams, that can be used to model business processes. Business process models expressed in UML activity diagrams can specify both the ordering of business activities as well as the flow of stateful business objects among these activities. State changes of these business objects are due to business activities. The actual object life cycles modeling these state changes are typically expressed as hierarchical object-oriented statecharts [5,15].

Business process models and object life cycles are used differently in the engineering process. Business process models are used in requirements engineering while object life cycles are used in the design phase, for instance to generate code or to configure an off-the-shelf software system. Next, business process models and object life cycles have a different scope. A business process model gives a global view of a business process, addressing different objects, while an object life cycle offers a local view, linked to one aspect of the global process view.

There is a need to relate both views, for instance to check consistency between a global and a local view, or to support traceability between requirements expressed in the global view and a design guided by the local view. However,

P. Atzeni, D. Cheung, and R. Sudha (Eds.): ER 2012, LNCS 7532, pp. 307–320, 2012.

an important obstacle is that object life cycles are only implicitly specified in business process models.

In this paper, we define an automated approach for synthesizing a hierarchical statechart from a UML activity diagram that specifies a business process model referencing a stateful object. This way, the approach discovers an object life cycle that is hidden in a business process model. The synthesized statechart can be used to generate or configure a software system supporting the business process, or to check consistency with existing statechart descriptions [2,11], either due to legacy systems or to specific industrial or governmental standards like ACORD (http://www.acord.org) and SCOR (http://supply-chain.org/scor).

The synthesis approach consists of two phases. In the first phase, nodes not relevant for the object life cycle are filtered from the activity diagram. In the second phase, the remaining part of the activity diagram is translated into a hierarchical statechart specifying the life cycle of the object referenced by the process model. The approach is fully automated and has been implemented using the graph transformation tool GrGen [4]. Section 5 gives more details on the prototype and our experiences in applying the prototype to case studies.

In this paper we consider UML2 activity diagrams that use object nodes [15], as explained in Section 2. Activity diagrams also support a pin-style modeling of object flows: then input and output objects of each activity are modeled with pins, which resemble parameters. We do not consider the pin style in this paper, but we plan to address it in future work.

The remainder of this paper is structured as follows. Section 2 gives an overview of the approach using a running example. Section 3 discusses the first phase of the synthesis approach, in which action nodes and irrelevant control nodes are filtered from the activity diagram. Section 4 details the second phase, in which a filtered activity diagram is translated into a hierarchical statechart. There we also explain that the translation may fail, since activity diagrams allow fine grained synchronization not expressible in statecharts. Section 5 presents a prototype that implements the approach and discusses our experiences in applying the prototype in case studies. Section 6 discusses related work. Section 7 ends the paper with the conclusion.

2   Overview

To introduce the salient features of the approach, we show in Fig. 1 an example business process model in a UML activity diagram. The process specifies handling an insurance claim. Atomic activities are represented by action nodes (ovals) like Receive. After receiving the claim, the policy and damage are checked in parallel, indicated with the bar symbols. Next, a decision (diamond symbol) is made to either reject the claim or to accept the claim. In that case, the cost are calculated and paid to the client, and in parallel the periodic contribution to be paid by the client is updated. Finally, the client is notified about the decision.

The process updates stateful object Claim. Each state of Claim is represented by an object node (rectangle). The Claim object can be in multiple states at

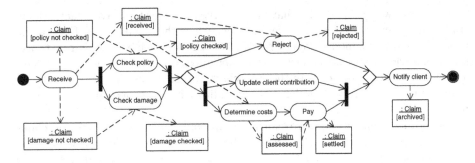

Fig. 1. Activity diagram specifying process for handling insurance claims

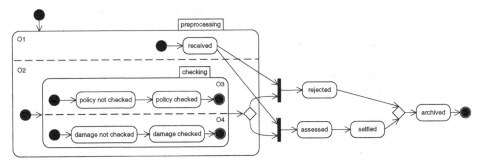

Fig. 2. Statechart specifying life cycle of object Claim in Fig. 1

the same time. For instance, after Receive has completed, the Claim object is in three parallel states: received, policy not checked, and damage not checked. Certain activities change the local state of the Claim object. For instance, Check policy changes the state from policy not checked to policy checked, but does not affect the other states.

Implicitly, the process model specifies the life cycle of the Claim object. Each object node in the activity diagram references a local state of the life cycle. That object life cycle can contain sequence, parallelism, choice, and loops. An example of a choice is the :Claim[received] object node in Fig. 1, which specifies according to the UML2 [15] semantics that object Claim in state received is either input to Reject or to Determine costs but not both.

It is desirable to automatically derive from a process model with an implicit object life cycle an explicit description of that life cycle. Such a description can be used to generate software code or to check consistency with an existing description of the life cycle. Since hierarchical statecharts are the default language for modeling object behavior, the life cycle description should use hierarchical statecharts as well. Fig. 2 shows a hierarchical statechart modeling the life cycle of a Claim object. Note that states received, policy not checked and damage not checked can be active in parallel, just as in Fig. 1. The statechart explicitly models the object life cycle specified implicitly in Fig. 1.

There are two problems that have to be solved in order to derive a hierarchical statechart description from a process model such as the one in Fig. 1. First, some parts of the process model are not relevant for the object life cycle. For instance, the Update client contribution activity does not affect any state of the Claim object, and therefore does not occur in the statechart. These irrelevant parts need to be removed from the process model, but the indirect flows between different object nodes need to be preserved.

Second, statecharts use hierarchical (compound) states, which have no counterpart in process models like activity diagrams. An example of a compound state is preprocessing in Fig. 2 that contains other states like O1 and received. To derive a hierarchical statechart, compound states need to be inferred from the activity diagram syntax. UML 1.5 [14] proposes to translate pairs of fork-join bars to AND states in order to map activity diagrams to statecharts. (A fork is a bar with one incoming and multiple outgoing edges while a join is a bar with multiple incoming and one outgoing edge.) However, in practice many activity diagrams do not satisfy this constraint but can be translated into a hierarchical statechart. For instance, the activity diagram in Fig. 5 contains a fork that matches three joins, but it can be translated to the statechart in Fig. 2.

We define an approach that solves these two problems. Input is an activity diagram specifying the behavior of a stateful object. The approach first filters irrelevant nodes from the activity diagram (Section 3), and next synthesizes a hierarchical statechart from the filtrate activity diagram (Section 4). Both steps are fully automated and do not require any user interaction. Applying the approach to the activity diagram in Fig. 1 results in the statechart shown in Fig. 2, modulo the names of the compound states. The approach may fail, since some activity diagrams cannot be translated into a hierarchical statechart, as we explain in Section 4.3. In that case, diagnostics can be provided giving precise feedback on which part of the process model causes the failure.

We focus in this paper on a single activity diagram with object nodes that references the same object type. If a single activity diagram references multiple object types, then for each object type a version of the activity diagram can be created that references only that object type, by removing from the original activity diagram those object flows and object nodes that do not refer to the object type. If multiple activity diagrams reference the same object type, they can be grouped into a single activity diagram by adding relevant control nodes to connect the different diagrams.

3   Filtering Activity Diagrams

In the first phase of the synthesis approach, nodes not relevant for the object life cycle are filtered from the activity diagram. For the input activity diagram, we require that every action node has one incoming and one outgoing control flow, and every object node has at least one incoming or outgoing object flow. An object node can have multiple incoming or outgoing object flows. The resulting activity diagram contains only object nodes and relevant control nodes and is,

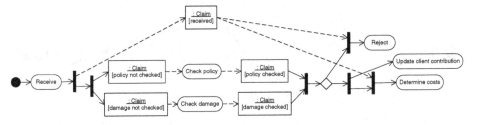

Fig. 3. Part of activity diagram in Fig. 1 after preprocessing

translated into a hierarchical statechart in the second phase, which is explained in the next section.

The filtering phase consists of two stages, which are explained in the sequel of this section. First, the activity diagram is preprocessed and transformed into a normal form. Second, several filtering rules are applied in arbitrary order to the activity diagram. The filtering stage stops if no filtering rule can be applied anymore to the activity diagram.

3.1 Preprocessing

In the preprocessing stage, we transform each activity diagram into a normal form by ensuring that each action node has one incoming and outgoing edge. An activity diagram in normal form has no dangling (object) nodes: each node is on a directed path from the initial to a final node. Preprocessing consists of two steps that performed iteratively in random order until the process model is not changed anymore.

In the first step, we ensure that each object node has at least one incoming and one outgoing edge. If an object node o has no incoming edge, then we take the action node a to which o is input. If a is not unique since o is input to multiple action nodes, this step fails. Otherwise, a is unique and there is a unique control flow that enters a. We change the target of the control flow from a to o. For instance, in Fig. 1 object :Claim[policy not checked] has no incoming object flow. The activity Check policy is targeted by one incoming control flow. The target of this control flow is changed to :Claim[policy checked], as shown in Fig. 3. A symmetrical rule is used for object nodes that have no outgoing edges, such as :Claim[policy checked] in Fig. 1.

However, if a has multiple object nodes as input or output, this preprocessing step is not applicable and the preprocessing fails. For instance, the activity diagram in Fig. 4 cannot be preprocessed; it is not clear whether in the final statechart, state S3 has to be in parallel with state S4 or not. If

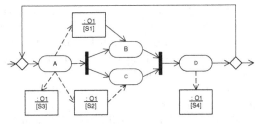

Fig. 4. Ambiguous activity diagram

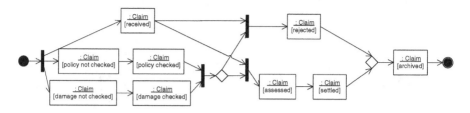

Fig. 5. Activity diagram of insurance claim process (Fig. 1) after filtering

the preprocessing fails for an object node, the user has to rewrite the model into an equivalent model, for instance by adding outgoing edges to the object node.

In the second step, we ensure that each action node gets exactly one incoming and one outgoing edge. By constraint, an action node has one incoming and one outgoing control flow. If an action node also has one or more incoming object flows, a bar is inserted that synchronizes the control flow and the object flows. This synchronization denotes that all inputs of activity need to be present before it can start, which is in line with the UML2 standard [15]. For instance, this step ensures that before both Reject and Determine costs in Fig. 3 bars are inserted that synchronize object flow and control flow. Similarly, if an action node has one or more outgoing object flows, a bar is inserted that splits the control flow and the object flows. For instance, in Fig. 3 after Receive an extra bar has been inserted.

Both steps may introduce control nodes that mix object flow and control flow, which is not allowed by the UML standard. However, this is harmless for the synthesis approach, since eventually object flows have to be translated into statechart control flows anyway.

3.2 Filtering

In the filtering stage, irrelevant nodes are removed. All action nodes are irrelevant, since they do not translate into any statechart construct. Furthermore, control nodes that do not influence object nodes are irrelevant. For instance, the rightmost pair of bars in Fig. 1 are irrelevant, since only one of the two parallel branches between the bars references object nodes, not both. Fig. 5 shows the activity diagram that results from filtering the activity diagram in Fig. 1. As Fig. 5 illustrates, control nodes are only included if they influence object nodes.

The actual filtering is realized by applying different filtering rules. A filtering rule eliminates irrelevant nodes from an activity diagram in normal form. The different filtering rules are applied iteratively in arbitrary order. The filtering step stops if no filtering rule can be applied anymore to the activity diagram. After the filtering rules have been applied, the activity diagram contains no action nodes, but only object nodes and control nodes. The resulting activity diagram is input to the translation from activity diagrams to hierarchical statecharts, detailed in the next section.

We use five filtering rules, R1–R5, which are graphically specified in Fig. 6. Reduction rules R4 and R5 resemble transformation rules defined by Hecht and

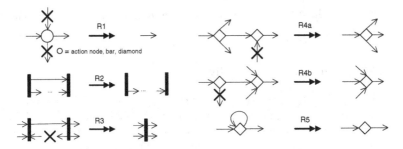

Fig. 6. Filtering rules for activity diagrams

Fig. 7. Example to illustrate different effect of filtering rules

Ullman [7] to test wether a flow graph is reducible [1], i.e. each loop has a single point of entry. Note that flow graphs are sequential, so they do not contain parallelism, as expressed with bars. Moreover, the objective of the reducibility test is to reduce a flow graph to a single node, while the filter approach needs to preserve relevant (object) nodes.

The most interesting feature of the rules is that diamonds and bars are treated in different ways, as illustrated by the example filterings shown in Fig. 7. The models on the right are the result of iteratively applying all filtering rules to the corresponding models on the left. Fig. 7(a) results by applying first rule R3 and next rule R1 two times. In Fig. 7(b), rule R4 is not applicable. We experimented with several other alternative rules for rule R4, for instance a rule for merging two diamonds similar to rule R2. However, such a rules merges the diamonds in Fig. 7(b), which is undesirable as explained before.

4   Synthesizing Statecharts

Output of the filtering phase is an activity diagram containing only object nodes and control nodes. In the next phase, a hierarchical statechart is synthesized from the activity diagram. Basis for the synthesis is an existing, formally defined translation from Petri nets to statecharts [3]. The syntax of Petri nets closely resembles that of activity diagrams. Key difficulty in synthesizing statecharts is the construction of hierarchical (AND and OR) states, which have no counterpart in activity diagram syntax. A synthesized statechart preserves both the structure and behavior of the input activity diagram.

We first explain how the state hierarchy, consisting of AND and OR states, is built from a filtered activity diagram. Next, we explain how the state hierarchy and the filtered activity diagram are used to construct the complete statechart.

Finally, we discuss how the synthesis can fail and how this influenced the definition of the approach.

4.1   Building the State Hierarchy

The state hierarchy is a tree of states. Leaves of the tree are BASIC states while internal nodes are AND and OR states. The tree is visualised by nesting child nodes inside parent nodes. For instance, in Fig. 2 BASIC node policy checked is child of OR node O3 which is in turn child of AND node check.

The AND/OR tree is built in a stepwise fashion, by applying transformation rules to a structure which is a hybrid of activity diagrams and statecharts. The structure contains states and edges resembling activity diagrams, but each state that is source or target of an edge is the root of an AND/OR tree; the states inside this tree are not incident to any edge. Two transformation rules (T1 and T2) are used: one for merging OR states (T1) and one for creating AND states (T2). Each transformation rule reduces edges from the structure but adds state hierarchy. The procedure stops if the structure contains no edges and one state that contains all other states. That state is the root of the state hierarchy. The procedure may fail, in which case the activity diagram cannot be translated into a statechart. Section 4.3 discusses the most common fail case and how this influenced the definition of the overall approach.

We now elaborate the initialization step, in which the initial hybrid structure is created, and the two rules T1 and T2. The next subsection explains how the rules are used to translate an activity diagram into a hierarchical statechart. To simplify the exposition, we do not show the formal specifications of the rules but we apply the rules to the claim processing example.

Initialization. The activity diagram is copied into a new structure. In this structure, each object node is replaced with the BASIC state to which it refers. Then for each BASIC state due to an object node and for each control node, except bars, an OR state is created. The OR state becomes parent of the node for which is created. Finally, the edge relation is for non-bar nodes lifted to their OR parents. So if the activity diagram contains an edge from a non-bar node to another non-bar node, then the structure contains an edge from the OR parent to the other OR parent. If the activity diagram contains an edge incident to a bar, then that part of the edge is not changed. If an edge connects two bars, there

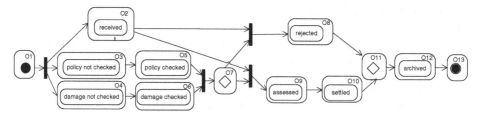

Fig. 8. Initialization of synthesis procedure for filtrate activity diagram in Fig. 5

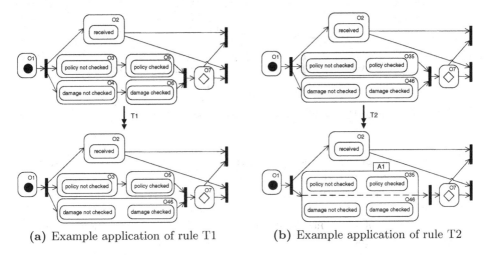

(a) Example application of rule T1   (b) Example application of rule T2

Fig. 9. Example applications of reduction rules on model in Fig. 8

is a cycle (since filtering rule R2 did not remove the edge) and so the activity diagram contains a deadlock, which is undesirable. Fig. 8 shows the initialization of the filtrate activity diagram in Fig. 5.

T1) Merging OR states. If an edge connects two OR states and if there is not a bar such that both states are predecessor or both successor of the bar, then the two OR states can be merged into a new OR state which becomes parent of all children of the two merged OR states. The new OR state replaces the old OR states and the edge connecting the two old OR states is removed from the structure. Note that this does not imply that the edge is not present in the final statechart, since statechart edges are defined separately (see Section 4.2). Figure 9a shows how the rule is applied in the synthesis of the state hierarchy for the structure in Fig. 8: OR nodes O4 and O6 are merged. New node O46 specifies a tree, so the two BASIC states it contains are not connected by an edge.

T2) Creating AND states. Each set of OR states that is input (output) to a bar translates into an AND state, which becomes parent of the OR states. Each pair of OR states in the set should have the same neighboring states. The set of OR states is replaced with a new OR state that becomes input (output) to the bar. The OR state becomes parent of the created AND state. Figure 9b shows an example application of this rule. OR nodes O35 and O46 have the same neighbors and can be grouped under parent AND node A1, which is new. Note that O2 cannot be grouped since it has different successors than O35 and O46. Furthermore, rule T2 is not applicable to the two activity diagram fragments shown in Fig. 9a, since for instance O3 and O4/O46 have different successors.

4.2   Constructing the Statechart

The previous step has resulted in an AND/OR tree of states. We now explain how a hierarchical statechart can be constructed from this tree plus the input

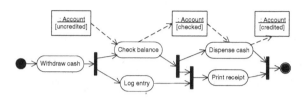

Fig. 10. Process of withdrawing money from ATM

activity diagram. States of the statechart are the states in the AND/OR tree plus additional fork and join pseudo states to represent bars.

Each node in the activity diagram has a unique counterpart in the statechart. If two nodes in the activity diagram are connected by an edge, in the statechart the counterparts of these nodes are also connected by an edge. For instance, object nodes :Claim[policy not checked] and :Claim[policy checked] are connected by an edge in the activity diagram in Fig. 5, so in the statechart BASIC states policy not checked and policy checked are connected by an edge; see Fig. 2.

Since compound statechart states have no counterpart in activity diagram syntax, there are no edges between compound states in the statechart in Fig. 2. We have defined postprocessing rules that rewrite the statechart edges into edges between compound states by eliminating interlevel edges and bars. Applying the translation to the activity diagram of Fig. 5 results in the statechart shown in Fig. 2. The initial and final states inside the AND state preprocessing are due to postprocessing. The names of compound states have been manually defined.

The synthesized statechart can be extended with events as follows (not shown in Fig. 2). Define for each activity A a completion event $cpl(A)$. If A modifies an object to a new state s in the activity diagram, then the corresponding edge that enters s in the statechart before postprocessing can be labelled with $cpl(A)$.

4.3   Discussion

As mentioned at the end of Sections 1 and 2, not every activity diagram can be translated into a hierarchical statechart. Fig. 10 shows a typical example of an activity diagram that cannot be translated directly into statecharts, assuming action nodes map to statechart BASIC states and no filtering rules are applied. After the Withdraw cash action, two parallel branches are started. However, there is a cross-synchronization between the two branches: Print receipt requires that both Check balance and Log entry have been completed. Such a synchronization cannot be expressed in UML 2.3 statechart control flow [15].

However, our approach is still able to synthesize a statechart, since the filtering rules remove the cross-synchronization construct that impedes the translation to statechart. The synthesized statechart contains a sequence of three BASIC states. Our definition of the approach—first filter irrelevant nodes, then synthesize the remaining part into a hierarchical statechart—is motivated by the possible failure cases explained above. An alternative approach would have been to first translate process models into statecharts and then filter irrelevant nodes from the statecharts, so reverse the phases of our approach. Advantage of our approach is that it succeeds in many situations where the other approach fails.

In future work, we plan to analyze problem points for remaining failures and define translation patterns for these cases to enlarge the translation scope.

5   Validation

To evaluate the feasibility of the approach, we have realized a prototype tool that implements the approach and we have tested the tool on several synthetic examples and on two process models that are based on real-life scenarios. In this section, we explain the architecture of the tool and discuss our experiences in applying the approach to the real-life scenarios.

5.1   Architecture

We have decided to implement the rules in Sections 3 and 4 as graph transformation rules using the general purpose graph transformation engine GrGen [4]. This engine provides a scalable implementation for state-of-the-art matching and rewriting constructs and provides especially useful support for visual debugging. Fig. 11 shows the overall architecture of the resulting implementation. The figure's left-hand side shows that that the tool reads activity diagrams ex-

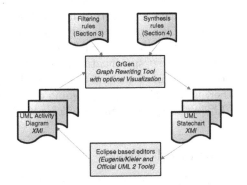

Fig. 11. Implementation architecture

pressed in XMI syntax according to the UML 2.3 standard. Such XMI code is generated by mainstream tools like MagicDraw or the Eclipse UML 2 plugin. The rectangle at the top of the figure represents the GrGen platform [4].

The right-hand and bottom side of Fig. 11 shows that our transformation implementation produces output models that can be consumed by many interesting tools, including Eclipe-based editors. The implementation produces hierarchical statecharts expressed in UML XMI that conforms to the official UML 2.3 standard. Moreover, the implementation produces XMI based on a very simple UML metamodel. The generated XMI can be used among others as input for a statechart simulator that we have developed using state-of-the-art tool Eclipse modeling technology. The architectural strength is that the rules from Sections 3 and 4 do not have to be changed for such extensions.

5.2   Case Studies

To further evaluate the feasibility of the approach, we applied the prototype to two real-life industrial processes that we modeled ourselves: ordering and delivery of bikes, and the handling of dermatology patients. As with the running example, the control flow of the two process models is block-structured (each bar that starts parallel branches has one matching bar that synchronizes the branches) but by preprocessing the object flows, extra bars are introduced that destroy the block-structure (cf. Fig. 3). Table 1 shows that the prototype constructs for each activity diagram a hierarchical statechart in a less than a second. We have shown before that the synthesis procedure defined in Section 4 runs in polynomial time [3] and that the GrGen-implementation scales well for large input models [16].

Table 1. Characteristics of the cases

Case study	# action nodes	# object nodes	# bars	# diamonds	# control flows	# object flows	preproc. & filter (ms)	synthesis (ms)
Bikeshop	15	12	7	4	33	15	297	577
Dermatology	17	5	4	10	39	6	250	280

It would be interesting to let trained professionals design activity diagrams with object flows. An obstacle might be the peculiar semantics of object nodes in combination with action nodes. For instance, inserting in Fig. 1 an object flow from :Claim[settled] to activity Notify client means according to the UML2 semantics that :Claim[settled] is needed for Notify client to start which contradicts the control flow specification that allows that activity Pay is not executed (if Reject is executed instead). This is the reason why many object nodes have either no incoming or no outgoing edge. Such UML modeling issues might make it difficult for professionals to construct activity diagrams with object nodes.

6   Related Work

Our work is most closely related to research on relating object life cycles and business process models (i) and to approaches for synthesizing statecharts (ii).

(i) There are a few approaches [8,9,13] that consider the relation between business process models that reference business objects and object life cycles of these objects. All these other approaches consider flat finite state machines whereas we consider hierarchical statecharts. Moreover, some approaches [9,13] generate a process model from a set of object life cycles, where each object life cycle is specified by a flat finite state machine. Whereas we study the reverse direction: how can an object life cycle be generated from a process model?

Kumaran et al. [8] give an algorithm for deriving sequential, flat state machines from a business process model with data flow. The algorithm focuses on discovering business entities of which the life cycles have to be derived, but in activity diagrams these entities are already modeled explicitly as objects. Next, the algorithm does not use filtering rules, which are essential to enable the discovery of implicitly specified object life cycles from business process models.

(ii) There are several works that study how to generate a statechart from a set of scenarios specified as either MSC or LSC, e.g. [6,17,18]. The constructed statechart satisfies the scenarios, i.e., each scenario is playable with the statechart. An important difference between scenarios and activity diagrams is that scenarios reference statechart events but not states, whereas activity diagrams reference statechart states (object nodes) but not events.

There are two important differences with our work. First, these approaches translate the complete control flow of the scenarios to a statechart. In our approach, we translate object flows to a statechart. Since an activity diagram is a mixture of object flow and control flow, the object flows need to be filtered from the activity diagram. This step is missing in the scenario-based synthesis approaches. The filtering phase is a key element of our approach to discover a hidden object life cycle from a process model.

Second, only a limited set of statecharts can be synthesized from scenarios, compared to the statecharts constructible with our approach. In most approaches [6,18], each scenario-based statechart consists of communicating sequential state machines, so there is one top-level AND state that contains sequential finite state machines. Whereas in our work, constructed statecharts can have concurrency at arbitrary levels of nesting, not just the top level.

A recent approach [17] studies synthesis of hierarchical state machines from UML 2.0 interaction diagrams, which contain activity diagram constructs to specify complex concurrent behavior. However, the interaction diagrams are required to be (block-)structured [10]: each fork matches with a join and pairs of matching nodes are properly nested and loops with multiple exits are not allowed. Consequently, the synthesized statecharts are also block-structured. Whereas the translation defined in Sect. 4 takes as input unstructured activity diagrams and constructs statecharts that can be unstructured, for instance containing loops with multiple exits or unbalanced forks and joins as in Fig. 5.

7   Conclusion

An approach for synthesizing an object life cycle from a business process model, making the implicit life cycle contained in the process model explicit, has been proposed. The approach is fully automated and has been implemented with the graph transformation tool GrGen [4]. Though the approach has been defined for UML activity diagrams, it is also applicable to models expressed in languages like the Business Process Model and Notation (BPMN) [12], provided the models use similar constructs as activity diagrams. Synthesized hierarchical statecharts can be used to generate or configure a software system supporting the business process, or to assess consistency with existing statechart descriptions [2,11].

Future work is to enlarge the scope of the translation in several ways. First, we will define translation patterns that can deal with frequently occurring activity diagram constructs that prohibit a translation into statecharts. Second, pin-style object flows in activity diagrams will be considered. Third, we will define a translation for BPMN models that cannot be expressed as activity diagrams.

References

1. Aho, A., Sethi, R., Ullman, J.: Compilers: Principles, Techniques, and Tools. Addison-Wesley (1986)

2. Engels, G., Küster, J.M., Heckel, R., Groenewegen, L.: A methodology for specifying and analyzing consistency of object-oriented behavioral models. In: Proc. ESEC / SIGSOFT FSE, pp. 186–195 (2001)
3. Eshuis, R.: Translating Safe Petri Nets to Statecharts in a Structure-Preserving Way. In: Cavalcanti, A., Dams, D.R. (eds.) FM 2009. LNCS, vol. 5850, pp. 239–255. Springer, Heidelberg (2009)
4. Geiß, R., Batz, G.V., Grund, D., Hack, S., Szalkowski, A.: GrGen: A Fast SPO-Based Graph Rewriting Tool. In: Corradini, A., Ehrig, H., Montanari, U., Ribeiro, L., Rozenberg, G. (eds.) ICGT 2006. LNCS, vol. 4178, pp. 383–397. Springer, Heidelberg (2006)
5. Harel, D., Gery, E.: Executable object modeling with statecharts. IEEE Computer 30(7), 31–42 (1997)
6. Harel, D., Kugler, H.: Synthesizing state-based object systems from LSC specifications. Int. Journal of Foundations of Computer Science 13(1), 5–51 (2002)
7. Hecht, M., Ullman, J.: Characterizations of reducible flow graphs. J. ACM 21, 367–375 (1974)
8. Kumaran, S., Liu, R., Wu, F.Y.: On the Duality of Information-Centric and Activity-Centric Models of Business Processes. In: Bellahsène, Z., Léonard, M. (eds.) CAiSE 2008. LNCS, vol. 5074, pp. 32–47. Springer, Heidelberg (2008)
9. Küster, J.M., Ryndina, K., Gall, H.: Generation of Business Process Models for Object Life Cycle Compliance. In: Alonso, G., Dadam, P., Rosemann, M. (eds.) BPM 2007. LNCS, vol. 4714, pp. 165–181. Springer, Heidelberg (2007)
10. Liu, R., Kumar, A.: An Analysis and Taxonomy of Unstructured Workflows. In: van der Aalst, W.M.P., Benatallah, B., Casati, F., Curbera, F. (eds.) BPM 2005. LNCS, vol. 3649, pp. 268–284. Springer, Heidelberg (2005)
11. Nejati, S., Sabetzadeh, M., Chechik, M., Easterbrook, S.M., Zave, P.: Matching and merging of statecharts specifications. In: Proc. ICSE, pp. 54–64. IEEE Computer Society (2007)
12. OMG. Business Process Model and Notation (BPMN) Specification, Version 2.0. Object Management Group (2011), http://www.bpmn.org
13. Redding, G., Dumas, M., ter Hofstede, A.H.M., Iordachescu, A.: Generating business process models from object behavior models. IS Management 25(4), 319–331 (2008)
14. UML Revision Taskforce. OMG UML Specification v. 1.5. Object Management Group, 2003. OMG Document Number formal/2003-03-01, http://www.uml.org
15. UML Revision Taskforce. UML 2.3 Superstructure Specification. Object Management Group, 2010. OMG Document Number formal (May 05, 2010)
16. Van Gorp, P., Eshuis, R.: Transforming Process Models: Executable Rewrite Rules versus a Formalized Java Program. In: Petriu, D.C., Rouquette, N., Haugen, Ø. (eds.) MODELS 2010, Part II. LNCS, vol. 6395, pp. 258–272. Springer, Heidelberg (2010)
17. Whittle, J., Jayaraman, P.K.: Synthesizing hierarchical state machines from expressive scenario descriptions. ACM Trans. Softw. Eng. Methodol. 19(3) (2010)
18. Whittle, J., Schumann, J.: Generating statechart designs from scenarios. In: ICSE, pp. 314–323 (2000)

Conceptual Modeling for Systems Integration

Narasimha Bolloju[1], Sandeep Purao[2], and Chuan-Hoo Tan[1]

[1] Department of Information Systems,
City University of Hong Kong, Hong Kong
{narsi.bolloju,ch.tan}@cityu.edu.hk
[2] College of IST, Penn State University, University Park, PA
spurao@psu.edu

Abstract. Conceptual modeling for systems integration is challenging because of two reasons. First, it must deal with a universe of discourse that includes human creations: software applications and interactions among them. Second, the knowledge needed to create appropriate conceptual models must be obtained from several users who can only possess partial views. The constructs and methods employed for conceptual modeling for systems integration should, therefore, facilitate representation as well as combination of partial knowledge captured from multiple organizational participants. We present an overview of an approach named systems integration requirements engineering (SIRE) to address this set of challenges. The approach includes a set of modeling constructs and an associated method that allows generating and then merging local conceptual models for systems integration. The paper presents formal descriptions of the constructs, a method for developing individual model fragments, an results from an initial empirical evaluation that compares our proposals against a benchmark provided by UML communication diagrams.

Keywords: Conceptual modeling, Systems Integration, Requirements.

1   Introduction

Although systems integration projects are becoming common in many organizations, there are few methods or (conceptual) modeling techniques devoted to systems integration concerns. Such methods and modeling techniques are important to surface and verify integration requirements that can lead to development and implementation of systems integration solutions. Systems integration [8, 11, 20] involves designing and deploying mechanisms for seamless exchange of information and control across diverse organizational units and information systems that have been separated by scale, specialization and geographical distribution. The magnitude of the problem has led to enduring escalation of IT expenditures for integration [1], and has been associated with a number of challenges such as technology platform differences and organizational behaviors [5, 15]. Research on system integration has been extensive, with the bulk of it *either* technical *or* conceptual in nature. Technical approaches proposed include multi-agent coordination [21,22], service-oriented platforms for integration [4, 14, 28], and data integration [6, 16]. Although useful, these approaches

P. Atzeni, D. Cheung, and R. Sudha (Eds.): ER 2012, LNCS 7532, pp. 321–330, 2012.
© Springer-Verlag Berlin Heidelberg 2012

have often focused on specific technologies and emphasized solution delivery instead of front-end activities such as conceptual modeling for the design of integration solutions. They have, instead, assumed *a-priori* knowledge of the systems to be integrated and the integration requirements. The few notable exceptions include work and information flow modeling (e.g., [2]), and repurposing existing modeling techniques (such as BPMN and UML) to capture integration requirements (e.g., [13, 26]. In spite of their potential, these approaches suffer from incomplete articulation, lack of clear conceptual foundations, and lack of validation.

The *objective* of this research is to develop and demonstrates an approach to facilitate conceptual modeling for systems integration. Our *approach* allows: (a) surfacing of partial and potentially incomplete integration requirements from functional managers and business analysts; (b) representing these requirements at varying levels of formalization so that functional managers and business analysts as well as solution developers can use the information; and (c) merging partial and incomplete integration requirements to develop actionable implementation plans. This paper focuses on the first two aspects our approach. Our *contributions* build on and synthesize a number of concepts from prior work including coordination [17], integration patterns [12], and the black-box technique [9]. We demonstrate utility of the approach following a summative evaluation in an authentic setting, and point out opportunities for further improvements.

2   Challenges in Conceptual Modeling for Systems Integration

A conceptual modeling language provides the set of constructs that provide the basis for this process, which often takes place during the early stages of a software development lifecycle. For conceptual modeling of systems integration solutions, a feasible approach is obtaining partial models from functional managers and business analysts who are familiar with specific information systems, before merging them. Important differences between conceptual modeling for individual systems vs. systems integration, however, remain, and produce new challenges. The *first* challenge is the characterization of conceptual modeling for systems integration as a bottom-up process. Unlike a software designer who may engage in top-down approach to extract concepts and relationships, conceptual modeling for systems integration must follow a bottom-up approach. Further, the domain of interest contains systems and interactions – both artificial constructs [18] that must be obtained from stakeholders instead of discovering them as immutable constructs. The *second* challenge deals with the varying level of detail and formalization that must be incorporated in the process. Unlike traditional conceptual modeling, which requires a progression from conceptual to logical models, modeling for systems integration cannot separate some of the logical details from a conceptual view because of the models must represent existing systems and interactions among them. In fact, to ensure participation from IT and functional managers, the modeling process must allow simultaneous use of detail and formalization.

3   An Approach for Conceptual Modeling for Systems Integration

Our approach, which we call SIRE (Systems Integration Requirements Engineering), consists of a set of modeling constructs and a method. The *modeling constructs* draw on several precursors such as: considering systems interactions as first-class citizens (van der Aalst et al. 2000), box-structured modeling principles [9], coordination theory [17], enterprise integration patterns [12], and task dependencies in process models as surrogates for interactions among software systems [25]. The *method* involves generating individual model fragments from the perspective of each system before combining them into a synthesized model fragment. This bottom-up method takes into account the fragmented nature of domain knowledge of modelers by facilitating capturing of integration requirements from the perspective of each system or a group of related systems as individual model fragments prior to the merging process required for synthesized model fragments. This process of combining model fragments including synthesized model fragments can be performed until we arrive at an enterprise-wide integration requirements model or a model representing integration requirements of a select group of systems.

To develop the modeling constructs and the method, performed multiple design-evaluate cycles [10] to generate: (a) *constructs* that build on the key abstraction of systems interactions, treating them as first-class citizens, and (b) a *method* to surface and merge integration requirements utilizing this construct, suggesting larger-granularity patterns that capture recurring integration structures [19]. We followed up with an empirical evaluation aimed at understanding effectiveness of the conceptual modeling constructs. The empirical assessment consisted of comparison of proposed modeling constructs with a closest alternative based on the Unified Modeling Language for representing integration.

4   Conceptual Modeling Constructs and Method

4.1   Overview

In developing the modeling constructs, our intent was to respond to several properties of good conceptual modeling constructs such as: faithfully representing the domain of interest [3], avoiding structural and implementation details [23, 24], and allowing communication with non-IT professionals. With these in mind, we developed three types of primitive constructs for conceptual modeling of systems integration solutions: two types of *nodes* for representing systems (notation: rectangles) and coordination mechanisms (notation: rectangles with rounded corners), and *directed links* representing one of more interactions between a pair of nodes (notation: directed arc). Figure 1 shows three examples of using these modeling constructs. The first shows that an interaction can be either a service invocation (e.g., getQuanityOnHand()) or an event with information exchange (e.g., newOrder). The second and third show that interactions describe dependencies across systems that may be coordinated through appropriate routing (e.g., DISTRIBUTE) and/or transformations (e.g., GROUP).

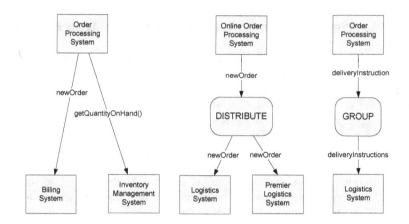

Fig. 1. Examples of systems interactions and coordination mechanisms

Malone and Crowston [17] define coordination as "managing dependencies between activities." For systems integration, we conceptualize interactions among systems as "dependencies" that must be coordinated. The coordination mechanisms address routing and transformation of resources available (services provided and events published) and resources required (services consumed and events subscribed) through common interactions. We propose nine types of mechanisms (shown in Figure 2) for this purpose. These represent the outcomes of multiple cycles of formative evaluation.

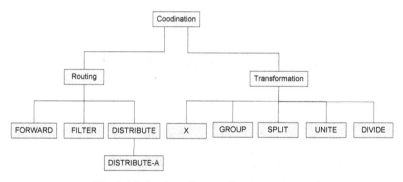

Fig. 2. Mechanisms for coordination of interactions

Each coordination mechanism processes incoming object instances from source system(s), and creates instances of outgoing interaction objects for destination system(s). Additional information through another event or service may be used to control coordination, and/or to enrich information present in the incoming objects. Examples include waiting for a signal before forwarding an object, using the results of a service request to identify a destination node, appending additional values to an incoming object, and discarding an incoming interaction object instance. Table 1 elaborates the semantics underlying each mechanism.

Table 1. Semantics of different coordination mechanisms

Coordination type (description)							
	Interaction types		Types Different?	Outward Processing	Destination Selection	Inward Processing	Role of External Control/ Input
	Incoming	Outgoing					
	1 or more	1 or more	Yes	Map using params/ event attributes	All	Map using params/ event attributes	Suspend/ Resume
FORWARD (incoming interactions are forwarded to all destination nodes)							
FORWARD	1 or more	1 or more	No		All	Return First Response	
FILTER (incoming instances not satisfying a given filter condition are discarded)							
FILTER	1	1	No	if condition is true	All	Return First Response	
DISTRIBUTE (incoming interactions are routed to one or more of destination nodes based on distribution specification)							
DISTRIBUTE	1	1	No		Based on spec	Return First Response	
DISTRIBUTE-A (applicable to synchronous service requests; responses received are aggregated and returned)							
DISTRIBUTE-A	1	1	Yes or No		Based on spec	Aggregate Responses	+Enrich
X (incoming interactions are translated according to mapping specification)							
X	1	1	Yes	+With mapping	All	Return First Response	+Enrich
GROUP (incoming interaction instances are grouped and the group forwarded)							
GROUP	1	1	Yes	+ Pack	All	Split Response	+Send Partial/ Discard; +Enrich
SPLIT (incoming interaction instances are split and forwarded)							
SPLIT	1	1	Yes	+ Unpack, Sort	All	Consolidate Response	+Enrich
UNION (incoming interactions of different types are joined and forwarded after all join components are received)							
UNION	2 or more	1	Yes	+ Join	All	Split Response	+Send Partial/ Discard; +Enrich
DIVIDE (incoming interactions are projected into different types of outgoing interactions)							
DIVIDE	1	2 or more	Yes	+ Project	All	Consolidate by time window	+Enrich

Note: "+" indicates functionality added to that described for the Abstract type

4.2 Primitive Constructs

The constructs described above are used to represent each system or data source as a Node. A system node (s_i) is defined as a 6-tuple <l, *descr*, SP, SC, PE, SE> where l, *descr*, SP, SC, PE and SE stand for label, description, set of services provided (SP), set of services consumed (SC), set of events published (PE), and set of events subscribed (SE) by s_i. As the integration requirements are identified and modeled the last four sets, initially set to null, will accumulate specific interactions involving s_i. An interaction o_k between a pair of nodes n_o and n_d is represented by a tuple <l, m, P_{in}, P_{out}, n_o, n_d> which represents a stream of interaction object instances with label l from node n_o to node n_d. A given type of interaction identifies either service request from n_o for a service provided by n_d or events published by n_o and subscribed by n_d. The mode of interaction, *m*, can be *sync* (synchronous request), *async* (asynchronous request), or *event* (event with attributes). The parameter set P_{in} represents either input parameters or event attributes associated with the interaction, and the parameter set P_{out} is applicable only for synchronous requests. Each $p \in P_{in}$ or P_{out} is defined a tuple $<l, t, descr>$ where *l*, *t*, *descr* indicate parameter name, type and description respectively.

A coordination node (c_j) is an intermediary node between two or more nodes of either system or coordination type, and it represents one of the several coordination mechanisms (described above) on the streams of incoming interaction object instances to produce streams of outgoing interaction object instances as defined by the object types in OI and OO. A coordination node c_i is defined as a tuple $<l, t, spec, OI, OO>$ where *l*, *t* and *spec* stand for its label, type and specification respectively, and OI and OO are sets of incoming and outgoing interaction object types respectively. Interactions among nodes are specified for each coordination mechanism identified in Table 1 above. For example, semantics for the "Forward" interaction can be stated as: all incoming and outgoing object instances are of the same type, i.e., $\forall o_i, o_j \in OI \cup OO$ ($o_i.t = o_j.t$). Similar specifications are created for each coordination mechanism to respect the semantics outlined.

4.3   Model Fragments

The primitive constructs described above are used to compose individual model fragments, and synthesized model fragments. An individual model fragment captures interactions for a system (see Figure 3). Individual model fragments for a set of systems can be combined to define a synthesized model fragment which depicts all the interactions for the set of systems.

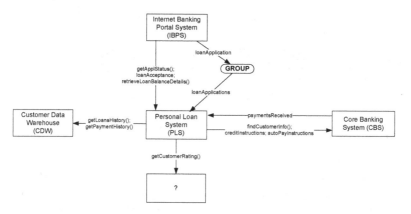

Fig. 3. Example of an individual model fragment

System interactions are represented as service invocation(s) or events exchanged. An unknown system is included for the purpose of representing required interactions but the providers of corresponding services or publishers of corresponding events are not known. Such interactions are resolved during the subsequent individual model fragment refinement and the model synthesizing process as the services and events offered by other systems becomes clearer. A model fragment is described formally in the following manner. Let S be the set of nodes representing various systems $\{s_1, s_2, ..., s_n\}$ to be integrated. Each system S_i S provides certain services to other systems, consumes services offered by other systems, publishes events for subscription by other systems, and subscribes to events published by other systems. Interactions for a given system $s_i \in$ S are represented by a model fragment imf_i which is a 3-tuple $<S_i, C_i, O_i>$ where S_i, C_i and O_i refer to sets of participating systems, coordination mechanisms and interactions respectively. Each model fragment imf_i satisfies the following conditions: (i) It is a fully connected directed graph with s_i as the central system node to which all other systems interacting with s_i are located at the periphery; the set of peripheral nodes SP for s_i defined as $(S_i - \{s_i\}) \cup \{s_\perp\}$; (ii) the system s_i interacts with the peripheral systems nodes directly or indirectly via one or more coordination nodes $c \in C_i$, in other words each interaction in O_i is either directly linked to s_i or via one or more coordination nodes from C_i; (iii) interactions of systems in SP are only with s_i or some c in C_i; i.e., no interactions among the systems in SP are represented in imf_i; and (iv) each interaction branch of s_i represents either outgoing or incoming interactions terminating at one of the nodes in SP (no combination of interactions along a path from S_i to peripheral nodes).

4.4   Constructing Model Fragments

Our proposal also includes a method to create individual and synthesized model fragments. The method captures the set of iterative tasks a designer is likely to engage in for creating each model fragment. Although the modeling constructs used for depicting systems, coordination mechanisms, and systems interactions do result in model fragments, sometimes these model fragments can be fairly large. A simple 'collection' node allows us to introduce the notion of layering. The notation for a collection can be either a list of systems, coordination mechanisms or interactions. Together, the conceptual modeling constructs (with semantics outlined in Table 1), the layering notion and the method for constructing model fragments (not presented in this paper) provide the conceptual modeling approach for systems integration. It represents the outcome of multiple refinement cycles and formative evaluation episodes including testing with scores of projects done by student teams over multiple years.

5   Evaluation

To assess the effectiveness of constructs in SIRE, we carried out a comparative evaluation against constructs in the UML communication diagram (UMLCD) approach. Established modeling constructs for system integration are absent and the closest, known approach that we can find is the UMLCD where objects, message flows and notes are used represent systems, interactions, and coordination mechanisms respectively. Though it may not necessarily be the most appropriate benchmark; however, the comparison with it could shed lights to our SIRE when in used. In preparation, the SIRE approach was incorporated into a graduate course on systems integration. Most students enrolled in this course were experienced IT professionals with an average experience of about 6 years. The students were taught the constructs and creation of model fragments with SIRE as well as UMLCD. After these introductions, all students completed an individual task of identifying integration requirement of a licensing system (a real world case) and created two versions of model fragments. In the following week, half of the students (randomly chosen) identified integration requirements and created model fragments using the SIRE constructs; the other half of the students performed the same task using UMLCD constructs. In the following week, each student used the other set of constructs for completing a similar task for a different case study. To ensure authenticity, both case descriptions were based on the open tender specification documentations issued by the Singapore government.

To prevent bias, the two techniques were taught and given equal weight to the assessment of the students' performance. The students were also given sufficient amount of time and training on the two sets of modeling constructs. The three weeks long assessment was necessary due to the inherent complexity of system integration domain. The research expectation was not shown and the training scripts were pre-written and verified. Upon the completion of the exercise, the students were asked to rate the SIRE constructs against UMLCD constructs for three aspects: usefulness, ease of use and satisfaction (7-Likert scale with 1 as least). Table 2 depicts the list of questions together with the statistical results.

Table 2. Our proposals compared to the UMLCD approach

Item	Mean (Std. dev.)	Cron-bach's Alpha	Factor Analysis (Component)		
			1	2	3
Perceived usefulness					
1. Is useful	5.040	.845	.195	.167	**.825**
2. Increases productivity	(.805)		.076	.158	**.873**
3. Is effective			.126	.232	**.831**
Perceived ease of use					
1. Is easy to use	4.560	.887	**.898**	.227	.065
2. Is simple to use	(1.091)		**.882**	.283	.137
3. Is understandable			**.775**	.275	.249
Satisfaction					
1. I am satisfied with it	4.600	.906	.308	**.875**	.128
2. I would recommend it to a friend	(.987)		.280	**.852**	.220
3. It is wonderful			.250	**.806**	.309

Independent sample t-tests were also conducted to determine whether the means of perceived usefulness, perceived ease of use and satisfaction from using SIRE approach were significantly higher than the neutral rating of 4. Cronbach's Alpha values indicate good consistencies of the items used for the three measures. Our analyses revealed that the use of SIRE approach (compared to the benchmarked UMLCD approach) lead to significantly higher than neutral of 4 for perceived usefulness of the approach (t = 11.184, p < .01), perceived ease of use (t = 4.446, p < .01) and satisfaction derived (t = 5.301, p < .01). The results clearly indicate that our proposals were favored over the UMLCD approach.

6   Discussion and Concluding Remarks

This research identifies and proposes a small set of conceptual modeling constructs – SIRE – for systems integration. This is a problem that is qualitatively *different* from conventional conceptual modeling because it must deal with artificial constructs, capturing information systems and interactions among these systems as part of the universe of discourse to be modeled. Our contribution hence mainly revolves around the derivation of the SIRE to support system integration activities, which have traditionally been performed less systematically, in the areas of: (a) surfacing integration requirements from individual system's perspective, (b) representing these through a combination of graphical and textual formats as model fragments, and (c) merging such fragments to arrive at a consolidated set of requirements (the last component is not part of this paper). Our proposals include a minimal set of constructs: systems, interactions and coordination mechanisms. The modeling constructs and method in its current version represents an outcome that reflects several design –evaluate cycles over a period of three years.

Findings from a comparative analysis of SIRE against UMLCD points to benefits such as perceived usefulness, perceived ease of use and satisfaction. We acknowledge earlier, however, that it is possible that the SIRE constructs exhibit a natural bias towards modeling integration requirements because there are no equivalent constructs in UMLCD. However, UMLCD is the closest known approach that can be applied to system integration thus serves as the closest sensible benchmarking tool. A useful next step would be to test our approach in real life settings.

Acknowledgements. The work by the second author was substantially supported by a grant from the Research Grants Council of the Hong Kong Special Administrative Region, China (Project No. CityU 110308).

References

1. Bass, C., Lee, J.M.: Building a Business Case for EAI. EAI Journal, 18–20 (2002)
2. Casati, F., Discenza, A.: Modeling and Managing Interactions Among Business Processes. Journal of Systems Integration 10, 145–168 (2001)
3. Chan, H.C., Wei, K.K., Siau, K.: User-Database Interface: The Effect of Abstraction levels on Query Performance. MIS Quarterly 17(4), 441–464 (1993)
4. Erl, T.: Service-Oriented Architecture: A Field Guide to Integrating XML and Web Services. Prentice-Hall, Upper Saddle River (2004)
5. Evgeniou, T.: Information Integration and Information Strategies for Adaptive Enterprise. European Management Journal 20(5), 486–494 (2002)
6. Foster, I., Grossman, R.L.: Data Integration in a Bandwidth-rich World. Communications of the ACM 46(11), 50–57 (2003)
7. Gregor, S.: The Nature of Theory in Information Systems. MIS Quarterly 30(3), 611–642 (2006)

8. Hasselbring, W.: Information System Integration. Communications of the ACM 43(6), 33–38 (2000)
9. Hevner, A.R., Mills, H.D.: Box-Structured Requirements Determination Methods. Decision Support Systems 13, 223–239 (1995)
10. Hevner, A., March, S., Park, J., Ram, S.: Design science research in information systems. MIS Quarterly 28(1), 75–105 (2004)
11. Hobday, M., Davies, A., Prencipe, A.: Systems Integration: A Core Capability of the Modern Corporation. Industrial and Corporate Change 14(6), 1109–1143 (2005)
12. Hohpe, G., Woolf, B.: Enterprise Integration Patterns: Designing, Building and Deploying Messaging Solutions. Addison-Wesley Longman Publishing Co. Inc., Boston (2003)
13. Jonkers, H., Lankhorst, M., Buuren, R.V., Hoppenbrouwers, S., Bonsangue, M., Torre, L.V.D.: Concepts for Modeling Enterprise Architectures. International Journal of Cooperative Information Systems 13(3), 257–287 (2004)
14. Krafzig, D., Banke, K., Slama, D.: Enterprise SOA: Service-Oriented Architecture Best Practices. Prentice Hall PTR, Upper Saddle River (2004)
15. Lee, J., Siau, K., Hong, S.: Enterprise Integration with ERP and EAI. Communications of the ACM 46(2), 54–60 (2003)
16. Li, H., Su, S.Y.W.: Business Object Modeling, Validation, and Mediation for Integrating Heterogeneous Application Systems. Journal of Systems Integration 10, 307–328 (2001)
17. Malone, T.W., Crowston, K.: The Interdisciplinary Study of Coordination. ACM Computing Surveys 26(1), 87–119 (1994)
18. March, S.T., Allen, G.N.: On the Representation of Temporal Dynamics. In: Siau, K. (ed.) Advances in Database Management, pp. 37–53. Idea Group Publishing (2003)
19. March, S.T., Smith, G.F.: Design and natural science research on information technology. Decision Support Systems 15(4), 251–266 (1995)
20. Markus, M.L.: Paradigm Shifts – e-Business and Business/Systems Integration. Communications of the Association for Information Systems 4(10), 1–44 (2000)
21. Sikora, R., Shaw, M.J.: A Multi-agent Framework for the Coordination and Integration of Information Systems. Management Science 44(11), S65-S78. (1998)
22. Sutherland, J., Heuvel, W.J.V.D.: Enterprise Application Integration and Complex Adaptive Systems. Communications of the ACM 45(10), 59–64 (2002)
23. Teo, H.H., Chan, H.C., Wei, K.K.: Performance Effects of Formal Modeling Language Differences: A Combined Abstraction level and Construct Complexity Analysis. IEEE Transactions on Professional Communication 49(2), 160–175 (2006)
24. Topi, H., Ramesh, V.: Human Factors Research on Data Modeling: A Review of Prior Research, An Extended Framework and Future Research Directions. Journal of Database Management 13(2), 3–19 (2002)
25. Umapathy, K., Purao, S., Barton, R.R.: Designing Enterprise Integration Solutions – Effectively. European Journal of Information Systems 17(5), 518–528 (2008)
26. Umapathy, K., Purao, S.: Systems Integration and Web Services. IEEE Computer 43(11), 91–94 (2010)
27. van der Aalst, W., et al.: Workflow Modeling Using Proclets. In: Scheuermann, P., Etzion, O., et al. (eds.) CoopIS 2000. LNCS, vol. 1901, pp. 198–209. Springer, Heidelberg (2000)
28. Vernadat, F.B.: Interoperable Enterprise Systems: Principles, Concepts, and Methods. Annual Reviews in Control (31), 137–145 (2007)

Ontology Evolution: Assisting Query Migration

Haridimos Kondylakis and Dimitris Plexousakis

Institute of Computer Science, FORTH
{kondylak,dp}@ics.forth.gr

Abstract. Information systems rely more and more on semantic web ontologies to share and interpret data within and across research domains. However, an important problem when dealing with ontologies is the fact that they are living artefacts and subject to change. When ontologies evolve, queries formulated using a past ontology version might become invalid and should be redefined or adapted. In this paper we propose a solution in order to identify the impact of ontology evolution on queries and to ease query migration. We present a module that receives as input the sequence of changes between the two ontology versions along with a set of queries and automatically identifies the specific change operations that affect the input queries. Besides the automatic identification of the affecting change operations, query migration is further aided by providing an explanation for the specific invalidation. This explanation is presented graphically by means of change paths that represent the evolution of the specific parts of the ontology that invalidate the query. We evaluate the time complexity of our approach and show how it can possibly reduce the human effort spent on query redefinition/adaptation.

1   Introduction

The development of new scientific techniques and the emergence of new high throughput tools have led to a new information revolution. The nature and the amount of information now available open directions of research that were once in the realm of science fiction. During this information revolution the data gathering capabilities have greatly surpassed the data analysis techniques, making the task to fully analyze the data at the speed at which it is collected a challenge. The amount, diversity, and heterogeneity of that information have led to the adoption of ontology-based data access systems [1, 2] in order to manage it and further process it.

In such systems, an ontology is used to establish an explicit formal vocabulary to share. Then, queries formulated using that ontology are used to access data. However, despite the great amount of work done in establishing such an architecture, an important problem which is usually ignored is that ontologies are living artifacts and subject to change [3]. Due to the rapid development of research, ontologies are frequently changed to depict the new knowledge that is acquired. When ontologies change, the problems that occur are the following: a) the data access layer should be reconfigured and b) the queries relying on the previous ontology version may become inconsistent and should be updated and adapted [4].

P. Atzeni, D. Cheung, and R. Sudha (Eds.): ER 2012, LNCS 7532, pp. 331–344, 2012.

Although a number of works (e.g. [5]) deal with the first problem we have seen little effort devoted on the second important one. The identification of the affected queries that should be adapted and the propagation of the changes to them is a time-consuming, labour-intensive and error-prone activity [5], and the lack of proper tools to help the developers make it even more difficult. In this paper we offer a solution in order to identify the impact of RDF/S ontology evolution on SPARQL queries and to ease query migration.

We address the problem of the automatic detection of invalid SPARQL queries (a core subset corresponding to union of conjunctive queries (UCQ)) that need to be adapted when ontology evolution occurs. To that direction, we present a module, named *Affected Queries Detection* (AQD) module that gets as input a) the sequence of changes between the two ontology versions and b) a set of queries and automatically detects the queries that will be *affected* in the new ontology version. In order to identify the sequence of changes between the ontology versions another module is employed, described extensively in [6], that gets as input the two ontology versions and produces automatically the sequence of changes between them.

Initially, we define the notion of *affecting change operations* for a query. We show that if such operations exist, the query cannot be correctly answered using the evolved ontology version. Besides providing the reasons for the failure, the AQD module can produce *the change paths* for all affecting change operations as well. A *change path* presents the history of the evolution of a specific part of the ontology. So, it can be used to drive developer's understanding on ontology evolution for the specific parts of the query that become invalid, thus minimizing the time they have to devote on updating the queries. Furthermore, we show that change paths are a valuable tool for ontology engineers as well.

Finally, we describe our implementation and we present our experimental analysis using queries from information systems relying on two well-known ontologies CIDOC-CRM [7] and Gene Ontology [8]. Experiments performed show the feasibility of our approach and the considerable advantages gained.

The rest of the paper is organized as follows: Section 2 presents related work and Section 3 provides preliminaries and introduces the problem by an example. Then, Section 4 describes shortly the adopted language of changes. Section 5 shows how to identify the affecting change operations with respect to input queries, and how to present an explanation for the evolution of the ontology that concerns those changes. Section 6 describes the implemented system and presents our experimental analysis. Finally, Section 7 provides a summary and an outlook for further research.

2   Related Work

The need for mechanisms to identify the consequences of ontology evolution with respect to the dependent artifacts was early identified and recognized. Although, former research mostly focused in the possible inconsistencies inside the ontology, the authors in [9] differentiate among invalidation of *data instances*, *dependent ontologies* and *applications*.

Approaches dealing with instance invalidation are [10] and [11]. In [10] the authors distinguish between structural and semantic validity of data instances and propose approaches to ensure them. To achieve that, they propose *semantic views* as a subset of the ontology which is then used to detect instance invalidation. In [8] on the other hand, the authors evaluate the evolution of the most popular life science ontologies and determine the impact of the evolution on the dependent mappings and instances. Other approaches like [12] develop a change detection and analysis method that predicts the effect of changes on the concept hierarchy and allows ontologies to evolve without unpredictable effects on other depending ontologies. In [13] the authors propose a generic approach to annotate generated ontology mappings. Then, as the ontology evolves, stability measures are calculated over the annotations to identify the quality of the mappings and the impact of ontology evolution on them. Other approaches such as MORE [14] employ *temporal logic* to detect the consequences raising from ontology changes with respect to the problem of compatibility to existing applications or restrict ontology evolution scenarios to maintain compatibility [15, 16].

Our system differs from all the above in terms of both goals and techniques. All of them are used to identify the impact of ontology evolution on instances, mappings and dependent ontologies whereas in our case we identify the impact of ontology evolution on queries. Although most of these systems measure the impact using metrics or act proactively to reduce that impact, in our case we help developers to adapt input queries *after* the ontology evolution has occurred.

Other approaches that try to assess the impact of ontology evolution based on the end-user's incoming queries, either lack solid formal foundations [17], or come from database schema evolution. The authors for example in [18] identify the inconsistencies in the application layer when database schema changes. A promising approach is PRISM [5], which tries to predict the effect of schema changes on current applications and to translate old queries to work on the new schema version. However, it requires the repeated manual mapping among the schema versions and disambiguation is usually needed in several places without offering strong guarantees. In our case, the changes among ontology versions are produced automatically, they are unique and they don't need disambiguation. Moreover, in several cases – presented later on – that no solution is provided by such systems, we actively assist developers to redefine queries, thus minimizing query migration time.

3   Preliminaries and Motivating Example

In this paper we restrict ourselves to *valid RDF/S knowledge bases*. Most of the Semantic Web Schemas (85,45%) are expressed in RDF/S [19] and RDF/S offers, in our case, an optimum trade-off between expressive power and efficient reasoning support. The validity constraints [20] that we consider in this work concern mostly the *type uniqueness*, i.e., that each resource has a unique type, the *acyclicity* of the *subClassOf* and *subPropertyOf* relations and that the subject and object of the instance of some property should be correctly classified under the domain and range of the property, respectively. Those constraints are enforced in order to enable *unique* and *non-ambiguous* detection of the changes among the ontology versions as we shall later discuss.

When using RDF/S ontologies, the natural language for formulating queries is SPARQL [21]. It is currently the standard query language for the semantic web, it is an official W3C recommendation and most of the current ontology-based data access systems [1] use it. A SPARQL query consists of three parts: the pattern matching part, the solution modifiers and the output. In this paper, we do not consider *OPT* and *FILTER* operators since we leave it for future work. The remaining SPARQL fragment we consider here corresponds to *union of conjunctive queries* [22].

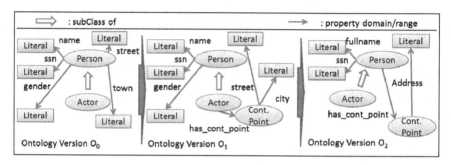

Fig. 1. Example Ontology Evolution

Now as an example, consider an ontology-based data access system that uses the ontology shown in the left of Fig.1 (ontology version O_0). This ontology is used as a point of common reference, describing persons and their contact points. Assume, furthermore, a query that asks for the "*name*", the "*town*" and the "*gender*" of all "*Persons*":

q_0: SELECT ?NAME ?TOWN ?GENDER WHERE   {
    ?X   type   Actor;   name   ?NAME;
              town   ?TOWN;   gender   ?GENDER. }

Using the semantics from [22] the algebraic representation of q_0 is equivalent to:

q_0: $\pi_{?NAME,?TOWN,?GENDER}$((?X, type, Actor) AND (?X, name, ?NAME) AND
    (?X, town, ?TOWN) AND(?X, gender, ?GENDER))

Assume now that at some point in time, the ontology evolves and we get O_1 by adding the class "*Cont.Point*" describing contact points and the property "*has_cont_point*" between the class "*Actor*" and the class "*Cont.Point*". Moreover, the domain of the literals "*street*" and "*city*" is changed to the class "*Cont.Point*".

Then the ontology designer decides to evolve again the ontology and to produce O_2. So, the domain of the "*has_cont_point*" property is moved from the class "*Actor*" to the class "*Person*", and the property "*gender*" is deleted. Moreover, the "*street*" and the "*city*" properties are merged to the "*address*" property as shown on the right of Fig. 1. In order to query the data sources, using the new ontology version, the underlying data access layer should be reconfigured.

However, besides reconfiguring the data access layer, the application layer should be changed as well. This is because the query that previously asked for the *"name"*, the *"gender"* and the *"town"* of all *"Persons"* will become *invalid*. This happens since the *"gender"* and the *"town"* properties that were available in the former ontology version no longer exist in the latter one and the *"name"* property has been renamed to *"fullname"*. Although some of the approaches already presented in the related work, such as PRISM [5], would try to automatically rewrite the query to the new ontology version - based on manually created mappings between the ontology versions - none of them would succeed, since the *"gender"* property no longer exists.

A mechanism that would automatically detect the invalid queries when the ontology evolves would be of great assistance to the information system's developers. Besides the detection of the invalid queries, an ideal system would report the reasons that led to the invalidation of a specific query, i.e., the deletion of the properties *"gender"* and *"town"* and the renaming of the property *"name"* in our example. This would reduce the downtime of the applications, and would ease query migration.

4   Modeling Ontology Evolution

For modeling ontology evolution we use the language of changes and the corresponding detection algorithm as proposed in [6]. The language contains over 70 types of change operations and three of them are described in Fig. 2. A change operation is defined as follows:

Definition 4.1 (Change Operation): *A change operation u over an RDF ontology O, is any tuple (δ_a, δ_d) where $\delta_a \cap O = \phi$ and $\delta_d \subseteq O$. A change operation u from O_1 to O_2 is a change operation over O_1 such that $\delta_a \subseteq O_2 \backslash O_1$ and $\delta_d \subseteq O_1 \backslash O_2$.*

Obviously, δ_a and δ_d are sets of triples. For simplicity we will denote $\delta_a(u)$ the *added* and $\delta_d(u)$ the *deleted* triples of a change *u*. From the definition, it follows that $\delta_a(u) \cap \delta_d(u) = \phi$ and $\delta_a(u) \cup \delta_d(u) \neq \phi$ if $O_1 \neq O_2$. The application of a change *u* over an ontology version *O*, denoted by *u(O)*, is defined as $u(O) = (O \cup \delta_a(u)) \backslash \delta_d(u))$. Moreover the application of a sequence of change operations *us* to an ontology, i.e. *us(O)*, is defined as the sequential application of the change operation in *us* to *O*. An important note for those change operations is that for any two changes u_1, u_2 in such a sequence it holds that $\delta_a(u_1) \cap \delta_a(u_2) = \phi$ and $\delta_d(u_1) \cap \delta_d(u_2) = \phi$. The interested reader is forwarded to [6] for more information on the aforementioned language of changes.

Change	Generalize_Domain(a,b,c)	Rename_Property(a,b)	Merge_Properties(A,b)
Intuition	Change the domain of property *a* to from *b* to *a* superclass *c*	Rename property *a* to *b*	Merge properties contained in *A* into *b*
δ_a	[(a, domain, c)]	[(b, type, property)]	(b, type, property)
δ_d	[(a, domain, b)]	[(a, type, property)]	$\forall a \in A : (a, type, property)$
Inverse	Specialize_Domain(a,c,b)	Rename_Property(b,a)	Split_Property(b, A)

Fig. 2. Example Change Operations

In our example the change log between O_0 and O_1, i.e. the E^{O_0,O_1}, consists of the following change operations:

u1: Add_Class(Cont.Point, ø, ø, ø, ø, ø,)
u2:Add_Property(has_cont_point, ø, ø ,ø ,ø, Actor, Cont.Point, ø, ø)
u3: Move_Property(town, Person, Cont.Point))
u4: Move_Property(street, Person, Cont.Point))
u5: Rename_Property(town, city)

Moreover, the change log E^{O_1,O_2} consists of the following change operations:

u6: Delete_Property(gender, ø, ø ,ø ,ø, Person, xsd:String, ø, ø)
u7: Generalize_Domain(has_cont_point, Actor, Person)
u8: Merge_Properties({street, city},address)
u9: Rename_Property(name, fullname)

We need to note that we selected the specific language of changes for several reasons. One of them is because it is a high-level language[1] of changes. In such a language change operations are more *intuitive, concise, closer* to the intentions of the ontology editors and capture more accurately the semantics of change. Moreover, the language possesses nice properties such as *uniqueness, composition* and *inversion. Uniqueness* is a pre-requisite for our system whereas *composition* and *inversion* are desirable but not obligatory properties. So, instead of the specific language of changes other languages (and the corresponding detection algorithm) could be also used as long as they preserve *uniqueness* in the constructed sequence of changes between two ontology versions.

5   Automatic Detection of Invalid Queries

5.1   Affecting Change Operations

Given the sequence of change operations between two ontology versions, the task is now to identify the past queries that are becoming invalid when using the new ontology version. To identify the change operations that lead to such a result we define the notion of *affecting change operation*.

Definition 5.1 (Affecting change operation). *A change operation $u \in E^{O_1,O_2}$ affects the query q - expressed using terms from O_1 -, denoted by $u \Diamond q$, if there exists a triple pattern[2] $t \in q$ that can be unified with a triple $t' \in \delta_d(u)$.*

The algorithm for identifying the affecting change operations, given as input a query and a sequence of changes, is presented in Fig. 3. It checks directly the change operations for the condition described above. *Unification* is the act of identifying two terms with a suitable substitution (a standard operation in Artificial Intelligence [23]) which we adopt here and it is used to detect that the change operation evolves a part from the ontology that the query currently uses. The time complexity of the algorithm is

---

[1] High level languages employ operators that can describe complex updates, as for instance theinsertion of an entire sub-sumption hierarchy.
[2] Triple patterns are like RDF triples except that each of the subject, predicate and object may be a variable.

Algorithm 5.1: *IdentifyAffectingOperations(q, $E^{O1,O2}$)*
Input: The query *q* formulated using ontology version O_1 and the the evolution log $E^{O1,O2}$
Output: The set *S* of the affecting change operations
1. $S := \emptyset$
2. For each $u \in E^{O1,O2}$
3.   if there exists $t \in q$, $t' \in \delta_d(u)$ such that t unifies t'
4. $S := S \cup u$
5. Return *S*

Fig. 3. The algorithm for identifying affecting change operations for a query q

$O(N*M*T)$ where N is the number of change operations in E^{O_1,O_2}, M the number of triple patterns in q and T the maximum number of triples in the $\delta_d(u)$, $u \in E^{O_1,O_2}$.

Proposition 1 (Correctness): *The algorithm IdentifyAffecting Operations identifies the affecting change operations for a given query q, over E^{O_1,O_2}.*

Proposition 1 is immediately proved by construction. In our example, instead of using E^{O_0,O_1} or E^{O_1,O_2} we will use E^{O_0,O_2}, without loss of generality – since they can be composed , in order to better highlight the advantages of our solution in environments with multiple ontology versions. So, the AQD module, will report that the query q_0 is affected by the following change operations:

1. Move_Property(town, Person, Cont.Point)) - moves the property *"town"*,
2. Delete_Property(gender, $\emptyset$, $\emptyset$,$\emptyset$,$\emptyset$, Person, xsd:String, $\emptyset$, $\emptyset$) - deletes the property *"gender"*
3. Rename_Property(name, fullname) - that renames the property *"name"* to *"fullname"*

The developers can use that knowledge in order to identify and merely correct the problems on their queries.

5.2   Explaining Ontology Evolution with Change Paths

However, presenting only the affecting change operations does not necessarily provide insights on the corresponding ontology evolution. For example, with respect to the initial query q_0, the developer can easily decide what to do with the deletion of the property *"gender"* and the renaming of the property *"name"* but more information should be provided for querying the location of a person using O_2. When drastic ontology evolution occurs, presenting only the change operations that affect the application queries is not enough. Ideally, we would like to know how the specific part of the ontology - that the input query previously asked for - has been changed. So, instead of providing the list of affecting change operations that affect input queries, our idea is to present the history of the evolution of the specific parts of the ontology that the query previously asked for.

For example, by checking the change log E^{O_0,O_2} presented on Section 4, we can easily identify that the operations shown in Fig. 4, describe the evolution of the *"town"* property that the query q_0 asked for.

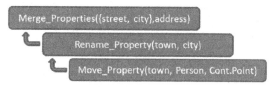

Fig. 4. The change path for u_3

Presenting such a graph to the developers, their understanding on the ontology evolution is focused on the specific parts that need to be redefined. In our example, the query can be updated using the "*address*" property to retrieve information about location. Such a sequence of change operations that depict the history of restructuring the ontology with respect to a specific change operation is called a *change path*.

Definition 5.2 (Change path for a change operation). *A change path for the change operation* $u \in E^{O_1,O_2}$, *denoted by* $us_{path}{}^u$, *is the minimal sequence of change operations in* E^{O_1, O_2} *such that* $u \in us_{path}{}^u$ *and that* $us_{path}{}^u (O_1) \subseteq O_2$.,

A change path is *minimal* in the sense that one cannot remove any of the change operations in it and still $us_{path}{}^u (O_1) \subseteq O_2$. The change path presents the history of the evolution of the specific part of the ontology for a specific change operation. For example, the *change path* for the change operation $u_3=Move\_Property(\ town,\ Person,\ Cont.Point))$ is $us_{path}{}^{u_3} =[u_3,\ u_5,\ u_8]$ as shown in Fig. 4 and $us_{path}{}^{u_3} (O_0) \subseteq O_2$.

Proposition 2 (Uniqueness): *The change path* $us_{path}{}^u$ *over* E^{O_1,O_2} *is unique.*

Proof: Assume $us_{path}{}^u$ is not unique. This would mean that we can have two change paths $us_{path1}{}^u$ and $us_{path2}{}^u$. Since they are both change paths it should hold that $size(us_{path1}{}^u)=size(us_{path2}{}^u)$ since they both have to be minimal. Now let $us_{path1}{}^u = [u_{k1}, ..., u_{kn}]$ and $us_{path1}{}^u = [u_{m1}, ..., u_{mn}]$. Since they are both change paths $u= u_{k1}= u_{m1}$. For $i>1$, each one of the u_{ki}, u_{mi} deletes a part of the ontology and adds another part. Since the two change paths have the same minimal size and $u= u_{k1}= u_{m1}$ in order to be different there must exist two change operations u_{ki}, u_{mj} such that $u_{ki} \neq u_{mj}$ and $\delta_d(u_{ki}) \cap \delta_d(u_{mi}) \neq \emptyset$ since they should delete a common part of the ontology. However, this is impossible since $\delta_d(u_1) \cap \delta_d(u_2)= \emptyset$ for our change operations∎

Now, we will present an algorithm that given a change log produces the change path for a change operation u. The algorithm is shown in Fig. 5. The idea is the following: The algorithm starts from the input change operation and identifies the triples that are added to the ontology, possibly by deleting other triples. Then it searches for the change operations that delete that added information in order to add new information and so on. After the execution of the algorithm the change path for u will be stored in us'.

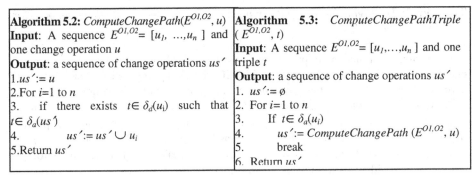

Algorithm 5.2: *ComputeChangePath*$(E^{O1,O2}, u)$	**Algorithm 5.3:** *ComputeChangePathTriple* $(E^{O1,O2}, t)$
Input: A sequence $E^{O1,O2} = [u_1, ...,u_n]$ and one change operation u	**Input**: A sequence $E^{O1,O2} = [u_1,...,u_n]$ and one triple t
Output: a sequence of change operations us'	**Output**: a sequence of change operations us'
1. $us' := u$	1. $us' := \emptyset$
2. For $i=1$ to n	2. For $i=1$ to n
3. if there exists $t \in \delta_d(u_i)$ such that $t \in \delta_a(us')$	3. If $t \in \delta_a(u_i)$
4. $us' := us' \cup u_i$	4. $us' := ComputeChangePath (E^{O1,O2}, u)$
5. Return us'	5. break
	6. Return us'

Fig. 5. Computing the change path for a given change operation (left) or a given resource(right)

Theorem 1: *The algorithm ComputeChangePath computes $us_{path}{}^u$ over $E^{O1,O2}$.*

Proof: In order to prove that algorithm *ComputeChangePath* computes the change path for a given change operation u over a change log $E^{O1,O2}$ we have to prove that (a) $u \in us'$, (b) $us'(O_1) \subseteq O_2$ and that (c) us' is minimal.

(a) From line 1 of the algorithm indeed $u \in us'$.

(b) Let's assume that $us'(O_1)$ is not a subset of O_2. This would mean that there exists at least one triple, assume t' in $us'(O_1)$ such that it does not exist in O_2. So, to reach O_2, there should be a change operation u' such that $t' \in \delta_d(u')$ such that $t' \in \delta_a(us')$ not identified by our algorithm. However this is impossible from line 3 of our algorithm.

(c) Now we prove minimality. Let's assume that us' is not minimal. This would mean that there is us_{path} with $size(us_{path}) < size(us')$. This would mean that there exist $u' \in us'$ such that $u' \notin us_{path}$. Of course this would mean from lines 3 and 4 that there exist t' such that $t' \in \delta_d(u')$ and $t' \in \delta_a(us')$. However, this would mean that t' does not belong to O_2, and should be deleted by another change operation. However for our change operations it holds that $\delta_d(u_1) \cap \delta_d(u_2) = \emptyset$ which makes the previous statement impossible. So us' is minimal as well∎

The time complexity of the algorithm is $O(N*M*S)$ where N is the number of change operations, M the maximum size of triples in a change operation u and S the number of triples in $\delta_a(us')$. Moreover, it is easy to change Algorithm 5.2 in order to retrieve the change path for a given *triple*. This will allow the developers to examine the evolution of the ontology concerning a specific triple:

Definition 5.3 (Change path for a triple). *The change path us_{path} over $E^{O1,O2}$ for the triple $t \in O_1$ is $us_{path}{}^u$, $t \in u$.*

The corresponding algorithm is shown in the right of Fig. 5. The idea is that we would like to retrieve the history of the evolution of the triple t. Since the triple t is inserted by at most one change operation, it is actually the change path of that change operation.

Theorem 2: *The algorithm ComputeChangePathTriple computes the change path for a given triple t over* E^{O_1,O_2}.

The algorithm is immediately proved by construction. Algorithm 5.3 needs to scan the change log one more time in order to identify the change operation that inserts the given triple and practically has the same time complexity with Algorithm 5.2.

Moreover, by exploiting the *invertibility* of our change operations (or by computing directly E^{O_2,O_0}), developers can use the Algorithm 5.3 to identify how a specific triple *t* from the *latest* ontology version O_2 has been produced. Thus, the *change path* produced can aid not only the developers in query redefinition, but also the ontology engineers in the identification of the modelling choices of the past.

6   Implementation and Evaluation

The AQD module described in this paper was implemented as a module of our *Exelixis* plarform[3]. The platform uses JAVA for the algorithms and HTML/jQuery for the presentation layer. Using the *Exelixis* platform, a user is able to load an RDF ontology to visualize and explore it. Furthermore, as more ontology versions become available, the change logs between them are automatically constructed and stored to the system. When a change log is available, the user is able to query ontology evolution, and to visualize the change path for a specific change operation or a triple. Moreover, a user can issue queries - denoting also the ontology version that those queries are using and the ontology version to which they need to evolve. The invalid queries are identified, and the corresponding affecting change operations as well. The demo of the entire platform was presented at [24] whereas the module for automatically generating the sequence of changes among two ontology versions was presented at [6]. Another module [4] tries to respond to massive number of queries that might need to be changed by producing possible rewritings as well. However, this is beyond the scope of this paper and is not presented here.

In order to evaluate our system we used a workstation with an Intel Core 2 Duo processor running at 3.0 Ghz, and 4GB memory, using Windows 7x64. Moreover, we used queries built using two well-known ontologies. One medium-size ontology (CIDOC-CRM [7]) from the cultural domain which is rarely changed and one large-size ontology (Gene Ontology [8]) from the bioinformatics domain which is heavily updated daily.

CIDOC-CRM is an ISO standard which consists of nearly 80 classes and 250 properties. For our experiments we used versions dated from 02.2002 (v3.2.1) to 06.2005 (v4.2). The detected change log that was automatically produced identified 726 total changes from v3.2.1 to v.4.2. To check the effectiveness of our system we used 21 template queries coming from hundreds of user queries (9 query templates from [25] and 12 query templates from project 3d-COFORM[4]).

---

[3] http://139.91.183.29:8080/exelixis/
[4] http://www.3d-coform.eu/

Gene Ontology (GO) on the other hand, is composed of about 28000 classes and 1350 property instances. GO is updated on a daily basis and for our experiments we used the change log from 25.11.08 to 26.05.09. The change log that was automatically produced contained 4175 changes. To perform our experiments we used the 38 most popular queries as they have been identified and provided from the AmiGO[5] search engine.

6.1   Identifying Invalid Queries

To illustrate the impact of our system we first present the percentage of the invalid queries detected by the AQD module as the number of change operations increased, in the two aforementioned cases. The results for CIDOC-CRM are shown on the left of Fig. 6 whereas for GO are shown on the right of Fig. 6. We can easily observe that the number of queries affected by change operations increases almost linear as the number of change operations increases as well. This is reasonable as more changes in the ontology result in the invalidation of more queries.

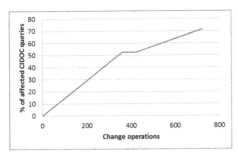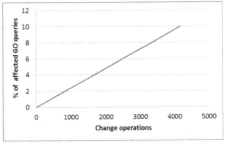

Fig. 6. Percentage of the affected queries for CIDOC-CRM (left) and GO (right) queries

For CIDOC-CRM, even after 726 change operations, 30% of the initial queries do not need to be checked when evolution occurs. But even then, from the remaining 70%, only the 14.35% (on average) of the triple patterns need to be examined, as shown in Fig. 7. By focusing directly on these triple patterns, the total time that developers have to devote for updating the queries is substantially reduced. For GO, we observe that after 4175 change operations just 10% of the total queries need to be checked for adaptation. This may seem to be peculiar because of the large number of change operations. However, if we carefully examine the change operations in each case we will identify that the 4175 change operations change only a small percentage of the GO ontology (10% of the entire ontology) whereas for the CIDOC-CRM the 726 change operations change 54% of the entire ontology. Moreover, queries formulated using GO involve one single term only. That's why in each case we have to change the entire query. Finally, GO is evolved in a backward-compatible manner. That is why most of the queries can be answered in the future versions as well.

---

[5] http://amigo.geneontology.org/cgi-bin/amigo/go.cgi

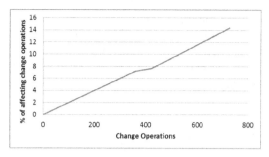

Fig. 7. The average number of affecting change operations per query wrt. change operations

6.2   Generating Change Paths

Next we present experiments concerning the scalability of the algorithms for generating the change paths. Initially, we measured the running time of our algorithm for the different sizes of the change paths. To do that, we retrieved the change path for all triples in the aforementioned ontologies. The average running time in each case is shown in Fig. 8 (left).

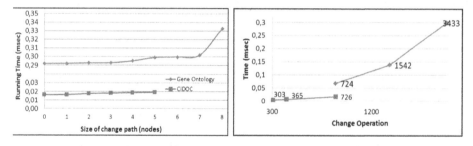

Fig. 8. The running time compared to the size of change path (left) and change log (right)

As shown in the figures, the time to compute a change path is greater for the Gene Ontology than for CIDOC-CRM. This is reasonable since for the Gene Ontology we have to search 4175 changes, whereas for CIDOC-CRM we only have to search 726 changes. Moreover, having a larger set of changes to reason on, we could identify change paths with more nodes for the Gene Ontology (8 nodes for GO vs. 5 nodes for CIDOC-CRM). In all cases, the running time was too small (maximum 0.34 msec), which shows the feasibility and the scalability of our approach even for ontologies that change very often. As shown in our experiments, that the larger the size of the change path, the greater the time which is required.

Then we performed experiments to check the time to construct one change path compared to the size of the logs. The results of our experiments are shown on the right of Fig. 8. For both ontologies we identified that the average time to produce a change path increased linear to the number of changes we had to search. Our results are in line with the complexity evaluation we presented previously.

Finally, trying to understand the change paths that were produced, we made several interesting observations. One of them for example, was the following: We identified that in the evolution of the CIDOC-CRM ontology from version v3.2.1 to version v3.3.2, one ontology engineer renamed the class *"E11 Modification"* to *"E11 Modification Event"*. A few years later another ontology engineer was employed to evolve the ontology. So in v4.2 we can see that the class *"E11 Modification Event"* was again renamed to *"E11 Modification"*. If the second ontology engineer had an indication of the previous renaming he would avoid cycles, he would be able to identify possibly the reasons behind each renaming since we are also able to show comments from the ontology evolution. Hence change paths can be used not only to drive query redefinition but they are also a valuable tool supporting ontology evolution. This enhances the practical value of our system.

7   Conclusion

In this paper, we argue that ontology evolution is a reality and information systems using ontologies should be aware and ready to deal with that. To that direction, we presented a novel module that assists query evolution as the reality that ontology model changes.

This is performed by the automatic detection of the queries that become invalid as the ontology used evolves. Besides the detection of the invalid queries our module can assist developers to redefine them by presenting the specific reasons that led to that invalidation. To that direction, the evolution of the specific part of the ontology that invalidates the query is presented as well, thus minimizing the total time for query redefinition. Experiments performed, show the potential impact of our approach. For example, for GO queries only 10% of them need to be revised. For CIDOC-CRM queries, although the 70% of them need to be adapted, our approach limits the triple patterns that should be examined to the ~14% of the total triples patterns.

As future work, several challenging issues need to be further investigated. An interesting topic would be to extend our approach for OWL ontologies or for the full SPARQL query language. Another interesting topic would be to present summaries of the evolved change path. Ontology evolution is becoming more and more important topic and several challenging issues remain to be investigated in near future.

Acknowledgement. This work has been supported by the eHealthMonitor, EURECA and p-Medicine projects and has been partially funded by the European Commission.

References

1. Poggi, A., Lembo, D., Calvanese, D., Giacomo, G.D., Lenzerini, M., Rosati, R.: Linking data to ontologies. Journal on data semantics X, 133–173 (2008)
2. Calvanese, D., De Giacomo, G., Lembo, D., Lenzerini, M., Poggi, A., Rodriguez-Muro, M., Rosati, R.: Ontologies and Databases: The DL-Lite Approach. Reasoning Web, 255–356 (2009)
3. Flouris, G., Manakanatas, D., Kondylakis, H., Plexousakis, D., Antoniou, G.: Ontology change: Classification and survey. Knowl. Eng. Rev. 23, 117–152 (2008)

4. Kondylakis, H., Plexousakis, D.: Ontology Evolution in Data Integration: Query Rewriting to the Rescue. In: Jeusfeld, M., Delcambre, L., Ling, T.-W. (eds.) ER 2011. LNCS, vol. 6998, pp. 393–401. Springer, Heidelberg (2011)
5. Curino, C.A., Moon, H.J., Ham, M., Zaniolo, C.: The PRISM Workwench: Database Schema Evolution without Tears. In: ICDE, pp. 1523–1526 (2009)
6. Papavassiliou, V., Flouris, G., Fundulaki, I., Kotzinos, D., Christophides, V.: On Detecting High-Level Changes in RDF/S KBs. In: Bernstein, A., Karger, D.R., Heath, T., Feigenbaum, L., Maynard, D., Motta, E., Thirunarayan, K. (eds.) ISWC 2009. LNCS, vol. 5823, pp. 473–488. Springer, Heidelberg (2009)
7. Doerr, M., Ore, C.-E., Stead, S.: The CIDOC conceptual reference model: a new standard for knowledge sharing. In: ER, pp. 51–56 (2007)
8. Gene Ontology Consortium: The Gene Ontology (GO) database and informatics resource. Nucl. Acids Res. 32, 258–261 (2004)
9. Klein, M., Fensel, D.: Ontology versioning on the semantic web. In: ISWC, pp. 75–91 (2001)
10. Qin, L., Atluri, V.: Evaluating the validity of data instances against ontology evolution over the Semantic Web. Inf. Softw. Technol. 51, 83–97 (2009)
11. Hartung, M., Kirsten, T., Rahm, E.: Analyzing the Evolution of Life Science Ontologies and Mappings. In: Bairoch, A., Cohen-Boulakia, S., Froidevaux, C. (eds.) DILS 2008. LNCS (LNBI), vol. 5109, pp. 11–27. Springer, Heidelberg (2008)
12. Klein, M., Stuckenschmidt, H.: Evolution Management for Interconnected Ontologies. In: ICSW-Workshop on Semantic Integration, pp. 55–60 (2003)
13. Thor, A., Hartung, M., Gross, A., Kirsten, T., Rahm, E.: An Evolution-based Approach for Assessing Ontology Mappings - A Case Study in the Life Sciences. GI-Fachtagung für Datenbanksysteme in Business, Technologie und Web, BTW (2009)
14. Huang, Z., Stuckenschmidt, H.: Reasoning with Multi-version Ontologies: A Temporal Logic Approach. In: Gil, Y., Motta, E., Benjamins, V.R., Musen, M.A. (eds.) ISWC 2005. LNCS, vol. 3729, pp. 398–412. Springer, Heidelberg (2005)
15. Xuan, D.N., Bellatreche, L., Pierra, G.: A Versioning Management Model for Ontology-Based Data Warehouses. In: Tjoa, A.M., Trujillo, J. (eds.) DaWaK 2006. LNCS, vol. 4081, pp. 195–206. Springer, Heidelberg (2006)
16. Wang, J., Miao, Z., Zhang, Y., Lu, J.: Semantic Integration of Relational Data Using SPARQL. In: IITA, vol. 01, pp. 422–426. IEEE Computer Society (2008)
17. Liang, Y., Alani, H., Shadbolt, N.R.: Changing Ontology Breaks Queries. In: Cruz, I., Decker, S., Allemang, D., Preist, C., Schwabe, D., Mika, P., Uschold, M., Aroyo, L.M. (eds.) ISWC 2006. LNCS, vol. 4273, pp. 982–985. Springer, Heidelberg (2006)
18. Maule, A., Emmerich, W., Rosenblum, D.S.: Impact analysis of database schema changes. In: ICSE, pp. 451–460. ACM, Leipzig (2008)
19. Theoharis, Y.: On Graph Features of Semantic Web Schemas. IEEE Transactions on Knowledge and Data Engineering 20, 692–702 (2007)
20. Papavassiliou, V.: Detecting Deterministically High-level Changes for RDF/S Knowledge Bases. Dept. of Computer Science. University of Crete, Heraklion (2010)
21. http://www.w3.org/TR/rdf-sparql-query/
22. Perez, J., Arenas, M., Gutierrez, C.: Semantics and complexity of SPARQL. ACM Trans. Database Syst. 34, 1–45 (2009)
23. Lloyd, J.W.: Foundations of logic programming. Springer-Verlag New York, Inc. (1987)
24. Kondylakis, H., Dimitris, P.: Exelixis: Evolving Ontology-Based Data Integration System. In: SIGMOD, pp. 1283–1286 (2011)
25. Theodoridou, M., Tzitzikas, Y., Doerr, M., Marketakis, Y., Melessanakis, V.: Modeling and querying provenance by extending CIDOC CRM. Distrib. Parallel Databases 27, 169–210 (2010)

Ontology of Dynamic Entities*

Lior Limonad[1,**], Pieter De Leenheer[2,3], Mark Linehan[4],
Rick Hull[4], and Roman Vaculín[4]

[1] IBM Haifa Research Lab, Haifa 31905, Israel
`liorli@il.ibm.com`
[2] Collibra nv/sa, Ransbeekstraat 230, 1120 Brussels 12, Belgium
`pieter@collibra.com`
[3] VU University Amsterdam, De Boelelaan 1081a, 1081HV Amsterdam,
The Netherlands
[4] IBM T.J. Watson Research Lab, Hawthorne, New York, USA
`{mlinehan,vaculin,hull}@us.ibm.com`

Abstract. This paper describes "dynamic business entities", ontological
classes that dynamically acquire and lose properties and relationships as
a function of other aspects of the entities. Specifically, we propose how
acquisition and loss instructions for such transient properties may be
inherent in the definition of the entities possessing them. We use SBVR
to demonstrate the specification of dynamic business entities showing
how our idea could be applied in practice. We illustrate with an example
drawn from the Flanders Research Information Space.

Keywords: Ontology, Conceptual Modeling, Business Entity, Data
Integration.

1 Introduction

Enterprise Application Integration (EAI) and *Service Interoperability* are aimed
at the realization of cross-party operations through the establishment of software
architecture and computer service links between originally independent units. In
business, these types of cross-party operations often implement supply channels.
A significant practical difficulty among service channel partners is semantic-
mismatch due to different understandings of the data communicated among
them. The design of such technological ties typically relies on the construction
of a *semantic layer*, serving as an underlying conceptualization to help design
concrete data and service adapters for overcoming the mismatch. Associated
with the business units that take part in the integration, the main product
typically being produced at the core of the semantic layer is a corresponding
ontology, typically expressed using a concrete knowledge representation language
(e.g., formal using OWL [12,7] or in controlled natural language expressed in

---

* The research leading to these results has received funding from the European
Community's 7th Framework Programme under grant agreement no. 257593.
** All authors have equally contributed to this work.

SBVR [11] or visual such as UML class diagrams), providing a shared and formal specification of key business artifacts [6]. Such *business artifacts* reflect mutually agreed upon conceptualized entities that are conceived by the parties as being central to the mutual domain of integration. The better aligned the specification of the semantic layer with the mutual glossary of the partners, the easier it becomes to attain interoperability. Hence, effectively and timely synchronizing data and operations across the enterprise [9,4]. Ensuring such an alignment is anchored in the ontological expressiveness of the language that is used to describe the semantic layer. That is, its capacity to accurately and faithfully describe the mutual domain. As previously theorized in [16,17] and adapted to the purpose of this work, such capacity is inherent in the adherence of the concrete grammar to the following requirements:

1. *Ontological Clarity and Completeness* - refers to the capacity of the grammar to faithfully and comprehensively describe all the business entities and their properties (intrinsic and mutual) as generally perceived by the corresponding community of users. This kind of representation has been traditionally the focus of ontologies, realized by the usage of conceptual modeling such as ER as the specification language to specify entities, attributes, relationships and governing rules.

2. *State tracking* - refers to the capability of the grammar to keep track with domain changes. In the context of the aforementioned business artifact, state tracking entails a unique need to attain *lifecycle-congruency*, which means coherently and unambiguously represent all business entities in different contexts and applications, synchronized with the timely perception of the entities' properties by the partners. For example, it is possible that a manufactured product (e.g., a car) is considered as possessing the property `price` only after having it passed all quality assurance tests during its manufacturing. Before that point in its lifetime, the property `price` has no meaning when associated with the `car` being produced.

The main problem with existing grammars that are used for the specification of the semantic layer is their lack of capability to handle lifecycle-congruency. Simply put, there is no existing ontological apparatus that can account for this need in the design grammars that should adequately describe *dynamic entities*. The latter are entities whose possession of some properties (i.e., transient) may be valid at different circumstances within the overall lifetime of the entities.

Therefore, as a solution for the need to account for the external validity of ontological specifications, in this paper, we motivate (in Sect. 2) and propose (in Sect. 3) the conceptual apparatus of an ontology that is designed to handle not only the conceptualization of dynamic entities and the notion of a transient property, but more importantly illustrates the design of a *property possession algebra* for conceptualizing the behavior of transient properties across the lifecycle of corresponding entities.

Next, in Sect. 4, we show how a concrete modeling language (i.e., IBM's Business Artifacts) that is equipped to express business domains that comprise artifacts possessing transient properties can provide the semantic foundation for a concrete grammar (i.e., SBVR) to improve its lifecycle congruency. The illustration is drawn from a realistic case study in the Flanders Research Information Space, and the implementation is shown in an SBVR-based collaborative ontology management tool, i.e., Collibra's Business Semantics Glossary. Finally, in Sect. 5 we reflect on our work and outline future directions.

2   Motivation and Related Work

In this paper, we introduce the conceptual apparatus that is necessary to precisely describe the dynamics of *transient properties* being part of an *ontology of dynamic entities*. An ontology of dynamic entities is a business ontology whose conceptualized elements are formed by a combination of persistent and transient properties.

1. A *persistent* property is a property whose possession is fixed across the lifecycle of the entity or thing possessing it.
2. A *transient* property is a property whose possession is transient across the lifecycle of the entity or thing possessing it.

Consequently, the specification of the latter type is inherently equipped to determine possession validity at any point in the lifetime of its possessing thing. *Possession* or "expression" is used here to mean that a property is a quality or trait of an entity or a thing. An *entity* is equivalent to the notion of an ontological *kind* (e.g. an SBVR [11] *noun concept* or an OWL *class*), determined by a set of (possessed) *properties* such that the members of a kind are all those and only those things that share all the properties in the given set [3]. Hence, a *thing* means an instance or a member of an entity.

2.1   State of the Art

The goal of this work is to describe all grammar elements being required to specify property *transience* in dynamic entities, while in prior work its has been solely its conceptualization being addressed. For example, previous literature refers to the need for ontological views that are suited for expressing and reasoning about domains that comprise perdurant entities [18] i.e., entities for which their possession of properties and relationships may change in time. Such transient properties have been termed *fluents* in this prior work. Furthermore, a plethora of realization approaches has been accumulated throughout the years, proposing various concrete ways for expressing the existence of transient properties. This prior work includes distinction between optional and mandatory properties in general [5], property negation (i.e., non-possession) [1], the conceptualization of relationships between entities [15], the roles assigned to entities [2], all which may be considered as various forms for the specification of transient properties.

More recently, property transience has been further confirmed and clarified by work in the area of classification [13], similarly distinguishing between baseproperties (i.e., persistent) in a class which determine the classification of a thing (i.e., whether it is a member in a given class), and derived-properties which can be inferred from its membership.

The most fundamental driver justifying the significance of transience is driven by the philosophical paradigm underlying social ontologies according to which there is a clear rational for expressing not only materialized and substantial aspects of the domain, but also aspects being the mere outcome of social intentionality [14]. For example, how would one associate between two individuals (e.g., `John`, and `Kelly`) with the possession of the property `in_love(John,Kelly)` without acknowledging its possession may be transient?

2.2 Novelty of This Work

All of the previous work simply establishes that the existence of property transience is an essence in any linguistic form that is intended to facilitate faithful domain representations. However, none of the previous work is focused on the exact linguistic instrumentation that is necessary to express possession dynamics - i.e., expressing property acquisition and loss in the context of the exact circumstances affecting it such as time, form, association etc. Hence, in this work, our effort is to illustrate the most fundamental machinery that is required to exist in any language that is expected to faithfully and accurately describe possession dynamics. As mentioned above, we find such capability as being most desired in the context of interoperability in which the capability to both describe and interpret entity structures must be synchronized across independent silos.

Note that the possession of a property, whether persistent or transient, may be perceived or viewed by an observer in a way that makes the perception itself transient. We therefore distinguish between the ontological level in which the notion of transience is a fundamental characteristic or trait of the entities being expressed by the ontology regardless of any external view, and the perceptional level in which the perception of property possession may be a function of various contextual and spatial dimensions such as: calendar time, participant's or role's perspective, geography and other. We acknowledge that the perceptional level has been somewhat approached in prior work (e.g., expressed in the form of access controls and views) while the focus of this paper is the ontological level.

Particularly, our proposed solution is designed to account for any case in which possession of properties is a function of phases in the *lifecycle* of the entities being conceptualized by the ontology. Despite the existing body of knowledge aforementioned being focused on various possible conceptualizations for transient properties, existing solutions are essentially different than the one proposed here for either one (or a combination) of the following reasons:

1. Most existing solutions are focused only on a single factor as the potential source for property possession. The most common factor considered is time.
2. Most existing solutions are missing the expressive power needed to explicitly describe how property possession changes as a function of the factor(s) being considered.

Our innovation lies in a solution that:

1. acknowledges the need to associate possession of properties with various factors (e.g., state, geography, perspective). Specifically, we account for situations in which possession should be determined by existential and contextual knowledge about the mere object for which the possession itself has to be determined.
2. suggests the exact processing instructions being required to describe how possession may alternate as a function of the various triggering factors, and specifically as a function of the target objects' lifecycle contexts.

Note, we consider the possession of a property or the attribution of a property to an entity in an inclusive form, uniformly considering the intrinsic traits of entities and also their relationships to other entities as being expressed by properties that one may attribute or predicate about the entity[1]. In case of a relationship, the property is designated by an n-ary predicate being attributed to all entities participating in it (namely, a *mutual property*). This way for example, "having a red color" as an intrinsic property may be attributed to "my car" through the predicate color(red, my_car) while associating my car with "myself" as the owner may be expressed as a mutual property through the owning(self, my_car) predicate. While the former property in this example is persistent across any point during the lifecycle of my_car, the latter may be transient, attributing it to an owner only after the completion of its manufacturing.

Unlike traditional ontologies being merely aimed to express static entities i.e., entities comprising persistent properties only, ontologies of dynamic entities need to express entities comprising combinations of persistent and transient properties. Currently there is no such apparatus. For the latter, the specification also needs to clarify in which circumstances entities' properties may be acquired and dismissed. The dynamics of property acquisition and loss may be expressed as a function of various changes throughout the lifecycle of the entity possessing it. The underlying machinery that is required for such purposes is explained in Sect. 3. In Sect. 4, we demonstrate the advantage of our solution and how the proposed apparatus may be applied to the benefit of business integration and the creation of corresponding shared vocabularies.

3 Description of the Ontology for Dynamic Entities

In this section we describe in detail the internal features of an ontology that is aimed for clarifying the dynamics of property transience in dynamic entities. As mentioned above, in its core, such an ontology distinguishes between the conceptualization of static and dynamic entities and specifically includes a corresponding linguistic capability for describing the behavior that underlie changes

---

[1] It is worth noting that although the focus in this work is on the dimension of property transience, our findings may be developed further to consider its integration with other previously theorized dimensions of properties (e.g., intrinsic vs. mutual, hereditary vs. emergent etc.)

in the possession of transient properties in dynamic entities. An ontology of dynamic entities should implement the following features: property specifications, possession formula for these properties, and a life cycle context.

Properties of entities are associated with a *possession formula* that is based on values of other properties of entities or can be inferred from lifecycle traces that include the application of a *property possession algebra* i.e., a set of atomic operations about the acquisition and loss of properties, each which may be structured as follows:

- Property specification e.g., `color(car)`, the specification of a property in the ontological description of a domain applies to any property, persistent or transient. In the case of the latter, the specification should also be associated with a possession formula that can be used to evaluate property possession at run-time. Such a formula may be realized as a set of acquisition and loss statements, each comprising two components:
  - A possession instruction i.e., an instruction to either acquire or lose the corresponding property.
  - A lifecycle context (e.g., "when") i.e., a combination of certain property values and certain lifecycle indicators being an antecedent condition to the execution of the possession instruction.

For example, the property `price(1000 €, car)` may be associated with the following possession formula: {(`acquire, on_completion_of_manufacturing`),
(`lose, on_total_loss`)}, i.e., `price` as a property of a `car` is determined as being possessed only after the `car` is fully manufactured. Similarly, in case of extreme damage, a `car` will no longer have a `price`.

It is worth stressing that the aforementioned machinery is stated on a relatively abstract level, keeping it agnostic to possibly more concrete realizations for entity lifecycle styles (e.g., a state-machine), in which the suggested terminology may need to be further specialized. For example, when indeed specified in a form of a state-machine, the concept of possession-formula may be interpreted as being part of functions that describe transition of states. Similarly, other concrete styles may entail different interpretations.

3.1   Designing Possession as a Function of Context

Note that in this example the possession of `price` as a property cannot be expressed intuitively as a function of time: indeed the `on_total_loss` time point is unknown at design time. Hence, from a designer's perspective it is essential (and one may argue also more useful) to provide a grammar (as we illustrate here) that enables formulating the truth of possession as a function of various contextual factors, including factors that stem from inherent information (e.g., materialized) about the possessing entities themselves. As an example, consider the case of the property `attractedTo(matter_1,matter_2)` for which the corresponding possession formula may be formulated as follows:
{(`acquire,opposing(charge(matter_1,value_1),charge(matter_2,value_2)))`,
(`lose, NOT opposing(charge(matter1,value_1),charge(matter2,value_2)))`}.

In this case the possibility to determine whether the possession of the property `attractedTo` holds may be inferred directly from the capability to determine the charges `value_1` and `value_2` of both matters.

3.2 Run-Time Evaluation of Possession along the Entity Life Cycle

For convenience purposes, one may use a tagging mechanism at run-time to annotate each property with a possession indicator that is modified each time an acquisition or a loss statement is triggered. This way, instead of needing to evaluate historical traces of acquisition and loss of properties at run-time, there will be an immediate indication for whether the property is possessed or not. Furthermore, in a data-centric approach, lifecycle indicators may themselves be specified as properties such that in the example above, `on_completion_of_manufacturing (true/false, car)` and `on_total_loss(true/false, car)` may both be specified as persistent properties (e.g., as opposed to events). We will illustrate this possibility in SBVR.

Implied from the realization of the above features, the possession of any entity's property can be determined at runtime based on evaluation of the property possession formula. In restricted cases, the possession of a property can be determined by a tool at design time. Given a set of "possession analysis" criteria, this tool can determine (statically) that the expression specified by the possession formula depends upon properties of the entity that have known values at specific stages of the entity lifecycle (i.e., the lifecycle's context). To make this possible, the lifecycle has to be modeled explicitly, specifying notions such as stages and milestones, and the relationships among them as in [10,8]. Static analysis is simplified when lifecycles avoid cycles (i.e. where an entity can return from a later stage to an earlier stage) but is possible in limited circumstances when there are cycles.

Examples given below use the SBVR "Structured English" grammar for convenience. However, the underlying ideas of this paper are agnostic to the concrete grammar that is used to specify an entity lifecycle model. When a property is accessed at some point in the lifecycle, and if the possession formula refers to either (a) persistent properties, or (b) transient properties that are themselves possessed, and if the referenced properties have known values, then a tool can determine that the accessed property is possessed at that point in the lifecycle.

3.3 Mutable Attributes

In addition to the above it is also acknowledged that in current state of the art, a typical realization for the ontological notion of a property is attained through the usage of valued attributes e.g., the persistent property `color(red, car)` may be represented by an attribute `color` being associated with the value "red". Since it is not expected that a `car` will ever change (the value of) its `color` (i.e., a persistent property), once represented as an attribute it may be inferred that the attribute's value is immutable. On the other hand, a transient property such as `owner(person, car)` when represented as a valued attribute may be determined as mutable, enabling the underlying need to replace the possession of the property with another that indicates a different ownership. Henceforth, the dichotomy of being either mutable or immutable is a direct outcome of using valued attributes as the realization mechanism for properties. Therefore, in the following example, we also indicate for each property whether it is immutable or not.

4   Demonstration

Concluding the feasibility of the apparatus described above, in this section we demonstrate how a concrete modeling grammar (i.e., SBVR) may be used to represent dynamic entities with an example from the Flanders Research Information Space. Finally, by implementing a prototype in the Business Semantics Glossary, we show how the design of ontology for dynamic entities and the creation of shared vocabularies and rules in general may benefit each other.

4.1   Flanders Research Information Space

The Web is a catalyst for *open innovation*. Enterprises and research institutions have come to realize that they no longer can rely on their own research to innovate, but instead share or trade ideas and results to achieve a greater benefit to themselves and others. The Flemish government has taken the lead at driving European open innovation through Flanders Research Information Space (FRIS[2]), an ambitious change program that publishes data on innovation-related entities such as research institutes, researchers, and funded projects.

Many of these FRIS entities share the characteristics of dynamic entities. E.g., take a research project: they usually have long life cycles (up to several years), requiring FRIS data relating to properties (i.e., publications, deliverables, consortium) of these entities to be updated regularly. Secondly, not all properties are intended for publication (e.g., periodic review reports). Thirdly, the data related to these properties have to be provided by different parties (such as principle investigator, consortium members, project officer) and according to a certain semantics (in the case of FRIS based on the Common European Research Information Standard (CERIF[3]). Finally, they exhibit transient properties: e.g., the start date of a project is only valid if the project has been formally initiated. Summarising: an ontology for the dynamic entity Project should declare and enforce: "what are the attributes of an entity in which stage of the entity's lifecycle?" Currently CERIF is formalized using the ER grammar, which does not allow to model transient properties or possession formula.

4.2   Two Associated Dynamic Entities: Proposals and Projects

We illustrate our approach in terms of two FRIS entities that acquire or lose possession of properties and relationships in function of other aspects of the

---

[2] http://www.researchportal.be
[3] http://www.eurocris.org/

entities. We follow the SBVR practice of underscoring nouns (SBVR noun concepts), showing relationships (SBVR verb concepts or fact types) using italics, and using bold face for keywords such as "if". This approach does not require any change to SBVR, which already provides for conditional necessity rules.

Consider an entity Proposal with the following properties (in terms of SBVR binary verb concepts).

1. Persistent + immutable:
   Proposal *isownedby* Principle Investigator / Principle Investigator *owns* Proposal
2. Persistent + mutable:
   Proposal *isdescribedby* Discipline Code/ Discipline Code *describes* Proposal
3. Transient + immutable:
   Proposal *has* Evaluation Score / Evaluation Score *of* Proposal
4. Transient + mutable:
   Proposal *defines* Work Plan / Work Plan *isdefinedby* Proposal

Consider an entity Project with the following properties (in terms of SBVR binary verb concepts):

1. Persistent + immutable: Project *executes* Proposal / Proposal *is executed by* Project
2. Transient + mutable: Project *has* Start Date / Start Date *of* Project

For the persistent properties of Proposal and Project we define integrity constraints that are true independent of the stage in which the entity is:

1. **It is necessary that each** Proposal *isownedby* **exactly one** Principle Investigator
2. **It is necessary that each** Proposal *isdescribedby* **at least one** Discipline Code
3. **It is necessary that each** Project *executes* **exactly one** Proposal

Now for the transient properties we define the possession formula in terms of the following assumed lifecycle stages.

– For Proposal: → Submitting → Evaluating → Notifying → Submitting

– For Project: Initating → Reviewing → Finishing

For our example, we define "milestones" (related to achieving the end of these stages) as special types of characteristic in SBVR. A characteristic is a unary verb concept with Boolean type. We have four milestone characteristic types for Proposal:

1. Proposal *has been submitted*

2. Proposal *has been evaluated*

3. Proposal *is accepted*

4. Proposal *is rejected*

We have three "milestone" characteristic types for Project:

1. Project *is initiated*

2. Project *has been reviewed*

3. Project *is finished*

We can express for every transient property P of an entity E a (dis-)possession formula that has to be true if one or more characteristics M_i is/are true. A possession formula may use an SBVR "necessity" modality:

– **It is necessary that each** Proposal *defines* **exactly one** Work Plan **if the** Proposal *has been submitted.*

A dispossession formula may use an SBVR "impossibility" statement:

– **It is impossible that a** Proposal *defines* Work Plan **if the** Proposal *has* **not** *been submitted.*

A combination possession formula for entity-property Proposal has Work Plan using "if and only if" allows a shorthand notation for the conjunction of the two previous "if" statements. This version uses "always" as an alternative way to express "necessity":

– **A** Proposal **always** *defines* **exactly one** Work Plan **if and only if the** Proposal *has been submitted.*

The following are some example combination formulae. Combination possession formula for Proposal *has* Evaluation Score:

– **It is necessary that each** Proposal *has* **at least one** Evaluation Score **if and only if the** Proposal *has been submitted* **and the** Proposal *has been evaluated.*

Combination possession formula for Proposal *is executed by* Project:

– **It is necessary that each** Proposal *is executed by* **exactly one** Project **if and only if the** Proposal *is submitted* **and the** Proposal *has been evaluated* **and the** Proposal *is accepted* **and the** Proposal *is* **not** *rejected* **and the** Project *is initiated.*

Combination possession formula for Project *has* Start Date:

– **Each** Project **always** *has* **exactly one** Start Date **if and only if the** Project *is initiated* **and the** Proposal *that is executed by* Project *is accepted.*

The two above possession formula illustrate that the possession of properties may also depend on aspects of another entity. E.g., in the latter example, the validity of property Project *has* Start Date depends on an aspect of another entity Proposal that is associated (through the verb concept "*executed by*'), this aspect being the milestone: Proposal *is accepted.*

4.3 Transitivity of Possession

The above formula may be simplified if we define production rules for each sequence of two milestones. E.g., "Proposal *has been evaluated*" implies "Proposal *has been submitted*". This also assumes that Proposal *is accepted* and Proposal *is rejected* are mutually exclusive, the latter which can be specified with SBVR's mutual exclusion contraints.

Therefore, from the following (repeated from above):

– **It is necessary that each** Proposal *is executed by* **exactly one** Project **if and only if the** Proposal *is submitted* **and the** Proposal *has been evaluated* **and the** Proposal *is accepted* **and the** Proposal *is* **not** *rejected* **and the** Project *is initiated.*

We can infer automatically:

– **It is necessary that each** Proposal *executed by* **exactly one** Project **if and only if the** Proposal *is accepted* **and the** Project *is initiated.*

4.4 Acyclic Lifecycles

As defined above, part of the lifecycle of a Proposal has no cycles, meaning that a Proposal cannot "go back" from "*has been accepted*" to "*is rejected*". Thus a tool is able to statically determine that some transient properties of Project are available once the "Proposal *has been accepted*" and "Project *has been initiated*".

4.5 Implementation in Business Semantics Glossary

Figure 1 shows a screenshot of the noun concept Proposal within the "Proposal" vocabulary managed by the "CERIF" speech community that is part of the "FRIS" semantic community. The Business Semantics Glossary is a tool that implements the *business semantics management* (BSM) methodology [4]; hence for collaboratively managing the semantics of persistent properties of CERIF entities. The screenshot shows the attributes we defined in Subsect. 4.2 for the dynamic entity Proposal: fact types to express properties, characteristics denoting milestones, integrity constraints for persistent properties, and possession formula for transient properties. The underscores define hyperlinks to other parts in the glossary showing the embedding of our approach in the broader context of creating shared vocabularies based on CERIF for service interoperation in FRIS.

The right hand panel shows the governance settings: in the bottom-right corner is indicated which member in the community (here "Pieter De Leenheer") carries the role of "steward", who bears final accountability. The status "candidate" indicates that the term is not yet fully articulated: in this case 37.5%. This percentage is automatically calculated based on the articulation tasks that have to be performed according to the BSM methodology. Tasks are related to defining attributes and are distributed among stakeholders and orchestrated using workflows.

Fig. 1. Collaborative designing dynamic entities in the Business Semantics Glossary

5 Discussion and Future Work

The fundamental contribution of this work is prescribing the essence of any language that is aimed to adequately describe dynamic entities. Particularly this includes the unique mechanism that is needed to handle transience through possession and loss instructions being inherent in the language. Currently, no commercial tool, aside from the prototype in Collibra's Business Semantics Glossary used to demonstrate the feasibility of our solution, exists. Hence, we find our effort in this work as paving the road towards the development of improved data-integration tools that are better equipped to facilitate inter-silos communication. This may include tools such as IBM Infosphere MDM, Oracle Master Data Management, and SAP Enterprise Master Data Management.

In addition, our contribution has similar applicability to existing knowledge representation standards (e.g., OWL) and software development languages which may be extended with the proposed capabilities as well. Preliminary penetration of the solution is starting to show its first signs in technologies such as Java JSR-305, enabling basic annotation of "nullable" properties. Yet, more expressive solutions as proposed here seem to further alleviate the need to mitigate both the burden in handling reference availability mismatch at runtime, and also enabling richer static analysis of code.

In order to accommodate for the exact features expected in the underlying tools to adequately reason about property transience and inference about possession validity, our most immediate intention is devoted towards further investigation of specification well-formedness and discovery of inconsistencies.

Aside from possible applications of the proposed solution, future research work may be aimed to extend the proposed linguistic machinery with the capability to express epistemic aspects of property possession. This may include the possibility for example to attach roles to every possession formula indicating who is responsible for what milestone. Such capability is essential in data governance.

When business partners exchange information in Enterprise Application Integration (EAI) and Service Interoperation scenarios, they should distinguish persistent versus transient attributes of that information. Traditional modeling and ontology standards do not enable these distinctions. The method we propose exploits existing capabilities of SBVR to explicitly identify which attributes are valid under what circumstances. This removes doubt about exchanged information, and should improve the success of interoperation scenarios.

References

1. Allen, G.A., March, S.T.: The Proper Role of Optionality and Negation in Conceptual Modeling. In: Proceedings of the Eighth Annual Symposium on Research in Systems Analysis and Design, Richmond, VA, May 21-23 (2009)
2. Bera, P., Burton-Jones, A., Wand, Y.: The effect of domain familiarity on modelling roles: an empirical study. In: Proceedings of PACIS 2009, p. 110 (2009)
3. Bunge, M.: Treatise on basic philosoph. In: Ontology I: The Furniture of the World, vol. 3. Reidel, Boston (1977)
4. De Leenheer, P., Christiaens, S., Meersman, R.: Business semantics management: a case study for competency-centric HRM. Computers in Industry 61(8), 760–775 (2010)
5. Gemino, A., Wand, Y.: Complexity and clarity in conceptual modeling: Comparison of mandatory and optional properties. Data and Knowledge Engineering 55(3), 301–326 (2005)
6. Gruber, T.: A translation approach to portable ontology specifications. Knowledge Acquisition 5(2), 199–220 (1993)
7. Hitzler, P., Krötzsch, M., Parsia, B., Patel-Schneider, P.F., Rudolph, S. (eds.): OWL 2 Web Ontology Language: Primer. W3C Recommendation (October 27, 2009), http://www.w3.org/TR/owl2-primer/
8. Hull, R., Damaggio, E., Fournier, F., Gupta, M., Heath III, F(T.), Hobson, S., Linehan, M., Maradugu, S., Nigam, A., Sukaviriya, P., Vaculin, R.: Introducing the Guard-Stage-Milestone Approach for Specifying Business Entity Lifecycles (Invited Talk). In: Bravetti, M. (ed.) WS-FM 2010. LNCS, vol. 6551, pp. 1–24. Springer, Heidelberg (2011)

9. de Moor, A., De Leenheer, P., Meersman, R.: DOGMA-MESS: A Meaning Evolution Support System for Interorganizational Ontology Engineering. In: Schärfe, H., Hitzler, P., Øhrstrøm, P. (eds.) ICCS 2006. LNCS (LNAI), vol. 4068, pp. 189–202. Springer, Heidelberg (2006)

10. Nigam, A., Caswell, N.S.: Business artifacts: An approach to operational specification. IBM Systems Journal 42(3), 428–445 (2003)

11. OMG: Semantics of Business Vocabulary and Business Rules (2008),
http://www.omg.org/spec/SBVR

12. OWL Working Group: OWL 2 Web Ontology Language: Document Overview. W3C Recommendation (October 27, 2009),
http://www.w3.org/TR/owl2-overview/

13. Parsons, J., Wand, Y.: Using cognitive principles to guide classification in information systems modeling. MIS Quarterly 32(4), 839–868 (2008)

14. Searle, J.R.: Social ontology. Anthropological Theory 6(1), 12–29 (2006)

15. Wand, Y., Storey, V.C., Weber, R.: An ontological analysis of the relationship construct in conceptual modeling. ACM Trans. Database Syst. 24(4), 494–528 (1999)

16. Wand, Y., Weber, R.: On the ontological expressiveness of information systems analysis and design grammars. Information Systems Journal 3(4), 217–237 (1993)

17. Wand, Y., Weber, R.: On the deep structure of information systems. Information Systems Journal 5(3), 203–223 (1995)

18. Welty, C., Fikes, R.: A Reusable Ontology for Fluents in OWL. In: Proceedings of FOIS, pp. 226–236. IOS Press (2006)

A Supervised Method for Lexical Annotation of Schema Labels Based on Wikipedia*

Serena Sorrentino, Sonia Bergamaschi, and Elena Parmiggiani

DBGROUP, DII
University of Modena and Reggio Emilia, Italy
{Serena.Sorrentino,Sonia.Bergamaschi,Elena.Parmiggiani}@unimore.it

Abstract. Lexical annotation is the process of explicit assignment of one or more meanings to a term w.r.t. a sense inventory (e.g., a thesaurus or an ontology). We propose an automatic supervised lexical annotation method, called ALA$_{TK}$ (Automatic Lexical Annotation -Topic Kernel), based on the *Topic Kernel* function for the annotation of schema labels extracted from structured and semi-structured data sources. It exploits Wikipedia as sense inventory and as resource of training data.

1 Introduction

Lexical Annotation represents a powerful means in a wide range of semantic applications including schema and ontology matching [7]: starting from the "hidden meanings" associated with schema labels (i.e. class and attribute/properties names) it is possible to discover semantic correspondences (e.g., location = area) among the elements of different schemas/ontologies.

A manual process of lexical annotation is a time consuming and not scalable task. To perform automatic or semi-automatic lexical annotation, a method for *Word Sense Disambiguation* (WSD) has to be devised. WSD is traditionally a NLP (Natural Language Processing) problem and it can be defined as the ability of identifying the meanings of a word (a schema label, in our case) in a context.

WSD approaches can be divided in *supervised* and *unsupervised* methods [16]. Supervised approaches usually are based on machine-learning techniques: they need to be trained on a set of hand-annotated *training data* (i.e. manually annotated occurrences of the target label in a context) in order to automatically disambiguate a set of previously unseen label occurrences (in the following *test data*). Supervised methods are able to outperform significantly in precision and recall the unsupervised ones [9,16]. However, their main drawback is the need of a large amount of accurately sense-annotated training data (which are very costly to be produced as they have to be manually created) in order to obtain good results in term of precision. On the contrary, unsupervised methods

---

* The research leading to this work was partially supported by the Biogest-Siteia projects http://www.biogest-siteia.unimore.it, funded by Emilia-Romagna (Italy) regional government. Our sincere thanks to Professor Sanda Harabagiu, and to the PhD students Bryan Rink and Kirk Roberts for their support to this research.

P. Atzeni, D. Cheung, and R. Sudha (Eds.): ER 2012, LNCS 7532, pp. 359–368, 2012.

do not require any manually sense-annotated training data, and usually exploit raw (i.e., non annotated) corpora and thesaurus (e.g., WordNet [15]). For these reasons, so far, unsupervised methods have been preferred in several automatic applications [16]. Moreover, in the area of schema/ontology matching, the cost of manually producing training data might affect significantly the advantages of an automatic mapping discovery process w.r.t. a manual one.

A possible solution to overcome the drawback of supervised methods is represented by Wikipedia[1]. Wikipedia is a freely available and accessible knowledge resource which covers a wide range of domains. Several works have shown the similarities between Wikipedia and traditional thesauri, thus allowing us to use it as a sense inventory for lexical annotation and as a resource of training data [13].

In this paper, we propose a new automatic supervised lexical annotation method[2], called ALA_{TK} *(Automatic Lexical Annotator - Topic Kernel)*, to automatically annotate schema labels extracted from structured and semi-structured data sources, by exploiting Wikipedia. Our method is based on the *Topic Kernel* function, which represents a composition of the more promising supervised approaches proposed in the NLP research area so far: (1) the *Domain Kernel WSD algorithm* proposed by Gliozzo et. al in [8]; and (2) the Latent Dirichlet Allocation (LDA) topic model [3].

The rest of the paper is organized as follows: in Section 2, we briefly describe Wikipedia and how to have access to its contents; in Section 3, we propose our ALA_{TK} method (extension to the Domain Kernel function). Section 4, describes preliminary experimental evaluations of our method. Finally, in Section 5, we make some concluding remarks and illustrate related work.

2 Wikipedia

Wikipedia is a free online multilingual document collection edited by a wide community of contributors. Wikipedia has been applied in several semantic applications [13,14,11], thanks to the many similarities between its internal organization and the structure of traditional thesauri:

- Wikipedia *Articles* can be considered as concepts of an ontology/thesaurus which contain a brief abstract representing the *gloss* (i.e. the description in natural language of its meaning). E.g., the concept "library" is associated to the article http://en.wikipedia.org/wiki/Library where it is defined as "a collection of sources, resources, and services, etc. ".
- Each article has a *title* representing a *term* of a traditional thesaurus. By exploiting Wikipedia "redirects" we can discover *Synonym Terms*. E.g., the term "Reading room" is redirected to the article of "Library".
- *Disambiguation Pages* collect all the different meanings of a term as in a traditional thesaurus. E.g., the term "library" has a disambiguation page which contains meanings such as "Library_(computing)", "Library_(biology)" etc.

---

[1] http://www.wikipedia.org/
[2] Even if supervised, the method is automatic as training data are automatically collected from Wikipedia.

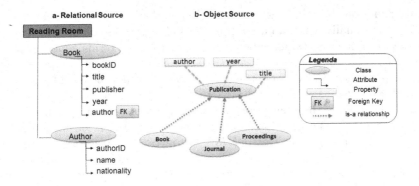

Fig. 1. Graph representation of two schemas to be annotated

- Each article contains a set of *Hyperlinks* which link it to other related articles. E.g., the article "Library" contains a redirect to the "Book" article.
- One or more *Categories* may be assigned to each article. Each category may have sub-category and so on. For example, "School Library" belongs to the "Academic libraries, School terminology, Types of library" categories.

Several projects tried to provide an easy and structured access to the Wikipedia information [13]. Among them, DBPedia [2] represents a valuable community effort to extract information from Wikipedia and to make it available on the Web by using RDF (Resource Description Framework)[3].We decided to use DBPedia for its querying capability on Wikipedia as a database: for instance, we can obtain the disambiguation pages by using the property *wikiPageDisambiguates*[4].

3   The ALA$_{TK}$ Method

The ALA$_{TK}$ method can be divided in two main phases: (1) *Extraction of Test and Training Data* and (2) *Automatic Supervised Disambiguation*. In the following, we briefly described both these phases.

3.1   Extracting Test and Training Data

Supervised methods are based on learning systems and require to collect test and training data sets: training data are used to train the learning system and permits to classify test data.

Test Data are automatically collected starting from the set of schema labels extracted from the structured and semi-structured data sources. All these labels are pre-processed, in order to remove morphological variations, expand acronyms

---

[3] http://www.w3.org/TR/rdf-primer/
[4] We can use DBPedia by the public SPARQL endpoint http://dbpedia.org/sparql or by downloading the datasets.

and abbreviations by using the normalization method described in [17]. In NLP applications, usually test data are represented by *instances* composed by a piece of text containing the target label and a set of others terms representing its context: e.g., the sentence *This <head>book</head> has been written by three authors.*, is a test instance where the tag *<head>* identifies the label to be disambiguated. However, as we deal with structured and semi-structured data sources, we decided to exploit the concept of class as aggregation of its properties, Foreign Key (FK) constraints in relational data sources and ISA relationships in ontologies/object schemas: given a *class label* its test instance is created by considering its attribute/property labels plus the classes and attributes labels connected to it by a FK, in the case of relational sources, or by the classes/properties in the case of ontologies/object schemas. The same hold for *attribute labels*. Let us consider for example, the relational schema shown in Figure 1-a: the test instance for the attribute label "author" of the "Book" class will be "Book, book_identifier, title, publisher, year, <head>author</head>, author_identifier, name, nationality"[5] as it exists a FK between the classes "Author" and "Book".

Training data are composed by a set of training instances. Each instance is represented by a piece of text containing an occurrence of the target label manually tagged with one or more meanings from a reference sense inventory (i.e., Wikipedia in our case). In our method, the "manual tagging" is automatically obtained by using Wikipedia: the Wikipedia hyperlinks, which connect a label occurrence to its corresponding article, are manually created by the Wikipedia contributors thus they can be considered as *manual lexical annotations*. For example, the disambiguation page of the term "book" contains eight different meanings. Within the article `http://en.wikipedia.org/wiki/ISBN`, we find an occurrence of the term "book" which is connected by an hyperlink to the article `http://en.wikipedia.org/wiki/Book` which represents its annotation. By using DBPedia, we automatically identify for each label the corresponding disambiguation page and extract the possible meanings[6]. Then, by exploiting the DBPedia property *wikiPageWikiLink*, for each label we ask for all its existing hyperlinks. All the sentences containing hyperlinks of the target label, are extracted and collected together with their annotations (i.e., Wikipedia article).

For example, if we consider again the label "book" and its hyperlink in the article `http://en.wikipedia.org/wiki/ISBN`, we extract the sentence "The International Standard Book Number (ISBN) is a unique numeric commercial book identifier based upon ...", from which we create a training instance by removing stop words and proper nouns.

3.2   Automatic Supervised Disambiguation

Supervised WSD approaches are based on two main aspects: the learning algorithm and the features to be considered during the learning task. Among the

---

[5] "bookID" and "authorID" have been previously expanded by normalization.

[6] If the disambiguation page does not exist, we interpret the label as a monosemic term and annotate it with the direct link to its article.

several supervised WSD approaches proposed, we focused on kernel methods. Kernel methods have already been successfully applied in several machine learning approaches [16]. Kernel methods work on the idea of embedding a generic set of object $S = x_1, x_2, ... x_i, x \in X$ in a features space F through a mapping function $\phi : X \rightarrow F$ and looking for linear patterns in such space. However, instead of directly using a mapping function they exploit a "kernel function" $k : X \times X \rightarrow \mathbb{R}$ where the set of object S is represented by a $n \times n$ matrix of pairwise comparisons $k_{i,j} = K(x_i, x_j)$. We focus on *linear kernel function*: the objects are vectors which are mapped from the input space X in the features space F by $\phi(x)$ and the kernel function K computes their inner product. The set of object is represented by the training/test instances (instances in the following) $S = i_1, i_2, ..., i_n$. Let $V = t_1, t_2, ..., t_k$ be the vocabulary including all the unique terms extracted from instances, S can be represented, by using a Vector Space Model (VSM) [18], as a $[k \times n]$ *term by instance matrix* in which each row i_j indicates an instance and each column t_i a unique term, and such that $f_{i,j}$ is the frequency of term t_i into the document d_j.

To define the Topic Kernel function K^T we decided to start from the Domain Kernel WSD method proposed by Gliozzo et. al in [8], which obtained the best performance in disambiguating texts in the evaluation of Senseval-3[7]. In particular, we decided to improve this function by employing the *Topic Model* [3], which has successfully been applied in several disambiguation approaches [4,5]. In the following, we briefly summarize both of them.

Domain Kernel WSD

The algorithm proposed in [8] uses the domain knowledge as the feature for meaning distinction (e.g., the term "virus" may mean "a malicious computer program" in the "Informatics" domain or "an infectious agent which spreads diseases" in the "Medicine" domain). The intuition is that the domain may be exploited to compute the similarity between training and test instances , and thus to infer the meaning of new term occurrences. They make use of a Support Vector Machine (SVM) learning system [6] based on the *Domain Kernel function*. The domain knowledge is represented by the *Domain Matrix* (DM) (with terms along the rows and domains along the columns) which is automatically computed by the using Latent Semantic Indexing (LSI) (also called Latent Semantic Analysis in the literature) [12], starting from text extracted by raw corpus. DM is used to define the mapping function ϕ^D that maps the document vector $\vec{t_j}$ into the vector $\vec{t^D}_j$ in the domain space.

The Topic Model

Topic models provide a simple way to analyze large volumes of raw text [3]. A "topic" consists of a cluster of terms that frequently occur together. The starting point of topic models is to decompose a conditional *term by the document probability distribution* $p(t|d)$ into two different distributions: *the term by topic*

---

[7] http://www.senseval.org/senseval3

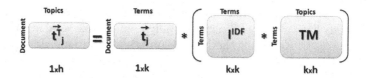

Fig. 2. The matrix representation of the Topic Kernel Mapping Function

distribution $p(t|z)$, and *the topic by document distribution* $p(z|d)$ as follow:

$$p(t|d) = \sum_z p(z|d)p(t|z)$$

this allows each semantic topic z to be represented as a multinominal distribution of terms $p(t|z)$, and each document d to be represented as a multinominal distribution of semantic topics $p(z|d)$. The model introduces a conditional independence assumption that document d and term t are independent conditioned on the hidden variable, topic z. LDA is a generative probabilistic model for document collection that represents one of the most common topic model [3].

The Topic Kernel Function
Starting from the Domain Kernel function and the Topic Model we define our Topic Kernel function. Our motivation is that the topic information can connect terms with similar meanings and distinguish between uses of terms with multiple meanings in different contexts as well as is done by the domain information. In [9] it has been shown that LDA outperforms LSI (used in the Domain Kernel function), in the representation of ambiguous words and in a variety of other linguistic processing and memory tasks.

By using LSI the term by documents matrix is decomposed into three matrices: UDV^T where U is a term by domains matrix, D is a diagonal matrix with singular values and V^T is the domain by documents matrix. In the Domain Kernel function the DM matrix is derived directly from the matrix U and the diagonal matrix D. As we previously describes, the topic model can be interpreted as matrix factorization where term by document probability distribution $p(t|d)$ can be split into two different distributions: the term by topic distribution $p(t|z)$ (that we call *TM* from now on), and the topic by document distribution $p(z|d)$. Thus, we can easily make a direct correspondence between the *DM* matrix obtained from LSI and the term by topic distribution *TM* obtained by using LDA. We define the mapping function that maps a document vector $\vec{t_j}$ in the standard VSM into the vector $\vec{t^T}_j$ in the *Topic Vector Space* by substituting the DM matrix with TM:

$$\phi^T(\vec{t_j}) = \vec{t_j}(I^{IDF}TM) = \vec{t^T}_j$$

where $\vec{t^T}_j$ is represented as a row vector, and I^{IDF} is a diagonal matrix to consider the document frequency of a term. Figure 2 shows the matrix representation of

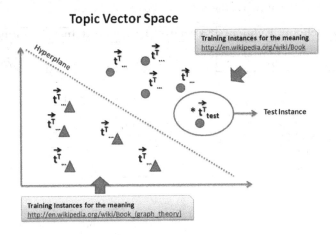

Fig. 3. The SVM classifier in the Topic Vector Space

the mapping function with the computation of the matrix products, where k is the number of terms, and h is the number of topics.

Both LDA and LSI permits to find a low dimensional representation for a set of documents w.r.t. the simple term by document matrix. This dimensionality (which represents for DM the number of domains and for TM the number of topics) in both cases has to be decided a priori. LDA has been demonstrated working well for a number of topics over 50 [3]. DM and TM can be built starting from arbitrary raw corpora or other knowledge resources. We built TM starting from Wikipedia, by extracting the abstracts of all the label meanings under consideration plus the abstracts of all their linked articles. Thus, the Topic Kernel function can be defined as:

$$K_{wsd}(t_i, t_j) = \langle \phi^T(\overrightarrow{t_i}), \phi^T(\overrightarrow{t_j}) \rangle$$

This function is used by the SVM binary classifier to learn the decision function for the classification (i.e., disambiguation) process. As shown in Figure 3, each training instance $\overrightarrow{t^T}_{...}$ is projected in the Topic Vector Space by using the mapping function $\phi^T(\overrightarrow{t_j})$. The *hyperplane* represents the linear decision function computed by using the Kernel method. It maximizes the distance between the training instances. Thus, SVM classifies each label by projecting its test instance in the Topic Vector Space. For example, the test instance $\overrightarrow{t^T}_{test}$ in Figure 3, for the label "book", will be classified (disambiguated) with the meaning http://en.wikipedia.org/wiki/Book.

4   Experimental Evaluation

We evaluated our method over the two relational schemas of the Amalgam integration benchmark for bibliographic data[8]. Our evaluation goals were: (1)

---

[8] www.cs.toronto.edu/~miller/amalgam

Table 1. Characteristics of test schemas

Number of	Schema Elements	Extracted Labels	Test Inst.	Training Inst.
Schema 1	117	149	149	12740
Schema 2	51	79	78	5782

Table 2. Lexical Annotation evaluation and comparison

WSD method	Precision	Recall	F-Measure
CWSD	0.70	0.69	0.70
ALA_{DK}	0.65	0.62	0.63
ALA_{TK}	0.68	0.65	0.67

investigating the use of Wikipedia as sense inventory and as resource of training data and (2) measuring the performance of our method based on ALA_{TK}.

Evaluating Wikipedia as Sense Inventory and Resource of Training Data
Table 1 summarizes the features of the test schemas[9]. Wikipedia was a very useful source of training data (see Table 1): for the Amalgam sources, we collected 18522 training instances with an average of 81 training instances for schema labels. With regards to coverage, Wikipedia contains all the schema labels present in WordNet. Moreover, it significantly helps in dealing with compound nouns and abbreviations: several labels such as "ISBN", "Technical Report", "Accession Number" that do not have an entry in WordNet have been found in Wikipedia. However, Wikipedia lacks several meanings for common labels: for instance, for the label "title" the WordNet meaning "the name of a work of art or literary composition" representing the correct meaning, it is not present in Wikipedia. The same holds for the label "series". Even if these cases affected a limited number of schema labels (12 labels occurrences) they represented an important limitation in the use of Wikipedia as sense inventory.

Evaluating ALA_{TK}
To evaluate the performance of our method, we used the Precision, Recall and F-Measure quality measures commonly used in the WSD area [16]. In particular, we compared the annotations returned by our method w.r.t. to a gold standard represented by manual annotations. To evaluate ALA_{TK}, we compared it with the lexical annotation results obtained by using the Domain Kernel function ALA_{DK} and the CWSD method [1], a WSD algorithm implemented in the MOMIS Data Integration System [7] and based on WordNet. The results of the annotation process are shown in Table 2: for the analyzed source our Topic Kernel function, even if slightly, outperforms the Domain Kernel function both in Precision and Recall.

---

[9] For compound nouns that do not exist in Wikipedia we extract their single term constituents: e.g., for "article_identifier" we extract "article" and "identifier". Thus, the number of extracted labels is higher than the one of schema elements.

Table 3. Evaluation of lexical annotation methods by excluding the schema labels without the correct meaning in Wikipedia/DBPedia

WSD method	Precision	Recall	F-Measure
CWSD (without labels with missing meanings)	0.70	0.69	0.70
ALA_{DK} (without labels with missing meanings)	0.72	0.71	0.71
ALA_{TK} (without labels with missing meanings)	0.74	0.73	0.74

As regards the comparison with CWSD, if we consider all the schema labels our method obtained worse performance. This result was caused by the presence of several labels that do not have a correct corresponding meaning in Wikipedia (e.g., "Title"). Other errors were caused by the fact that DBPedia is not complete: e.g., for the label "Manual" in Wikipedia there exists the meaning *Owner_manual*, however, it is not returned by the DBPedia property *wikiPageDisambiguates*. Other Wikipedia terms are completely missing: e.g., the term "Descriptor" exists in Wikipedia but not in DBPedia.

To evaluate, the proposed Topic Kernel function, independently from the gaps of Wikipedia and DBPedia, we excludes the labels that do not have the corresponding meaning in the reference sense inventory. As shown in Table 3, ALA_{TK} and ALA_{DK} improved significantly both in Precision and Recall, and thus overcome the CWSD performance (which are unchanged as WordNet contains all the correct meanings for the schema (expanded) labels).

5   Related Work and Conclusions

Recently, a number of systems that make use of topic models for sense disambiguation have been proposed. In [4], the authors proposed a topic model for unsupervised WSD, applied on textual source and based on WordNet. However, they conclude that their method is not comparable to the state-of-the art for errors due to the WordNet hyponymy structure. In [5], LDA is exploited in a supervised system based on traditional features such as part-of-speech of terms, local collocation etc. which cannot be used in structured data sources.

Several works have investigated the use of Wikipedia as sense inventory or as source of training data [13]. In [10], the authors concluded that there does not exist any other vocabulary for annotations providing such a large and broad coverage for concepts and instances as Wikipedia. In [14] a supervised WSD algorithm using Wikipedia as resource for training data, is proposed. It exploits features extracted from the syntax of text. However, these features cannot be exploited in the schema matching area where there is no syntactic structure.

In this paper, we presented ALA_{TK} a new automatic supervised method, based on the Topic Kernel function for automatically annotating schema labels w.r.t. Wikipedia. We performed a preliminary evaluation on relational data sources which shows promising results: the Topic Kernel function outperforms both the Domain Kernel function and the unsupervised CWSD method. The use of Wikipedia permits to improve the coverage of labels especially for compound nouns and abbreviations, but it lacks of several meanings for common labels w.r.t. WordNet. On the basis of our experiments, we concluded that both Word-Net and Wikipedia have some limitations as sense inventories, and that they should be exploited together, for instance, by the integrated ontology YAGO [11].

References

1. Bergamaschi, S., Po, L., Sorrentino, S.: Automatic Annotation in Data Integration Systems. In: Meersman, R., Tari, Z. (eds.) OTM-WS 2007, Part I. LNCS, vol. 4805, pp. 27–28. Springer, Heidelberg (2007)
2. Bizer, C., Lehmann, J., Kobilarov, G., Auer, S., Becker, C., Cyganiak, R., Hellmann, S.: Dbpedia - a crystallization point for the web of data. J. Web Sem. 7(3), 154–165 (2009)
3. Blei, D.M., Ng, A.Y., Jordan, M.I.: Latent dirichlet allocation. Journal of Machine Learning Research 3, 993–1022 (2003)
4. Boyd-Graber, J.L., Blei, D.M., Zhu, X.: A topic model for word sense disambiguation. In: EMNLP-CoNLL, pp. 1024–1033 (2007)
5. Cai, J., Lee, W.S., Teh, Y.W.: Improving word sense disambiguation using topic features. In: EMNLP-CoNLL, pp. 1015–1023 (2007)
6. Cristianini, N., Shawe-Taylor, J.: An Introduction to Support Vector Machines and Other Kernel-based Learning Methods. Cambridge University Press (2010)
7. Beneventano, D., Bergamaschi, S., Guerra, F., Vincini, M.: Synthesizing an integrated ontology. IEEE Internet Computing 7(5), 42–51 (2003)
8. Gliozzo, A.M., Giuliano, C., Strapparava, C.: Domain kernels for word sense disambiguation. In: The Association for Computer Linguistics, ACL (2005)
9. Griffiths, T., Steyvers, M., Tenenbaum, J.: Topics in semantic representation. Psychological Review 114(2), 211–244 (2007)
10. Hepp, M., Siorpaes, K., Bachlechner, D.: Harvesting wiki consensus: Using wikipedia entries as vocabulary for knowledge management. IEEE Internet Computing 11(5), 54–65 (2007)
11. Hoffart, J., Suchanek, F.M., Berberich, K., Lewis-Kelham, E., de Melo, G., Weikum, G.: Yago2: exploring and querying world knowledge in time, space, context, and many languages. In: WWW (Companion Volume), pp. 229–232 (2011)
12. Landauer, T.K., Foltz, P.W., Laham, D.: An introduction to latent semantic analysis. Discourse Processes 25(2), 259–284 (1998)
13. Medelyan, O., Milne, D., Legg, C., Witten, I.H.: Mining meaning from wikipedia. Int. J. Hum.-Comput. Stud. 67, 716–754 (2009)
14. Mihalcea, R.: Using wikipedia for automatic word sense disambiguation. In: HLT-NAACL, pp. 196–203 (2007)
15. Miller, A.: Wordnet: A lexical database for english. Communications of the ACM 38(11), 39–41 (1995)
16. Navigli, R.: Word sense disambiguation: A survey. ACM Comput. Surv. 41(2) (2009)
17. Sorrentino, S., Bergamaschi, S., Gawinecki, M., Po, L.: Schema label normalization for improving schema matching. Data Knowl. Eng. 69(12), 1254–1273 (2010)
18. Wong, S.K.M., Ziarko, W., Wong, P.C.N.: Generalized vector space model in information retrieval. In: SIGIR, pp. 18–25 (1985)

Optimising Schema Evolution Operation Sequences in Object Databases for Data Evolution

Tilmann Zäschke, Stefania Leone, and Moira C. Norrie

Institute for Information Systems, ETH Zurich
CH-8092 Zurich, Switzerland
{zaeschke,leone,norrie}@inf.ethz.ch

Abstract. We propose an approach to optimising schema evolution operation sequences in object databases. The approach separates operations that add structures from those that remove structures so that all additions are performed before any removals. This separation ensures that there is always a state during schema evolution where data can be evolved from structures that are to be deleted to structures that are new or already exist. Our approach also reduces and groups the schema operations to simplify implementation of data evolution functions by developers. We present a case study used as a first evaluation of the approach.

Keywords: object database, schema evolution, data evolution.

1 Introduction

Agile development techniques [8] are used increasingly in software development projects, as indicated by a recent study in Ireland [2]. For many types of projects, agile development has an excellent track record in aspects such as delivery on time and customer satisfaction. However, for projects involving the development of information systems, the agile practice of frequent software releases means that software users and developers face frequent database evolution tasks. In every release cycle, the developer realises new requirements and improves and refactors the existing application data model, based for example on user feedback or performance measurements. Since development is always in progress, the resulting changes per cycle are also numerous. For every release, the databases of existing information system installations must be evolved to comply with the new model. Given the frequency and amount of changes, the effort for individual releases should be minimised, both in terms of impact of evolution workload on running systems and development effort.

Such database evolution consists of both *schema* and *data* evolution, where schema evolution refers to changes to class definitions and data evolution involves initialising and updating corresponding data in the databases, which often has a semantic aspect. Our experience with the Herschel project of the European Space Agency [7] showed that the complexity of data evolution usually exceeds schema

P. Atzeni, D. Cheung, and R. Sudha (Eds.): ER 2012, LNCS 7532, pp. 369–382, 2012.

evolution and that tools with support for data evolution are still scarce [17]. Here, we focus on object databases (ODBMS) because the case has been made that they are well-suited to agile development since, unlike relational databases with object-relational-mapping layers, they usually do not have a mapping layer that requires refactoring [1].

In this paper, we introduce a new and innovative solution for optimising sequences of schema modification operations (SMOs) [6]. This solution is specifically targeted at minimising development effort for data modification operations (DMOs) and therefore suitable for agile information system development. Crucially, the approach separates operations that add structures from those that remove structures and performs all structure additions before any removals. This separation ensures that there is always a state during schema evolution that allows data evolution, in particular from structures that are to be deleted to structures that are to be created.

To validate our approach, we use a model-driven development environment that facilitates frequent evolution by regenerating database-related application code rather than requiring manual refactoring after each change to the application model. The application models that are used for application code generation can also be used to generate database evolution code that evolves databases from one model version to another. The generated code contains all required SMOs as well as insertion points for manually written DMOs. The latter have to be written manually because semantic aspects of data evolution, for example changing the physical unit of a measured value, cannot be derived from standard models.

The paper is organised as follows. First, we will present the background to this work in Sect. 2 before describing our approach in Sect. 3. The algorithm is described in detail in Sects. 4 and 5 with a description of the implementation in Sect. 6. A case study used as a first evaluation of the approach is presented in Sect. 7. Concluding remarks are given in Sect. 8.

2   Background

While schema evolution is frequently discussed in the research literature [15], we are aware of only [4] who discuss the optimisation of SMO sequences in ODBMS. The authors propose CHOP, an iterative approach to optimising SMO sequences and prove that any resulting sequence is correct and minimal. CHOP is integrated into the associated SERF framework [16], which builds on the query language OQL/ODMG [13] to support developers in schema and object migration. The framework provides an editor to construct database evolution code from evolution primitives and templates [3].

However, as described in [5], CHOP only considers dependencies of affected schema elements, but not dependencies of associated data evolution code. Thus, CHOP may invalidate data evolution code during optimisation. While the authors marked this as work in progress, we could not find any subsequent associated publication. Besides, when using model-driven development, it would be advantageous if database evolution code could, at least partly, be generated from application models, rather than having to be entered manually.

The DB-MAIN [11] framework is mainly aimed at relational databases, but claims to support other models as well [10]. They process *histories* (sequences) of manually entered schema change operations on models and databases. In [9], they sketch how such *histories* could be minimised by cancelling out operations with the inverses, but this approach has not been detailed since. Data evolution functions can be generated, but need manual intervention for conflict resolution and do not consider semantic changes.

Another framework that generates SMO sequences is called ESCHER [14]. It follows a state-based versioning approach [12], where changes to the source code of an application are analysed to extract SMOs. This pure state-based approach does not allow complete generation of class evolution operations, because it suffers from the classic ambiguity of create/delete operations. However, the class invariants available in their programming language allow certain semantic changes to be detected and corresponding conversion code to be generated.

None of the research that we found discusses the optimisation of SMO sequences for the insertion of DMOs. What we want to achieve is to streamline the process for creating and implementing database evolution sequences. The SMO sequence should avoid redundant operations and be optimised for simple insertion of manually created DMOs. This means that insertion points should be clearly marked and that developers should not need to reorder other SMOs for all required schema structures to be available.

3   Approach

Our approach is based on the assumption that a developer uses a modelling environment, such as a graphical UML editor, and evolves an application model from one version to the next using a number of schema evolution operations. The output of such an evolution cycle, is a new application model version and database evolution code, which is used to evolve running applications accordingly.

Such evolution cycles are of exploratory nature and characterised by long sequences of SMOs, where a user may, for example, explore different ways of extending the application model to cater for new requirements. Consequently, the number of evolution operations may be high, and some of the evolution operations may be cancelled out by others or may be redundant with respect to the new version. To optimally support data evolution, we therefore seek a pragmatic way of avoiding redundant operations.

In the top part of Fig. 1, we illustrate a simplified version of an evolution cycle. In this example, a user evolves the first version of a `Publication` class into a second version. First, the user creates a new class `Title`. The `Publication` class is then extended with an additional attribute `title` of type `Title`. The user then changes the `title` attribute type from `Title` to `String` and removes the previously created class `Title`. Finally, the attribute `text` is removed from the `Publication` class and a new attribute `body` added. From this rather short sequence of user operations, it is obvious that some operations, such as the creation and deletion of the `Title` class, cancel each other or are redundant with respect to the evolution of the class from version one to version two.

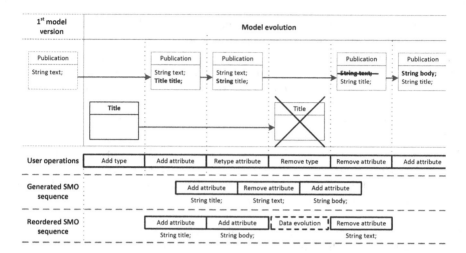

Fig. 1. Evolution example with user operations and optimized SMO sequence

In contrast to previous work, where SMO sequences were directly optimised [4], we adopt a state-based approach for versioning, where we store the model versions instead of evolution operations. In order to prevent create/delete ambiguities, all model elements have a unique identity that does not change between model versions. After the new version of the schema has been created and the corresponding database is to be evolved, an SMO sequence is generated by comparing the two model versions. This constructive approach to the generation of SMO sequences always yields a complete SMO sequence, and, in contrast to operation-based versioning approaches, avoids the detection and elimination of redundant operations.

However, such an SMO sequence may not be optimal with respect to support for data evolution. This is illustrated by the generated SMO sequence in Fig. 1. While this sequence is correct for schema evolution, it is not suitable for data evolution: Let us assume that, for each instance of `Publication` in the database, the initialisation values of the new attributes `title` and `body` attributes should be derived from the `text` attribute. The generated SMO sequence, however, deletes the `text` attribute before the `body` attribute is created and can be initialised.

Therefore, our approach reorders generated SMOs so that classes and attributes are only removed after all new ones have been created and initialised. This separation of create and delete operations allows developers to insert any required DMOs, such as initialisation functions to create instances of a new class or to initialise new attributes of existing classes, between create and delete operations. Given that no elements have been removed at this point, they may all serve as potential input for the said initialisation functions. Also, we group create operations for further optimisation: The creation of new classes and their attributes are grouped, as well as the creation of multiple attributes for an existing class. This procedure facilitates data evolution in that the initialisation of new class instances and their attributes can be handled together. The resulting reordered and optimised SMO sequence is shown at the bottom of Fig. 1.

Database evolution code is generated from this sequence of SMOs that has been enriched with placeholders for DMOs. In the generated code, the placeholders mark insertion points where the developer may then implement DMO operations. The code can then be executed to evolve databases from the previous to the new model version.

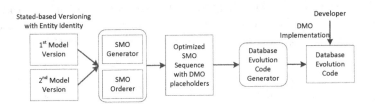

Fig. 2. Overview of the approach

Figure 2 provides an overview of our approach. On the left are versions 1 and 2 of an application model. The model versions are stored using a state-based versioning model, where all model elements have unique IDs. To the right of the model versions, the diagram shows the *SMO generator*, which compares the two models and generates the SMOs that evolve one model version into the other. During generation, the set of operations is ordered by the *SMO orderer* in a way that all new schema elements are first created, then placeholders for data evolution operations inserted, thereby yielding a database evolution sequence that is optimised with respect to schema and data evolution. The SMO & DMO placeholder sequence is then passed to the *database evolution code generator* that generates the database evolution program code. The order of SMOs and DMO placeholders in the optimised sequence is directly reflected in the order of the statements in the generated code.

4   SMO Generator

We now provide more details of the SMO generator which generates the set of SMOs needed to evolve one model version into the next.

The left part of Figure 3 illustrates the SMO generation process. The SMO generator first inspects the two model versions, compares the schema elements

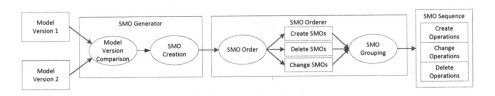

Fig. 3. Detailed SMO sequence generation process

between the two versions and generates a set of necessary SMOs. As already mentioned, the introduction of unique IDs for all model elements allows us to adopt a state-based model for the model versions, while preventing common create/delete as well as rename ambiguities. This concept is illustrated in Fig. 4. In the example on the left, a first model version contains an element *class A* and the second an element *class B*. The generator can compare the identities of the elements to tell that the element has simply been renamed. In the example on the right, an identity check could indicate that, although the name of a class is preserved, it is actually the case that *class X* should be deleted and recreated.

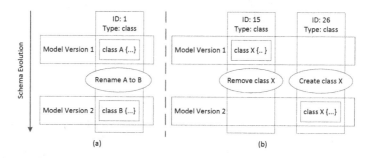

Fig. 4. Identity in state-based model versioning

The SMO generator generates SMOs of different types. In general, we distinguish between SMOs that create a new schema element, SMOs that delete a schema element and those that change a schema element. Table 1 gives an overview of SMOs for an object model as defined by a model such as UML. We have mapped more complex schema changes to these basic schema evolution primitives in order to also comply with ODBMS that only support simple schema evolution operations. Retyping an attribute, for example from `float` to `String` is mapped to *remove-* and *add-attribute* operations. Changes in the class-hierarchy result in the affected classes and their sub-classes being replaced by a set of new classes with the updated hierarchy. Finally, we treat UML associations as attributes. Changing the source or target of an association is therefore mapped to creating and removing attributes from the respective classes. Changing the

Table 1. Schema evolution operation primitives

Operation	Short	Description
add-class(c, d)	CA	Add new class c with super class $\{d \mid \oslash\}$.
delete-class(c)	CD	Delete class c if $subclasses(c) = \oslash$.
rename-class(c, d)	CR	Rename class c to d.
add-attribute(a, c)	AA	Add new attribute a to class c.
delete-attribute(a, c)	AD	Delete attribute from class c.
rename-attribute(a, b, c)	AR	Rename attribute a in class c to b.

multiplicity constraints of an association may also be mapped to create and remove attribute operations. For example, if the multiplicity constraints change from (0,1) to (0,n), the attribute of the target class needs to be changed from a reference to a set.

5   Sequence Orderer

While the SMO generator is responsible for generating a set of required SMOs, the SMO orderer reorders them to optimally support both schema and data evolution. The ordering process combines *SMO ordering* and *SMO grouping* so that a consistent, optimised SMO sequence is created.

As shown on the right of Fig. 3, the set of generated SMOs are ordered into three distinct sets, i.e. *create SMOs*, *delete SMOs* and *change SMOs*. Note that, while we have so far only addressed the ordering of create and delete operations, we also create a set of change operations. Given that most change operations are broken down into create and delete operations, this set only contains renaming operations, which are always performed before data evolution, and data evolution placeholders, for example for evolving multiplicity changes of associations. Rename operations are sometimes interleaved with creation operations, as discussed below, since a renaming may be required before a new element can be added. By having all renamings before data evolution, a developer can rely on the element names available in the new model version during data evolution.

The SMO orderer makes use of an element-existence tracker to prevent so-called *existential conflicts*. This type of conflict occurs when the creation or deletion of one schema element depends on the prior creation or deletion of another. For example, when a class B with a superclass A is created, many ODBMS require that A is created before B, meaning that the SMOs, from which the code is generated, must be ordered accordingly. The element-existence tracker serves as a virtual model of the database and simulates schema evolution, showing which elements would currently be available and which not. During the ordering process, each generated SMO is tracked by the element-existence tracker: for each *create* SMO an element is added to the tracker and for each *delete* SMO an element is removed from the tracker. The tracker informs the SMO orderer about existential conflicts. For example, if an SMO for creating a subclass has not been preceded by an SMO for creating its superclass, the orderer will be informed and reorders the two SMOs accordingly.

Similar to the elements-existence tracker, the SMO orderer makes use of a name tracker to detect and resolve naming conflicts. The example in Fig. 4(b) illustrates such a naming conflict due to class X being removed and a new class X created. To allow for data evolution, the general policy is to move *create* operations before any *remove* operations. Therefore, *create class* X (for ID 26) is moved forward in the sequence to occur before *remove class* X, allowing data to be migrated from the old X to the new X. Obviously, this causes a name clash, which is detected by the name tracker. The SMO orderer then inserts an additional *renaming* SMO, which renames the old X to, for example, X_to_be_removed, before creating the new

X. In parallel, a *delete* SMO is created to remove the renamed class X_to_be_removed. The name-tracker also recognises *circular naming conflicts*. Assume that two or more elements swap their names. Normally, if a conflict with an element is detected which is not to be deleted, the SMO orderer always renames the element first to its correct final name. However, in the case of name swaps, such renamings create conflicts due to the circular dependencies between the elements involved in the name swap. If such a circle is detected, the SMO orderer introduces a temporary name for one of the elements, then renames all other elements in the cycle, and finally renames the element with the temporary name to its correct final name. For such circles, the SMO orderer introduces $n+1$ additional renaming SMOs for n elements in the circle.

In the next step, the SMOs are grouped to minimise data evolution effort. This means that SMOs for the creation of attributes declared in the same class are grouped with the *class creation* SMO if the class is new. This grouping simplifies data evolution to a single data evolution function for the initialisation of the class instance and all its attributes. If they were not grouped, an unwary developer might implement more than one initialisation function, causing all instances of the class to be loaded several times to initialise attributes separately.

However, grouping may lead to *existential cross-dependency* conflicts, where two schema elements depend on each other, as depicted in Fig. 5, where a class A defines an attribute b of class B, which has not yet been created. Grouping would cause class A to be created along with an undefined attribute b of class B. Creating class B first is not a solution, because its attribute depends on class A. Fortunately, most ODBMS are transactional, also for schema management. Therefore, creating attributes with unknown value type is typically not a problem, as long as all required classes are available when the transaction is committed.

Fig. 5. Cross-dependency between two classes A and B

The element-existence tracker is also used to support the *grouping* process. For example, for a *class creation* SMO, the SMO orderer first checks whether there exists any *attribute creation* SMOs for that class. If so, the SMOs for *attribute creation* are grouped with the *class creation* SMO.

6   Implementation

The SMO generator and orderer have been implemented as an integral part of AgileIS [18], an agile information system development environment that supports model-driven development through graphical model editors. Fig. 6 illustrates the relevant components of AgileIS. Users define application models using a graphical model editor, which, for each agile development cycle, stores a new version of the model in the model repository using a state-based versioning approach with element identity as discussed previously. Database evolution is triggered by the user through the model editor. Upon activation, the SMO generator accesses the two relevant model versions in the repository, compares them and generates a set of SMOs. The SMOs are then ordered and serve as input for the database evolution code generator which generates database evolution code based on the database evolution API. This API is a generic database evolution API that has been implemented for a number of database systems, including db4o and Versant, as well as SQLite.

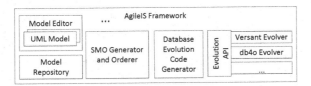

Fig. 6. AgileIS system architecture

The SMO generator and SMO orderer have been realised as a single class `SMOGenerator`, illustrated in Fig. 7, that executes both tasks in parallel. The method `generateSMOSequence` takes two model versions as input and performs the SMO sequence generation and ordering. It first generates two maps of model elements `v1` and `v2` representing the old and new model versions. From these two maps, SMOs are generated and stored in lists, one for every type of SMO. These lists are then ordered and returned as a optimised sequence of SMO instances.

The SMO sequence is then passed to the database evolution code generator, which generates a Java source file with a `main` method that can be executed to evolve the database to the new version.

The generated evolution code is independent of the underlying DBMS and uses a generic database evolution API. For each supported DBMS, we have implemented an *evolver* which maps the database evolution API to calls on the DBMS API. Figure 8 shows a simplified model of the evolution API implemented by database evolver implementations. The `Metadata` interface in the top left provides access to the metadata in a database. Two typical methods of it are `locateType`

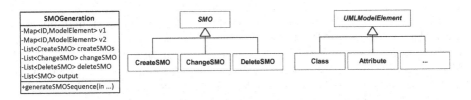

Fig. 7. UML of SMO sequence generation classes

and `defineType` to locate and define classes. They return `Type` instances that allow operations such as adding attributes and accessing all instances of that type. Instances of `Attribute` have methods to rename or remove them from their respective type. Instances of `Handle` provide handles on individual objects in the database. The associated methods allow, for example, individual attributes of an instance to be read or set. The methods on `Handle` are mainly called from within implementations of `AttributeInitializer`. The `AttributeInitializer` is a class that is typically implemented manually by the developer and defines database evolution code to define how a particular attribute has to be initialised.

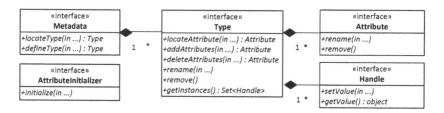

Fig. 8. Simplified model of the evolution API

The code example below shows generated Java code for creating a new class `Person` with two attributes. The /* TODO */, which will be automatically highlighted in most development environments, indicates to the developer that an initialisation function should be provided.

```
AttributeList attributes_Person = new AttributeList();
attributes_Person.addAttribute("firstName", "String");
attributes_Person.addAttribute("lastName", "String");
meta.defineType("Person", attributes_Person, null /* TODO */ );
```

Below is an example of a user-defined data evolution operation. The class `PersonInitializer` initialises the attribute `lastname` with the value `"Unknown"`.

```
class PersonInitializer implements AttributeInitializer {
    @Override public initialize(Iterator<Handle> handles) {
        while (handles.hasNext()) {
            handles.next().setValue("lastName", "Unknown");
        }
    }
}
```

The implementation of the database evolution API provides a mapping between the API and the API of the underlying database system. For most methods in the database evolution API, the mapping of operations to the DBMS APIs is straightforward, but there are of course differences between the APIs of these systems. In db4o, for example, some schema changes, such as adding attributes, are performed automatically, which means that the implementation of the `addAttributes` method only schedules the initialisation of new attributes.

```
public class Db4oTypeImpl implements Type {
    @Override public void addAttributes(AttributeList attributes,
                                        AttributeInitializer ai) {
        meta.addDataEvolver(ai, this);
    }
    ...
}
```

Note that while data evolution operations such as initialisations are defined alongside the element definition, the execution of database evolution code is bundled and only performed after schema elements have been created.

While some mappings, such as attribute retyping, are currently done in the SMO generator, we plan to move them completely into the evolvers to more efficiently use databases with native support for such features.

7 Case Study

We have validated our approach with a small experiment where a developer was asked to design an information system for managing scientific publications using AgileIS. The developer started with a very simple initial model, shown on the left of Fig. 9, which was evolved over the course of 14 model versions to a publication management system with rich means for organising and managing publications, shown on the right of Fig. 9. We simulated an agile development process, where, for every evolution cycle, the developer was provided with new user requirements.

For each evolution cycle, we generated optimised database evolution code and executed it on a running instance of our application. We tracked the number of edit operations, the optimised SMO sequence, and, for the sake of the experiment, we also generated the non-optimised SMO sequence that would have resulted from an operation-based versioning approach with no optimisation. The results are shown in Tbl. 2. The first column shows the version number. The first evolution starts with a new database instance and ends with the schema for the first model version. The second column shows the number of edit operations. In the third column, we give the number of operations in the SMOs that would be generated in an operation-based approach. The number of edit operations and the number in the generated SMOs may differ slightly due to the fact that UML specific operations are mapped to ODBMS compatible primitive operations, e.g. an attribute retyping operation is mapped to create and remove attribute operations. The fourth column shows the optimised SMO sequence generated by our implementation. Comparing the third and fourth columns shows the strength of the optimisation, without being diluted by operation mapping when comparing results with the editor operations in the second column. In the fifth column, the ratio of optimised vs. non-optimised number of operations indicates how well the SMOs could be optimised. Note that the high number of SMO operations compared to editor operations for version 14 results from a class hierarchy refactoring that is mapped to add/delete class operations.

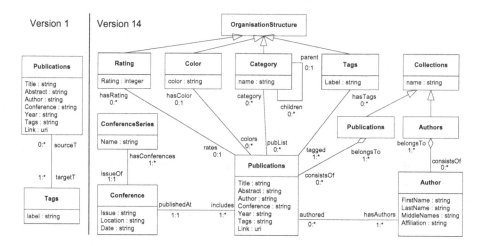

Fig. 9. Version 14 of the example model

Table 3 lists the optimisation ratios by operation type. The table lists only non-optimised and optimised SMO sequences, but not the editor operations. The results show that the developer chose to create classes rather carefully within an evolution cycle, since *add-class* and *delete-class* operations were hardly optimised. *Rename-class*, however, was performed more often within one evolution cycle, and the number of operations could be optimised. Most of the changes occurred on the attribute level, where the developer created, renamed and deleted attributes often within an evolution cycle, as shown by the ratio between optimised and non-optimised operations. *Pure data evolution*, which only occurs in the case of changes to an association's cardinality constraints, could always be grouped with schema evolution operations.

In Table. 4, we show results for the case where a running system is only evolved from time to time, i.e. not every version is released. The first three rows simulate what happens if only the versions 5, 10 and 14 were released. In this example, the database was directly evolved to version 5, then from 5 to 10 and from 10 to 14. The last row shows what happens if version 14 is the first one to be released. The ratios show that the optimiser is even more efficient when versions are combined.

In summary, the results show that the optimisation is efficient. It is not a surprise that such optimisation is possible, because developers using model editors usually modify elements multiple times in an exploratory manner before the version is accepted and released. The numbers also show that the potential for optimisation depends on the number of operations.

Table 2. Operations count per model version

V.#	Editor	Non-Opt.	Opt.	Ratio
1	30	32	2	0.06
2	12	13	3	0.23
3	13	14	4	0.29
4	6	6	6	1.00
5	9	10	2	0.20
6	12	13	7	0.54
7	9	11	6	0.55
8	20	22	2	0.09
9	11	10	3	0.30
10	8	14	7	0.50
11	28	35	11	0.31
12	18	22	2	0.09
13	9	11	6	0.55
14	9	23	15	0.65
Total	194	236	76	0.32

Table 3. Optimisation per operation type

Operation	Non-Opt.	Opt.	Ratio
add-class	20	18	0.90
delete-class	8	6	0.75
rename-class	21	8	0.38
add-attributes	74	13	0.18
delete-attributes	27	4	0.15
rename-attributes	70	27	0.39
pure data evolution	16	0	–
Total	236	76	0.32

Table 4. Multiple versions per release

V.#	Editor	Non-Opt.	Opt.	Ratio
0-5	70	75	6	0.08
5-10	60	70	12	0.17
10-14	64	91	18	0.20
0-14	194	236	12	0.05

8   Conclusion

We have presented an approach to optimising schema evolution sequences for data evolution which enables developers to implement data evolution without having to reorder schema evolution sequences manually. It also minimises and groups the evolution operations so that the resulting code is more readable and the execution time for data evolution is reduced.

We have shown how our pragmatic approach of using a state-based versioning approach with identity enables efficient generation of the set of required SMOs, which are optimised to support data evolution. While we have verified our approach with a small case study, the approach and implementation has been strongly influenced by our experience with the Herschel project of the European Space Agency [17,7]. This project went through 36 schema evolution cycles with about 250 persistent classes in the course of 6+ years. Database evolution was regularly required for over 100 customer databases. Schema and data evolution code was mostly implemented manually, and, if the project had used a model-driven approach, we believe that an approach such as the one we proposed could have considerably reduced the development efforts.

References

1. Ambler, S.W.: Agile Techniques for Object Databases (September 2005), http://www.db4o.com/about/productinformation/whitepapers/
2. Bustard, D.W., Wilkie, G., Greer, D.: Agile Software Development Diffusion: Insights from a Regional Survey. In: Proc. Intl. Conf. on Information Systems Development, ISD 2011 (2011)

3. Claypool, K.T., Jin, J., Rundensteiner, E.A.: SERF: Schema Evolution Through an Extensible, Re-Usable and Flexible Framework. In: Proc. 7th Intl. Conf. on Information and Knowledge Management, CIKM 1998 (1998)
4. Claypool, K.T., Natarajan, C., Rundensteiner, E.A.: Optimizing Performance of Schema Evolution Sequences. In: Proc. Intl. Symp. on Objects and Databases (2001)
5. Claypool, K.T., Rundensteiner, E.A.: Flexible Database Transformations: The SERF approach. IEEE Data Engineering Bulletin 22 (1999)
6. Curino, C.A., Moon, H.J., Zaniolo, C.: Graceful Database Schema Evolution: the PRISM Workbench. In: Proc. 38th Intl. Conf. on Very Large Databases, VLDB 2008 (2008)
7. European Space Agency. Herschel (2000), http://www.esa.int/herschel
8. Fowler, M., Highsmith, J.: The Agile Manifesto. Software Development 9(8) (2001)
9. Hainaut, J., Henrard, J., Hick, J., Roland, D., Englebert, V.: Database Design Recovery. In: Constantopoulos, P., Vassiliou, Y., Mylopoulos, J. (eds.) CAiSE 1996. LNCS, vol. 1080, pp. 272–300. Springer, Heidelberg (1996)
10. Hick, J.-M., Hainaut, J.-L.: Database Application Evolution: A Transformational Approach. Data and Knowledge Engineering 59(3) (2006)
11. LIBD Laboratory – University of Namur. DB-MAIN 9 – The Modeling Framework (2011)
12. Mens, T.: A State-Of-The-Art Survey on Software Merging. IEEE Transactions on Software Engineering 28(5), 449–462 (2002)
13. Object Data Management Group. ODMG 3.0 (2000), http://www.odmg.org
14. Piccioni, M., Oriol, M., Meyer, B., Schneider, T.: An IDE-based, integrated solution to Schema Evolution of Object-Oriented Software. In: Proc. Intl. Conf. on Automatic Software Engineering, ASE 2009 (2009)
15. Ram, S., Shankaranarayanan, G.: Research Issues in Database Schema Evolution: The Road Not Taken. Technical report, Boston University School of Management (2003)
16. Rundensteiner, E.A., Claypool, K., Li, M., Chen, L., Zhang, Z., Natarajan, C., Jin, J., De Lima, S., Weiner, S.: SERF: ODMG-based Generic Re-structuring Facility. In: Proc. Intl. Conf. on Management of Data, SIGMOD 1999 (1999)
17. Zäschke, T., Norrie, M.C.: Revisiting Schema Evolution in Object Databases in Support of Agile Development. In: Dearle, A., Zicari, R.V. (eds.) ICOODB 2010. LNCS, vol. 6348, pp. 10–24. Springer, Heidelberg (2010)
18. Zäschke, T., Zimmerli, C., Leone, S., Nguyen, M., Norrie, M.C.: Adaptive Model-Driven Information Systems Development for Object Databases. In: Proc. Intl. Conf. on Information Systems Development, ISD 2011 (2011)

Capturing Variability of Law with *Nómos* 2

Alberto Siena[1], Ivan Jureta[2], Silvia Ingolfo[1], Angelo Susi[3],
Anna Perini[3], and John Mylopoulos[1]

[1] University of Trento, via Sommarive 14, Trento, Italy
{a.siena,silvia.ingolfo,jm}@unitn.it
[2] University of Namur, 8, Rempart de la Vierge, 5000 Namur, Belgium
ivan.jureta@fundp.ac.be
[3] FBK-Irst, via Sommarive 18, Trento, Italy
{susi,perini}@fbk.eu

Abstract. Regulatory compliance is increasingly viewed as an essential element of requirements engineering. Laws, but also regulations and policies, frame their provisions through complex structures made of conditions, derogations, exceptions, which together generate a high number of alternative compliance solutions. This paper addresses the problem of modeling, exploring and selecting among alternatives in a variability space defined by laws. Our proposal includes a conceptual modeling framework for laws and reasoning techniques, called *Nómos* 2. The proposal is evaluated with a fragment of the Health Insurance Portability and Accountability Act (HIPAA).

Keywords: requirement engineering, variability, regulatory compliance.

1 Introduction

Socio-technical systems — complex systems consisting of software, human and organizational actors as well as business processes, running on an open network of hardware nodes — are increasingly gaining attention of governmental agencies for their tremendous potential to fulfill social functions but also the risks they introduce in case of mishap. Legislators are challenged to lay down comprehensive laws, abstract enough to avoid dependance on specific technological configurations, but able to foresee conceivable adverse conditions and prevent possible workarounds. The challenge for organizations is to understand and analyze the various ways their business goals can be achieved, while complying with applicable laws. The cost to ensure compliance is high: in the financial domain the U.S. Government Accountability Office estimates a $2.9 billion expense over five years[1] to implement the Wall Street Reform and Consumer Protection Act; in the Healthcare domain, organizations have spent $17.6 billion over a number of years[2] to align their systems and procedures with a single law, the Health Insurance Portability and Accountability Act (HIPAA).

---

[1] http://blogs.wsj.com/economics/2011/03/28/gao-implementing-dodd-frank-could-cost-2-9-billion/

[2] Medical privacy - national standards to protect the privacy of personal health information. Office for Civil Rights, US Department of Health and Human Services, 2000.

P. Atzeni, D. Cheung, and R. Sudha (Eds.): ER 2012, LNCS 7532, pp. 383–396, 2012.

The cost of ensuring compliance can be mitigated by understanding better variability in laws (including regulations and policies). For an organization, this means understanding that there are various ways to comply with the same law, and each of them can result in different costs. The purpose of this paper is to provide a language for modeling and analyzing variability in laws. A law consists of atomic elements (clauses) together with *conditions for their applicability*. A clause generally applies to a legal subject *if and only if the legal subject finds herself in a situation that satisfies the applicability conditions of the clause*. The legal subject *complies* with a clause if that clause applies to that legal subject and the subject satisfies that clause. The observation that clauses are conditional has an important consequence on the design of methods for assuring the compliance of requirements to laws. Namely, *by choosing the requirements that the system-to-be should satisfy, the requirements engineer chooses the conditions that the system-to-be will satisfy, and therefore also the clauses that the system-to-be should comply to*. Not all systems-to-be are therefore subject to the same clauses. And this is not only because they have different purposes, but also because the same purpose can be achieved in different ways. Each of these alternative sets of requirements that a system-to-be should satisfy will result in different properties and behavior of the system-to-be, and thus a different set of clauses which the system-to-be must comply to. Separating the identification, analysis and selection of requirements from the analysis of laws is therefore a mistake.

Choosing requirements should be done together with the identification and evaluation of the consequences of these choices on the *applicability* of clauses (i.e., whether the system-to-be should comply with the clause) and *satisfiability* of clauses (whether the system-to-be satisfies applicable clauses). In other words, the problem and solution space to explore during Requirements Engineering (RE) should also include representations of antecedents (i.e., preconditions) and consequents (postconditions) of clauses and analysis of requirements and laws should help evaluate both applicability and satisfiability of clauses. This is done by answering questions such as:

1. Given a law, containing a set of clauses articulated though conditions, exceptions and so on, which of them apply to a given requirements model?
2. Given a requirement, how to select a compliant way to achieve it, that is, a way which satisfies all applicable clauses?

This paper extends the *Nòmos* framework [1] for representing legal knowledge in requirements engineering, to (i) include antecedents and consequents of clauses in models of law, and (ii) use this information to evaluate the applicability and satisfiability of clauses for given sets of requirements, and thereby the compliance of these requirements. This new framework is called *Nòmos 2*.

The rest of the paper is organized as follows. We introduce the concepts and relations used in *Nòmos 2*, i.e., the ontology of the framework (§2). We define a modeling language used to represent the instances of the concepts and relations, i.e., for creating models of norms (§3). We define the algorithms for deciding about the applicability and satisfiability of norms (§4). We illustrate models using HIPAA norms (§5). We end with a discussion of related work (§6) and a summary of limitations and conclusions (§7).

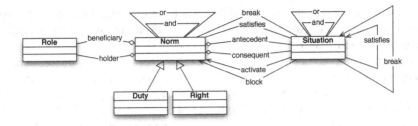

Fig. 1. A meta-model for the *Nòmos* 2 modeling language

2   Primitive Concepts and Relationships

The ontology of our proposed framework is founded on the assumption that *legal clauses are not requirements*. Stakeholders may or may not want compliance with a law, but they certainly want their requirements to be fulfilled by the system-to-be. Figure 1 presents the metamodel of our proposal in the form of a UML class diagram. The rest of this section defines the concepts and relations, providing a description of all the concepts presented in the meta-model.

Norm.   A *norm* is an atomic fragment of law with deontic status. Each norm consists of a 5-tuple (t, h, b, a, c), where:

- t is the *type* of the norm;
- h is the *holder* of the norm, the role having to satisfy the norm, if that norm applies;
- b is the *beneficiary*, the role whose interests are helped if the norm is satisfied;
- a is the *antecedent*, the conditions to satisfy for the norm to apply;
- c is the *consequent*, the conditions to satisfy for the norm to be satisfied.

A norm only takes effect under some condition. The condition, under which the norm takes effect, is its *antecedent*. If the norm takes effect, it is said to be **applicable**. For example, "Taxi drivers must exhibit their taxi driving license while driving the taxi" – In this norm, "driving the taxi" is the antecedent and, when it holds, the prescription of exhibiting the driving license also holds (i.e., it must be satisfied). If the norm has a antecedent, but the antecedent does not hold, then the norm is not applicable. In the taxi example, when not driving, there is no need to exhibit the license. If the norm has no antecedent, it is always applicable (wrt the holder role). A norm is **satisfied** under some conditions. The condition under which the norm is satisfied is its *consequent*. In the taxi driving example, "exhibiting the taxi driving license" is the norm's consequent. Satisfying a norm is different from complying with it. If the taxi driver exhibits the license while not actually driving the taxi, there is no compliance The norm's *holder* specifies the one(s) in charge of complying with the norms, whilst the *beneficiary* specifies the one(s) in the interests of whom the norm applies. A norm's *type* allows us to specialize the norm concept according to a legal ontology of choice.

We assume that at least two specializations of a norm are provided by the ontology: duty and right. A duty entails that if the antecedent holds, then the holder *must* bring about the consequent. A right states that if the antecedent holds, then the holder *is entitled to, but can choose not to* bring about the consequent of the right. A duty thus requires that the consequent necessarily holds if the antecedent does, while the right allows the consequent to hold when the antecedent does. Rights are used to exclude the need to comply with a duty.

The ultimate purpose of our proposed language is to evaluate *compliance*. In natural language, we say that duties are *complied with*, whereas rights can be *exercised*. In this paper, *we use the phrase comply with a norm to mean both "comply with a duty", if the norm is a duty, and "exercise a right", if the norm is a right*. A duty can be violated by its holder; a right can only be violated by those who are in charge of guaranteeing the right, but not by its holder.

Situation.   A situation is a state-of-affairs where some propositions are true, others false, and some neither true nor false. For example, when the situation "Christmas season" occurs, propositions such as "Falls within December" are true, while "Falls within summer" is false, and "Paolo likes Marta" is neither true nor false. Different norms are applicable in different situations, depending on whether the situation satisfies the norm's antecedent. A norm is satisfied in a given situation, if the situation satisfies the consequent of the norm. The concept of situation allows us to relate models of law to models of requirements independently from the specific language used for modeling requirements: consistent sets of requirements satisfy one or more situations, which make certain norms applicable. It is then unimportant if one sees requirements as use cases, goals, tasks, or otherwise, as long as one logically relates these requirements to situations.

Role.   Roles define the subjects of a norm. Instances of the role concept appear as the holder and beneficiary as parts of instances of the norm concept. Both holder and beneficiary can be a single role, a set of roles who together must comply with the norm, or a set of roles of which at least one must comply with the norm. In the present paper we focus on compliance problems addressing a single role, simplifying the reasoning to only situations and norm type.

Relations.   We define six types of relations between norms and situations:

- **Activate** from situations to norms: if situation is satisfied, then norm is applicable;
- **Block** from situations to norms: if situation is satisfied, norm does not apply;
- **Satisfy:**
  - From situations to situations: if the former is satisfied, the latter is too;
  - From situations to norms: if the former is satisfied, the latter is too;
  - From norms to situations: if the former is satisfied, the latter is too;
- **Break:**
  - From situations to situations: if the former is satisfied, the latter is not;
  - From situations to norms: if the former is satisfied, the latter is not;
  - From norms to situations: if the former is satisfied, the latter is not;
- **Conjunction** (And), between two or more norms to be satisfied together, or two or more situation instances to be satisfied together;
- **Disjunction** (Or), between two or more norms of which at least one ought to be satisfied, or two or more situation instances of which at least one ought to be satisfied.

$$S ::= p \mid S \wedge q \mid S \vee r \tag{1}$$

$$R ::= R \mid \wedge_{i=1..n, n>1} R_i \mid \vee_{i=1..n, n>1} R_i \tag{2}$$

$$N ::= (T, R_h, R_b, S_a, S_c) \tag{3}$$

$$\text{And-S} ::= S \mid \wedge_{i=1..n, n>1} S_i \tag{4}$$

$$\text{Or-S} ::= S \mid \vee_{i=1..n, n>1} S_i \tag{5}$$

$$\text{And-N} ::= N \mid \wedge_{i=1..n, n>1} N_i \tag{6}$$

$$\text{Or-N} ::= N \mid \vee_{i=1..n, n>1} N_i \tag{7}$$

$$\text{X-S} ::= \text{And-S} \mid \text{Or-S} \tag{8}$$

$$\text{X-N} ::= \text{And-N} \mid \text{Or-N} \tag{9}$$

$$\phi ::= S \mid N \mid \text{X-S} \mid \text{X-N} \tag{10}$$

$$Rel ::= \text{X-S} \xrightarrow{\text{activate}} N \mid \text{X-S} \xrightarrow{\text{block}} N \mid \text{X-S} \xrightarrow{\text{satisfy}} S \mid \text{X-S} \xrightarrow{\text{satisfy}} N$$
$$\mid \text{X-N} \xrightarrow{\text{satisfy}} S \mid \text{X-S} \xrightarrow{\text{break}} S \mid \text{X-S} \xrightarrow{\text{break}} N \mid \text{X-N} \xrightarrow{\text{break}} S \tag{11}$$

Fig. 2. BNF rules defining the grammar of the language

The applicability and satisfaction of a norm, as well as the satisfaction of a situation, are decided based on the relations that they participate in. The rules for deciding applicability and satisfaction are defined in the next section. Applicability and satisfaction values let us check if we need to comply with a particular norm, if we do comply with it, and in which situations we comply or fail to do so.

3   Syntax and Semantics

We describe below the modeling language through its syntax – symbols and grammar – and semantics – semantic domain and semantic mapping. The modeling language has a symbolic and a corresponding graphical syntax.

Symbolic Syntax. The concepts from the ontology give sorts for the language. For simplicity (to avoid decorating symbols with sort symbols), we use N for instances of norm, T for instances of a concept specializing norm (e.g., for duty instances, we can write T_D), S for instances of situation, and R for instances of role. Expressions in the language, the well-formulas, are denoted by Greek lowercase letters. Symbols for relations are $\wedge$ for Conjunction, $\vee$ for Disjunction, $\xrightarrow{\text{activate}}$ for Activate, $\xrightarrow{\text{block}}$ for Block, $\xrightarrow{\text{satisfy}}$ for Satisfy, and $\xrightarrow{\text{break}}$ for Break. Expressions of the language are generated via the BNF rules in Figure 2.

Semantics. The semantic domain includes two triples of values:

- **Applicability** values: AT reads *Applicable*, AF reads *Not applicable*, and AU reads *Applicability undefined*; applicability values *are assigned to norms;*
- **Satisfiability** values: ST reads *Satisfied*, SF reads *Not satisfied*, and SU reads *Satisfiability undefined*; satisfiability values *are assigned to situations and to norms.*

Conjunction and Disjunction obtain Applicability and Satisfiability value as shown in Table 1. The Applicability and Satisfiability values are interesting in the following sense.

Given a set of expressions in the language of the present framework, we want to make assumptions about the value of some of these formulas, and compute from there the value of others. For example, we want some situations to be ST and others SF, and we want to know what Applicability and Satisfiability values this propagates across other formulas.

Activate, Block, Satisfy, and Break relations do not have a value. We see them as propagating a value to their target. The value they propagate depends on whether the origin of the relation is a norm or situation, and on the relation type. Table 2 defines how Applicability and Satisfiability values propagate values over the Activate, Block, Satisfy, and Break relations. If there are several Activate, Block, Satisfy, Break relations targeting the same situation or norm, the values that they propagate to that target are treated as being in disjunction, and the following rules apply; let ψ be the target:

- If ψ receives both ST and SF, then value of ψ is ST;
- If ψ receives both ST and SU, then value of ψ is ST;
- If ψ is receives both SF and SU, then value of ψ is SU;
- If ψ receives both AT and AF, then value of ψ is AT;
- If ψ receives both AT and AU, then value of ψ is AT;
- If ψ is receives both AF and AU, then value of ψ is AU.

Compliance Value. Applicability and Satisfiability values are relevant to the extent that they tell us whether requirements, through situations, satisfy applicable norms, that is, are compliant. We call this *Compliance value*, and compute it as shown in Table 3. The informal meaning for Compliance values is the following:

- **Compliance**: The norm instance applies and is satisfied;
- **Non-compliance**: The norm instance applies, but is not satisfied;
- **Tolerance**: The norm instance does not apply, making it unnecessary to satisfy it;
- **Inconclusiveness**: It is not known if the norm instance applies. This can intuitively be interpreted as there being a need to go back to the model to determine if the norm applies, and if yes, how to satisfy it in order to comply.

Shortcuts. What we call a *shortcut* is a template defining a frequently used combination of primitive relations. Shortcuts allow us to define relations which are useful for the modeling of norms, but are themselves not primitive. Rather, such relations are defined

Table 1. Satisfiability and Applicability tables for Conjunction and Disjunction

		Satisfiability values		Applicability values			
ϕ	ψ	$\phi \wedge \psi$	$\phi \vee \psi$	ϕ	ψ	$\phi \wedge \psi$	$\phi \vee \psi$
ST	ST	ST	ST	AT	AT	AT	AT
ST	SF	SF	ST	AT	AF	AF	AT
ST	SU	SU	ST	AT	AU	AU	AT
SF	ST	SF	ST	AF	AT	AF	AT
SF	SF	SF	SF	AF	AF	AF	AF
SF	SU	SF	SU	AF	AU	AF	AU
SU	ST	SU	ST	AU	AT	AU	AT
SU	SF	SF	SU	AU	AF	AF	AU
SU	SU	SU	SU	AU	AU	AU	AU

Table 2. Propagation of satisfiability values across Activate, Block, Satisfy, and Break relations. The table reads as follows: If a situation ϕ has value ST and there is an Activate relation from ϕ to ψ, then that relation assigns AT to ψ.

$\phi \xrightarrow{\text{activate}} \psi$		$\phi \xrightarrow{\text{satisfy}} \psi$		$\phi \xrightarrow{\text{block}} \psi$		$\phi \xrightarrow{\text{break}} \psi$	
ϕ	ψ	ϕ	ψ	ϕ	ψ	ϕ	ψ
ST	AT	ST	ST	ST	AF	ST	SF
SF	AU	SF	SU	SF	AU	SF	SU
SU	AU	SU	SU	SU	AU	SU	SU

from primitive relations. For example, the Derogate relation from a norm N_1 to another norm N_2 is intended to convey that complying with N_1 (i.e., having N_1 applicable and satisfied) makes it unnecessary to satisfy N_2, that is, makes N_2 non-applicable. Derogate is what we call a shortcut, because $N_1 \xrightarrow{\text{derogate}} N_2$ can be rewritten using primitive relations, as follows: (i) there is a situation S_a such that $S_a \xrightarrow{\text{activate}} N_1$, and (ii) there is a situation S_s such that $S_s \xrightarrow{\text{satisfy}} N_1$, and (iii) $S_a \wedge S_s \xrightarrow{\text{block}} N_2$. We use the following shortcuts (but others could be defined):

- **Imply**: $N_1 \xrightarrow{\text{imply}} N_2$ means there is S_a, S_b such that $S_a \xrightarrow{\text{activate}} N_1$, $S_b \xrightarrow{\text{satisfy}} N_1$ and $S_a \xrightarrow{\text{activate}} N_2$, $S_b \xrightarrow{\text{satisfy}} N_2$.
- **Derogate**: $N_1 \xrightarrow{\text{derogate}} N_2$ means there is S_a, S_b such that $S_a \xrightarrow{\text{activate}} N_1$, $S_b \xrightarrow{\text{satisfy}} N_1$ and $S_a \wedge S_b \xrightarrow{\text{block}} N_2$.
- **Endorse**: $N_1 \xrightarrow{\text{endorse}} N_2$ means there is S_a, S_b such that $S_a \xrightarrow{\text{activate}} N_1$, $S_b \xrightarrow{\text{satisfy}} N_1$ and $S_a \wedge S_b \xrightarrow{\text{activate}} N_2$.

Graphical Syntax.   Mapping from linear syntax to graphical syntax is shown in Figure 3. The graphical syntax is illustrated with an excerpt of the (Italian) traffic rules. The figure can be read as follows: It is forbidden (norm N_1 of type duty, represented through the symbol △) to stop the car on a motorway (situation S_1, represented through the symbol ▢). In case of a car failure (S_3) it is permitted (norm N_2 of type right, represented through the symbol ▷) to stop the car on the motorway (S_1), provided that blinking lights are switched on (S_2); in this case, the first prohibition is not applicable. In case of heavy snow (S_5), it becomes mandatory (N_3) to stop the car (S_1) and install snow chain on the car (S_4). Notice that shortcut relations (e.g., $N_2 \xrightarrow{\text{derogate}} N_1$) connect two norms but have the semantics described above. Norms, situations and the relations among them

Table 3. Compliance values of norms

Norm is a duty			Norm is a right		
Applicability	Satisfiability	Compliance value	Applicability	Satisfiability	Compliance value
AT	ST	Compliance	AT	ST	Compliance
AT	SF	Non-compliance	AT	SF	Tolerance
AT	SU	Non-compliance	AT	SU	Tolerance
AF	ST	Tolerance	AF	ST	Tolerance
AF	SF	Tolerance	AF	SF	Tolerance
AF	SU	Tolerance	AF	SU	Tolerance
AU	ST	Inconclusiveness	AU	ST	Inconclusiveness
AU	SF	Inconclusiveness	AU	SF	Inconclusiveness
AU	SU	Inconclusiveness	AU	SU	Inconclusiveness

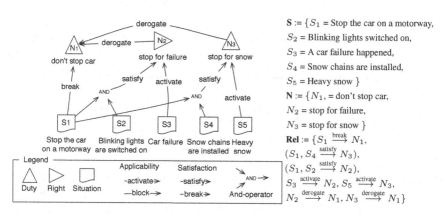

Fig. 3. Graphical syntax for the modeling language

define the three sets N, S and Rel that form the norm model L. In the example, all the norms in L are held by a single role (say, "Driver"); since in the present paper we only reason on one-role models, the holder (and beneficiary) are not depicted here.

Compliance to a Law. On a norm model — intended as a set of norms, situations and the relations between them — is defined Law compliance as a global property. A law as a whole is complied with if (i) each norm is complied with; or (ii) one or more norms are not complied with, but they are derogated by others, and the derogating ones are complied with; or (iii) one or more norms are not applicable; or, finally, (iv) one or more norms are rights, applicable but not exercised by their holder. A set of situations make a norm model compliant, if $\forall N \in L, tol(N) \vee com(N)$, where $tol(N)$ means that a *tolerance* condition (as of Table 1) is inferred on the norm N, $com(N)$ means that a *compliance* condition is inferred for that norm. In a model with multiple roles — in which one's duty is to ensure another's right — compliance with a right can also be interpreted in the following more constraining way: a right is complied with if and only if that right (i) is applicable, (ii) is satisfied, and (iii) there are no situations which are satisfied, and which block or break that right. We can verify if a right is satisfied in this sense if its compliance value is Compliant, and there are no block or break relations to that right, originating in a satisfied situation.

4   Reasoning with Norm Models

Norm models can be analyzed by means of forward or backward reasoning algorithms.

Forward Reasoning involves giving applicability/satisfiability labels to some nodes of a *Nòmos* 2 network and then propagating these labels to all other nodes of the network, as described in [2]. Specifically, to each situation (and to each norm) we associate a variable $Sat(S)$ (and $Sat(N)$) with values ST, SF, SU, representing the current evidence of satisfiability of situation S (and norm N). To each norm is associated a variable $App(N)$ with values AT, AF, AU, representing the current evidence of applicability of norm N. Default values are SU and AU. Given an initial values assignment to situations (input situations from now on) forward reasoning focuses on the forward propagation of these initial values to all other situations and to the norms of the graph according to the rules described

Algorithm:ComplianceAnalysis

Input: L, initialAssignment
Output: complianceValues
1 assignment = ForwardReason(L,initialAssignment);
2 **foreach** $N_i \in L$ **do**
3   **if** *assignment.appValue(N_i) = AT* **then**
4   **if** *assignment.satValue(N_i) = ST* **then**
5   solution.set(N_i, compliance);
6   **else**
7   **if** N_i *instanceOf Duty* **then**
8   solution.set(N_i, non_compliance);
9   **else**
10   solution.set(N_i, tolerance);
11   **if** *assignment.appValue(N_i) = AU* **then**
12   solution.set(N_i, inconclusiveness);
13   **else**
14   solution.set(N_i, tolerance);
15 **end**
16 **return** solution;

Algorithm: ComplianceSearch

Input: L, initialAssignment
Output: selectedAssignment
1 assignment = ForwardReason(L,initialAssignment);
2 **while** *ComplianceAnalysis(L,assignment)* **contains**
   non_compliance **or** *inconclusiveness* **do**
3   attempt = BackwardReason(L,assignment);
4   **if** *attempt* = $\emptyset$ **then**
5   **return** $\emptyset$;
6   **else**
7   assignment = attempt
8   **end**
9 **end**
10 **return** assignment;

Algorithm 1. Compliance analysis. This algorithm evaluates the compliance status of a set of norms belonging to a law (L) with respect to an initial set of situations.

Algorithm 2. Compliance search. This algorithm evaluates the situations to satisfy in order to comply with the norms applicable from an initial set of situations.

in Table 1. So for example, if $Sat(S_1)$ = ST and $S_1 \xrightarrow{\text{activate}} N_1$, then the algorithm propagates $App(N_1)$ = AT. Initial values represent the evidence available about the satisfaction of the situations, namely evidence about the state of the situation. After the forward propagation of the initial values, the user can look the final values of the norms of interest. In other words, the user observes the effects of the initial situation over the norms.

Backward Reasoning involves giving applicability/satisfiability labels to some target nodes of a *Nòmos* 2 network and then searching back through the network for possible input values leading to some desired final value, as described [3]. We set the desired final values of a target norm, and we want to find possible initial assignments to the input situations, which would cause the desired final values of the target norms by forward propagation. For example, if $S_1 \xrightarrow{\text{activate}} N_1$, $S_2 \xrightarrow{\text{satisfy}} N_1$, $S_3 \xrightarrow{\text{satisfy}} N_1$, and in the initial assignment we specify that we want N_1 to be complied with (i.e., we want it to be $Sat(N_1)$ = ST and $App(N_1)$ = AT), then the algorithm may find the assignment $Sat(S_1)$ = ST, $Sat(S_2)$ = SF, $Sat(S_3)$ = ST. Backward and forward reasoning algorithms also take care of possible cycles in graphs.

Forward and backward reasoning algorithms can be applied to norm models to answer questions such as those raised in section 1.

Applicability Analysis is intended to find the (sub)set of norms applicable to a given (sub)set of situations. The analysis consists in a straightforward application of the forward reasoning algorithm. The output of the propagation is that subset L^a of the norms L such that $\forall N \in L^a, App(N)$ = AT.

Compliance Analysis aims at providing evidence of compliance (or violation) of a norm model. The algorithm for inferring compliance of a given norm model L (comprised by a set of norms, a set of situations, and a set of relations) and an assignment to their applicability and satisfiability values, is illustrated in Algorithm 1. Given an initial assignment to situations, compliance analysis propagates forward applicability and satisfiability values (line 1). When the propagation terminates, each norm of the model is evaluated (line 2). If applicability can not be inferred (AU), inconclusiveness is concluded (line 12); if applicability is inferred to be false, tolerance is concluded (line 14); if applicability is true, then satisfiability has to be checked (line 3). If it is also true (line 4), then we are surely in a compliance case; otherwise, we have to evaluate the type of the norm: if duty, we have a violation (a duty has to be satisfied but it's not the case); if right, tolerance is concluded, as right are discretional. When the algorithm terminates *solution* contains the compliance values of the norms.

Compliance Search aims at finding a compliance solution for a given set of situations, possibly identifying other situations that complement the given set. In other words, supposing that some situation S_k is desired to be legally satisfied (i.e., satisfied and compliant), this analysis answers the question: what else must we do if we want the satisfaction S_k to be compliant?

Searching for a compliant way to satisfy a desired situation is done through an iterative algorithm, as listed in Algorithm 2. Initially, input satisfaction values of situations are propagated to norms applicability and satisfiability values (line 3). The resulting assignment is evaluated for compliance (line 2). If any non-compliance is found, an iteration is performed, until either an assignment is found, which does not include violations, or no further assignment can be made. For each iteration, the backward reasoning algorithm searches for an assignment (*attempt*, line 3) that makes the current one satisfied. If such assignment does not exist, the algorithm exits. Otherwise, the current one is checked for compliance and the algorithm exits with success or iterates again, according to whether compliance is found, or not.

5   Illustrative Case Study

The U.S. Health Insurance Portability and Accountability Act (HIPAA) is a well-known law to researchers and practitioners. Although years already passed since it has been laid down, the effort for engineering HIPAA-compliant requirements is still expensive because of its complexity. For example, section §164.502[3] of HIPAA lays down general rules of uses and disclosures of protected health information for covered entities. The section contains 174 paragraphs, all of them directly or indirectly concerning one single action, abstractly denoted as "use or disclosure of protected health information", which under some circumstances is forbidden, under other circumstance is permitted or even required. The paragraphs regulate such circumstances.

Figure 4 depicts a model extracted from HIPAA section §164.502. The model contains 10 situations (S_1, ..., S_{10}; situation S_{11}, S_{12}, S_{13} and S_{14} have been extracted from cross-referenced fragments and depicted for explanation purposes, but not counted here) and 12 norms, linked by 35 relations. Considering that a situation can be known to hold (ST), known to not hold (SF) or uncertain, due to insufficient knowledge, ambiguity or else (SU), we have a total of $3^{10} = 59.049$ possible assignments. On the other hand each norm can be not applicable, applicable but not complied with, or applicable and complied with. Overall, the situation generate therefore a total of $3^{12} = 531.441$

---

[3] http://www.hipaasurvivalguide.com/hipaa-regulations/164-502.php

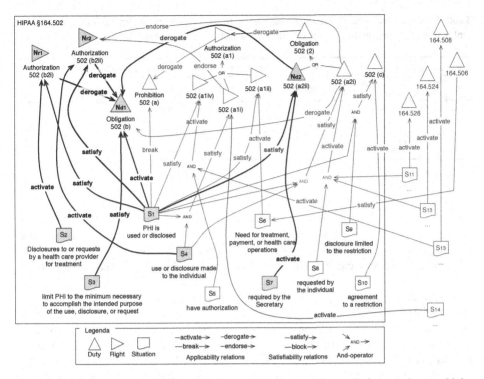

Fig. 4. HIPAA paragraphs 502 (1) to (2vi) law model. The situations, relations, and norms highlighted in the model are the one considered in the illustrative case study.

possibilities for norms. Out of these, we want to search for (i) the (sub)set of norms applicable to a given (sub)set of situations, and (ii) the set of situations that need to be satisfied in order to comply with the norms activated by an initial set of situation. To the first question we answer through *Applicability analysis*; to the latter question we answer through *Compliance analysis* and *Compliance search*.

We now show how these three analyses work in an example taken from the model depicted in Figure 4. In the following example we consider the sub-model composed of five situations S_1, S_2, S_3, S_4, S_7 (highlighted in grey in Figure 4) and the effects of these five situations on four norms $N_{r1}, N_{r2}, N_{d1}, N_{d2}$.[4] The relations considered in the example are highlighted in bold.

Applicability Analysis. We are interested in *finding the applicability values for the norms*. We start the analysis with the initial assignment of satisfaction values for the situations, $Sat(S_1, S_2, S_3, S_4, S_7) = $ (SF, SF, ST, ST, ST). The four relations in S_1 propagate values to all four norms (AU or SU respectively), the relation in S_2 assigns AU to N_{r1}, S_3 assigns ST to N_{d1}, S_4 propagates AT to N_{r2}, while S_7 sets N_{d2} to AT. The result of this forward analysis is summarized in Table 4 (second column). Analyzing the shortcut-relation $N_{r1} \xrightarrow{\text{derogate}} N_{d1}$, we have that it was as if in the model there was a relation

---

[4] The subscript r indicates a norm of type right (N_r), while d is for duties (N_d).

Table 4. Values returned by the three analyses with the given input assignment (left part of the table) and after one iteration of the compliance search (right part of the table)

InitialAssignment: $Sat(S_1,S_2,S_3,S_4,S_7)$=(SF, SF, ST, ST, ST)			$Sat(S_1,S_2,S_3,S_4,S_7)$=(ST, SF, ST, ST, ST)		
Norm	Forward reasoning	Compliance analysis	Compliance search	Forward reasoning	Compliance analysis	Compliance search
N_{r1}	AU, SU	inconclusiv.		AU,ST	inconclusiv.	
N_{r2}	AT, SU	tolerance	$Sat(S_1, S_2, S_3, S_4, S_7)$	AT,ST	compliance	$Sat(S_1, S_2, S_3, S_4, S_7)$
N_{d1}	AU, ST	inconclusiv.	= (ST, SF, ST, ST, ST)	AF,ST*	tolerance	= (ST, ST, ST, ST, ST)
N_{d2}	AT, SU	non-compliance		AT,ST	compliance	

* The relation $N_{r1} \xrightarrow{\text{derogate}} N_{d1}$ now holds as both S_1 and S_4 are now ST.

$S_1 \wedge S_2 \xrightarrow{\text{block}} N_{d1}$. Since the two situations are not satisfied, the block relation does not hold, and therefore the norm is not derogated. Similarly in the other derogation relation, the applicability of N_{d1} is not altered. The applicability analysis therefore returns that N_{r2} and N_{d2} are applicable, while the other two norms have undefined applicability.

Compliance Analysis. Given the labeling propagation identified in the previous step (line 1 of Algorithm 2), this analysis *evaluates all norms and returns their compliance status* (see Table 4, third column). Norm N_{r1} is evaluated as inconclusive because — as we can see also from Table 3 — a right with undefined applicability and satisfiability has status of inconclusiveness (line 12 of Algorithm 1). Norm N_{r2} is tolerated because an applicable right with undefined satisfiability (evaluated by line 10 of the algorithm), is assigned value "tolerance". Norm N_{d1} is evaluated as inconclusive because when a duty has undefined applicability but is satisfied, then a precise compliance value cannot be drawn. Finally, norm N_{d2} is identified in a status of non-compliance because it is an instance of a duty that is applicable and is not satisfied (line 8).

Compliance Search. We are now interested in *finding the situations* (and their satisfiability value) *needed to comply with the applicable norms* of our initial situation. Following Algorithm 2 (line 1) we perform forward reasoning analysis and Compliance Analysis, obtaining the results described in the left side of Table 4. As the Compliance Analysis of our example has both non-compliance and inconclusiveness instances (line 2), the algorithm performs backward reasoning (line 3). Tracing backward the relation $S_1 \xrightarrow{\text{satisfy}} N_{d2}$, the algorithm identifies that both issues can be solved by satisfying S_1, which is now set to ST. A new iteration is started (with $Sat(S_1)$=ST) and Compliance Analysis is performed again, revealing that now N_{r1} is evaluated as inconclusive (its label are now AU,ST) (see right side of Table 4). A new iteration of the algorithm finds that the final assignment $Sat(S_1, S_2, S_3, S_4, S_7)$ = (ST, ST, ST, ST, ST) leads to a compliant status where no case of inconclusiveness or non-compliance are present.

6   Related Work

In Tropos [4] requirements of the system-to-be are analyzed in terms of goal models, represented by means of goal graphs composed of goal nodes and goal relations. It is possible to analyze goal graphs with both forward reasoning and backward reasoning. In Tropos approach however, the notion of applicability is absent and so the inference rules to reason about compliance. In Kaos [5] requirements are modeled and analyzed through goals, requirements, scenarios, and responsibility assignments. Goal models are formalized using temporal logic to achieve correctness. Such formalism has several advantages, but they lack the concepts of norm and situation. Goal oriented techniques have been used to represent legal prescriptions. For example, Darimont and Lemoine have used Kaos to represent objectives extracted from regulation texts [6]. Ghanavati et al. [7] use URN (User Requirements Notation) to model goals and actions prescribed by laws. Likewise, Rifaut and Dubois use *i** to produce a goal model of the Basel II regulation [8]. Goal-oriented approaches to norm modeling are useful when the complexity of norms is little enough to be reduced to goal relations. When the notion of applicability is needed, goal-oriented techniques fail in capturing its effects on reasoning about alternatives.

AI techniques have been used to build formal models of regulations. Among recent works, [9] presents a formal framework for reasoning about derogation in law. In [10] conditional elements of norms are formalized into an operational framework. [11] provides a strong characterization for a formal elaboration of the notion of applicability. We depart from these approaches, in that our approach is less expressive but better suited for the purpose of supporting requirements engineering. Legal reasoning approaches are hardly usable by non-experts, so that they are rendered useless during the requirements engineering phase, which generally involves a continuous interaction with stakeholders to check that the work of the analysts has captured their needs and preferences.

Breaux & Antón focus on information extraction from natural language texts of privacy policies [12]. A set of heuristics is created to systematically convert unstructured legal texts into structured artifacts. Artifacts are then combined into a frame-based method for manually acquiring legal requirements from regulations. Such approach has been used as a basis for tool-supporting the identification of requirements in legal documents [13]. The Privacy APIs framework converts regulatory privacy rules in natural language into formal expressions, which are automatically analyzed to identify problematic statements [14]. The SALEM system adapts a linguistic approach for the extraction of semantic concepts from Italian legal documents, such as actors, actions and properties. It also classifies different types of provisions using a text categorization algorithm [15]. We depart from such works in that our analysis starts after the information extraction from legal text has been concluded.

7 Conclusions

In this paper we propose *Nòmos* 2, a framework for modeling preconditions and post-conditions of legal norms and their relationships into early representations of the requirements problem and solution space. Moreover, we describe how to traverse norm models to evaluate the applicability and satisfiability of norms for given sets of requirements, and thereby the compliance of these requirements. Automatic reasoning is accomplished by using extensions of existing forward and backward reasoning algorithms. The scope of the present work is limited to analyze situations that concern a single role. Further work is needed to reason on multiple roles and delegation of responsibilities.

Acknowledgements. This research has been partially funded by the European Research Council (ERC) through advanced grant 267856, titled "Lucretius: Foundations for Software Evolution" (04/2011-03/2016). http://www.lucretius.eu.

References

1. Siena, A., Mylopoulos, J., Perini, A., Susi, A.: Designing Law-Compliant Software Requirements. In: Laender, A.H.F., Castano, S., Dayal, U., Casati, F., de Oliveira, J.P.M. (eds.) ER 2009. LNCS, vol. 5829, pp. 472–486. Springer, Heidelberg (2009)
2. Giorgini, P., Mylopoulos, J., Nicchiarelli, E., Sebastiani, R.: Formal Reasoning Techniques for Goal Models. In: Spaccapietra, S., March, S., Aberer, K. (eds.) Journal on Data Semantics I. LNCS, vol. 2800, pp. 1–20. Springer, Heidelberg (2003)
3. Sebastiani, R., Giorgini, P., Mylopoulos, J.: Simple and Minimum-Cost Satisfiability for Goal Models. In: Persson, A., Stirna, J. (eds.) CAiSE 2004. LNCS, vol. 3084, pp. 20–35. Springer, Heidelberg (2004)
4. Bresciani, P., Perini, A., Giorgini, P., Giunchiglia, F., Mylopoulos, J.: Tropos: An agent-oriented software development methodology. Autonomous Agents and Multi-Agent Systems 8(3), 203–236 (2004)
5. van Lamsweerde, A.: Requirements Engineering: From System Goals to UML Models to Software Specifications. Wiley (2009)
6. Darimont, R., Lemoine, M.: Goal-oriented analysis of regulations. In: Laleau, R., Lemoine, M. (eds.) CAiSE 2006. CEUR Workshop Proceedings, CEUR-WS.org, vol. 241 (2006)
7. Ghanavati, S., Amyot, D., Peyton, L.: Towards a Framework for Tracking Legal Compliance in Healthcare. In: Krogstie, J., Opdahl, A.L., Sindre, G. (eds.) CAiSE 2007 and WES 2007. LNCS, vol. 4495, pp. 218–232. Springer, Heidelberg (2007)
8. Rifaut, A., Dubois, E.: Using goal-oriented requirements engineering for improving the quality of iso/iec 15504 based compliance assessment frameworks. In: Proceedings of RE 2008, pp. 33–42. IEEE Computer Society, Washington, DC (2008)
9. Dinesh, N., Joshi, A., Lee, I., Sokolsky, O.: Reasoning about Conditions and Exceptions to Laws in Regulatory Conformance Checking. In: van der Meyden, R., van der Torre, L. (eds.) DEON 2008. LNCS (LNAI), vol. 5076, pp. 110–124. Springer, Heidelberg (2008)
10. Sartor, G.: The structure of norm conditions and nonmonotonic reasoning in law. In: Proceedings of the 3rd International Conference on Artificial Intelligence and Law, ICAIL 1991, pp. 155–164. ACM, New York (1991)
11. Boella, G., Governatori, G., Rotolo, A., van der Torre, L.: *Lex Minus Dixit Quam Voluit, Lex Magis Dixit Quam Voluit*: A Formal Study on Legal Compliance and Interpretation. In: Casanovas, P., Pagallo, U., Sartor, G., Ajani, G. (eds.) AICOL-II/JURIX 2009. LNCS, vol. 6237, pp. 162–183. Springer, Heidelberg (2010)
12. Kiyavitskaya, N., Zeni, N., Breaux, T.D., Antón, A.I., Cordy, J.R., Mich, L., Mylopoulos, J.: Automating the Extraction of Rights and Obligations for Regulatory Compliance. In: Li, Q., Spaccapietra, S., Yu, E., Olivé, A. (eds.) ER 2008. LNCS, vol. 5231, pp. 154–168. Springer, Heidelberg (2008)
13. Breaux, T., Antón, A.: Analyzing regulatory rules for privacy and security requirements. IEEE Trans. Softw. Eng. 34, 5–20 (2008)
14. May, M.J., Gunter, C.A., Lee, I.: Privacy apis: Access control techniques to analyze and verify legal privacy policies. In: Proceedings of the 19th IEEE Workshop on Computer Security Foundations, pp. 85–97. IEEE Computer Society, Washington, DC (2006)
15. Biagioli, C., Francesconi, E., Passerini, A., Montemagni, S., Soria, C.: Automatic semantics extraction in law documents. In: Proceedings of the 10th International Conference on Artificial Intelligence and Law, ICAIL 2005, pp. 133–140. ACM, New York (2005)

Deriving Variability Patterns in Software Product Lines by Ontological Considerations

Mohsen Asadi[1], Dragan Gasevic[2], Yair Wand[3], and Marek Hatala[1]

[1] Simon Fraser University, Canada
[2] Athabasca University, Canada
[3] University of British Columbia, Canada
{masadi,mhatala}@sfu.ca, dgasevic@acm.org
Yair.wand@ubc.ca

Abstract. Variability modeling is widely used in software product line engineering to support reusability. Specifically, it is used in the derivation of concrete software products from a reusable solution within a family of products. To help manage variability, several modeling languages have been proposed for representing variability within a family of products. The study and evaluation of languages to model variability has so far focused on practical aspects of such languages. Less attention has been paid to more theoretical approaches to the analysis of variability modeling languages. In developing such approaches it would be of particular interest to explore the ability of variability modeling to represent the information about the real world (application) domain for which the product family is designed. In information systems research, evaluation of expressiveness of conceptual modeling languages has been done based on onto-logical theories. This paper describes a framework for general analysis of types of variability based on Bunge's ontology and derives a variability framework which is used to evaluate variability modeling languages.

Keywords: Variability modeling, ontological theory, software product lines.

1 Introduction

Software Product Line Engineering (SPLE) is an approach to develop a set of software systems which satisfy requirements of a specific domain and share common features [1]. The key factor for the success of SPLE is variability which can be categorized into [3]: *essential variability* (i.e., variability from the system usage pers-pective) and *technical variability* (i.e., variability related to the realiza-tion/implementation of essential variability and/or variability related to the IT-infrastructure)[2]. Essential variability is variability in domain knowledge and can be represented by *conceptual models*. Conceptual models are representations of static (e.g., things and their properties) and dynamic (e.g., process and events) phenomena in a domain of interest [7]. Variability languages should be able to represent both essential variability (manifested in the conceptual models) and technical variability (manifested in the implementation models). Several variability languages have

P. Atzeni, D. Cheung, and R. Sudha (Eds.): ER 2012, LNCS 7532, pp. 397–408, 2012.

been proposed which use approaches such as feature modeling [1] and orthogonal variability modeling [3]. The evaluation and improvement of expressiveness of variability languages have attracted some attention in SPLE research [4][5]. However, theoretical analysis of variability modeling languages with respect to their ability to represent variability in real world domain models has yet not been studied in detail. Theoretical analysis of variability languages help in better understanding of expressiveness of these languages and consequently changes in these languages for proper representation of variability between conceptual models.

In this paper, we investigate the use of ontological theories for theoretical analysis of variability modeling languages. An ontological theory defines constructs required for describing the structure and processes of the world in general. Ontological theories have been used to evaluate modeling languages in terms of correspondence of ontological concepts to modeling constructs. Bunge's ontology (as adapted by Wand and Weber [7]) has been used to evaluate several conceptual modeling languages [7][8]. This ontology includes a set of high level constructs for representing real world phenomena. The evaluation of modeling languages is based on the assumption that an information system is an artifact that represents a real-world domain. Since Bunge's ontology provides concepts for representing real world phenomena, we presume that it is appropriate for analyzing the *essential* variability of software systems which represents variability of real-world domains. Therefore, in this paper, we analyze *essential* variability of products in a software product line by investigating variability between real-world domains they are intended to represent. More specifically, we employ concepts from Bunge's ontology (as adapted for information systems [7]) as the theoretical framework for identifying possible variability patterns among products based on real-world domains variability. Afterward, we develop a theoretical framework for variability and apply the framework for evaluating expressiveness of variability languages.

2   Backgrounds

2.1   Bunge's Ontology

In Bunge's ontology [6], the "world is made up of substantial *things* which possess *properties*". Examples of things are specific instances of person, car, or book and properties are name, color, or weight. A *property* can be either *intrinsic* (i.e., is possessed by one thing) or *mutual* ("meaningful only in the context of two or more things" [7]). "Properties in general are represented by *state functions*, the values of which express individual properties" ([8] p.4). A *functional schema* is formed by a set of state functions that are used to describe similar things. A *state* is "the vector of values for all property functions of a thing" [8]. An event is "a change of a state of a thing" [8]. A state *law* defines any restriction on the individual properties (or attributes) of a thing. The *process* of a thing comprises the events that the thing might undergo and is determined by a set of transition laws [7]. A set of things that possess a common property is termed a *class*. A set of things having several properties in common is called a *kind*. A *natural kind* is a kind of things adhering to the same laws. Bunge's ontology defines more concepts that will be mentioned when are needed in the paper.

2.2   Variability in Software Product Line Engineering

Variability is the central concept in SPLE [2] and have been investigated with different perspectives and assumptions. Mainly, existing points of views on the variability notion can be categorized into two classes:

- Bosch et al. [9] defines variability as "the ability to be changed, customized, configured, or extended for use in a specific context".
- Weiss et al. [10] define variability as "an assumption about how members of a family may differ from each other ".

Mentzger et al. [11] refer to software variability as the first class of variability and product line variability as the second class. Product line variability differs from software variability; product line variability is an explicit decision of product management about what should vary between the systems in a product line and what should not [3]; software variability is an inherent property of the software under development and represents different behavior of the software. Product line variability originates from differences among real-world domains which are represented by the products of a product line. For example, in a shopping domain, real-world shopping systems may vary in their types of payment methods where one may include *credit card* and *debit card* and another may have *debit card* and *cash*. This variation leads to product line variability for their corresponding information systems. But, in a shopping domain, if all shopping systems have all three types of payment methods and hence, the payment methods cannot be varied among different members of the product line; that is, there may be software variability but there is no product line variability.

Since software products are representation of real-world domains, one can analyze the sources and types of variability which exist among real-world domains by using an ontological theory. In this paper, we concentrate on the ontological analysis of essential variability to identify variability patterns that can exist among different products.

2.3   An Illustrative Example

To exemplify the concepts in the remainder of the paper, we use a standard case study commonly used in comparative analyses of information system methodologies [12]. This is the IFIP working conference case study ([12] pp. 8-9). We assume three conference examples named conference A, conference B, and conference C, and we analyze their variability. Table 1 shows these three conferences.

Table 1. Three imaginary conferences based on the IFIP working conference example

	Conference A	Conference B	Conference C
Things	Specific instances of Program Committee, Organizing Committee, Participants, Papers	Specific instances of Program Committee, Organizing Committee, People Involved, Papers, Demos	Specific instances of Program Committee, Organizing Committee, Attendees, Papers, Art Works,
Properties (attributes)	Paper (Submission ID, Title, Author(s) Name, Quality, Type, Status), Authors(Name, Papers(s), Affiliation, Role)	Paper (Paper ID, Title, Author(s) Name, Quality, Category, Status), Authors (AName, Paper(s), Affiliation, Role, Conflict)	Paper (Paper ID, Title, Author(s) Name, Quality, Category, Status), Authors(AName, Papers(s), Affiliation, Role)
Lawful State Spaces	Paper-Status (Submitted, Accepted, Rejected, Short Papers) Paper-type (Experience, Research, Evaluation. Ideas)	Paper-Status (Submitted, Accepted, Rejected, Conditionally Accepted) Paper Category (Theoretical Foundation)	Paper-Status (submitted, Accepted, Rejected, Conditionally Accepted) Paper-Category (Theory, Practice)
Lawful Events	Paper{(Submitted→Accepted) (Submitted → Rejected), (Submitted →Short Paper Accepted)	Paper{(Submitted→Accepted) (Submitted → Rejected), (Submitted → Conditionally Accepted), (Conditionally Accepted→Accepted), (Conditionally Accepted →Rejected)}	Paper {(Submitted→Accepted) (Submitted → Rejected), (Submitted → Conditionally Accepted), (Conditionally Accepted→Accepted), (Conditionally Accepted →Rejected) }

3   Ontological Analysis of Essential Variability

According to the representation premise of Bunge's ontology, we can conclude that every product (i.e. information system) in a product line family is a representation of a real domain. For example, a software system developed for managing conference A is a representation of conference A in the real-world. Thus, the software system's conceptual model represents static and dynamic phenomena of the real-world domain of the system under study (i.e., conference A). As already indicated, essential variability represents variability among knowledge of real-world domains, and can be represented by different conceptual models of the products. Hence, we investigate essential variability by exploring variability of domains which these products are intended to represent. To investigate variability between different real world domains corresponding to different products, we assume that we can map the phenomena in one domain to the phenomena in another domain.

Mapping Premise: We assume that we can establish corresponding mappings between *things, attributes, states,* and *events* in one domain and *things, attributes, states,* and *events* in another domain.

For example, for the IFIP conference context, we can establish mappings between property *Submission ID* of thing *Paper* ∈ *Conference A* with property *Paper ID* of thing *Paper* ∈ *Conference B*. The mapping is denoted as $A(Paper. Submission ID) \leftrightarrow B(Paper. Paper ID)$.

To investigate variability among different domains, first we analyze similarity patterns among things in these domains. Next, we consider variability as opposite to similarity and define variability patterns between different real-world systems.

3.1   Variability Patterns among Sets of Phenomena

Before introducing variability classes in terms of Bunge's concepts (i.e., thing, property, state, and event), we introduce a set of general variability patterns between two sets of phenomena (By phenomena we refer to any possible observation that can be made about the domain or part of it). Afterwards, considering these variability patterns and Bunge's concepts, a set of variability classes among two domains are defined.

Assume $S = \{s_1, s_2, \ldots, s_m\}$ is a set of phenomena belonging to domain D_1 and $T = \{t'_1, t'_2, \ldots, t'_n\}$ is a set of phenomena belonging to domain D_2. Now, we have one of following situations with respect to similarity between these two sets.

Definition 1 (Equivalent Sets of Phenomena): S is *equivalent to* T (denoted as $S \equiv T$), if and only if there is a mapping between elements in S and elements in T.

Definition 2 (Similar Sets of Phenomena): S is *similar to* T with respect to p (denoted as $S \cong_p T$) if and only if there is a subset of S (i.e., $S' \subset S$) and of T (i.e., $T' \subset T$) which are equivalent $S' \equiv T'$. p is equivalent subset i.e. $p = S' = T'$.

Definition 3 (Completely Dissimilar Set of Phenomena): S is *completely dissimilar to* T with respect to (denoted as $S \neq T$) if and only if there are no subsets of S (i.e., $S' \subset S$) and of T (i.e. $T' \subset T$) that are equivalent.

Based on Definitions 1-3, we can define the following similarity patterns between two different sets of phenomena:

- *Full similarity double side* – when a set of phenomena S and set of phenomena T are equivalent (i.e., $S \equiv T$).
- *Full similarity one side* – when a set of phenomena S and a set of phenomena T are *similar* (i.e., $S \cong_p T$) *and* when we have either $S' \subset S$ and $S' \equiv T$ or $T' \subset T$ and $S \equiv T'$.
- *Partial similarity* – when a set of phenomena S and a set of phenomena T are *similar* (i.e., $S \cong_p T$) and there is no subset of one functional schema that is equivalent to the other functional schema.
- *Complete Dissimilarity* – when two sets *of* phenomena are *completely dissimilar*.

One of the above similarity patterns may happen between attribute sets of phenomena in different real-world domains. All the above patterns, except *full similarity double side*, represent possible variability between two sets of phenomena. To investigate variability between more than two sets which belong to different real-world domains, we can explore the variability patterns between each set and the rest of the sets.

3.2   Variability among Things of Different Real-World Domains

According to Bunge's ontology, a real-world domain is comprised of things. To investigate variability between domains, we need to explore variability among things in the product domains in terms of the variability of their structure and processes. The structure and processes of things is defined in terms of a combination of their properties, states, laws and events. We analyze variability among things by analyzing variability among their *attributes (properties representation), lawful state space* and *lawful event space*. To investigate variability among two things in terms of their attributes, we form functional schemas which are sets of attributes for defining those things. Table 2 shows different similarity classes of two things in different real-world domains.

Table 2. Similarity classes between things of different product domains

Patterns	Class name	Description
Full similarity	Full similarity among functional schemas	Equivalent functional schemas
	Full similarity among lawful state spaces	Equivalent lawful state spaces
	Full similarity among lawful event spaces	Equivalent event spaces
Full similarity one side	Full similarity one side among functional schemas	Similar functional schemas and a subset of one functional schema is equivalent of the other functional schema
	Full similarity one side among lawful state spaces	Similar lawful state spaces and a subset of one lawful state space is equivalent of the other lawful state space.
	Full similarity one side among lawful event spaces	Similar lawful event spaces and a subset of one of the lawful event spaces is equivalent of the other lawful event space.
Partial similarity	Partial similarity among functional schemas	Similar functional schema and there is no subset of one functional schema that is equivalent to the other one.
	Partial similarity among lawful state spaces	Similar lawful state spaces and no subset of one lawful state space that is equivalent to the other lawful state space.
	Partial similarity among lawful event spaces	Similar lawful event space and there is no subset of one lawful event space that is equivalent to the other event space.
Complete dissimilarity	Complete dissimilarity among functional schemas	Dissimilar functional schemas
	Complete dissimilarity among lawful state spaces	Dissimilar lawful state space
	Complete dissimilarity among lawful event spaces	Dissimilar Lawful event space

For instance for IFIP conferences, we have a full similarity double side among functional schema defining paper in conference A and functional schema defining paper in conference C. Moreover, there is a subset of the functional schema defining authors of conference B which is equivalent to the functional schema describing authors of conference A. Hence, there is a full similarity one side between the functional schemas of the authors in the two conferences.

3.3   Variability among Real-World Domains

After exploring variability among *things* of different product domains, we can define variability classes between two *product domains* belonging to a product line. Variability among product domains may occur in both static (structure) and dynamics (processes) of the domains. The structure of a domain is defined based on things in the domain. Considering Bunge's ontology, the structure of a domain can be shown using functional schemas and a lawful state space of the whole domain. Hence, to investigate structural variability between two domains, we need to consider combinations of variability classes defined for functional schemas and lawful state spaces of these two domains. A difference between the functional schemas implies a difference between the conceivable state spaces and lawful state spaces of the two product domains. Therefore, variability between functional schemas of the domains leads to the variability of the state spaces of the domains. However, an equivalence of the two functional schemas means'the equivalence of their conceivable state space, but does not mean an equivalence of their lawful states because different laws may govern their properties. Hence, there may be structural variability between domains in terms of their lawful state spaces, even though there is a full-similarity double side between their functional schemas. For instance for IFIP conferences, we have a full similarity double side among the functional schema defining *paper* of *conference A* and the *functional schema* defining the *paper* in the *conference C*, but possible values for *status* of the paper in conference A are *Submitted, Accepted, Rejected, Short Papers* and in the conference C are *Submitted, Accepted, Rejected, Conditionally Accepted.* This shows variability in the lawful state spaces, even though their functional schemas are completely similar.

On the other hand, when considering dynamics of the domain – a process can be defined in terms of a *sequence of changes* (i.e. events) [7]. These changes may happen within things (internal events of the things) and between things (i.e. interactions between things) of a domain. The processes of a domain can be shown using the *lawful event space* of the domain and the *ordering* between these events. Hence, to investigate variability between processes of different product domains, we need to explore variability classes between their lawful event spaces of the domains and their sequences. As an example for variability in ordering of events, assume that determining the program of conference A involves an order *Workshop→ Tutorial→Main Conference* and for conference B *Main Conference → Workshop→ Tutorial.* Although both the conferences contain the same set of events, the ordering of events is different.

Table 3 shows similarity classes between two different product domains. All of the above similarity classes, except full similarity double side, show variability classes among domains. When investigating more than two domains, combination of these classes may exist between one product domain and the rest of product domains.

3.4   Variability Framework Derived from Bunge's Ontology

Having identified variability classes in section 3.3, we describe a framework for evaluating variability languages by using the ontological theory. The evaluation framework is based on the assumption that variability languages must be ontologically expressive and must be able to represent all the variability classes which may happen among the elements of conceptual models, represented in Table 3.

By investigating the variability classes in section 3.3, we specify two main concepts for variability framework: *variability sources* and *variability patterns.* A *variability source* shows elements in which variability may happen (c.f. sections 3.1 and 3.2). A *variability pattern* shows a different recurring type of variability (see section 3.1) between sets of phenomena of different product domains. Considering Bunge's ontology, *structure of a domain* including *things, properties (attributes),* and *lawful state space* and *processes of the domain* including *lawful event space* and *time of occurrence* are variability sources. The common variability patterns for both the structure and process of domains are *full-similarity one side, partial similarity,* and *complete dissimilarity.* Additionally, the *ordering* variability pattern is dedicated to the processes of domain and shows another aspect of difference between the sets of lawful event spaces.

Similar to ontological analysis for conceptual modeling languages [7], we identify two evaluation criteria for assessing the capability of languages to model variability – *variability completeness* and *variability clarity*. Variability completeness is concerned with investigating if modeling languages have constructs for representing all the variability patterns and consider variability in all possible sources. Variability clarity means that there is a one-to-one mapping among variability constructs in variability languages and variability patterns of the framework.

Table 3. Similarity Classes between two product domains

Similarity among structures of Domains			Similarity among Processes of Domains		
Functional Schema	Lawful State Space	Class Name	Lawful Event Space	Sequence Difference	Class Name
Full Similarity – Double Side	Full Similarity - Double side	Completely similar structures	Full Similarity – Double Side	No	Full Similarity-Double side among Process
	Full Similarity – One Side	Complete similar macro structure, different micro structure		Yes	Full Similarity-Double side among events and different order
	Partial Similarity			No	Full Similarity –one Side among Process
	Dissimilarity	Complete similar macro structure and complete dissimilar micro structure	Full Similarity – One Side	Yes	Full Similarity –one side among events and different order
Full Similarity – One Side	Full Similarity – one side	High similar macro and micro structure	Partial Similarity	No	Partial Similarity among Process
	Partial Similarity			Yes	Partial Similarity among Process and different order
	Dissimilarity	High similar macro and complete dissimilar micro structures	Dissimilarity	NA	Complete Dissimilarity among process
Partial Similarity	Partial Similarity	Medium similar macro and micro structure			
	Dissimilarity	Medium similar macro and complete dissimilar micro structure			
Dissimilarity	Dissimilarity	Complete Dissimilar Structure			

4   Analysis of Variability Languages with Variability Framework

In this section, we analyze two variability languages, i.e. feature models [1] and Orthogonal Variability Models (OVMs) [3] using the proposed variability framework.

Feature models are widely employed in SPLE to model variability and provide representations for variability relations. A central notion in identifying and modeling variability in feature-oriented software product lines is *feature*, which is defined as follows:

"Important distinguishing aspects, qualities, or characteristics of a family of systems" (Kang et al. [1]); and "a logical unit of behavior specified by a set of functional and non-functional requirements" Bosch [13]. In the feature model, variability is represented through the following variability relations: *Optional feature, Alternative feature group,* and *Or feature group.* OVMs represent variability using the variation points (VP) and variants (V) constructs [3]. A variation point is a representation of variability *subject,* "a variable item of the real-world or a variable property of such item" [3]. A variant is a representation of a variability *object,* a particular instance of a variability subject [3]. For example, a variable subject *color* is a property of a real-world item (e.g. Car) and variable objects for *color* are *green, red*, and *blue*. Then, color is a variation point and green, red, and blue are variants. In OVM, variability is specified using relations defined between variation points and variants. These relations are *optional* and *alternative choice* with cardinality *min..max [11]*.

In order to analyze these two languages, based on the evaluation criteria defined in our framework, we derive a research question: *What are the representational shortcomings of feature models and OVM in light of the theoretical variability framework?*

Table 4. Analysis results of feature models and OVM using variability framework (√) match, (×) not match, (±) ambiguity

Variability Languages			Feature models		OVM	
Concepts in the Framework				Explanation		Explanation
Variability Source	Structure	Things	√	relating features to natural kind in Bunge's Ontology	√	relating variation point (subject) and variants (object) to natural kind in Bunge's Ontology
		Properties	√		√	
		Lawful state Space	√		√	
	Process	Event Space	√		√	
		Time	√		√	
Variability Patterns		Full Similarity One-side	√	Optional relations	√	Optional relations
		Partial Similarity	±	OR relation	±	Cardinality [m..n]
		Dissimilarity	√	Alternative Relation	√	Cardinality [1..1]
		Ordering Variability	×	Not Considered	×	Not Considered

To answer the research question, we established a mapping between constructs of these languages and the concepts in the variability framework (see Table 4).

With respect to completeness criteria, the variability languages should encompass constructs for presenting variability sources and variability patterns. Hence, we investigate if feature models and OVM can represent variability among all different sources of variability (i.e., things, properties, lawful state space, and lawful event space).

Variability Sources: Variability in feature models is represented in terms of difference among features and in OVM in terms of variation points and variants. According to definitions for features, we can conclude a feature is a particular set of properties or processes of one or more products in a product family. To interpret features based on Bunge's ontology, we can relate features to *natural kinds* because natural kinds are used to define things with a set of common properties that adhere to the same laws including both transition and state laws [8]. Hence, the natural kinds similar to features can be used to represent both the processes (lawful event spaces and time) and structure (things, properties, and lawful state space) of the domain. Considering the ways features and natural kinds can be related, we propose that a "good" set of features is required to represent all sources of conceptual variability including things, properties, lawful state space, and lawful event space. Similarly in OVM, a variation point and a variant can represent both processes (lawful event spaces and time) and structure (things, properties, and lawful state spaces) of the real-world domain. Therefore, the variation point and variant can be related to natural kinds. This means that OVM can represents all sources of conceptual variability.

Variability Patterns: To interpret variability relations in feature models and OVM, using our variability patterns, for simplicity we consider a product line with three products (e.g., conferences A, B, and C). However, the overall discussion and interpretation is applicable for product lines with any number of products. We make two assumptions. First, if a product has a feature (in feature models) or a variant (in OVM)f, then there are things in the real-world domain of the product belonging to natural kind nk_f. Second, we assume that all products involved in a variability relation, contain *thing(s)* in their corresponding real-world domain which are mapped to *thing(s)* in the real-world domain of the other involved products.

Optional Relation: If a feature (variant in the OVM) f is optional then one or two of the products in the example product line contain f. Consequently, things in the real-world domain of products containing feature (variant in OVM) f belong to natural kind nk_f. However, things in the real-world domains of the other products (i.e., products without optional feature or variant f) do not have a set of properties and processes defined by nk_f. This means there is a full similarity one side with respect to nk_f among things which have mapping to the real-world domains of the products belonging to the product line.

Alternative Relation (Alternative choice with cardinality 1..1): In an alternative relation (i.e. XOR group f_1, f_2, f_3), from a group of alternative features (variants in the OVM), each product contains only one of the features (variants in the OVM). This means that there are things in the domain of products involved in the alternative group such that these things belong to natural kinds nk_{f1}, nk_{f2} and nk_{f3}. The mapped things of different products involved in the alternative relation are completely dissimilar with respect to the properties and laws defined in $nk_{f1}, nk_{f2},$ and nk_{f3}.

OR Relation (alternative choice with cardinality *min..max*): A group of features connected by the OR relation means that products in the group have one or more features. Similarly, in OVM, a group of variants with cardinality *min..max* refers the selection of at least *min* and at most *max* number of variants. To interpret these patterns, all the variability patterns, including complete dissimilarity, partial similarity, and full-similarity one side may happen (but do not have to) among mapped things of each pair of products. Therefore, an OR group in feature models and alternative choice with cardinality *min..max* in OVM can be mapped into complete dissimilarity, partial similarity, and full-similarity one side variability patterns.

Ordering Variability: Neither feature models nor OVM has constructs for representing ordering variability between features and variants, respectively. Consequently, feature models and OVM cannot represent part of process variability classes shown in table 3.

Based on our analysis, we conclude feature models and OVM have the same representational expressiveness for modeling conceptual variability. Also, both languages have the lack of variability completeness as they do not have any construct for representing *ordering* variability. Finally, due to the ambiguity in OR and alternative choice with cardinality *min..max*, we cannot establish one–to-one mapping between variability patterns in our framework and these variability relations. This is a lack of *variability clarity* with respect to our ontological framework.

5   Related Work

Similar to our work, Reinhartz-Berger et al. [14] conducted an ontological analysis of variability modeling using Bunge's ontology. They analyzed process variability and identified a set of variability patterns in the behavior of products. They used their variability framework to perform a formal operational feasibility analysis. In our study, we have investigated both structural and behavior variability which need to be modeled using variability languages. Moreover, our framework is used to evaluate expressiveness of variability languages for representing types of differences among products.

Several frameworks have been employed for evaluating variability languages from practical points of view. Specifically, two main works have been done to evaluate feature models [4][5]. Diebbi et al. [5] performed a study in an company and established a set of criteria by studying software engineers' main expectations for a variability notation. Heymans et al. [4] developed a formal quality method based on SEQUAL (SEmioticQUALity) for evaluating the expressiveness of feature models. In comparison to these works, instead of practical consideration we employed a theory-based approach which can better reveal complete and non-redundant set of variability types.

6   Conclusion and Future Works

Variability plays a pivotal role in systematic reusability in SPLE. This prompts a need for new methods for detailed and formal analysis of the variability notion. However, existing variability formalizations are mainly practice-oriented and lack theoretical foundation. Trying to address this challenge, our work explores the notion of product line variability using the ontological theory and provides a theoretical analysis of variability modeling. Ontological understanding of variability is beneficial from several aspects. First, based on the results of ontological analysis, software developers can clearly identify sources of essential variability in information systems and consider those sources during domain engineering lifecycle. Second, the possible patterns of variability (i.e., types of differences) among products in terms of structure and processes can be identified. This can support the evaluation of variability modeling languages and the improvement of their expressiveness in representing variability. This work is part of continuing work for developing theoretical foundations for variability modeling and configuration decisions in software product lines. We intend to enrich the variability framework with patterns and sources of technical variability.

References

1. Kang, K.C., Cohen, S.G., Hess, J.A., Novak, W.E., Kang, A.: Feature-Oriented Domain Analysis (FODA) Feasibility Study. Technical Report CMU (1990)
2. Pohl, K., Metzger, A.: Variability management in software product line engineering. In: Proc. 28th Int'l Conf. Software Engineering, pp. 1049–1050 (2006)

3. Pohl, K., Böckle, G., van der Linden, F.: Software Product Line Engineering: Foundations, Principles and Techniques. Springer-Verlag New York, Inc., Secaucus (2005)
4. Heymans, P., Schobbens, P.Y., Trigaux, J.C., Matulevicius, R., Classen, A., Bontemps, Y.: Towards the comparative evaluation of feature diagram languages. In: Proc. Software and Services Variability Management Workshop (2007)
5. Djebbi, O., Salinesi, C.: Criteria for Comparing Requirements Variability Modeling Notations for Product Lines. In: 4th Int'l WShComparative Evaluation in Req. Eng., pp. 20–35 (2006)
6. Bunge, M.: Treatise on Basic Philosophy. Ontology I: The Furniture of the World, vol. 3. Reidel, Boston (1977)
7. Wand, Y., Weber, R.: On the deep structure of information systems. Information Systems Journal 5(3), 203–223 (1995)
8. Evermann, J., Wand, Y.: Ontology based object-oriented domain modelling: fundamental concepts. Requirements Engineering 10(2), 146–160 (2005)
9. Van Gurp, J., Bosch, J., Svahnberg, M.: On the notion of variability in software product lines. In: Proc. Working IEEE/IFIP Conf. Software Architecture, pp. 45–54 (2001)
10. Weiss, D.M., Lai, C.T.R.: Software Product-Line Engineering, A Family-Based Software Development Process. Addison Wesley (1999)
11. Metzger, A., Pohl, K., Heymans, P., Schobbens, P.-Y., Saval, G.: Disambiguating the Documentation of Variability in Software Product Lines: A Separation of Concerns, Formalization and Automated Analysis. In: RE 2007, Delhi, India, pp. 243–253 (2007)
12. Olle, T.W., Stuart, A.A.V., Sol, H.G.: Information Systems Design Methodologies. In: A Comparative Review. Proceedings of the IFIP WG 8.1 Working Conference on Comparative Review of Information Systems Design Methodologies (1982)
13. Bosch, J.: Design and Use of Software Architectures: Adopting and Evolving a Product-Line Approach. ACM Press / Addison-Wesley (2000)
14. Reinhartz-Berger, I., Sturm, A., Wand, Y.: External Variability of Software: Classification and Ontological Foundations. In: Jeusfeld, M., Delcambre, L., Ling, T.-W. (eds.) ER 2011. LNCS, vol. 6998, pp. 275–289. Springer, Heidelberg (2011)

A Preference-Aware Query Model
for Data Web Services

Soumaya Amdouni[1], Djamal Benslimane[1], Mahmoud Barhamgi[1],
Allel Hadjali[2], Rim Faiz[3], and Parisa Ghodous[1]

[1] LIRIS Laboratory, Claude Bernard Lyon1 University
69622 Villeurbanne, France
{Soumaya.Amdouni,Djamal.Benslimane,Mahmoud.Barhamgi,
Parisa.Ghodous}@liris.cnrs.fr
[2] Enssat, University of Rennes 1
22305, Lannion, France
hadjali@enssat.fr
[3] University of Carthage-IHEC
2016 Carthage, TUNISIA
Rim.Faiz@ihec.rnu.tn

Abstract. Data Mashup is a special class of applications mashup that combines data elements from multiple data sources (that are often exported as data web services) to respond to transient business needs on the fly. In this paper, we propose a semantic model for data services along with a declarative approach for creating data mashups without any programming involved. Given a query formulated over domain ontologies, and a set of preferences modeled using the fuzzy set theory, our approach selects the relevant data services based on their semantic modeling using an RDF query rewriting algorithm. Selected services are then orchestrated using a ranking-aware algebra to rank their returned results based on users preferences at the data mashup execution time.

Keywords: Data Mashup, Query Models, Data Ranking, Fuzzy Preferences.

1 Introduction

Data Mashups are *Situational Applications* (i.e., applications that come together for solving some immediate business problems) that combine data elements from different data sources to provide value-added information for immediate business data needs. Typically, the access to these data sources is carried out through Web services. This type of services is known as DaaS (Data-as-a-Service) [17], or data services simply [4]. Data mashup has become so popular over the last few years; its applications vary from addressing transient business needs in modern enterprises [7,12] to conducting scientific research in e-science communities [21,3]. However, in spite of its popularity, current data mashup applications are still limited to very primitive information integration. This is due to many challenges

P. Atzeni, D. Cheung, and R. Sudha (Eds.): ER 2012, LNCS 7532, pp. 409–422, 2012.
© Springer-Verlag Berlin Heidelberg 2012

introduced by data mashup. Indeed, mashing-up data involves carrying out many challenging tasks including: selecting the data services that are relevant to user's needs, mapping their inputs and outputs to each other (and probably applying some mediation functions when inputs/outputs don't fit each other) and performing some processing on intermediate results. In addition, data mashup are usually written in some procedural programming languages such as JavaScript, and the code rarely separates the user interface layer (dynamic HTML) from the data integration layer. These challenging tasks hinder average users from building data mashup applications at large.

1.1   Motivating Example

Consider a Web user *Melissa* planning to buy a new apartment. Melissa would like to find an apartment in a clean city, with an affordable price and located near to high schools with cheap tuition fees and good reputation. Melissa needs to retrieve cheap schools along with their tuitions fees and addresses from *nces.ed.gov* and their ratings from *psk12.com*. She needs then to connect to some e-commerce sites (e.g., *apartments.com*) to locate cheap apartments near to the schools found. She needs also to connect to the Outdoor City Pollution Database on *who.int* to filter out apartments located in polluted cities. Assume that these information are provided by the data services in Table 1. The service S_1 returns the schools, along with their tuitions fees, reputations and addresses at a given country; S_2 returns the apartments for sale along with their prices at a given city, and S_3 returns the pollution level at a given city. Input and output parameters are proceeded by "$" and "?" respectively. Obviously, Melissa can answer her query by composing the following services.

Table 1. Available Web Services

Service	Functionality
$S_1(\$c, ?s, ?t, ?r, ?a)$	Returns the schools s along with their tuition fees t, reputation r and addresses a in a given country c
$S_2(\$a, ?ap, ?p)$	Returns the apartments for sale ap, their prices p at a given address a
$S_3(\$a, ?po)$	Returns the pollution level po for a given city a

1.2   Challenges

Mashing-up data services presents many challenges for the mashup creator (i.e., Melissa):

- *Understanding the Semantics of Data Services*: Melissa needs to delve into the data service space and understand the semantics of each individual service in order to identify the services that may contribute to the resolution of her query. The semantics of a data Web service resides not only in its input's and output's types, but also in "how" inputs and outputs are related to each other; i.e., many services may have the same input's and output's types, but completely different semantics. In the lack of a clear definition of such input/output relationships inside the service description files, mashup creators will miss much of the services that are relevant to their queries; even worse they may wrongly select services that are irrelevant to their needs.

– *Selecting and Mashing-up Data Services*: Let us assume now that Melissa is able to understand the semantics of available data services, the next step would be to select participant services, figure out their execution order, build the data mashup plan, define the mapping between input and output messages of interconnected services and drop some programming code to carry out the intermediate data processing inside the mashup. Unfortunately non-expert users like Melissa (an ordinary Web user) cannot conduct the previous tasks that require important technical and programming skills.

– *Ranking the results and selecting the best ones*: Let us assume now that Melissa was able to create and execute the data mashup. Each of the composed services may return a huge number of business objects (e.g., apartments, schools) that may more or less match the user's preferences. As a result, Melissa will be overwhelmed with a great number of answers and may miss the ones that are most relevant to her needs.

1.3   Contributions

In this paper we propose a declarative approach for mashing up data web services. We summarize below our major contributions in this paper:

– *A Semantic Model for Data Services*: We propose to model data services as *RDF Views* over domain ontologies. An RDF view allows capturing the semantics of the associated service in a "declarative" way based on concepts and relations whose semantics are formally defined in domain ontologies. The semantic model allows characterizing the returned data and the non-functional aspects of a service using the fuzzy set theory.

– *A Declarative Model for Composing Data Services*: We propose to use the mature query rewriting techniques for mashing-up data services. Specifically, we devised a composition algorithm that given a query and a set of services (described as RDF views), it selects the relevant services and rewrites the query in terms of calls to selected services. Our model enables average users to mashup data as all what they need to do is just specifying their data needs declaratively.

– *A Data Ranking Model to select the best Answers*: We propose a ranking-aware composition algebra (and implementation thereof) that allows ranking the returned results based on the user preferences at the data mashup execution time.

The remainder of this paper is organized as follows. In Section 2, we present our declarative approach to construct data mashups. We show also our semantic modeling to data services and queries; present the composition algorithm through an example, and introduce our ranking-aware composition algebra. In Section 3, we evaluate our approach. In section 4, we compare our approach to related works. Finally, in section 5, we summarize and conclude the paper.

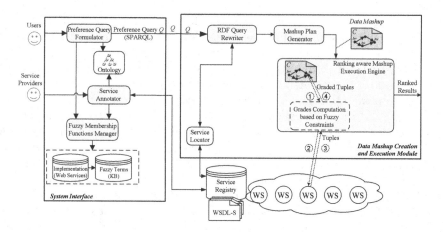

Fig. 1. An overview of the proposed approach

2   A Preference-Aware Query Model

In this section we present a "declarative" approach for data mashup that addresses the challenges discussed in the previous section. We show the different phases involved in data service mashup, starting from the service modeling to the generation of the final data mashups that will be returned to users.

2.1   Approach Overview

Figure 1 gives an overview of our proposed data mashup approach. Our approach to data mashup is "declarative"; i.e., mashup creators are relieved from having to select services and build manually the mashup plan, a task that would generally require important programming skills. They need only to formulate their declarative data mashup queries over domain ontologies using the do facto ontology query language SPARQL. They can also specify their preferences in the form of fuzzy constraints that are concretely interpreted according to functions (called *membership functions*) that are implemented as Web services. The components *Preference Query Formulator* and *Fuzzy Membership Functions Manager* assist the users in that task. The first step towards the automation of data mashup is to semantically represent the capabilities of data services. In our approach, we model data services as *RDF Views* over domain ontologies. The RDF views can be used (by the *Service Annotator* component) then to annotate the service description files (e.g., WSDL files, SA-Rest, etc). Our approach exploits the mature query rewriting techniques [1] to fully automate the data mashup process. Declarative mashup queries are rewritten (by the *RDF Query Rewriter* component) in terms of available data services based on our proposed modeling to data services (i.e., RDF views). The mashup system will then orchestrate the selected services using a ranking-aware mashup algebra that we have devised for that purpose. The mashup will be then displayed to users, who will be able to execute it with their

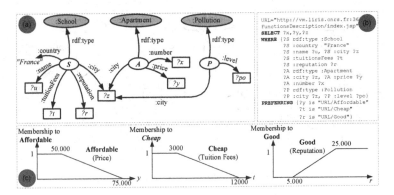

Fig. 2. (a) a Graphical Representation of the Query, (b) the Formulated Query in SPARQL, (c) the Associated Membership Functions

inputs. The data mashup plan can be then executed by a ranking-aware mashup execution engine implementing our defined algebra. The execution engine assigns *grades* to results returned from services' calls based on their matching to users' preferences. We detail all of the previous steps in the subsequent subsections.

2.2 A Semantic Model for Data Services and Preference Queries

Preference Queries: Users express their preference queries over domain ontologies using a slight modification of SPARQL. Figure 2 gives the formulated query for the running example in (b), and its graphical representation in (a). The user's preferences are expressed in the "PREFERRING" clause. We model user's preferences using the fuzzy sets theory [20]. Formally, preferences are a set of constraints of the form C_i: x is $FuzzyTerm$, where x is a variable, and the $FuzzyTerm$ is a fuzzy term (e.g., "Cheap") interpreted according to a membership function $\mu_F : X \rightarrow [0,1]$, specifying for each value of x the grade (i.e., $\mu_F(x)$) to which x belongs to $FuzzyTerm$. Note that $\mu_F(x) = 1$ reflects full membership of x to $FuzzyTerm$ and $\mu_F(x) = 0$ absolute non-membership. For example, the fuzzy terms used in Figure 2 (part-b) (i.e., *Affordable* price, *Cheap* tuition fees, and *Good* reputations) are interpreted using the membership functions shown in Figure 2 (part-c).

Data Services: The functionalities of data services, as opposed to traditional Web services that encapsulate software artifacts, can be only captured by representing the semantic relationship between inputs and outputs [19,1]. Therefore, we model data services as RDF Parameterized Views (RPVs) over domain ontologies Ω. Each view captures the semantic relationships between input and output sets of a data service using concepts and relations whose semantics are formally defined in ontologies. Formally, a data service S_i is described over Ω as a predicate $S_i(\$X_i, ?Y_i) :- < F(X_i, Y_i, Z_i), C_i >$, where:

- X_i and Y_i are the sets of input and output variables of S_i, respectively. Input and output variables are also called as distinguished variables. They are prefixed with the symbols "\$" and "?" respectively.

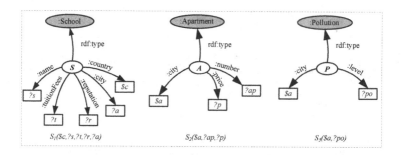

Fig. 3. RDF Parameterized Views - Examples

- $F(X_i, Y_i, Z_i)$ is the functionality of the service and represents the semantic relationship between input and output variables. Z_i is the set of existential variables relating X_i and Y_i. $F(X_i, Y_i, Z_i)$ has the form of RDF triples where each triple is of the form $subject.property.object$.
- C_i is a set of value constraints expressed over the X_i, Y_i or Z_i variables. C_i may include fuzzy constraints to characterize the data manipulated by S_i.

For example, the service S_2 is described by the following RDF view, where "*Apartment*" is an ontological concept and "*city*", "*number*" and "*price*" are its properties.

```
S2($a,?ap,?p):-
          F:{ ?A rdf:type :Apartment,
              ?A :city     $a,
              ?A :number   ?ap,
              ?A :price    ?p}
          Constraints: {?p is CHEAP}
```

Figure 3 gives graphical representations of the RDF Parameterized Views of our sample services. *School, Apartment,* and *Pollution* are ontological concepts. Note that, the current Web service description standards (e.g., WSDL) can be extended straightforwardly with our proposed modeling to data services, as RPVs can be incorporated within the description files (e.g., WSDL) as annotations.

2.3   Mashup Query Rewriting

In a previous work [1] we proposed an efficient RDF query rewriting algorithm. We exploit that algorithm to "mash up" data services. Given a data mashup query Q and a set of data services represented by their corresponding *RPV*s $V = v_1, v_2, v_i$, the algorithm rewrites Q as a composition of data services whose union of RDF graphs (denoted to by G_V) covers the RDF graph of Q (denoted to by G_Q). The rewriting algorithm has two phases:

Phase-I: Finding Relevant Sub-Graphs. In the first phase, our data mashup system compares G_Q to every *RPV* v_i in V and determines the *class-nodes* (i.e., the variables in Q whose types are ontological classes, e.g., "A", "S" and "P" in Figure 2) and object properties in G_Q that are covered by v_i. The system stores information about covered class nodes and object properties as a partial

Table 2. Mapping Table: the covered sub graphs by sample data services

Service	Covered classnodes & properties
$S_1(\$c, ?u, ?t, ?r, ?z)$	$Q.S(\text{"france"}, u, t, r, z)$
$S_2(\$z, ?x, ?y)$	$Q.A(z, x, y)$
$S_3(\$z, ?po)$	$Q.P(z, po)$

containment mapping in a *mapping table*. The mapping table points out the different possibilities of using an RPV to cover parts of G_Q.

Example: The service S_1 has a class node $S_1.S$ that can be matched with $Q.S$. All the data-type properties of $Q.S$ that bound to distinguished variables in Q also bound to distinguished variables in S_1. Furthermore, $Q.S$ is involved in a join over the variable $?z$ with the class-nodes $Q.A$ and $Q.P$. Even though S_1 does not cover the class-nodes $Q.A$ and $Q.P$, the join over $?z$ can be still enforced as $?z$ is a distinguished variable in S_1. Therefore, S_1 can be used to cover $Q.S$, and thus inserted in the Table 2. The same discussion applies to S_2 and S_3.

Phase-II: Generating data service Compositions. After the construction of the mapping table in the previous phase, the mashup system explores the different combinations from that table. It considers the combination of disjoint sets of covered object properties and class nodes. A combination is said to be a valid rewriting of Q (also a valid composition) if (*1*) it covers the whole set of class-nodes and object-properties in Q, and (*2*) it is executable. A composition is said to be executable if all input parameters necessary for the invocation of its component services are bound or can be made bound by the invocation of primitive services whose input parameters are bound.

Example- Continuing with the running example, there is only one possible combination $C_1 = \{S_1, S_2, S_3\}$. Only $S_1(\$c, ?u, ?t, ?r, ?z)$ can be invoked at the beginning as its input parameter is bound. After the invocation of S_1, the variable z become available; hence, the services S_2, S_3 become invokable. Consequently C_1 is executable and is considered as a valid composition.

2.4   A Ranking-Aware Algebra for Data Mashup

In this section we propose an algebra to orchestrate the data services selected in the previous steps. The proposed algebra allows ranking the returned results based on their relevances to user's preferences. Results ranking is important as the results number may be very large which may cause the users to miss the ones that are most relevant to their needs. To enable ranking-aware query processing, our proposed algebra relies on the mature fuzzy database foundations [2]. This new algebra enables and determines our query execution model and operator implementations.

We describe below two sets of ranking-aware data mashup operators that follow two approaches to rank data: (*i*) scalar grades based ranking, and (*ii*) vector grades based ranking.

Table 3. Implemented norms and conorms

Name	$TNorm : \top(x,y)$
Zadeh	min(x,y)
Probabilistic	xy
Lukasiewicz	$max(x+y-1,0)$
Hamacher	$\frac{xy}{\gamma+(1-\gamma)(x+y-xy)}$
Weber	$\begin{cases} x & \text{if } y=1 \\ y & \text{if } x=1 \\ 0 & \text{else} \end{cases}$

Name	$Conorm :\bot\ (x,y)$
Zadeh	$max(x,y)$
Probabilistic	$min(x+y,1)$
Lukasiewicz	$max(x+y-1,0)$
Weber	$\begin{cases} x & \text{if } y=0 \\ y & \text{if } x=0 \\ 1 & \text{else} \end{cases}$

Scalar Grade based Results Ranking Algebra : The operators in this set assume that each manipulated tuple is associated with a grade computed as the aggregation of the different grades associated to its attributes that are involved in fuzzy preferences. We define the following operators:

- *The Grade-aware Invocation $Invoke^g(S, t_{in}^g, O^g)$:* Let S be a service, t_{in}^g the graded input tuple with which S is invoked, O^g the graded output, and $S.O$ be the output returned by S. The $Invoke^g$ operator relays the tuples from $S.O$ to O^g, and for each relayed tuple t_i it computes the grade $g(t_i)$ as follows. First, assume t_i is involved in n preference fuzzy constraints P_j(where $1\leq j \leq n$), the operator computes $g_1(t_i)= \top(\mu_{P_1(t_i)}, \mu_{P_2(t_i)}, ..., \mu_{P_n(t_i)})$ where $\top$ is a t-norm operator (that generalizes the conjunction operation) and μ_{P_i} the membership function associated with P_i. We implemented the T-norms presented in Table 3. The Zadeh t-norm is the greatest t-norm, thus leading to an optimistic aggregation strategy. The Lukasiewicz and Weber t-norm yield a pessimistic aggregation strategy. Second, it computes $g(t_i)$ as follows: $g(t_i)=\top(g(t_{in}), g_1(t_i))$.
- *Graded Join:* $\infty^g(I_1^g, I_2^g)$, where I_1^g and I_2^g are two graded data sets. The grade of an outputted tuple is given by: $g(\infty^g(t,t')) = \top(g(t), g(t'))$ where $\top$ is a t-norm, and t and t' are tuples from I_1^g and I_2^g respectively.
- *Graded Projection* $\prod_A^g$. The projection is an operation that selects specified attributes $A=\{a_1, a_2, ...\}$ from a results set. The grade of an outputted tuple t is: $g(t) =\bot\ (g(t_1'), .., g(t_i'), .., g(t_n'))$ where $t = \prod_A(t_i')_{i=1:n}$ and $\bot$ is the co-norm corresponding to the t-norm $\top$ used in the graded join.
- *Graded Union* $\cup^g$. The grade of an outputted tuple t is: $g(t)= \bot\ (g(t_1'), .., g(t_i'), .., g(t_n'))$, where $t_i' = t$ and $i = 1 : n$
- *Graded Rank $Rank^g$:* the rank operator orders all outputted tuples according to assigned grades. Let t_1, t_2 be two tuples and g_1, g_2 be the grades respectively. If $g_2 \leq g_1$ so t_1 appears before t_2.

Example : We explain the previous operators based on our motivating example. The services S_1, S_2 and S_3 can be composed to find the apartments for sale located in cities with low pollution levels and near to schools with good reputations in a given country as shown in Figure 4. First, S_1 is invoked with

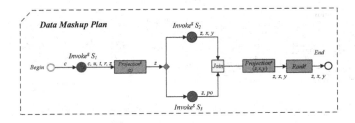

Fig. 4. Data Mashup Plan

the desired country (e.g., France). The invocation operator $Invoke^g(S_1)$ computes the grades of obtained tuples. The $Projection^g(z)$ operator projects the obtained tuples on the city attribute (i.e., z). Then, the obtained cities are used to invoke S_2 to retrieve nearby apartments along with their prices. In parallel, S_3 is invoked to retrieve the pollution levels of obtained cities. The results of S_2 and S_3 are joined over the variable z. Then, the $Projection^g(z, x, y)$ operator retains only the apartments information (i.e., numbers, prices and cities). All of these operators compute the tuples' rankings according to our defined equations presented earlier. Figure 5 shows the results (along with their rankings) at the output of each of these operators, and the final results at the mashup's output. The $Invoke^g(S_1)$ operator invokes S_1 with the value "France", and computes the different grades. For instance the grades of the school "s3" are computed as follows: based on the membership functions associated with the tuition fees and the reputation fuzzy predicates, the grade of the *tuition fees* attribute is 0.22 and the grade of the *reputation* attribute is 0.05. Hence, $Grade_{zadeh}(s3) = min(0.22, 0.05) = 0.05$, $Grade_{probabilistic}(s3) = 0.22 * 0.05 = 0.011$, and $Grade_{Lukasiewicz}(s3) = max(0.22 + 0.05 - 1, 0) = 0$.

The $Projection^g(z)$ operator projects the obtained tuples on the city attribute (i.e., z) and computes the grades of obtained tuples. For example, the grade of the outputted tuple corresponding to "Lyon" is computed as follows:
$Grade_{zadeh}(Lyon) = max(1, 0.4) = 1$
$Grade_{probabilistic}(Lyon) = min(1 + 0.311, 1) = 1$
$Grade_{Lukasiewicz}(Lyon) = max(1 + 0.178 - 1, 0) = 0.178$
The $Invoke^g(S_2)$ operator invokes, for each input tuple, the service S_2 and computes the grades of obtained tuples. For example, the grade of the apartment "a1" (at the output of $Invoke^g(S_2)$) is computed as follows: given that the apartment "a1" accessed by S_2 has a grade of 1, and that the grades of the city "Lyon" at the input of $Invoke^g(S_2)$ are shown above, then the grades of "a1" at the output of $Invoke^g(S_2)$ are: $Grade_{zadeh}(a1) = min(1, 1) = 1$, $Grade_{probabilistic}(a1) = 1 * 1 = 1$, and $Grade_{Lukasiewicz}(a1) = max(1 + 0.178 - 1, 0) = 0.178$. The *join* operator joins S_2 and S_3 outputted tuples and computes the associated grades. For example, the grade of the tuple corresponding to "a3" is computed as follows:
$Grade_{zadeh}(a3) = min(0.6, 0.52) = 0.52$
$Grade_{Probabilistic}(a3) = 0.312 * 0.36 = 0.112$
$Grade_{Lukasiewicz}(a3) = max(0.12 + 0.15 - 1, 0) = 0$

Invoke S_1

?u	?t	?r	?z	$c	Grade		
					T_Z	T_P	T_L
s1	2800$	30000	Lyon	France	1	1	1
s4	6600$	21000	Nice	France	0.6	0.48	0.4
s5	2900$	16000	Nancy	France	0.55	0.55	0.55
s2	5000$	13000	Lyon	France	0.4	0.311	0.178
s3	10000$	6000	Paris	France	0.05	0.011	0

Project(z)

z	Grade		
	T_Z	T_P	T_L
Lyon	1	1	0.178
Nice	0.6	0.48	0.4
Nancy	0.55	0.55	0.55
Paris	0.05	0.011	0

Invoke S_2

?x	?z	?y	Grade		
			T_Z	T_P	T_L
a1	Lyon	50000	1	1	0.178
a4	Nancy	60000	0.55	0.33	0.15
a3	Nice	62000	0.52	0.312	0.12
a2	Paris	120000	0	0	0

Invoke S_3

?z	?po	Grade		
		T_Z	T_P	T_L
Lyon	20	1	1	0.178
Nice	35	0.6	0.36	0.15
Nancy	65	0.25	0.14	0
Paris	80	0	0	0

Join(Invoke S_2, Invoke S_3)&Rank

Final Results

?x	?z	?y	?po	Grade		
				T_Z	T_P	T_L
a1	Lyon	50000	20	1	1	0
a3	Nice	45000	35	0.52	0.112	0
a4	Nancy	60000	65	0.25	0.046	0
a2	Paris	120000	80	0	0	0

Fig. 5. The intermediate and final results along with their grades

Finally, the $Rank^g$ orders results in ascending order(from the most satisfactory to the least satisfatory).

Vector Grade Based Results Ranking Algebra: Merging different grades in one aggregated scalar grade is interesting but presents two main drawbacks. First, it does not allow users to know why a given tuple is a good or a bad result. Details on how fuzzy user preferences match data are not kept. Second, the tuples ordering may vary from one t-norm to another. To overcome these drawbacks, we propose to associate to each tuple a vector of grades. One may not always prefer to aggregate the different computed partial grades. In this case, each tuple t is associated with a vector of grades (instead of a scalar grade). To rank query results, one should revisit the above graded algebraic operators. For instance, The following set of revised graded operators are defined.

- Graded Join $\infty^g(I_1^g, I_2^g)$, where I_1^g and I_2^g are two graded data sets. The revised grade of an outputted tuple is given by:
  $g(\infty^g(t, t')) = (\top(g_1(t), g_1(t')), \cdots, \top(g_d(t), g_d(t')))$
  where $\top$ is a t-norm and t (resp. t') is a tuple of the set I_1^g (resp. I_2^g), and $g_j(t)$ is the grade of the tuple t relative to a fuzzy predicate P_j.
- Graded Projection $\prod_A^g$. The grade of an outputted tuple t is: $g(t) = \{\bot (g_1(t_1'), .., g_1(t_n')), .., \bot (g_j(t_1'), .., g_j(t_n')), .., \bot (g_m(t_1'), .., g_m(t_n'))\}$ where $t = \prod_A(t_i')_{i=1:n}$ and $\bot$ is the co-norm corresponding to the t-norm $\top$ used in the graded join, and $g_j(t')$ is the grade of the tuple t' relative to a fuzzy predicate P_j. The implemented co-norm are presented in Table 3.
- Graded Union $\cup^g$. The grade of an outputted tuple t is:
  $g(t) = \{\bot (g_1(t_1'), .., g_1(t_n')), .., \bot (g_j(t_1'), .., g_j(t_n')), .., \bot (g_m(t_1'), .., g_m(t_n'))\}$
  where $t_i' = t$, $i=1:n$ and $g_j(t')$ is the grade of the tuple t' relative to a fuzzy predicate P_j.

Table 4 shows the final answers along with the different grades (for the fuzzy constraints *Cheap*, *Good*, *Affordable* and *Low*).   •

Table 4. Vector Ranking results

ap	a	p	po	Cheap	Good	Affordable	Low
a1	Lyon	50000$	20	1.00	1.00	1.00	1.00
a2	Paris	120000$	80	0.22	0.05	0.00	0.00
a3	Nice	45000$	35	0.60	0.80	0.52	0.75
a4	Nancy	60000$	65	1.00	0.55	0.60	0.25

3   Implementation and Evaluation

To evaluate and validate our approach, we implemented all of the components shown in Figure 1 in Java. The ranking-aware mashup execution engine was implemented to allow for both scalar and vector grades computations and with any of the three T_Z, T_P, and T_L norms. We conducted a series of experiments to evaluate the efficacy of our approach. The experiments covered many queries from the real-estate domain with a rich set of fuzzy preferences over a set of services returning synthetic data about Apartments, Lands, Restaurants, etc. Our experiments shown that the overhead incurred by computing the rankings is negligible compared to the time necessary to execute the same generated compositions without any ranking at all. In addition, the returned top-k tuples were always correct, proving the soundness of our proposed operators. Figure 6 shows the interface of our data mashup system:

1. The user uses the window (a) to enter his/her sparql query with fuzzy preferences. This query is formulated over an existing ontology. Fuzzy terms are those stored in the fuzzy terms knowledge base of our system. Users can edit and test them via the window (b) to identify the relevant fuzzy terms. Users can also define their own fuzzy terms.

2. The user can execute the query and chose any of the displayed compositions (window (a)). The mashup execution plan is then displayed on window (c) and the user is allowed to choose an execution strategy. If the user wants to aggregate the grades of its fuzzy preferences, he/she chooses the scalar grades computing; otherwise if he/she wants to keep an eye on the grades of all of its fuzzy preferences, he/she chooses the vector grades computing. The user has also to set his/her strategy: optimistic (T_Z norm), reinforcement (T_P norm), and pessimistic (T_L norm).

3. Ranked results are displayed (in windows d and e). In window (d) results are ordered by their aggregated grades. In window (e), results are ranked according to their vectors of grades (each grade corresponds to a degree of satisfaction of a fuzzy user preference). To do so, we make use of the leximin ordering which leads to a total order. This ordering is borrowed from the multicriteria decision field [5].

4. In case of the user is not satisfied by the obtained query results, he/she can choose another service composition from window (a) and execute it.

5. In case of empty (resp. too few) results, users can relax (by introducing some tolerance) their fuzzy constraints present in the initial query. The relaxation operation allows for enlarging the support of the membership functions

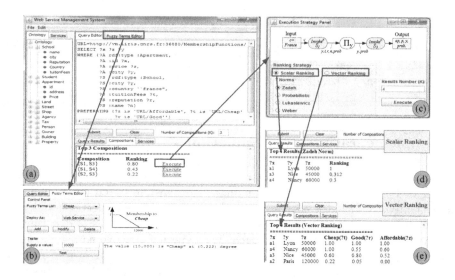

Fig. 6. Demo - Mashups and Results Ranking

associated with each constraint, thus making the query less selective. It is worthy to note that this operation requires to re-execute the grades computation step. For example, relaxing the affordable constraint of Q will return more results: some of the tuples previously ranked to 0 in scalar grades, or ranked to 0 in all dimensions in vector grades.

4   Related Works

Due to space limitation, in this section, we analyze only the closest works and discuss their limitations. Several mashup editors have been introduced by the industry with the objective of making the process of mashups creation as simple and "programmable-free" as possible. Examples include Yahoo Pipes [18], Google Mashup Editor [6], Intel Mash Maker [11]. These products allow average users to create mashups without any programming involved; the users need just to drag and drop services, operators and/or user inputs and to visually connect them. However, the knowledge required from users is not trivial because they are still expected to know exactly what the mashup inputs and outputs are, and to figure out all the intermediate steps needed to generate the desired outputs from the inputs. This includes selecting the needed services/data sources, mapping their inputs and outputs to each other and probably adding some mediation services/functions when inputs/outputs don't fit each other. The Web Service Management System in [15] models data services as relations and allows users to mashup data services by expressing their mashup queries directly in terms of these relations. Along the same lines, the Web Service Mediator System WSMED [13] allows users to mashup data services by defining relational views on top of them (called the WSMED Views). Unfortunately, users in these systems are assumed to have an understanding of the semantics of the data services that are available to them to be able to formulate their queries. Furthermore, users are supposed to import the services relevant to their needs; define views on top of imported services and enhance the views with primary-key constraints. These tasks are difficult and hinder average users from mashing up data services at large. These systems model data services as relations with inputs and outputs as the relations' attributes. This modeling is poor in semantics, as opposed to the use of domain ontologies, and may lead to ambiguity when data services have similar inputs and outputs (attributes), but different semantics. Compared to these works and to other academic mashup systems (e.g., [16]), users of our system are not required to select the services manually, connect them to each other and drop code (in JavaScript) to mediate between incompatible inputs/outputs of involved services. This is completely carried out by the system in a transparent fashion. That is, our approach is *declarative*; users need just to specify the information they need without specifying how this information is obtained. Furthermore, these systems do not provide any effective means to rank the data results returned by the mashups.

Our work is also related to the works around top-k queries and data ranking. Ranking queries are becoming dominant in many domain applications such as multimedia databases, middelwares, and datamining. The increasing of ranking works support ranking mainly in relational database management system and recently pay an attention of the research community. The previous research works [10] in that area adopt one of two ranking models: *(i) top-k selection* and *(ii) top-k join*. In the top-k selection model, scores are attached to tuples in one single relation, and the query reports the k tuples with the highest scores [8]. Another research work attempted to integrate these two models [9]. In contrast, the ranking in our system is complete: it computes the grades of individual data sets returned by data services and the integration thereof to compute the rankings of the final results based on the users fuzzy preferences. In addition, our approach adopts a flexible approach for preference modeling (i.e., a fuzzy approach) and provides two ranking models: scalar and vector models.

5   Conclusion

In this paper, we proposed a declarative approach for mashing up data Web services on the fly. We proposed to model data services as RDF Views over domain ontologies to represent their semantics declaratively. Our semantic model allows characterizing the returned data using the fuzzy set theory. Our approach is based on the usage of the query rewriting techniques to automate the mashup creation process, and allows to rank-order the mashups results based on the users preferences, which are modeled as fuzzy constraints. As a future work, we intend to define a cost model for data ranking, and explore the mashup optimization issues based on the defined model.

References

1. Barhamgi, M., Benslimane, D., Medjahed, B.: A query rewriting approach for web service composition. IEEE Transactions on Services Computing (TSC) 15(5), 795–825 (2010)
2. Bosc, P., Buckles, B.B., Petry, F.E., Pivert, O.: Fuzzy Databases. Fuzzy Sets in Approximate Reasoning and Information System. Kluwer Publishers (1999)
3. Butler, D.: Mashups mix data into global service. Nature 439, 6–7 (2006)
4. Carey, M.J.: SOA What? IEEE Computer 41, 92–94 (2008)
5. Dubois, D., Prade, H.: Beyond min agregation in multicriteria decision: (ordered) weighted mean, discri-min, leximin, pp. 181–192. Kluwer Publishers (1997)
6. Google Inc. Google Mashup Editor, http://code.google.com/gme/
7. Guinard, D., Trifa, V., Karnouskos, S., Spiess, P., Savio, D.: Interacting with the SOA-Based Internet of Things: Discovery, Query, Selection, and On-Demand Provisioning of Web Services. IEEE T. Services Computing 3(3), 223–235 (2010)
8. Fagin, R., Lotem, A., Naor, M.: Optimal Aggregation Algorithms for Middleware. J. Comput. Syst. Sci. 66(4), 614–656 (2003)
9. Ilyas, I.F., Aref, W.G., Elmagarmid, A.K.: Supporting Top-k Join Queries in Relational Databases. VLDB Journal 13(3), 207–221 (2004)
10. Ilyas, I.F., Beskales, G., Soliman, M.A.: A survey of topk query processing techniques in relational database systems. ACM Comput. Surv. 40(4) (2008)
11. Intel Inc. Intel Mash Maker, http://mashmaker.intel.com/web/
12. Jhingran, A.: Enterprise Information Mashups: Integrating Information, Simply. In: VLDB, Seoul, Korea, pp. 3–4 (2006)
13. Sabesan, M., Risch, T.: Adaptive Parallelization of Queries to Data Web Service Operations. T. Large-Scale Data-and Knowledge-Centered Systems 5, 49–69 (2012)
14. Soliman, M.A., Ilyas, I.F., Saleeb, M.: Building Ranked Mashups of Unstructured Sources with Uncertain Information. PVLDB 3(1), 826–837 (2010)
15. Srivastava, U., Munagala, K., Widom, J., Motwani, R.: Query Optimization over Web Services. In: VLDB 2006, pp. 355–366 (2006)
16. Tatemura, J., Chen, S., Liao, F., Po, O., Agrawal, D.: UQBE: uncertain query by example for web service mashup. In: SIGMOD Conference, pp. 1275–1280 (2008)
17. Truong, H.L., Dustdar, S.: On Analyzing and Specifying Concerns for Data as a Service. In: The 2009 Asia-Pacific Services Computing Conference (IEEE APSCC 2009), Singapore, pp. 7–11 (2009)
18. Yahoo Inc., Yahoo Pipes, http://pipes.yahoo.com/pipes/
19. Yu, Q., Liu, X., Bouguettaya, A., Medjahed, B.: Deploying and managing Web services: issues, solutions, and directions. VLDB J. 17(3), 537–572 (2008)
20. Zadeh, L.A.: Fuzzy Sets. Information and Control 8, 338–353 (1965)
21. Zhao, Z., Fang, J., Cheng, J.: CAFISE-S- An Approach to Deploying SOA in Scientific Information Integration. In: ICWS 2008, pp. 425–432 (2008)
22. http://www.oasis-open.org/committees/wsbpel/

User Feedback Based Query Refinement by Exploiting Skyline Operator

Md. Saiful Islam[1,2], Chengfei Liu[1], and Rui Zhou[1]

[1] Swinburne University of Technology, Melbourne, VIC 3122, Australia
{mdsaifulislam,cliu,rzhou}@swin.edu.au
[2] University of Dhaka, Dhaka-1000, Bangladesh

Abstract. This paper presents FlexIQ, a framework for feedback based query refinement. In FlexIQ, feedback is used to discover the query intent of the user and skyline operator is used to confine the search space of the proposed query refinement algorithms. The feedback consists of both unexpected information currently present in the query output and expected information that is missing from the query output. Once the feedback is given by the user, our framework refines the initial query by exploiting skyline operator to minimize the unexpected information as well as maximize the expected information in the refined query output. We validate our framework both theoretically and experimentally. In particular, we demonstrate the effectiveness of our framework by comparing its performance with decision tree based query refinement.

Keywords: Skyline Operator, User Feedback and Query Refinement.

1 Introduction

Database query models such as SQL implement the binary retrieval technique. In this technique, a tuple is present in the result set only if it satisfies each constraint set by the user. Therefore, a user may miss some expected results as well as receive some unexpected results in the query output if conditions in the query predicates are not set appropriately. There are two possible solutions to address this problem: (1) modify the original tuple values in the database so that unexpected/expected tuples dissatisfy/satisfy the query predicates and (2) modify the user submitted query by fixing the initial conditions so that refined query excludes/includes unexpected/expected tuples from/in the new query output. Each of these two solutions has different application scope. The first solution is suitable for applications where static queries are used and data is untrusted. On the other hand, the second solution is suitable for applications where query modifications are allowed and data is trusted. The techniques proposed by Huang et al. [6] and Herschel et al. [5] fall into the first category. However, their works are limited to missing (expected) tuples only. The techniques proposed by Tran et al. [15] and He et al. [4] fall into the second category and again this work is limited to missing (expected) tuples only. We believe that it would be more helpful for users if we can fix the initial query conditions, not only for expected

P. Atzeni, D. Cheung, and R. Sudha (Eds.): ER 2012, LNCS 7532, pp. 423–438, 2012.
© Springer-Verlag Berlin Heidelberg 2012

User Query: Query Output:
SELECT PubID FROM Publication
    WHERE CitationCnt ≥ 80 AND
        PubYear ≥ 1986;

PubID	CitationCnt	PubYear
P1	96	1989
P2	128	1986
P3	100	1989
P4	90	1993
P5	148	1986
P6	148	1990
P7	81	1996
P8	103	1986
P9	82	1994
P10	117	1987

Expected Tuples:

PubID	CitationCnt	PubYear
P11	60	1995
P12	72	1996
P13	64	1996
P14	67	1995

Fig. 1. Motivating example

tuples, but also for unexpected tuples in trusted data applications where query modifications are allowed. For example, consider a selection query which is issued by a user to a publication database and the corresponding result set as shown in Fig. 1, assume that the user has a different image of the query output in her mind and the result set does not match her expectation, she may then ask "How can I exclude P1, P3 and P10 from my query output? How can I include P11 and P12 into my query output?"

Recently, Islam et al. [7] proposed a decision tree (DT) based query refinement technique for minimizing unexpected information and maximizing expected information in the refined query output. To the best of our knowledge, there is no other work that makes an unified (considering both expected and unexpected tuples) attempt to solve the above problem. In this paper, we propose a **Flex**ible Interactive **Q**uerying (FlexIQ) framework to address the mentioned problem by exploiting the skyline operator.

Why Skyline Operator? Why-Not DT? The skyline operator is well-known for its intuitive query formalization and easy to understand semantics [1], [14]. This operator can also be used to draw a boundary between answers and non-answers (explained in Section 2). The advantage of this skyline operator based boundary is that it is data-driven and its formulation does not depend on any underlying data distribution. On the other hand, DT is basically an information-gain-theory based linear classifier and its success relies heavily on underlying data distribution [13]. In real life, relational data does not guarantee any statistical data distribution (though there are functional dependencies and sometimes correlations between attributes). Therefore, the DT technique is not very successful in relational data settings (experiments suggest this too).

The organization and contributions of this paper are as follows: after defining the data-driven query output semantics which serves as a summarized explanation of query output (Section 2); (a) we discuss the conflict and redundancy in user feedback and related properties (Section 3); (b) we show how to construct a new boundary for the refined query output by exploiting skyline operator (Section 4); (c) we propose a solution for the mentioned query refinement problem (Section 5); (d) we validate our approach with experiments and demonstrate its effectiveness by comparing our results with DT based query refinement (Section 6); we discuss related works (Section 7) and conclude the paper (Section 8).

2 Background

Let Q be the query, D be the database tuples (universe of discourse for Q), $Q(D)$ denote all tuples satisfied by the predicates given in Q, $Q'(D)$ denote all tuples not satisfied by the predicates given in Q, where $D = Q(D) \cup Q'(D)$. We use R and $Q(D)$ alternatively in this paper. Let G be the predicate preference in Q which consist of d atomic predicates i. e., $\{g_1, g_2, ..., g_d\}$. Each g_k is a triple $< a_k, op_k, v_k >$, where a_k be the attribute in D (i.e., column_name), op_k be the operator and v_k be the binding value. We assume that each atomic predicate g_k is equally important to the user. We consider G^d be the d-dimensional space where each dimension represents a particular preference $g_k \in G$. Therefore, each tuple $t_i \in D$ represents a d-dimensional point in G^d. We use $t_i.v_k$ to denote the kth dimensional value of t_i so that t_i can be represented as $t_i = < t_i.v_1, t_i.v_2, ..., t_i.v_d >$. We use $Q \vdash t_i$ to indicate that t_i satisfies the preference G in Q.

Definition 1. *A tuple t_i is said to dominate another tuple t_j, denoted by $t_i \succ_G t_j$, iff $\forall k \in \{1, 2, ..., d\}, t_i \geq_{g_k} t_j$ and $\exists l \in \{1, 2, ..., d\}, t_i >_{g_l} t_j$; t_i is said to be as good as t_j, denoted by $t_i \succeq_G t_j$, iff $\forall k \in \{1, 2, ..., d\}, t_i \geq_{g_k} t_j$.*

The relation $t_i \geq_{g_k} t_j$ holds if $t_i.v_k \otimes t_j.v_k$, where $\otimes$ is op_k that appears in g_k with equality ('=') added. The relation $t_i >_{g_k} t_j$ holds if $t_i.v_k \otimes' t_j.v_k$, where $\otimes'$ is $\otimes$ with equality ('=') dropped. If relation $t_i \succ_G t_j$ holds for tuples t_i and t_j in D, then the relation $t_i \succeq_G t_j$ holds implicitly. The relation $t_i \succ_G t_j$ is known as *weak pareto-dominance* [8],[16]. It should be noted that $t_i \succeq t_i$, but $t_i \nsucc t_i$.

Example 1. Consider the dataset given in Fig. 1. We can see that tuple $P3$ dominates tuple $P1$ in terms of CitationCnt and is as good as $P1$ in terms of PybYear. Therefore, we say $P3 \succ_G P1$ (also $P3 \succeq_G P1$). Similarly, $P12 \succ_G P11$ and $P3 \succeq_G P3$.

Lemma 1. *For all $t_i, t_j \in D$, if $t_i \succeq_G t_j$ and $Q(D)$ includes t_j (i.e., $Q \vdash t_j$), then $Q(D)$ must include t_i (i.e., $Q \vdash t_i$). Similarly, if $t_i \succeq_G t_j$ and $Q(D)$ does not include t_i (i.e., $Q \nvdash t_i$), then $Q(D)$ must not include t_j (i.e., $Q \nvdash t_j$).*

Definition 2. *A tuple t_j is said to be in space(t_i) iff $t_j \succeq t_i$. That is, space(t_i) includes each tuple t_j that dominates or is as good as t_i.*

Definition 3. *A tuple $t_i \in Q(D)$ is said to be a boundary tuple for $Q(D) = \{t_1, t_2, ..., t_n\}$ in terms of G iff $\forall m \in \{1, 2, ..., n\}, t_i \nsucc_G t_m$.*

Similar to the skyline tuples, if t_i and t_j are two boundary tuples, then $t_i \nsucc_G t_j$ and $t_j \nsucc_G t_i$. Let Γ be the *boundary* of $Q(D)$. Then, we compute Γ as follows: (1) Γ is initialized to $Q(D)$ and (2) for each $t_i \in \Gamma$, if $\exists t_j \in \Gamma$ such that $t_i \neq t_j$ and $t_i \succeq_G t_j$, then we remove t_i from Γ (steps 2 through 4 in Algorithm 1). There is a subtle difference between boundary tuples and traditional skyline tuples [1]. Boundary tuples capture the contour of query output (as shown in Fig. 2) and are dominated by other tuples. On the contrary, skyline tuples generally dominate other tuples [1], [14]. We twist the definition of traditional skyline tuples from "dominate" to "be dominated" to serve our purpose in this paper.

Algorithm 1. Computing Boundary

Input: $Q(D)$ **Output:** Γ

1: $\Gamma \leftarrow Q(D);$ //initialization
2: **for** each $t_i \in \Gamma$ **do**
3:   **if** $\exists t_j \in \Gamma$ such that $t_i \neq t_j$ and $t_i \succeq_G t_j$ **then**
4:     Remove t_i from Γ;

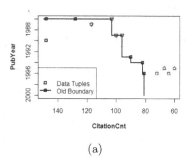
(a)

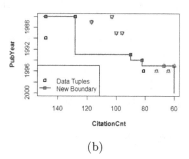
(b)

Fig. 2. (a) Data tuples and Old boundary (b) Data tuples and New boundary: unexpected and expected tuples are marked with lower and upper triangles, respectively

Example 2. According to Definition 3 and Algorithm 1, we compute Γ for the example query output given in Fig. 1 as {P1, P4, P7, P8, P9}. These boundary tuples separate $Q(D)$ and $Q'(D)$ as shown in Fig. 2(a). We also see from Fig. 2(a) that this boundary includes the unexpected tuples (marked with red colored lower triangles) in $Q(D)$, but misses the expected tuples (marked with green colored upper triangles) .

Definition 4. *Let Ω be the semantics of $Q(D)$. We then define Ω as follows:*

$$\Omega = \{s_i\} \text{ where } s_i = \{< a_1, \otimes_1, t_i.v_1 >, ..., < a_d, \otimes_d, t_i.v_d >\}, \forall t_i \in \Gamma \quad (1)$$

where a_k and $\otimes_k$ ($\otimes_k$ is op_k with equality ('=') added) come from $g_k = <a_k, op_k, v_k >$ and $d = |G|$. The s_i describes all tuples t_j that are in $space(t_i)$.

Example 3. According to Definition 4, we compute Ω of $Q(D)$ given in Fig. 1 as $\Omega=\{s_1=\{<\text{CitationCnt}, \geq, 96>, <\text{PubYear}, \geq, 1989>\}, s_4=\{<\text{CitationCnt},\geq, 90>,<\text{PubYear},\geq,1993>\}, s_7=\{<\text{CitationCnt},\geq,81>, <\text{PubYear},\geq,1996>\}, s_8= \{<\text{CitationCnt}, \geq, 103>,<\text{PubYear}, \geq, 1986>\}, s_9=\{<\text{CitationCnt}, \geq, 82>, <\text{PubYear}, \geq, 1994>\}\}$. The s_1, s_4, s_7, s_8 and s_9 describe the tuple set {P1, P3, P6}, {P4}, {P7}, {P2, P5, P6, P8, P10} and {P9}, respectively.

Lemma 2. *Semantics Ω precisely describes the query output $Q(D)$.*

Proof. We know from the definition of the semantics Ω of $Q(D)$ that for every tuple $t_i \in \Gamma$ there exists an $s_i \in \Omega$. We also know from the definition of Γ that for all $t_l \in Q(D)$ there exists a boundary tuple $t_i \in \Gamma$ such that $t_l \succeq t_i$. Therefore, we can say that Ω precisely describes the query output $Q(D)$.

2.1 Problem Statement

Let U be the set of unexpected tuples, $U \subseteq R$ and E be the set of expected tuples, $E \subseteq Q'(D)$. We use $Q \vdash U$ to indicate that every tuple $t_i \in U$ satisfies $G \in Q$ and $Q \nvdash E$ to indicate that every tuple $t_j \in E$ does not satisfy $G \in Q$. We then formally define our query refinement problem as follows: *"Given query Q, result set R, set of unexpected tuples U and set of expected tuples E, modify the initial query Q to Q^f in a way so that $t_i \in U$ does not satisfy the modified preference $G^f \in Q^f$ but $t_j \in E$ satisfies the modified preference $G^f \in Q^f$"*. In other words, find the refined query Q^f such that $Q^f \vdash (R \setminus U) \cup E$.

The discovery of the refined query Q^f relies heavily on the discovery of a new boundary that can separate unexpected and expected tuples. This can be easily observed from Fig. 2(a) and Fig. 2(b). This suggests that the query refinement problem is eventually transformed to the adjustment of boundary tuples for the refined query. The next challenge is how we can transform this adjusted boundary to the refined query, Q^f. We explain this transformation step in Section 5.

2.2 Overview of FlexIQ

In FlexIQ, we collect both unexpected and expected tuples as feedback from the user. Then, we analyze this feedback to find a minimal representation. We also find the boundary that separates answers (i.e., $Q(D)$) from non-answers (i.e., $Q'(D)$), as shown in Fig. 2(a). Then, we form a new boundary that excludes unexpected tuples and includes expected tuples in $Q^f(D)$, as shown in Fig. 2(b). Then, we offer a baseline algorithm for constructing the refined query Q^f in relation to the new boundary (see Section 5.1). Finally, we offer a trade-off algorithm to minimize the number of clauses in the refined query so that it becomes semantically as close as possible to the original query (see Section 5.2).

3   User Feedback

In FlexIQ, a user actively participates in the solution process. That is, once the initial query result is presented to the user, she then identifies the portion of the current result that is unexpected (U) and the part of the new information that is expected (E). In collecting these feedback, there is a possibility of *conflict* and *computational redundancy*.

Definition 5. *Tuples t_i and t_j conflict with each other iff any of the following holds: (a) $t_i \in U$ and $t_j \in Q(D) \setminus U$ such that $t_i \succeq_G t_j$; (b) $t_i \in E$ and $t_j \in Q'(D) \setminus E$ such that $t_j \succeq_G t_i$ and (c) $t_i \in U$ and $t_j \in E$ such that $t_i \succeq_G t_j$.*

We resolve conflicts as pre-processing in our framework. The basic idea of this pre-processing is checking the pairwise dominance between tuples of U and $Q(D) \setminus U$; E and $Q'(D) \setminus E$; finally between tuples of U^+ and E^+, where U^+ and E^+ are the extended version of U and E respectively. The construction of U^+ is done as follows: (1) U^+ is initialized to U (2) for all $t_i \in U$ if there exists

---

Algorithm 2. Resolving User Conflict

Input: U and E **Output:** U^+ and E^+

1: $U^+ \leftarrow U$;//initialization
2: **for** each $t_i \in U$ **do**
3:     **if** $\exists t_j \in Q(D) \setminus U$ such that $t_i \succeq_G t_j$ **then**
4:         Add t_j to U^+; // t_j is no better than t_i
5: $E^+ \leftarrow E$;//initialization
6: **for** each $t_i \in E$ **do**
7:     **if** $\exists t_j \in Q'(D) \setminus E$ such that $t_j \succeq_G t_i$ **then**
8:         Add t_j to E^+; // t_j is as good as t_i
9: **for** each $t_i \in U^+$ **do**
10:     **if** $\exists t_j \in E^+$ such that $t_i \succeq_G t_j$ **then**
11:         Remove t_j from E^+;

---

any $t_j \in Q(D) \setminus U$ such that $t_i \succeq t_j$, then t_j is added to U^+ (see steps 1 through 4 in Algorithm 2). A similar approach is followed for the construction of E^+ (see steps 5 through 8 in Algorithm 2). Finally, we resolve user conflict by deleting tuples from E^+ that are dominated by tuples in U^+ (see steps 9 through 11 in Algorithm 2). The rationale of this deletion is that expected tuples are believed to be better than unexpected tuples in terms of G.

Example 4. Consider the user feedback $U=\{P1, P3, P10\}$ and $E=\{P11, P12\}$ as given in Fig. 1. Then, we compute U^+ as $\{P1, P3, P8, P10\}$ and E^+ as $\{P11, P12, P13, P14\}$. The tuple $P8$ is added to U^+ as $P10 \succeq_G P8$. Similarly, $P13$ and $P14$ are added to E^+ as $P13 \succeq_G P11$ and $P14 \succeq_G P11$.

As user feedback may have conflict, we restate the query refinement problem given in Section 2 as follows: "*Given query Q, result set R, set of unexpected tuples $U \subseteq R$ and set of expected tuples $E \subseteq Q'(D)$, find the refined query Q^f such that $Q^f \vdash (R \setminus U^+) \cup E^+$*".

The above problem statement allows users to incompletely define the feedback. That is, the user does not need to mention all of her feedback as long as other members of feedback are no better than the currently provided unexpected tuples and no worse than the currently provided expected tuples.

Definition 6. *Computational redundancy*

(a) *A tuple $t_i \in U^+$ is said to be redundant wrt G iff $\exists t_j \in U^+$ such that $t_j \succeq_G t_i$.*
(b) *A tuple $t_i \in E^+$ is said to be redundant wrt G iff $\exists t_j \in E^+$ such that $t_i \succeq_G t_j$.*

In the above definition, we say t_i is computationally redundant because the exclusion of t_j ensures the exclusion of t_i in $Q^f(D)$ (see Lemma 1). Similarly, the inclusion of t_j ensures the inclusion of t_i in $Q^f(D)$ (Lemma 1). Let ζ and ξ be the *redundancy-free* feedback for unexpected and expected information, respectively. Then, we define the properties of these sets as follows:

Definition 7. $\forall t_i \in \zeta$ *the following holds: (a) $\neg \exists t_j \in U^+ \setminus \zeta$ such that $t_j \succ_G t_i$ and (b) $\neg \exists t_j \in \zeta$ such that $t_j \succ_G t_i$.*

Definition 8. $\forall t_i \in \xi$ *the following holds: (a)* $\neg \exists t_j \in E^+ \setminus \xi$ *such that* $t_i \succ_G t_j$ *and (b)* $\neg \exists t_j \in \xi$ *such that* $t_j \succ_G t_i$.

The computation of ζ is done as follows: (1) ζ is initialized to U^+ and (2) for each $t_i \in \zeta$, if $\exists t_j \in \zeta$ such that $t_i \neq t_j$ and $t_j \succeq_G t_i$, then we remove t_i from ζ. Similarly, ξ is computed as follows: (1) ξ is initialized to E^+ and (2) for each $t_i \in \xi$, if $\exists t_j \in \xi$ such that $t_i \neq t_j$ and $t_i \succeq_G t_j$, then we remove t_i from ξ. We say that minimal user feedback consists of ζ and ξ.

Example 5. Consider the U^+ and E^+ given in Example 4. According to the definition of ζ and ξ given above, we get ζ={P3, P10} and ξ={P11}.

Proposition 1. *The ζ and ξ are the necessary and sufficient information needed for updating G to exclude U^+ and include E^+ into $Q^f(D)$.*

Proof. We know that $\zeta \subseteq U^+$ and $\xi \subseteq E^+$. Definition 7 ensures $\neg \exists t_i \in \zeta$ and $\neg \exists t_j \in U^+ \setminus \zeta$ such that $t_j \succ_G t_i$ and $\forall t_i \in \zeta$, $\exists t_j \in U^+$ such that $t_i \succeq_G t_j$. Therefore, ζ is the necessary and sufficient information needed for updating G (i.e., the initial query Q) for the exclusion of U^+ from $Q^f(D)$ (follows from Lemma 1). Similarly, Definition 8 ensures $\neg \exists t_i \in \xi$ and $\neg \exists t_j \in E^+ \setminus \xi$ such that $t_i \succ_G t_j$ and $\forall t_j \in E^+$, $\exists t_i \in \xi$ such that $t_j \succeq_G t_i$. Therefore, ξ is the necessary and sufficient information needed for updating G (i.e., the initial query Q) for the inclusion of E^+ into $Q^f(D)$ (follows from Lemma 1). Therefore, we say that ζ and ξ are the necessary and sufficient information needed for updating G to exclude U^+ and include E^+ into $Q^f(D)$, respectively.

Lemma 3. *The refined query Q^f can be constructed from the result set R and the minimal feedback ζ and ξ such that $Q^f \vdash R \setminus U^+ \cup E^+$.*

4   Boundary Adjustment

We describe in Section 2 that the query refinement problem is essentially transformed to the adjustment of boundary tuples of $Q(D)$. That is, the adjusted boundary captures the semantics of the refined query output $Q^f(D)$ (as the original boundary captures the semantics of $Q(D)$). We prove this later in this section. Now, let Γ be the boundary between $Q(D)$ and $Q'(D)$, ζ and ξ be the minimal feedback for the unexpected and expected tuples respectively. Then, we define our boundary adjustment problem as follows: *"Given boundary Γ of the original query Q and minimal feedback ζ and ξ, find the new boundary Γ' for refined query Q^f such that $Q^f \vdash R \setminus U^+ \cup E^+$".*

The basic idea of our boundary adjustment algorithm is checking the pairwise dominance between feedback tuples and boundary tuples of the original query Q. That is, if any unexpected tuple t_i dominates any boundary tuple $t_j \in \Gamma$ then we adjust the boundary in a way so that t_i will not dominate any boundary tuple in Γ' again. We also apply similar idea to include expected tuples in the new boundary Γ'. Algorithm 3 implements the above and computes the new boundary Γ' given Γ, ζ and ξ for the refined query Q^f. The following proposition

---

Algorithm 3. Boundary Adjustment

Input: original boundary, Γ and minimal feedback, unexpected ζ and expected ξ
Output: new boundary Γ' for the refined query Q^f
 1: $\Gamma' \leftarrow \Gamma$;
 2: **for** each $t_i \in \zeta$ **do**
 3:     **if** $\exists t_j \in \Gamma'$ such that $t_i \succeq_G t_j$ and $t_i \in space(t_j)$ **then**
 4:         Remove t_j from Γ';
 5:         $temp\_tuple\_set \leftarrow space(t_j)$;
 6:         **if** $\exists t_m \in temp\_tuple\_set$ such that $t_i \succeq t_m$ **then**
 7:             Remove t_m from $temp\_tuple\_set$;
 8:         $temp\_boundary \leftarrow$ Compute boundary for $temp\_tuple\_set$;//Algorithm 1
 9:         Add $temp\_boundary$ to Γ';
10: **for** each $t_i \in \xi$ **do**
11:     **if** $\exists t_j \in \Gamma'$ such that $t_j \succeq_G t_i$ and $t_j \in space(t_i)$ **then**
12:         Remove t_j from Γ'; Add t_i to Γ'; // $space(t_j) \subseteq space(t_i)$
13:     **else**
14:         Add t_i to Γ';

---

proves that the new boundary Γ' precisely captures the semantics of the refined query output $Q^f(D)$.

Proposition 2. *The new boundary Γ' precisely captures the semantics of the refined query output $Q^f(D)$.*

Proof. Let Ω' be the semantics of the refined query output $Q^f(D)$ and be constructed from Γ' according to the definition of query output *semantics* given in Section 2. Algorithm 3 ensures $\neg \exists t_i \in \Gamma'$ and $\neg \exists t_j \in U^+$ such that $t_j \succeq t_i$. Algorithm 3 also ensures $\forall t_j \in E^+$, $\exists t_i \in \Gamma'$ such that $t_j \succeq t_i$. In other words, Γ' defines the contour for $R \setminus U^+ \cup E^+$. Therefore, we can say that Γ' precisely captures the semantics of $Q^f(D)$.

Example 6. Given dataset in Fig. 1, $\zeta=\{P3, P10\}$, $\xi=\{P11\}$ and $\Gamma=\{P1, P4, P7, P8, P9\}$, we compute the new boundary Γ' as $\{P2, P4, P9, P11\}$. This new boundary clearly separates E^+ and U^+ as shown in Fig. 2(b).

5   Query Refinement

This section presents the final step of our framework. The basic idea of our query refinement framework is constructing the new boundary for the refined query Q^f by analyzing the user feedback and then transforming the new boundary to the refined query Q^f. In Section 4, we show how to construct the new boundary Γ' given the original boundary Γ and the minimal feedback ζ and ξ. Now, we provide two algorithms for constructing the refined query Q^f from the adjusted boundary Γ': a baseline algorithm and then a trade-off algorithm.

5.1   The Baseline Algorithm

The baseline algorithm (TBA) converts the adjusted boundary Γ' to disjunction of conjunctions for the refined query Q^f. That is, the constituents s_i of the semantics Ω' constructed from Γ' are transformed directly to the disjunctive query. The transformation formula is given below:

$$Q^f \leftarrow \vee_{i=1}^{|\Omega'|} conjunct(s_i), \forall s_i \in \Omega' \tag{2}$$

The function $conjunct(s_i)$ returns the corresponding conjunction for $t_i \in \Gamma'$ as follows:

$$conjunct(s_i) \leftarrow a_1. \otimes_1 .t_i.v_1 \wedge a_2. \otimes_2 .t_i.v_2 \wedge ... \wedge a_d. \otimes_d .t_i.v_d \tag{3}$$

The size of the refined query Q^f returned by TBA is $|\Omega'|$ (also $|\Gamma'|$). That is, it consists of $|\Omega'|$ conjunctions (i.e., subqueries).

Example 7. The baseline algorithm computes the refined query for the example dataset given in Fig. 1 as follows: SELECT pubid FROM publication WHERE ((citationcnt $\geq$ 128 AND pubyear $\geq$ 1986) OR (citationcnt $\geq$ 90 AND pubyear $\geq$ 1993) OR (citationcnt $\geq$ 82 AND pubyear $\geq$ 1994) OR (citationcnt $\geq$ 60 AND pubyear $\geq$ 1995)).

Lemma 4. *The refined query Q^f obtained by following eqns. (2) and (3) optimally separates unexpected and expected tuples. That is, eqns. (2) and (3) ensures $Q^f \vdash R \setminus U^+ \cup E^+$.*

5.2   The Trade-Off Algorithm

The TBA converts each boundary tuple $t_i \in \Gamma'$ to a subquery for Q^f. The number of subqueries is equal to $|\Omega'|$, which is a major disadvantage of TBA. We can reduce this number of subqueries by combining them (i.e, approximation). Let us consider two tuples t_i and t_{i+1} from Γ'. We propose four types of combination strategies for these two tuples: (a) *maximal* (b) *minimal* (c) t_i and (d) t_{i+1}.

Definition 9. *(maximal combination) Two tuples t_i and t_{i+1} are said to be maximally combined to t_{max} only if $space(t_i) \cup space(t_{i+1}) \subseteq space(t_{max})$. The definition of $space(t_i)$ is the same as we define it in Section 2. The construction of t_{max} is done as follows:*

$$t_{max}.v_k \leftarrow \begin{cases} min(t_i.v_k, t_{i+1}.v_k) \text{ if } op_k \in \{`>=`,`>`\} \\ max(t_i.v_k, t_{i+1}.v_k) \text{ if } op_k \in \{`<`,`<=`\} \end{cases} \tag{4}$$

Definition 10. *(minimal combination) Two tuples t_i and t_{i+1} are said to be minimally combined to t_{min} only if $space(t_i) \cap space(t_{i+1}) \neq \emptyset$ and $space(t_i) \cap space(t_{i+1}) \subseteq space(t_{min})$. The construction of t_{min} is done as follows:*

$$t_{min}.v_k \leftarrow \begin{cases} max(t_i.v_k, t_{i+1}.v_k) \text{ if } op_k \in \{`>=`,`>`\} \\ min(t_i.v_k, t_{i+1}.v_k) \text{ if } op_k \in \{`<`,`<=`\} \end{cases} \tag{5}$$

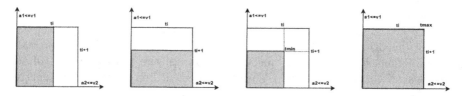

Fig. 3. The four combinations of tuples t_i and t_{i+1} are t_i, t_{i+1}, t_{min} and t_{max}. The shaded area indicates the portion covered by the corresponding combination.

Let G represents a two-dimensional space with predicates $a_1 \leq v_1$ and $a_2 \leq v_2$. Then, we can visualize t_i, t_{i+1}, t_{min} and t_{max} as we see it in Fig. 3. We replace t_i and t_{i+1} by any of them. The selection of t_{max} depends on whether any unexpected tuple $t_u \in U^+$ dominates any tuple $t_j \in space(t_{max})$ or not (i.e., valid combination). We define this condition as follows:

Definition 11. *if $\neg \exists t_u \in U^+$ such that $t_u \succeq t_{max}$, then replace t_i and t_{i+1} with t_{max}.*

If the condition given above does not hold, then we select the one among t_i, t_{i+1}, t_{min} and t_{max} which retains the highest fraction of positive tuples from $space(t_i) \cup space(t_{i+1})$ and introduces fewer false positives. We term the fraction of positive tuples retained by any combination as *fitness score* (FS). Let δ_i denotes $|space(t_i)|$, δ_{i+1} denotes $|space(t_{i+1})|$, δ_{min} denotes $|space(t_{min})|$ and δ_{max} denotes $|space(t_{max})|$. We define FS for t_i, t_{i+1}, t_{min} and t_{max} as follows:

$$FS(t_i) = \frac{\delta_i}{\delta_i + \delta_{i+1} - \delta_{min}}, FS(t_{i+1}) = \frac{\delta_{i+1}}{\delta_i + \delta_{i+1} - \delta_{min}} \tag{6}$$

$$FS(t_{min}) = \frac{\delta_{min}}{\delta_i + \delta_{i+1} - \delta_{min}}, FS(t_{max}) = \frac{\delta_i + \delta_{i+1} - \delta_{min}}{\delta_{max}} \tag{7}$$

To minimize the loss, we further maintain a predefined threshold, δ. If the selected combination does not maintain its FS above δ then we leave t_i and t_{i+1} as they were before. Finally, we establish the order by which we can traverse the tuples in Γ'. We propose to sort the tuples in Γ' along any dimension and then consider successive pair of tuples for combination. The pseudo-code given in Algorithm 4 implements all these approximations described above. We call this algorithm as trade-off algorithm (TOA) and its complexity is $O(|\Gamma'|\log|\Gamma'| + C|\Gamma'|)$.

Approximated Query:
SELECT pubid FROM publication WHERE
((citationcnt $\geq$ 128 AND pubyear $\geq$ 1986) OR
(citationcnt $\geq$ 60 AND pubyear $\geq$ 1993));

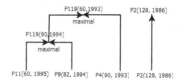

Fig. 4. An example of computational steps made by TOA

Example 8. The refined query returned by TOA for dataset given in Fig. 1 is: SELECT pubid FROM publication WHERE ((citationcnt $\geq$ 128 AND pubyear $\geq$ 1986) OR (citationcnt $\geq$ 60 AND pubyear $\geq$ 1993)). The construction of this query is illustrated in Fig. 4.

Algorithm 4. Trade-Off Algorithm

Input: New Boundary Γ' **Output:** Refined Query Q^f

1: Sort Γ' in any dimension;
2: **for** $\forall t_i, t_{i+1} \in \Gamma'$ **do**
3:     $t_{max} \leftarrow$ maximal combination of t_i and t_{i+1};//According to eqn. (4)
4:     **if** $\neg\exists t_u \in U^+$ such that $t_u \succeq t_{max}$ **then**
5:         Replace t_i, t_{i+1} with t_{max}; //t_{max} is a valid combination
6:     **else**
7:         $t_{min} \leftarrow$ minimal combination of t_i and t_{i+1};//According to eqn. (5)
8:         Compute $FS(t_i), FS(t_{i+1}), FS(t_{min})$ and $FS(t_{max})$;//Accord. to eqn. (6)-(7)
9:         $t_c \leftarrow$ argmax $\{FS(t_i), FS(t_{i+1}), FS(t_{min}), F(t_{max})\}$;
10:       **if** $FS(t_c) \geq \delta$ **then**
11:          Replace t_i and t_{i+1} with t_c;// t_i and t_{i+1} is approximately replaced with t_c
12: Compute Ω' for Γ';//According to eqn. (1)
13: $Q^f \leftarrow \vee_{i=1}^{|\Omega'|} conjunct(s_i), \forall s_i \in \Omega'$// According to eqns. (2) and (3)

---

6 Experiments

6.1 Setup

Environment: We run all experiments using an Intel(R) Core(TM) Duo E8400 3.0 GHz Windows XP PC with 3.49 GB RAM. The refinement algorithms are implemented in Java along with MySQL server 5.1 and SDK is Eclipse 3.5.

Dataset and Queries: We use DBLP dataset (download: http://dblp.uni-trier.de/xml/) of size 456 MB. We convert the XML based DBLP dataset into relational data format which consists of six tables. We create a set of base queries that returns 15-200 tuples in the initial result set. The base queries consist of both uni-point queries (no disjunction in query condition) and multi-point queries (disjunction of several subqueries).

User Feedback: We randomly pick a set of tuples from the initial result set as unexpected tuples. To get expected tuples, we pick tuples from non-answers (i.e., $Q'(D)$). Once we get expected tuples and unexpected tuples, we then construct *redundancy-free feedback* and resolve the *conflict* (described in Section 3).

Decision-Tree Based Query Refinement: A *decision tree* (DT) is a tree-like model of decisions. Given an input with well-defined attributes, the DT can classify the input entirely based on making choices about each attribute [13]. Tran et al. [15] and Ma et al. [12] propose DT based query refinement techniques to handle expected tuples and unexpected tuples in the refined query output, respectively. To compare the effectiveness of our framework, we use the DT based query refinement approach described in [7] for managing both unexpected and expected tuples in the refined query output. We use the best-known and most widely-used C4.5 decision tree learning algorithm and WEKA implementation of it [3]. To boost the precision of DT, we set WEKA parameters as follows: *minNumObj* to 1, *unpruned* to False and *subtreeRaising* to True.

Evaluation Process: We evaluate our algorithms proposed in this paper in two different aspects: (1) Quality of Results: false positive rate (FPR) ($=n_{fp}/(n_{tp} + n_{fp})$), false negative rate (FNR) ($=n_{fn}/(n_{tn} + n_{fn})$) and accuracy (ACC) ($=(n_{tp}+n_{tn})/(n_{tp}+n_{fp}+n_{tn}+n_{fn})$) measures [13]; where n_{tp} is the number of truly positive tuples, n_{fp} is the number of false positive tuples (i.e., unexpected tuples), n_{tn} is the number of truly negative tuples and n_{fn} is the number of false negative tuples (i.e., expected tuples); and (2) Query Size: number of subqueries in the refined query. We compare the performance of the proposed query refinement algorithms as well as the DT based query refinement [7] as follows: (1) We make 100 different input instances for each base query by randomly picking the result tuples as the unexpected tuples and setting the virtual tuples (randomly relaxing the predicates in the given query) from non-answers as the expected tuple set; and (2) Finally, we calculate the average of FPR and FNR (in %) and the refined query size (i.e., number of subqueries(#NSUB)) of the proposed algorithms and the DT based query refinement.

Table 1. Average FPR(%)

Experiments	TBA	TOA	DT
Exp#1$_{(\delta=0.65)}$	0.00	26.27	9.72
Exp#2$_{(\delta=0.70)}$	0.00	20.94	9.66
Exp#3$_{(\delta=0.75)}$	0.00	15.32	8.99
Exp#4$_{(\delta=0.80)}$	0.00	11.25	8.79
Exp#5$_{(\delta=0.85)}$	0.00	5.74	14.34
Exp#6$_{(\delta=0.90)}$	0.00	2.18	9.57
Exp#7$_{(\delta=0.95)}$	0.00	0.38	9.79
Exp#8$_{(\delta=0.97)}$	0.00	0.11	9.54
Exp#9$_{(\delta=0.99)}$	0.00	0.00	8.06

Table 2. Average FNR(%)

Experiments	TBA	TOA	DT
Exp#1$_{(\delta=0.65)}$	0.00	6.39	30.47
Exp#2$_{(\delta=0.70)}$	0.00	6.75	29.20
Exp#3$_{(\delta=0.75)}$	0.00	7.71	25.97
Exp#4$_{(\delta=0.80)}$	0.00	6.83	30.05
Exp#5$_{(\delta=0.85)}$	0.00	6.19	27.99
Exp#6$_{(\delta=0.90)}$	0.00	4.54	30.74
Exp#7$_{(\delta=0.95)}$	0.00	1.35	28.59
Exp#8$_{(\delta=0.97)}$	0.00	1.23	30.47
Exp#9$_{(\delta=0.99)}$	0.00	0.04	28.96

6.2   Results

TBA vs Other Methods: The baseline algorithm (TBA) performs well compared to TOA and DT based query refinement in terms of quality of the refined query results. The FPR and FNR measures for TBA is 0% as we see these measures in Table 1 and Table 2. This means that TBA optimally separates answers from non-answers, as we prove in section 5.1 (Lemma 4). Therefore, the ACC measure of TBA is 100% (see Table 3). To achieve this quality of results, TBA offers complex formulation of the refined queries compared to other methods as we see it in Table 4 (#NSub) which is a major disadvantage of TBA.

TOA vs DT: We conduct nine different experiments for TOA by setting δ to nine different values from 0.65 to 0.99 (though we can set δ to any value in the range 0.0 to 1.0 and the effect of δ in TOA is explained later in this section) to demonstrate its effectiveness for UPQs. The observed results are shown in Table 1-4. The TOA offers less FPR (for $\delta \geq 0.85$) and FNR (any value of δ) compared to DT based query refinement as we see in Table 1 and Table 2. The TOA also offers better ACC measure compared to DT based query refinement (see it in Table 3). For example, we achieve 99.32% ACC when we set δ to 0.95. The offered complexity of TOA depends on a particular setting of δ. We achieve less complex queries for TOA compared to DT by setting δ to ≤ 0.80 in our experiments (the first four rows in Table 4).

Table 3. Average ACC(%)

Experiments	TBA	TOA	DT
Exp#1$_{(\delta=0.65)}$	100.00	77.05	78.19
Exp#2$_{(\delta=0.70)}$	100.00	81.33	78.23
Exp#3$_{(\delta=0.75)}$	100.00	86.00	81.80
Exp#4$_{(\delta=0.80)}$	100.00	89.57	78.09
Exp#5$_{(\delta=0.85)}$	100.00	93.85	78.47
Exp#6$_{(\delta=0.90)}$	100.00	97.04	78.19
Exp#7$_{(\delta=0.95)}$	100.00	99.32	78.41
Exp#8$_{(\delta=0.97)}$	100.00	99.67	78.59
Exp#9$_{(\delta=0.99)}$	100.00	99.99	78.28

Table 4. Average #NSub

Experiments	TBA	TOA	DT
Exp#1$_{(\delta=0.65)}$	13.04	1.35	2.84
Exp#2$_{(\delta=0.70)}$	12.93	1.66	2.86
Exp#3$_{(\delta=0.75)}$	13.01	2.05	2.87
Exp#4$_{(\delta=0.80)}$	13.04	2.63	2.88
Exp#5$_{(\delta=0.85)}$	12.94	3.95	2.92
Exp#6$_{(\delta=0.90)}$	13.13	5.02	2.94
Exp#7$_{(\delta=0.95)}$	12.97	6.70	2.91
Exp#8$_{(\delta=0.97)}$	12.97	7.03	2.82
Exp#9$_{(\delta=0.99)}$	12.82	7.22	2.91

To demonstrate the effectiveness of TOA for multi-point queries, we run four different experiments for TOA by setting δ to four different values from 0.80 to 0.95. Table 5 presents the average performance of TOA and DT for multi-point queries. We observe that TOA outperforms DT from all perspectives. This outcome suggests that TOA performs better for multi-point queries compared to uni-point queries. For example, at $\delta = 0.95$, average FPR for multi-point queries is 0.21% whereas for uni-point queries this score is 0.38%.

Table 5. Multi-Point Query Results: TOA vs. DT

Experiments	TOA				DT			
	FPR(%)	FNR(%)	Acc(%)	#NSUB	FPR(%)	FNR(%)	Acc(%)	#NSUB
Exp#1$_{(\delta=0.80)}$	5.62	10.87	91.84	2.34	7.21	47.02	69.92	4.54
Exp#2$_{(\delta=0.85)}$	2.99	11.57	94.26	2.79	7.06	51.35	68.94	4.80
Exp#3$_{(\delta=0.90)}$	0.89	7.05	97.25	3.45	6.12	51.51	68.63	4.78
Exp#4$_{(\delta=0.95)}$	0.21	0.59	99.61	4.49	6.98	48.66	68.55	4.68

Effect of δ on FPR, FNR, ACC and #NSUB in TOA: The effect of δ on FPR, FNR, ACC and #NSUB in TOA can be observed from graphs shown in Fig. 5 (a)-Fig. 5 (d). The FPR and FNR tend to zero in TOA if δ tends to one and the complexity of the refined queries tends to be minimum if δ tends to zero. The ACC measure for TOA tends to 100% as δ tends to 1.0.

Trading-off δ in TOA: We can trade-off δ in TOA to achieve different quality metrics (quality of results and refined query size). If we are interested in low false positive rates and low false negative rates (as well as better ACC measure)

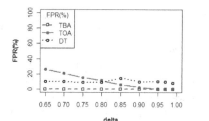
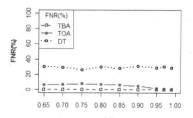

Fig. 5(a). Effect of delta (δ) on FPR(%)   **Fig. 5(b).** Effect of delta (δ) on FNR(%)

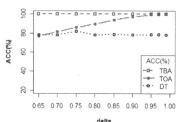
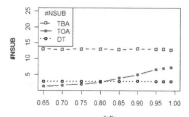

Fig. 5(c). Effect of delta (δ) on ACC(%)   **Fig. 5(d).** Effect of delta (δ) on #NSUB

we can set δ close to 1.0. On the other hand, if we need less complex refined queries we can set δ to $\ll 1$.

7   Related Work and Discussion

This work is inspired by the previous works found in [6], [5], [15], [4], [12], [10], [9] and [11]. In [6] Huang et al. and in [5] Herschel et al. propose to modify the original tuple values in the database so that missing tuples become part of the query output. However, their techniques are not applicable in trusted data applications. In [2] Chapman et al. propose to identify the culprit operator(s) that filters out expected tuple(s). As a next step, Tran and Chan [15] model query refinement by collecting missing tuples as feedback from the user. These authors exploit the idea of *skyline queries* to report the closest refined query wrt the original one to minimize the distance between refined and original query. In a very recent work [4], He et al. propose an approach to answer why-not questions (missing tuples) on top-k queries. This work also does not consider why (unexpected) tuples. In [12], Ma *et al.* model query refinement as both learning the structure of the query as well as learning the relative importance of query components. But they do not consider what new information a user expects to see (i.e. what is missing). In [10], Liu *et al.* collect *false positives* (which we call unexpected information in this paper) which are identified by users to modify the initial rules in information extraction settings. Koudas et al. [9] propose relaxation skyline as a solution for the empty answers problem. In [11], Ma et al. propose a framework that combines the positive aspects of both similarity retrieval and skyline retrieval into one single technique so that the user can retrieve results in the order of relevance.

None of the above models treats both unexpected and expected feedback. Recently, Islam et al. [7] proposed a decision tree based query refinement for minimizing unexpected information as well as maximizing expected information in the refined query output. The disadvantage of DT based query refinement is that it is based on information gain theory and relies heavily on underlying data distribution. In this paper, we propose FlexIQ, a framework for feedback based query refinement exploiting skyline operator. The advantage of FlexIQ is that it is independent of the underlying data distribution and the user can also trade-off different quality metrics in the refined queries. To the best of our knowledge, this is the first attempt ever made where skyline operator is exploited for controlling both unexpected and expected information in the refined query output.

8   Conclusion

This paper presents a novel feedback based query refinement framework which exploits the skyline operator. In this framework, we use feedback to discover the query intent of the user. In addition, the skyline operator is exploited to confine the search space of the query refinement algorithms and render the approach to be more intuitive to the user. The experimental results demonstrate that the proposed framework can effectively minimize the number of unexpected tuples as well as maximize the number of expected tuples in the refined query output. In our framework, the user can also trade-off different quality metric such as quality of results and number of subqueries in the refined query.

Acknowledgments. This work is partially supported by the grant of Australian Research Council Discovery Project No. DP120102627.

References

1. Börzsönyi, S., Kossmann, D., Stocker, K.: The skyline operator. In: ICDE (2001)
2. Chapman, A., Jagadish, H.V.: Why not? In: SIGMOD (2009)
3. Hall, M., Frank, E., Holmes, G., Pfahringer, B., Reutemann, P., Witten, I.H.: The weka data mining software: an update. SIGKDD Explorations 11(1) (2009)
4. He, Z., Lo, E.: Answering why-not questions on top-k queries. In: ICDE (2012)
5. Herschel, M., Hernández, M.A.: Explaining missing answers to spjua queries. PVLDB 3(1) (2010)
6. Huang, J., Chen, T., Doan, A., Naughton, J.F.: On the provenance of non-answers to queries over extracted data. PVLDB 1(1) (2008)
7. Islam, M.S., Liu, C., Zhou, R.: On modeling query refinement by capturing user intent through feedback. In: Australasian Database Conference (2012)
8. Kießling, W.: Foundations of preferences in database systems. In: VLDB (2002)
9. Koudas, N., Li, C., Tung, A.K.H., Vernica, R.: Relaxing join and selection queries. In: VLDB (2006)
10. Liu, B., Chiticariu, L., Chu, V., Jagadish, H.V., Reiss, F.: Automatic rule refinement for information extraction. PVLDB 3(1) (2010)

11. Ma, Y., Mehrotra, S.: Integrating Similarity Retrieval and Skyline Exploration Via Relevance Feedback. In: Kotagiri, R., Radha Krishna, P., Mohania, M., Nantajeewarawat, E. (eds.) DASFAA 2007. LNCS, vol. 4443, pp. 1045–1049. Springer, Heidelberg (2007)
12. Ma, Y., Mehrotra, S., Seid, D.Y., Zhong, Q.: RAF: An Activation Framework for Refining Similarity Queries Using Learning Techniques. In: Li Lee, M., Tan, K.-L., Wuwongse, V. (eds.) DASFAA 2006. LNCS, vol. 3882, pp. 587–601. Springer, Heidelberg (2006)
13. Mitchell, T.M.: Machine learning. McGraw-Hill (1997)
14. Su, I.-F., Chung, Y.-C., Lee, C.: Top-k Combinatorial Skyline Queries. In: Kitagawa, H., Ishikawa, Y., Li, Q., Watanabe, C. (eds.) DASFAA 2010. LNCS, vol. 5982, pp. 79–93. Springer, Heidelberg (2010)
15. Tran, Q.T., Chan, C.Y.: How to conquer why-not questions. In: SIGMOD (2010)
16. Voorneveld, M.: Characterization of pareto dominance. Oper. Res. Lett. 31(1) (2003)

A NFR-Based Framework
for User-Centered Adaptation

Fabiano Dalpiaz[1], Estefanía Serral[2], Pedro Valderas[2],
Paolo Giorgini[1], and Vicente Pelechano[2]

[1] DISI, University of Trento
{dalpiaz,pgiorgio}@disi.unitn.it
[2] PROS, Universitat Politècnica de València
{eserral,pvalderas,pele}@dsic.upv.es

Abstract. Pervasive environments support users' daily routines in an invisible and unobtrusive way. To do so, they include a technical pervasive infrastructure, which is aware of and adaptive to both the operational context and the users at hand. Non-Functional Requirements (NFRs) have been effectively used to inform decision-making in software engineering: functional alternatives are compared in terms of their contribution to NFRs satisfaction. In this work, we consider user preferences over NFRs as a key driver for the adaptation of a pervasive infrastructure. We devise a model-driven framework for building pervasive systems that maximize fitness with the context and the user. Our contributions are: (i) *adaptive task models*, a conceptual model to describe user routines that accounts for user preferences over NFRs; and (ii) an adaptation framework, which uses our models at runtime to guide a pervasive infrastructure in adapting its behaviour to user preferences and context.

Keywords: pervasive environments, self-adaptation, NFRs.

1 Introduction

In pervasive computing [17], technical systems are deployed in the environment—the so-called *pervasive environment*—so as to support humans in their daily activities. Crucially, pervasive environments have to remain, while executing, invisible and unobtrusive to users. The technical infrastructure of pervasive environments (*pervasive infrastructure*) effects changes in the environment and suggests appropriate activities to the users. While being guided by this system, the user should not realize that the system is "thinking" on her behalf.

Task models [14] are a modelling language to represent user routines [18] (sets of habitually performed tasks). They are an example of executable conceptual models, as they hierarchically specify and temporally relate the tasks a system should execute for supporting a user in the conduction of her daily routines. These models have been successfully adopted in model-driven pervasive infrastructures [18].

However, task models provide limited adaptation to user preferences about non-functional properties. In order to be unobtrusive and invisible, the system

P. Atzeni, D. Cheung, and R. Sudha (Eds.): ER 2012, LNCS 7532, pp. 439–448, 2012.

has to execute routines that support courses of action the user finds natural to her (i.e., that match her preferences). If the user is interested in carbon emissions reduction, the system should minimize heating usage and suggest going to work on foot. However, if she has scheduled early meetings and is late, the system should recommend fast transportation means such as driving.

We investigate how user preferences over non-functional properties can be taken into account by a pervasive infrastructure. Our approach relies upon Non-Functional Requirements (NFRs) [13]. NFRs have been successfully used to inform decision-making by choosing the alternative that maximizes the satisfaction of qualities. Based on successful applications in requirements models [23], architectures [5], and business processes [15], we investigate their effective usage NFRs with task models.

In this paper, we extend task models and propose a model-driven framework that enables a pervasive infrastructure to adapt its behaviour to the user preferences and the current context. Our contributions are as follows:

- *adaptive task models*, a modelling language that enriches task models with NFRs. The model is created at design-time, and used by the system at runtime to decide upon how to adapt its behaviour. User preferences over NFRs are captured by our proposed *contextual preference model*: each user specifies, in a context-dependent way, the priority she assigns to each NFR;
- *an adaptation framework* for building pervasive infrastructures that exploit adaptive task models and contextual preference models at runtime.

The paper is organised as follows. Sec. 2 presents our baseline. Sec. 3 introduces adaptive task models and the contextual preference model. Sec. 4 proposes an adaptation framework that exploits our proposed models at runtime. Sec. 5 discusses related work, draws our conclusions and outlines future directions.

2   Research Baseline: Task Models

We build on top of the task models by Serral et al. [18], which specify how a pervasive infrastructure can support its users in carrying out everyday activities. Task models are inspired by Hierarchical Task Analysis (HTA) [20], which constructs a task tree that refines a high-level task into a set of executable ones.

Running Example. A smart-home pervasive environment supports the daily routines of the home inhabitants through a set of pervasive services [18] that are interfaced with sensors and effectors. Consider the following routine: "Every working day, the system turns on the bathroom heating at 7:50 a.m. to make it warm enough for Bob to take a shower. At 8:00 a.m., the system makes a wake-up call, repeating it until Bob wakes up. Then, the room is illuminated and Bob is notified about the weather. Afterwards, when Bob enters the kitchen, the system makes a coffee, and suggests him the best way to go to work."   □

Fig. 1 illustrates the "Waking Up" routine. The root task is broken down into simpler tasks by means of two task refinement constructs: *exclusive refinement*

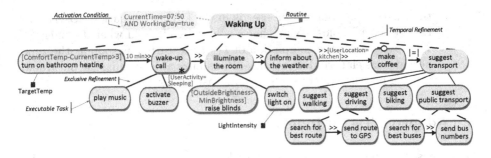

Fig. 1. Task model representing the "Waking Up" user routine

and *temporal refinement*. Exclusive refinement (graphically, a solid line) decomposes a task into a set of subtasks so that exactly one subtask will be executed. Temporal refinement (graphically, a dashed line) also decomposes a task into subtasks; however, all the subtasks shall be performed following a specific order which is depicted by the arrows between sibling tasks. Temporal constraints employ Concurrent Task Trees (CTT) operators [14]. For example, in Fig. 1:

- *Enablement* $(T_1 \gg T_2)$: task T_2 is triggered when task T_1 finishes. For instance, the system has to illuminate the room after waking up the user;
- *Task Independence* $(T_1 |=| T_2)$: T_1 and T_2 can be performed in any order. For instance, "make coffee" and "suggest transport" are temporally independent.

Task refinement ends when every leaf task is linked to a pervasive service (controlled by the pervasive infrastructure), which will execute the task. For example, "raise blinds" is executed by a pervasive service controlling the blinds engine.

A routine can be carried out through alternative sets of tasks depending on the current state of the context (the "situation" [11]). Situations are used to indicate the relationship between context and routine execution (see Fig. 1):

- *Activation condition*: it is associated with the root task of each routine. It indicates the situation in which the routine is activated. For instance, the "Waking Up" routine is to be executed every working day at 7:50 a.m.;
- *Task precondition*: it can be associated with a task to indicate that its execution depends on whether a situation (between square brackets) holds. For instance, the bathroom heating shall be turned on only if the difference between comfort and current temperature is greater than three degrees.
- *Iterative task*: it is executed repeatedly while the situation associated with the task holds. These tasks are graphically marked with an asterisk. For instance, task "wake-up call" is iterated while the user sleeps, and the iteration stops as soon as the user wakes up;
- *Temporal constraints:* the following relationships indicate that the execution of two tasks (linked by an arrow) is subject to a temporal constraint:
  - $T_1 \gg [s] \gg T_2$: after the completion of T_1, T_2 is started as soon as situation s holds. In Fig. 1, the system makes coffee after informing about

the weather, as soon as the user is in the kitchen (situation "UserLocation=kitchen" holds). Note that s could already hold when T_1 ends;

- $T_1 \; t \gg T_2$: after the completion of T_1, T_2 is started as soon as the time period t has elapsed. For instance, 10 minutes after turning the bathroom heating on, the system shall execute task "wake-up call".

3 Adaptive Task Models

We propose an extended version of task models (*adaptive task models*) to describe system behaviour that takes into account both the *personal context* [22], i.e., the individual requirements and preferences of specific users, and the *physical and social context*, i.e., observable characteristics of the environment that the system can monitor (e.g., who is in the room, temperature, closed and open doors).

Our extension enables not only adaptation to the preferences of different users—while one may be more concerned with energy efficiency, another may give priority to user comfort—, but also to the changing preferences of a specific user. For instance, if a user is in a hurry, she may favour efficiency over comfort; when not at home, instead, she may be more interested in energy saving.

Fig. 2 depicts the meta-model of adaptive task models. The red-coloured classes show our proposed contextual preference model (Sec. 3.1), which indicates user preferences over NFRs. The white-coloured classes represent the extended task model itself: we introduce optional and parametric tasks (Sec. 3.2), as well as task contributions to NFRs (Sec. 3.3). The green-coloured classes represent the context model (from previous work [18]).

3.1 Contextual Preference Model for NFRs

Each user has different preferences, which depend on the situation and vary over time. To represent users preferences over NFRs, we propose the contextual preference model (which extends [11]). Our model enables analysts to define the relevant NFRs and the priority assigned by each user in different contexts.

Table 1. Partial contextual preference model for user Bob

User Comfort (UC) :	⟨UserLocation≠Home, 0⟩
	⟨UserLocation=Home ∧ UrgentTasks=false, 0.7⟩
	⟨UserLocation=Home ∧ UrgentTasks=true, 0.5⟩
User Efficiency (UE) :	⟨UserLocation≠Home, 0⟩
	⟨UserLocation=Home ∧ UrgentTasks=true, 1⟩
	⟨UserLocation=Home ∧ UrgentTasks=false, 0.3⟩
Energy Efficiency (EE) :	⟨UserLocation≠Home, 0.9⟩
	⟨UserLocation=Home, 0.4⟩

For each NFR, a set of couples consisting of a situation s and a weight w (a real number in the range [0,1]) is specified. Each couple indicates that, when situation

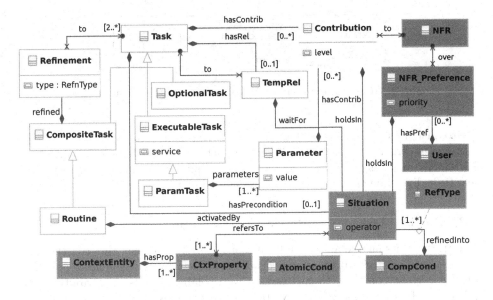

Fig. 2. Overview of adaptive task models

s holds, the NFR has priority w. Value 0 indicates minimum importance, while value 1 indicates maximum importance. The situations for a given NFR have to be mutually exclusive. Table 1 shows part of the contextual preference model for Bob. The weight of NFR user comfort is 0 when he is not at home; if Bob is at home, it is 0.7 if he has no urgent tasks, 0.5 otherwise.

3.2   Task Models with Optional and Parametric Tasks

Our approach to adaptation is model-driven in general, and NFR-driven in particular. The system adapts by choosing an adequate course of action from its task model, on the basis of contextual factors and user preferences over NFRs. To such extent, we enrich task models with optional and parametric tasks.

Optional Tasks are not essential to accomplish a specific routine. Their execution depends on user preferences over NFRs. For instance, task "make coffee" is optional. If the user has a meeting early in the morning, the NFR user efficiency will have a high priority, and task "make coffee" will not be executed. Note that optional is different from contextual: a contextual task is executed only if the context precondition holds, while an optional task relies on preferences. Graphically (see Fig. 3), optional tasks are represented by drawing a hollow circle to the incoming refinement link (like optional features in feature models [10]).

Parametric Tasks are leaf tasks in the task model whose execution can be tuned by adjusting the values of their parameters. This tuning is intended to maximise satisfaction of user preferences. For instance, task "turn on bathroom heating" can be tuned by adjusting the target temperature, while "switch

Fig. 3. Graphical representation of parametric and optional tasks

bedroom light on" by setting light intensity. Depending on the value of their parameters, the tasks will have different impacts on the NFRs. Graphically (see Fig. 3), parameters are labels—representing the parameter name—connected to leaf tasks by a line ending with a square-shaped arrow.

3.3   Linking Tasks to NFRs via Contributions

Variation points enable adaptability in a model-driven system: they are the loci in the model wherein alternative decisions are taken by the system (depending on the current context). In this section we explain how task contributions to NFRs can be exploited in adaptive task models to drive system adaptation so as to choose the alternative that fits best with user preferences.

Like in goal analysis [9], analysts indicate the contribution of each task to individual NFRs in the range [-1,+1]. Let c be such contribution. A task can be neutral ($c=0$), provide a negative contribution ($c<0$), or a positive contribution ($c>0$). If no value is specified, we consider a neutral contribution relation.

While expressing contributions, analysts have to distinguish between tasks where the system automates an activity (automation) and tasks in which the system suggests the user a specific course of actions (recommendation):

– *automation*: the contribution quantifies the direct impact of task execution by the system. For example, contributions for task "raise blinds" refer to the system action of turning on the blinds engine;
– *recommendation*: the contribution evaluates the indirect impact of having the user following the suggestion. For example, contributions for task "suggest walking" refer to the impact of accepting the suggestion and walking to work, and not to the action of recommending the user.

We require contributions to be specified in correspondence of all variation points in a routine. We suppose that the running system explores the task model for a routine in a top-down fashion, and takes decisions about which alternative to choose whenever it encounters a variation point.

Adaptive task models include three variation point types: (i) exclusive refinement: one subtask is to be selected and executed; (ii) optional tasks: can be either executed or skipped; and (iii) parametric tasks: parameters can be tuned, leading to different runtime behaviours. We detail each variation point type:

Exclusive Refinement: The system has to choose the best alternative subtask by comparing their contribution to NFRs. For a non-executable task, the contribution approximates the level of contribution of the abstract task, irrespective of the specific executable tasks that refine it. Contributions can be context-dependent. For instance, consider NFR user comfort and task "suggest driving" in the "Waking Up" routine. The contribution of this task could be +0.5 if the user lives and works in the city outskirts, -0.5 if the user either works or lives in the city centre, where traffic jams are very likely to occur.

Optional Task: The decision is whether to execute or skip the task, depending on the contribution to NFRs and user preferences. Contribution to NFRs is expressed as explained for the exclusive refinement variation point. The rule of thumb is that the task is carried out if the weighted contribution to NFRs is positive (>0), and is skipped otherwise. Take, for instance, task "make coffee", and suppose its contribution to NFR user comfort is +0.6 if the user has no early meetings (situation "UrgentTasks" does not hold), -0.8 otherwise; and its contribution to NFR User Efficiency is -0.4. Take Bob's preferences from Table 1. Depending on the current context, the task is executed or skipped:

- "UrgentTasks=true": user comfort has weight 0.5, user efficiency 1.0. The average contribution value is $\frac{(-0.8*0.5)+(-0.4*1)}{0.5+1} = -0.53$; being negative, the task is skipped;
- "UrgentTasks=false": user comfort has weight 0.7, user efficiency 0.3. The average contribution value is $\frac{(0.6*0.7)+(-0.4*0.3)}{0.7+0.3} = +0.3$; being positive, the task will be executed by the system.

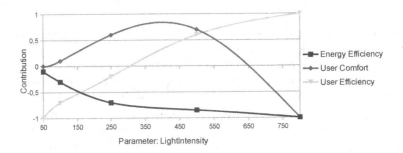

Fig. 4. Interpolation functions for NFR contributions of the task "switch light on"

Parametric Task: The system has to tune the parameters for optimizing NFRs. If the task depends on multiple parameters, so as to simplify the specification of contributions, we suppose the analysts will merge these parameters into a single numeric parameter in a discrete interval. Contribution values are assigned for a set of known values, obtained either by expertise, through interviews, from data sheets, or via measurements. The system will determine the contribution for the missing values using interpolation functions [19] (e.g., polynomial, spline, cubic).

Take, for instance, task "switch light on". Depending on the light intensity, NFRs energy efficiency, user comfort, and user efficiency receive different contributions. In Fig. 4, the analysts have specified contributions to the three NFRs for different light intensity values (50, 100, 250, 400, 800 lux), based on her own experience and the light bulb data sheet. A spline interpolation has been applied to compute the contribution values for the missing light intensity values.

When a task is both parametric and a subtask in an exclusive refinement, parametric task contributions are considered instead of contextual contributions.

4   Executing Adaptive Task Models

Adaptive task models are machine-processable and are executable models. At runtime, they drive the adaptive behaviour of a pervasive infrastructure. In this way, all the efforts invested at design time are reused at runtime providing new opportunities for adaptation capabilities without increasing development costs.

The adaptation process is activated by triggering events, which define *when* the system should adapt. These triggering events are: changes in user preferences, task execution faults, plan failures or context evolution.

The pervasive infrastructure will use the information about the occurrence of a trigger in the next execution (instance) of a routine. In such next instance, the plan that supports the user best—i.e., maximizes NFRs—will be selected. A plan consists of a set of executable tasks in the routine and ordering constraints between those tasks that, together, carry out the routine.

For instance, consider it's a hot Monday of September, and Bob has urgent tasks at work. In this context, NFRs weights are: user comfort = 0.5, user efficiency = 1, energy efficiency = 0.4. The pervasive infrastructure executes the adaptive task model of Fig. 1 as follows: the root task is temporally refined, so its subtasks are examined. The precondition of task "turn on the heating" does not hold, thus the task is skipped. After ten minutes, a wake up call is made. The task is exclusively refined. The buzzer option is chosen by comparing the weighted contributions to NFRs. Since user efficiency has priority over comfort, task "activate buzzer" is executed. Bob awakes immediately. The room is illuminated by raising the blinds, since the contribution of this task to user efficiency is higher than the contribution of "switch light on", no matter its tuning. Bob is informed about the weather by executing such task. When Bob enters in the kitchen, driving is suggested, as this option has the best contribution to user efficiency. Note that the optional task "make coffee" is not executed because Bob has urgent tasks and is in a hurry (see Sec. 3.3).

5   Discussion

We have proposed adaptive task models, an executable modelling language that a pervasive infrastructure can use to support users in their daily routines. These models do not only support alternatives to carry out a routine, but also include the decision-making rationale. Moreover, by handling preferences over NFRs as a separated model, a specific routine can be presented to the user in different ways just by changing the contextual preference model, without altering the routine.

We have also devised a user-centric adaptation framework (more details in [7]) that uses adaptive task models at runtime to adapt system behaviour. While automating user routines, the system adapts its behaviour by choosing a course of action that maximizes user preferences over NFRs.

The use of NFRs to drive decision-making has been widely explored in Goal-Oriented Requirements Engineering (GORE). Most GORE approaches rely upon (variants of) the NFR framework [13] or the i^* framework [23], and exploit the concept of soft-goal to represent NFRs and reason about NFR satisfaction.

Chung et al. [5] use a NFR graph to select among alternative architectural designs. Adaptive software systems (e.g., [12]) use soft-goals to choose the configuration that maximizes the satisfaction of a set of NFRs. We also rely on optimizing NFRs; however, unlike those approaches, we account for the priorities of each user, and we allow for contextual contributions.

Brown et al. [3] exploit a goal-oriented specification to define adaptation requirements, i.e. how the system switches from one configuration to another. In a similar spirit, Souza et al. [21] define awareness requirements as meta-requirements to drive adaptations. Our framework embodies an adaptation requirement, i.e. the optimization of user preferences over NFRs.

Some approaches [16,8] explore contextual variations in business process models. Interestingly, in [8], context analysis [2] is used to specify context. Such approach could be exploited to define situations. Another interesting direction is assessing the suitability of BPM languages for representing user routines.

Other approaches use feature models [4,1] to describe architectural configurations, each consisting of components that are activated depending on the current context. In addition to contextual factors, we consider user preferences over NFRs so as to adapt the system. Also, we do not focus on individual components, but on the adaptation of complex system behaviours (user routines).

Future work includes the development of a pervasive software infrastructure based on our adaptation framework. We intend to rely upon previous work on pervasive infrastructures [18] and goal-oriented adaptation [6]. The effectiveness of the infrastructure in adapting to users will be assessed through case studies.

Acknowledgement. This work has been developed with the support of: the European Union Seventh Framework Programme (FP7/2007-2013) under grants no 257930 (Aniketos) and 256980 (Nessos); and MICINN under the project EVERYWARE TIN2010-18011.

References

1. Acher, M., Collet, P., Fleurey, F., Lahire, P., Moisan, S., Rigault, J.P.: Modeling Context and Dynamic Adaptations with Feature Models. In: Proc. of Models@run.time 2009. LNCS (2009)
2. Ali, R., Dalpiaz, F., Giorgini, P.: A Goal-based Framework for Contextual Requirements Modeling and Analysis. Requirements Engineering 15(4), 439–458 (2010)
3. Brown, G., Cheng, B.H.C., Goldsby, H., Zhang, J.: Goal-oriented Specification of Adaptation Requirements Engineering in Adaptive Systems. In: Proc. of SEAMS 2006, pp. 23–29. ACM (2006)

4. Cetina, C., Giner, P., Fons, J., Pelechano, V.: Autonomic Computing through Reuse of Variability Models at Runtime: The Case of Smart Homes. IEEE Computer 42, 37–43 (2009)
5. Chung, L., Nixon, B., Yu, E.: Using Non-Functional Requirements to Systematically Select among Alternatives in Architectural Design. In: Proc. of IWASS 1995, pp. 31–43. ACM (1995)
6. Dalpiaz, F., Giorgini, P., Mylopoulos, J.: Adaptive Socio-Technical Systems: a Requirements-driven Approach. Requirements Engineering (to appear, 2012)
7. Dalpiaz, F., Serral, E., Valderas, P., Giorgini, P., Pelechano, V.: A NFR-based Framework for User-Centered Adaptation. TR DISI-12-022, DISI, University of Trento (2012)
8. de la Vara, J.L., Ali, R., Dalpiaz, F., Sanchez, J., Giorgini, P.: COMPRO: A Methodological Approach for Business Process Contextualisation. In: Proc. of CoopIS 2010, pp. 132–149 (2010)
9. Giorgini, P., Mylopoulos, J., Nicchiarelli, E., Sebastiani, R.: Reasoning with Goal Models. In: Spaccapietra, S., March, S.T., Kambayashi, Y. (eds.) ER 2002. LNCS, vol. 2503, pp. 167–181. Springer, Heidelberg (2002)
10. Griss, M.L., Favaro, J.M., d'Alessandro, M.: Integrating Feature Modeling with the RSEB. In: Proc. of ICSR 1998, pp. 76–85 (June 1998)
11. Henricksen, K., Indulska, J.: Developing Context-aware Pervasive Computing Applications: Models and Approach. In: Pervasive and Mobile Computing, vol. 2, pp. 37–64 (2004)
12. Lapouchnian, A., Yu, Y., Liaskos, S., Mylopoulos, J.: Requirements-driven Design of Autonomic Application Software. In: Proc. of CASCON 2006 (2006)
13. Mylopoulos, J., Chung, L., Nixon, B.: Representing and Using Nonfunctional Requirements: A Process-Oriented Approach. IEEE Transactions on Software Engineering 18(6), 483–497 (1992)
14. Paternò, F.: ConcurTaskTrees: An Engineered Approach to Model-based Design of Interactive Systems. In: The Handbook of Analysis for Human-Computer Interaction, pp. 483–500. Lawrence Erlbaum Associates (2002)
15. Pavlovski, C.J., Zou, J.: Non-Functional Requirements in Business Process Modeling. In: Proc. of APCCM 2008, pp. 103–112 (2008)
16. Santos, E., Pimentel, J., Dermeval, D., Castro, J., Pastor, O.: Using NFR and Context to Deal with Adaptability in Business Process Models. In: Proc. of RE@RunTime (2011)
17. Satyanarayanan, M.: Pervasive computing: Vision and challenges. IEEE Personal Communications 8(4), 10–17 (2001)
18. Serral, E., Valderas, P., Pelechano, V.: Supporting Runtime System Evolution to Adapt to User Behaviour. In: Pernici, B. (ed.) CAiSE 2010. LNCS, vol. 6051, pp. 378–392. Springer, Heidelberg (2010)
19. Shepard, D.: A Two-dimensional Interpolation Function for Irregularly-spaced Data. In: Proc. of the ACM National Conference, pp. 517–524 (1968)
20. Shepherd, A.: Hierarchical Task Analysis. Taylor & Francis, London (2001)
21. Silva Souza, V.E., Lapouchnian, A., Robinson, W.N., Mylopoulos, J.: Awareness Requirements for Adaptive Systems. In: Proc. of SEAMS 2011, pp. 60–69 (2011)
22. Sutcliffe, A., Fickas, S., Sohlberg, M.M.: Personal and Contextual Requirements Engineering. In: Proc. of RE 2005, pp. 19–28 (2005)
23. Yu, E.: Modelling Strategies Relationships for Process Reengineering. PhD thesis, Department of computer science, University of Toronto (1995)

Concept-Based Web Search

Yue Wang[1], Hongsong Li[2], Haixun Wang[2], and Kenny Q. Zhu[3,*]

[1] University of Massachusetts, Amherst
[2] Microsoft Research Asia
[3] Shanghai Jiao Tong University

Abstract. Traditional web search engines are keyword-based. Such a mechanism is effective when the user knows exactly the right words in the web pages they are looking for. However, it doesn't produce good results if the user asks for a concept or topic that has broader and sometimes ambiguous meanings. In this paper, we present a framework that improves web search experiences through the use of a probabilistic knowledge base. The framework classifies web queries into different patterns according to the concepts and entities in addition to keywords contained in these queries. Then it produces answers by interpreting the queries with the help of the knowledge base. Our preliminary results showed that the new framework is capable of answering various types of concept-based queries with much higher user satisfaction, and is therefore a valuable addition to the traditional web search.

1 Introduction

Keyword based search works well if the users know exactly what they want and formulate queries with the "right" keywords. It does not help much and is sometimes even hopeless if the users only have vague concepts about what they are asking. The followings are some examples of such "conceptual queries":

Q1. database conferences in asian cities
Q2. tech companies slogan

Although the intentions of these queries are quite clear, they are not "good" keyword queries by traditional standard. In the first query, the user wants to know about the database conferences located in Asian cities, without knowing the names of the conferences or cities. In the second query, the user wants to find out the various slogans of tech companies.

Fig. 1 and 2 show the top results returned by Google and Bing respectively. None of these results render meaningful answers to the actual questions implied by the queries. Most of them are likely to be considered irrelevant by the user. For example, the top two results in Fig. 1 have nothing to do with Asian cities, or database conferences. Fig. 2 happens to return a page with a list of company slogans, not because these are tech companies but because one of the company names contains the word "tech".

Apparently, *database conferences*, *asian cities* and *tech companies* are abstract concepts while *slogan* can be regarded as an attribute of tech companies. None of these are

* Kenny Q. Zhu was partially supported by NSFC grants No. 61033002 and 61100050.

P. Atzeni, D. Cheung, and R. Sudha (Eds.): ER 2012, LNCS 7532, pp. 449–462, 2012.

Fig. 1. Google results for Q1   **Fig. 2.** Bing results for Q2

keywords in the traditional sense. Traditional search engines are good at handling entities in these concepts (e.g. VLDB as an entity of database conferences), But in reality, quite a significant percentage of web queries are not *entity only* queries. Our statistics from the search log of a major search engine in last two years suggests that about 62% of the queries contain at least one conceptual class term (see Fig. 6). To better serve such conceptual queries, we need to understand concepts in web pages.

In this paper, we present a framework that leverages a knowledge base and query interpretation techniques to improve web search on concept-related web queries. This knowledge base, named Probase [1,2,3,4,5], is a general-purpose probabilistic taxonomy, automatically constructed by integrating information from web pages and other more reliable data sources such as Freebase [6] and Wordnet [7]. It contains large number of concepts, entities and attributes, organized in a hierarchical structure with subsumption, similarity and other relations. Here, an *entity* refers to a specific object, a *concept* refers to a collection of things, and an *attribute* refers to a property of one or more objects. With the help of Probase, we can identify concepts and attributes in queries and interpret them by replacing the concepts with their most likely entities, and hence formulate more accurate keyword-based queries. The results of these new queries from the traditional search engines can then be ranked and presented to users, in addition to normal keyword queries.

Fig. 3 shows the top search results of the query "database conferences in asian cities" from our prototype system. These results do not contain the keywords "database conferences" or "asian cities", but instead directly gives information about three recent VLDB conferences that were actually hosted in Asia. This information is a lot more targeted and relevant from the user perspective. When serving the query "tech companies slogan", our prototype system realizes that the information is immediately available from the taxonomy, and hence presents all the slogans it knows directly in a table, which is shown in Fig. 4.

It is important to note that the lack of a concept-based search feature in all mainstream search engines has, in many situations, discouraged people from expressing their queries in a more natural way. Instead, users are forced to formulate their queries as keywords. This makes it difficult for people who are new to keyword-based search to effectively acquire information from the web.

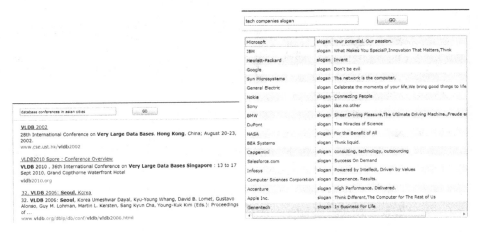

Fig. 3. Our top results for Q1   **Fig. 4.** Our top results for Q2

The proposed framework is not meant to replace the existing keyword based search, but complements keyword search as we can now handle some of "non-keyword" queries as well.

In the remainder of this paper, we will first introduce Probase the knowledge base in Section 2, and then present our framework in Section 3 and some of our experimental results in Section 4, before discussing some related work (Section 5) and concluding the paper (Section 6).

2 The Knowledge Base

In order to better understand queries, the search engine needs to have access to a knowledge base, which knows that, for example, *VLDB* is a database conference, *Hong Kong* is an Asian city, many companies have their *slogans*, and the slogan of Google, a well known tech company, is "Don't be evil." Furthermore, we also need certain meta information, for example, how entities are ranked by their representativeness within a same concept (e.g., What are the top 5 Internet companies?), or how plausible is a claim (e.g., Is Pluto a planet, or a dwarf planet?)

In this work, we take advantage of the Probase taxonomy [1] for rewriting queries. The backbone of Probase is constructed using linguistic patterns such as Hearst patterns [8]. Fig. 5 illustrates a snapshot of the Probase taxonomy which includes the concept "politicians", plus its super-concepts, sub-concepts, entities and similar concepts.

Probase is unique in two aspects. First, the Probase taxonomy is very *rich*. The core taxonomy alone (which is learned from 1.68 billion web pages and 2 years' worth of search log from a major search engine) contains 2.7 million concepts. With 2.7 million concepts obtained directly from Web documents, the knowledge base has much better chance to encompass as many concepts in the mind of humans beings as possible. As shown in Fig. 6, at least 80% of the search contains concepts or entities in Probase.

Second, the Probase taxonomy is *probabilistic*, which means every claim in Probase is associated with some probabilities that model the claim's plausibility, ambiguity, and

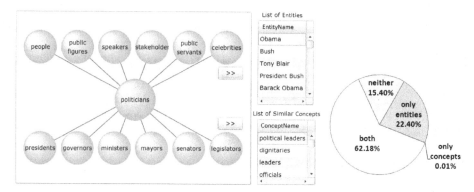

Fig. 5. A fragment of Probase taxonomy

Fig. 6. Queries with concepts or entities

other characteristics. It enables Probase to rank the information it contains. For example, it can answer questions such as "What are the top 5 Internet companies?", or "How likely is Pluto a planet vs. a dwarf planet?"

3   Our Approach

Our concept-based search framework contains three main modules. the *parser*, the *interpreter* and the *processor*, illustrated in Figure 7. When a user issues a query, the *parser* uses Probase to decompose the query into possible term sequences, which consist of terms of 4 different types: concepts, entities, attributes and keywords. The *interpreter* identifies the intent or the semantics of the term sequence based on a set of query patterns. It then rewrites the parsed queries into a set of candidate queries by substituting abstract concepts with their specific entities. The *processor* ranks the candidate queries based on their likelihood, which is estimated by word association probabilities, and then it submits top queries either to Probase or to the search engine index to obtain a list of raw results. These results are ranked before being returned to the user. Next, we discuss these modules in more details.

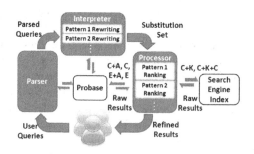

Fig. 7. The framework of concept-based search

3.1 Query Parsing

We regard a search query as a sequence of *terms*. We are interested in four kinds of terms: *entity* terms, *concept* terms, *attribute* terms, and *keyword* terms. Keywords are non-trivial types of words other than concepts and entities. Table 1 gives some examples of each type of terms which are highlighted.

Table 1. Query Terms and Examples

Term Type	Examples
Entity	**Citigroup** **ICDE** in Hong Kong **Hong Kong** area
Concept	**companies** **big financial companies** campaign donation **database conferences** in **asian cities**
Attribute	tech companies **slogan** Hong Kong **area** movies **director**
Keyword	Oracle **acquire** Sun big financial companies **campaign donation** **what's the date today**

The first three types are usually noun phrases, while the keyword terms are often verbs, adjectives or any combinations of other terms that are not recognized as one of the first three types.

Formally, we represent a raw query by an array of words, that is, $q[1,n] = (w_1, \cdots, w_n)$. We parse it into a sequence of terms, where each term is an entity, a concept, an attribute, or an attribute value in the Probase taxonomy, or simply a keyword otherwise. Specifically, we represent a parsed query as $p[1,m] = (t_1, \cdots, t_m)$, where each t_k is a consecutive list of words in the raw query, i.e., $t_k = q[i,j] = (w_i, w_{i+1}, \cdots, w_j)$.

Clearly, there may exist many possible interpretations of a query, or multiple different parses. For example, *"president george bush fires general batiste"* can be parsed as

[president] (george bush) fires [general] (batiste)

[president] george [bush fires] [general] (batiste)

(president) (george bush) fires [general] (batiste)

where () denotes an entity, [] a concept, <> an attribute. The reason of multiple parses is because both *george bush* and *bush fires* are valid Probase terms. Further more, *president* can either be a concept, which refers to all presidents, or a specific entity in "political leader" concept. So the parser needs to return all meaningful parses.

For an n-word query, there are at most $2^{(n-1)}$ possible parses which is expensive to compute. We first introduce a greedy algorithm to solve this problem. Then we improve this algorithm using a dynamic programming approach. The greedy algorithm contains three steps: (1) find all possible terms; (2) find all correlations among terms; (3) use a scoring function to find one meaningful parse in a greedy manner.

First, we find all terms in a query. For a sequence of n word, there are $n(n+1)/2$ possible subsequences. We check each of them to see if they are concepts, entities or attributes. We give a term t a *score* according to its type and length:

$$s_{term}(t) = w_{term}(t) \cdot w_{len}(|t|) \tag{1}$$

where $|t|$ is the number of words in t, $w_{term}(t)$ is the weight function defined as:

$$w_{term}(t) = \begin{cases} w_e, & \text{if t is an entity} \\ w_c, & \text{if t is a concept} \\ w_a, & \text{if t is an attribute} \\ 0, & \text{otherwise} \end{cases}$$

and $w_{len}(x) = x^\alpha$, where w_e, w_c and w_a are constants, and $w_e > w_c$. We let $\alpha > 1$ to bias toward longer terms.

Next, we consider the correlations among terms. Currently we focus on three kinds of correlations: Entity-Attribute, Concept-Attribute, and Concept-Entity. We use R_1-R_2 to denote the correlation between one R_1 term and several R_2 terms. For example, "<population> of (china)" is an instance of Entity-Attribute correlation, and "[tech companies] <slogan> and <founder>" is an instance of Concept-Attribute correlation. Note that terms in a correlation do not have to be physically adjacent to each other, which means keywords can be mixed with correlated terms, e.g. "[presidents] and their <wives>".

Based on terms and term scores, we define *block* and *block score*. A block is either a correlation or a single term. The block score is defined as:

$$s_{block}(q[i,j]) = \max_{i',j'}\left\{w_{block}(p[i',j']) \cdot \sum_{k=i'}^{j'} s_{term}(t_k)\right\} \tag{2}$$

where $p[i', j']$ is a term sequence parsed from $q[i, j]$, and

$$w_{block}(p[i',j']) = \begin{cases} w_{e-a}, & \text{if p[i',j'] is an E-A correlation} \\ w_{c-a}, & \text{if p[i',j'] is a C-A correlation} \\ w_{c-e}, & \text{if p[i',j'] is a C-E correlation} \\ 1, & \text{otherwise} \end{cases}$$

where w_{e-a}, w_{c-a} and w_{c-e} are all greater than 1. The above formula rewards blocks with a term correlation, and if there is no correlation, the block score is equal to the sum of term scores. (Now we empirically decide w_e, w_c, w_a, w_{e-a}, w_{c-a} and w_{c-e}, which could be obtained in a more precise way, e.g. machine learning methods, in the future.)

Finally, after finding all possible terms and blocks, we greedily select the block with the highest score. Once a block is selected, all blocks it overlaps with are removed.

We show the three-step greedy algorithm for parsing query "*president george bush fires general batiste*" in Fig. 8. In step (1), we identify all terms in the query and score them according to Eq. 1. In step (2), we generate blocks based on these terms. Each term becomes a block and their block scores equal to their term scores. At the same time, we notice that (george bush) is an entity of concept [president], so we build a C-E correlation block. Same goes for (batiste) and [general]. In step (3), we perform greedy selection on blocks. We identify "[president] (george bush)" as the best block among all, and remove other overlapping blocks such as "(president)" and "[bush fires]". Similarly we keep block "[general] (batiste)" and remove its overlapping blocks. Finally, the algorithm returns "[president] (george bush) fires [general] (batiste)" as the best parse.

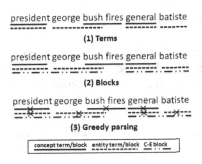

Fig. 8. Example of greedy parsing

However, the greedy algorithm does not guarantee an optimal parse, and it cannot return a list of top parses. As an improvement, we propose the following dynamic programming algorithm. We define a *preference score* $s_{pref}(n)$ to represent the quality of the best parse of a query of n words:

$$s_{pref}(n) = \begin{cases} \max\limits_{i=0}^{n-1}\{s_{pref}(i) + s_{block}(q[i+1,n])\}, & \text{if } n > 0 \\ 0, & \text{if } n = 0 \end{cases}$$

By memoizing the sub-solutions of $s_{pref}(n)$, one can produce the parse with highest preference score. Moreover, when defining $s_{pref}(n)$ as a score set of top parses, one can also obtain the top k parses.

3.2   Query Interpretation

In this module, we classify the input parsed queries into different patterns. Our analysis on the search log during the period of Sep. of 2007 to Jun. of 2009 of a major search engine (see Fig. 6) shows that about 62% of the queries contain at least one concept term. More detailed analysis revealed that common web queries can be classified into a number of different patterns. The following six basic patterns account for the majority of all the queries during that period: *Single Entity (E), Single Concept (C), Single Entity + Attributes (E+A), Single Concept + Attributes (C+A), Single Concept + Keywords*

Table 2. Query Patterns and Examples

Patterns	Queries
E	(VLDB) (Citigroup)
C	[database conferences] [big financial companies]
E+A	(Apple) <slogan> [Hong Kong] <country> <area>
C+A	[tech companies] <slogan> [films] <language> <tagline>
C+K	[big financial companies] campaign donation [species] endangered
C+K+C	[database conferences] in [asian cities] [politicians] commit [crimes]

(C+K), and *Concept + Keywords + Concept (C+K+C)*. Table 2 lists some example queries of each pattern.

Once the system determines the pattern of each parsed query, it starts interpreting them using the following strategies. The general approach is substituting the abstract concepts in a query with more specific search terms such as their associated entities which are more suitable for traditional keyword search.

For *E* and *E+A* queries, no further interpretation is necessary since this type of queries are already specific and can be searched directly in both Probase and the search engine index.

For a *C* or *C+A* query, it substitutes a list of top entities associated with that concept in Probase for the concept term to form a list of E or E+A queries.

For a *C+K* query, it replaces the concept with its associated entities to form a list of Entity + Keywords queries which require no further interpretation.

For a *C+K+C* query, it is considered as an extended form of C+K queries. We replace both concepts with their associated entities to form a list of Entity + Keywords + Entity queries. Note that the number of such queries can be very large but we will show in the next subsection how to reduce them to only relevant queries.

Finally, the output of the Interpreter module is a set of substituted queries of the following 4 patterns: E, E+A, E+K and E+K+E.

3.3   Query Processing

The Processor module takes as input a set of substituted candidate queries, submits some or all of these queries to Probase or a search index, and presents a final set of ranked results to the user.

For E and E+A pattern queries, the processor queries the Probase taxonomy for all the detailed information about this particular entity. This information is returned as a table which will eventually be presented to the user as an info-box (e.g. Fig. 4).

In the rest of this subsection, we will focus on E+K and E+K+E queries which are more interesting and challenging to process. One naive approach is to submit all these substituted queries to a traditional search engine index, combine the results, and present ranked results to the user. However, number of such queries can be prohibitively large because many concepts are associated with large number of the entities. For example, Probase contains hundreds of *politicians* and thousands of *crimes*. For query "*politicians commit crimes*", the system would generate millions of candidate substitutions even though most of these, such as "*obama commit burglary*", are not relevant at all.

Our proposed technique to address the above problem is to compute the *word association* values between an entity and a sequence of keywords, and multiply that with the typicality score of this entity in the concept it belongs to, to get a *relevance score* for a given query. The typicality score is the conditional probability $P(e|c)$, where e is an entity and c is the parent concept.

A word association value is used to measure the associativity of a set of words. To compute the word association value between two words, which we call *two-way association*, we measure the frequency of *co-occurrence* of the two words in a document among all documents or in a sentence among all sentences in the all documents.

In this paper, we approximate multi-way association by combining the values of two-way association. In general, given a set of m words: $W = \{w_1, w_2, ..., w_m\}$, it is impossible to compute the exact word association $wa(W)$ based only on two-way association $wa(\{w_i, w_j\})$, where w_i and w_j are any two distinct words in the W. However, because an co-occurrence of W together implies a co-occurrence of (w_i, w_j), for any $w_i, w_j \in W$, we have $wa(W) \leq \min wa(\{w_i, w_j\})$.

In other words, the minimum two-way association value provides an upper bound to the m-way association value. In this computation, we approximate the m-way association, by the minimum value of the two-way association between a word in the entity and the key word, or

$$wa(\{e, w_{key}\}) \approx \min_{w_i \in e} wa(\{w_i, w_{key}\}). \tag{3}$$

This technique is based on the notion of *pivot words*. A pivot word is the most informative and distinguishing word in a short phrase or term. Given that W contains the words from an entity term and a keyword, if two words w_i and w_j from W co-occur the minimum number of times, we argue that there is a high probability that one of them is the pivot word of the entity term and the other is the keyword. This is because the pivot word appears less frequently than the other more common words in the entity. It is even rarer for the pivot word to appear with an arbitrary keyword than with the other words in the entity. Therefore $wa(\{e_{pivot}, w_{key}\})$ is likely to be minimum, and can be used to simulate $wa(\{e, w_{key}\})$.

We can further extend (3) to compute the association of an entity term and a sequence of keywords:

$$wa^+(\{e, k_1, ... k_n\}) \approx \min_{i \in [1,n]} wa(\{e, k_i\}). \tag{4}$$

To get the word association values, we first obtain a list of most frequently used words on the web (excluding common stop words) as our word list and then compute the sentence-level pairwise co-occurrence of these words. We chose to count the co-occurrence within sentences rather than documents because this gives stronger evidence of association between two words.

We first group the E+K and E+K+E queries by their prefix E+K. As a result, each E+K query form a group by itself; and E+K+E queries with the same prefix form a group. Next, we compute a relevance score for each group G as

$$\max_{q \in G}(wa^+(q_{e1}, q_k, q_{e2}) \times rp(q))$$

where q_{e1}, q_k and q_{e2} are the first entity, keywords and the second entity of query q (q_{e2} may be null if q is an E+K query), and $rp(q)$ is the representativeness score of q which can be obtained from Probase or any other possible sources.

We then select the top n groups with the best relevance scores. Finally we collect all results from the top n groups and rank them according to some common search engine metric such as PageRank before returning to the user.

4   Evaluation

In this section, we evaluate the performance of online query processing and offline pre-computation (word association). To facilitate this evaluation, we create a set of benchmark queries that contain concepts, entities, and attributes, for example, "politicians commit crimes" (C+K+C), "large companies in chicago" (C+K), "president washington quotes" (E+A), etc. For a complete list of the queries and their results, please see [9].

4.1   Semantic Query Processing

The experiment is done on a workstation with a 2.53 GHz Intel Xeon E5540 processor and 32 GB of memory running 64-bit Microsoft Windows Servers 2003. In all search quality related experiments, we asked three human judges to rate each search result as being "relevant" or "not relevant", and we record the majority vote.

Quality of C+K & C+K+C Queries. In the first experiment, we compare the user ratings of the top 10 search results returned by our prototype system with that of Bing, Google, Hakia [10], Evri [11] and Sensebot [12], as illustrated in Fig. 9. The last three are popular semantic search engines. It shows that our prototype has a clear advantage at answering concept related queries. In addition, it also shows that results for C+K+C queries have lower relevance than C+K queries across the three systems, because C+K+C queries often involve more complicated semantic relations than C+K queries. The semantic engines have even fewer relevant results than Google and Bing do, because they index limited number of documents. For example, we found Sense-Bot's top 10 results are often extracted from 3~5 documents.

In the second experiment, we show the ability of our prototype system to pick up relevant results that keyword based search engines would miss. To see this, we focus on relevant results in the top 10 results returned by our prototype system. Among the top 10 results for the 10 C+K queries we use, 84 are judged relevant. The number for C+K+C queries is 74. We check whether these results will ever show up in Google or Bing. Fig. 10 shows that at least 77% of the relevant C+K results and 85% of the relevant C+K+C results in our top 10 could not be found in Bing or Google even after scanning the first 1000 results.

Quality of E, C, E+A & C+A Queries. For E, C, E+A, and C+A queries (see [9]), we return tables containing relevant entities instead of hyperlinks. We ask human judges

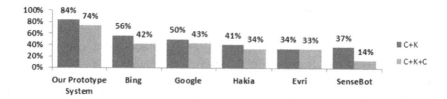

Fig. 9. % of relevant results for C+K & C+K+C queries

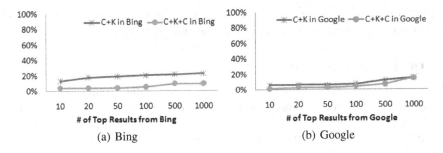

Fig. 10. Bing/Google miss the relevant results in our top 10

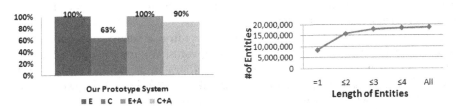

Fig. 11. % of relevant results

Fig. 12. Number of words in Probase entities

to rate the relevancy of each returned table. Fig. 11 shows the percentages of relevant results. All results are relevant for E and E+A queries while some results are wrong for C and C+A queries, because Probase contains some erroneous concept-entity pairs. For example, it doesn't know if "George Washington" is a book or a person.

Time Performance. We next evaluate the efficiency of our prototype system. Table. 3 shows the average running time of the benchmark queries in different categories. Note that the system is configured to return only the top 10 results for C+K, C+K+C, C and E patterns, and all results for the other two patterns.

Table 3. Execution Time (secs)

Pattern	Pr.	It.	Pc.	First Result
E	0.06	0.16	0.16	0.38
C	0.33	0.23	0.32	0.88
E + A	0.15	0.16	0.08	0.39
C + A	0.16	0.66	0.15	0.97
C + K	0.12	0.50	5.24	0.62
C + K + C	0.36	1.22	13.21	2.83

* Pr. = Parsing, It. = Interpretation, Pc. = Processing.

We currently use only one machine to communicate with Bing's public API to support our system. The API accepts only 2 queries per second. So we can see that C+K & C+K+C queries take much more time on processing than other queries. In this paper, we focus on improving user experience by presenting the first result as soon as it becomes available instead of showing all results at the end. [1] We can also find that C+K+C

---

[1] Accessing the document index directly and concurrently, or offering the user the rewritten queries to let he/she click on them, would definitely improve timing.

queries take more time to process than C+K ones because two round trips to Bing are required for each query – one for filtering and one for final results.

4.2 Offline Precomputation

We precompute a word association matrix using a map-reduce framework and a distributed storage system on a cluster with 30 machines. Each machines has an 8-core 2.33 GHz Intel Xeon E5410 CPU and 16 GB of memory, and is running 64-bit Microsoft Windows Servers 2003. However, our program is single threaded and hence uses only one core at a time.

One assumption we made in this paper is that we can estimate the association between an entity term and a keyword using simple two-way word association. The following experiments verify this assumption.

We sampled 9,154,141 web pages (25GB on disk) from a web corpus snapshot of 366,185,148 pages for counting co-occurrences of words. We first examine the length of all the entities in Probase. Fig. 12 shows that 44.36% of the entities contains just one word, almost 84% have 2 or fewer words, and over 95% have 3 or fewer words. One-word entities, which account for almost half of all entities, are straight-forward to process using 2-way word association with another keyword. For the other half of the entities which have more than 1 word, the next experiment indicates that there indeed exists pivot words in many such entities and 2-way association results are very similar to the exact results using multi-way association.

We take the 10 benchmark C+K queries [9], and for each query generate a set of E+K candidate queries by substituting the concept with its associated entities. We next rank this candidate set using two different methods and compare the results. In the first method, which serves as a baseline, we rank the E+K queries by the actual association values of all the words in these queries. In other words, we compute the exact multi-way association of the entity term and the keywords In the second method, we rank the same set of E+K queries by computing 2-way association of the pivot word and the keywords in each query using (4) in Section 3.3. To find out the effectiveness of pivot words and 2-way association, we compute the similarity between the two rankings for each of the 10 benchmark queries using *normalized Kendall's Tau* [13]:

$$\bar{K}(\tau_1, \tau_2) = 1 - \frac{K(\tau_1, \tau_2)}{n(n-1)/2}$$

where $K(\tau_1, \tau_2)$ is Kendall's tau distance between two equi-length sequence τ_1 and τ_2.

Fig. 13 shows that the two rankings are similar enough across all the benchmark queries (see [9] for the the pivot words we discovered in these queries). Fig. 14 shows that the algorithm for multi-word association scales well with the size of the word list.

5   Related Work

There have been some attempts to support semantic web search using some form of a knowledge base. A well-known example is PowerSet [14] which maintains huge indexes of entities and their concepts. This approach, however, will not scale because updates on the entities can be extremely expensive. Another noteworthy general semantic

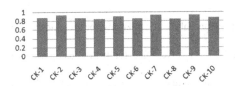

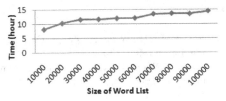

Fig. 13. Normalized similarity between ideal and simulated rankings

Fig. 14. Word association scaling

search engine is Hakia [10] which leverages a commercial ontology and the QDex technology. The QDex technology is unique in that it indexes paragraphs by the embedded frequent queries (sequence of words). Hakia's search results on many of our benchmark queries were similar to keyword search results, which suggest the coverage of their ontology is too limited to help understanding the queries. Other semantic engines include WolframAlpha [15], Evri [11], SenseBot [12] and DeepDyve [16], etc. These engines exploit human curated or automatically extracted knowledge in specific domains such as science, news and research documents. Qiu and Cho proposed a personalized topic search [17] using topic-sensitive PageRank [18], which emphasizes on the analysis of user interests and the disambiguation of entities.

Understanding users' queries and helping users formulate "better" queries is important to document retrieval and web search. Work in this space can be collectively referred to as *query rewriting*. The relevant techniques include query parsing [19], query expansion [20,21,22], query reduction [23,24], spelling correction [25] and query substitution [26,27].

Computing a two-way association matrix for a set of N words generally takes $O(N^2)$ time, while computing a multi-way matrix is more costly. Li [28] proposed a method to estimate multi-way association. It uses an improved sketch method when sampling inverted index list of the words and then use maximum likelihood estimation to solve the problem. Consequently, one only needs to store the sampled inverted list of each word, instead of all the association values of different word combinations. However, scanning the inverted list remains to be costly if the number of documents is too large, especially at the scale of the entire web.

6 Conclusion

In this paper, we propose an idea to support concept-based web search. Such search queries contains abstract or vague concepts in them and they are not handled well by traditional keyword based search. We use Probase, an automatically constructed general-purpose taxonomy to help understand interpret the concepts in the queries. By concretizing the concepts into their most likely entities, we can then transform the original queries into a number of entity-based queries which are more suitable for traditional search engines.

References

1. Wu, W., Li, H., Wang, H., Zhu, K.Q.: Probase: a probabilistic taxonomy for text understanding. In: SIGMOD (2012)
2. Song, Y., Wang, H., Wang, Z., Li, H., Chen, W.: Short text conceptualization using a probabilistic knowledgebase. In: IJCAI (2011)
3. Lee, T., Wang, Z., Wang, H., Hwang, S.: Web scale taxonomy cleansing. In: VLDB (2011)
4. Zhang, Z., Zhu, K.Q., Wang, H.: A system for extracting top-k lists from the web. In: KDD (2012)
5. Liu, X., Song, Y., Liu, S., Wang, H.: Automatic taxonomy construction from keywords. In: KDD (2012)
6. Bollacker, K., Evans, C., Paritosh, P., Sturge, T., Taylor, J.: Freebase: a collaboratively created graph database for structuring human knowledge. In: SIGMOD (2008)
7. Fellbaum, C. (ed.): WordNet: an electronic lexical database. MIT Press (1998)
8. Hearst, M.A.: Automatic acquisition of hyponyms from large text corpora. In: COLING, pp. 539–545 (1992)
9. Wang, Y.: Supplementary material for topic search on the web (2010), http://research.microsoft.com/en-us/projects/probase/topicsearch.aspx
10. Hakia: Hakia (2011), http://www.hakia.com
11. Evri: Evri (2011), http://www.evri.com
12. SenseBot: Sensebot (2011), http://www.sensebot.net
13. Fagin, R., Kumar, R., Sivakumar, D.: Comparing top k lists. SIAM Journal on Discrete Mathematics 17, 134–160 (2003)
14. Powerset: Powerset (2011), http://www.powerset.com
15. Alpha, W.: Wolfram alpha (2011), http://www.wolframalpha.com
16. DeepDyve: Deepdyve (2011), http://www.deepdyve.com
17. Qiu, F., Cho, J.: Automatic identification of user interest for personalized search. In: WWW, pp. 727–736 (2006)
18. Haveliwala, T.H.: Topic-sensitive pagerank: A context-sensitive ranking algorithm for web search. IEEE Trans. Knowl. Data Eng. 15(4), 784–796 (2003)
19. Guo, J., Xu, G., Cheng, X., Li, H.: Named entity recognition in query. In: SIGIR, pp. 267–274 (2009)
20. Lesk, M.E.: Word-word associations in document retrieval systems. American Documentation 20, 27–38 (1969)
21. Salton, G., Buckley, C.: Improving retrieval performance by relevance feedback. JASIS 41(4), 288–297 (1990)
22. Fonseca, B.M., Golgher, P.B., Pôssas, B., Ribeiro-Neto, B.A., Ziviani, N.: Concept-based interactive query expansion. In: CIKM, pp. 696–703 (2005)
23. Jones, R., Fain, D.C.: Query word deletion prediction. In: SIGIR, pp. 435–436 (2003)
24. Kumaran, G., Carvalho, V.R.: Reducing long queries using query quality predictors. In: SIGIR, pp. 564–571 (2009)
25. Durham, I., Lamb, D.A., Saxe, J.B.: Spelling correction in user interfaces. Commun. ACM 26(10), 764–773 (1983)
26. Radlinski, F., Broder, A.Z., Ciccolo, P., Gabrilovich, E., Josifovski, V., Riedel, L.: Optimizing relevance and revenue in ad search: a query substitution approach. In: SIGIR, pp. 403–410 (2008)
27. Antonellis, I., Garcia-Molina, H., Chang, C.C.: Simrank++: Query rewriting through link analysis of the click graph. In: VLDB (June 2008)
28. Li, P., Church, K.: A sketch algorithm for estimating two-way and multi-way associations. Computational Linguistics 33(3), 305–354 (2007)

A Linear Algebra Technique for (de)Centralized Processing of SPARQL Queries

Roberto De Virgilio

Dipartimento di Informatica e Automazione
Universitá Roma Tre, Rome, Italy
dvr@dia.uniroma3.it

Abstract. We are witnessing the evolution of the Web from a worldwide information space of linked documents to a global knowledge base, composed of semantically interconnected resources (to date, 25 billion RDF triples, interlinked by around 395 million RDF links). RDF comes equipped with the SPARQL language for querying data in RDF format. Although many aspects of the challenges faced in large-scale RDF data management have already been studied in the database research community, current approaches provide centralized hard-coded solutions, with high consumption of resources; moreover, these exhibit very limited flexibility dealing with queries, at various levels of granularity and complexity (*e.g.* so-called *non-conjunctive queries* that use SPARQL's UNION or OPTIONAL). In this paper we propose a general model for answering SPARQL queries based on the first principles of linear algebra, in particular on *tensorial calculus*. Leveraging our abstract algebraic framework, our technique allows both quick decentralized processing, and centralized massive analysis. Experimental results show that our approach, utilizing recent linear algebra techniques—tailored to performance and accuracy as required in applied mathematics and physics fields—can process analysis efficiently, when compared to competitors.

1 Introduction

Today, many organizations and practitioners are all contributing to the "Web of Data", building RDF repositories either from scratch or by publishing in RDF data stored in traditional formats. RDF comes equipped with the SPARQL language for querying data in RDF format. Using so-called *triple patterns* as building blocks, SPARQL queries search for specified patterns in the RDF data. For example, to retrieve all persons in the RDF data that have family name "Lee" together with their homepage we would write:

```
SELECT ?x ?y  WHERE ?x foaf:familyName "Lee" . ?x foaf:homepage ?y .
```

Many aspects of the challenges faced in large-scale RDF data management have already been studied in the database research community, including: native RDF storage layout and index structures [4,8]; SPARQL query processing and optimization [3,9]; as well as formal semantics and computational complexity of SPARQL [11,13]. In particular, current research in SPARQL pattern processing (e.g., [14]) focuses on optimizing the class of so-called *conjunctive* patterns (possibly with filters) under the assumption that these patterns are more commonly used than the others. A *conjunctive pattern*

P. Atzeni, D. Cheung, and R. Sudha (Eds.): ER 2012, LNCS 7532, pp. 463–476, 2012.

with filters (CPF pattern for short) is a pattern that uses only the operators AND and FILTER. Although a sizable amount of SPARQL queries (*i.e.* 52%) consists of CPF patterns, non-CPF patterns (*i.e.* so-called *non-conjunctive* patterns with filters) that use SPARQL's UNION and OPTIONAL operators are more than substantial (*i.e.* 48%) [7]. While UNION and OPTIONAL correspond to the relational database operators OUTER UNION and LEFT OUTER JOIN, respectively, it has been noted by Perez et al. [11] that optimization techniques developed for the latter in relational DBMSs cannot be applied to SPARQL patterns involving the former. Unfortunately, Perez et al. [11] and Schmidt et al. [13] show that query evaluation quickly becomes hard for patterns including the operators UNION or OPTIONAL: given the importance of UNION and OPTIONAL in practical patterns, more research for processing non-conjunctive patterns is required. Moreover, DBMS-based solutions require complex ad-hoc indexing and administration tasks for data maintenance.

Contribution. Leveraging such background, this paper proposes a general model of RDF graph, mirrored with a formal tensor representation and endowed with specific operators, allowing both quick decentralized and centralized massive analysis on large volumes of data (*i.e.* billions of triples). We allow high flexibility dealing with queries, at various levels of granularity and complexity (*i.e.* CPF and no-CPF). Our model and operations, inherited by linear algebra and tensor calculus, are therefore theoretically sound, and their implementation may benefit of several numerical libraries developed in the past. Additionally, due to the properties of our tensorial model, we are able to attain two significant features: the possibility of conducting computations in memory-constrained environments such as on mobile devices, and exploiting modern parallel and distributed technologies, owing to the possibility for matrices, and therefore tensors, to be dissected into several chunks, and processed independently. Finally we provide TENSORRDF, a Java tool implementing our approach.

Related Work. Existing systems for the management of Semantic-Web data can be discussed according to two major issues: *storage* and *querying*. Considering the storage, two main approaches can be identified: the first focuses on developing native storage systems to exploit ad-hoc optimisations, while the second makes use of traditional DBMSs (such as relational and object-oriented). Generally speaking, native storage systems (such as OWLIM or RDF-3X [8]) are more efficient in terms of load and update time, whereas the adoption of mature data management systems has the advantage of relying on consolidate and effective optimisations. Indeed, a drawback of native approaches consists in the need for re-thinking query-optimization and transaction-processing techniques. Early approaches to Semantic-Web data management (e.g., Sesame, Jena, Virtuoso) and the semantic extensions implemented in Oracle Database 11g R2 [3]) focused on the triple *(subject, predicate, object)* or quads *(graph, subject, predicate, object)* data models, which can be implemented as one large relational table with three, or four, columns. Queries are formulated in RDF(S)-specific languages (e.g., SPARQL or SeRQL), translated into SQL and sent to the RDBMS. However, the number of required self-joins makes this approach infeasible in practice and the optimisations introduced to overcome this problem have proven to be query-dependent [4] or to introduce significant computational overhead. On the querying side, in the earlier systems the efficiency of query processing depends only on the logical and physical

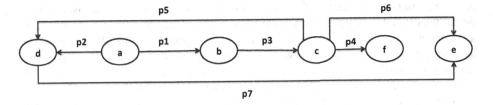

Fig. 1. An example of RDF graph

organization of the data and on the query language complexity. Current research in SPARQL pattern processing (e.g., [8,9,14]) focuses on optimizing the class of so-called *conjunctive* patterns (possibly with filters) under the assumption that these patterns are more commonly used than the others. In particular, a *conjunctive pattern with filters* is a pattern that uses only the operators AND and FILTER. Nevertheless, Möller et al [7] were the first to analyze a log of SPARQL queries, studying (1) the types of SPARQL queries posed (SELECT, ASK, CONSTRUCT, DESCRIBE) and (2) the forms of triple patterns posed in practice. In the same way, Arias et al. [1] performed a similar analysis. Both works have observed that *non-conjunctive* queries that use SPARQL's UNION or OPTIONAL operators appear non-negligibly often practically, providing statistics on the types of joins appearing in practical queries. As we will demonstrate in Section 5, all the above approaches are optimized for centralized massive analysis, in particular focusing on SELECT SPARQL queries, requiring significant amount of resources (*i.e.* memory and space consumption). The efficiency on such proposals is strictly depending on the logical and physical organization, and on the complexity of the SPARQL query.

Outline. Our manuscript is organized as follows. Section 2 will be devoted to a brief recall of RDF and SPARQL. The general model of an RDF graph, accompanied by a formal tensorial representation is supplied in Section 3, subsequently put into practice in Section 4, where we provide the reader a method of analyzing RDF data within our framework. We benchmark our approach with several test beds, and supply the results in Section 5. Finally, Section 6 sketches conclusion and future work.

2   Preliminary Issues

The following paragraphs will be devoted to a brief recall of RDF and SPARQL syntax, which is preliminary to the introduction of our model in the ensuing sections.

RDF. All information in RDF is represented by triples of the form $\langle s, p, o \rangle$, where s is called the *subject*, p is called the *predicate*, and o is called the *object*. Each set of RDF triples can easily be represented graphically as an edge-labeled graph in which the subjects and objects form the nodes, and edges are labeled by the corresponding predicates. For this reason, sets of RDF triples are also called *RDF graphs*. Fig. 1 shows an example of RDF graph (for the sake of simplicity we use simplified labels).

SPARQL. Abstractly speaking, a SPARQL query Q is a 5-tuple of the form $\langle qt, RC, DD, G_P, SM \rangle$, where qt is the *query type*, RC the *result clause*, DD the *dataset definition*, G_P the *graph pattern* and SM the *solution modifier*. At the heart of Q there lies the *graph pattern* G_P that searches for specific subgraphs in the input RDF dataset. In particular the official SPARQL syntax considers operators UNION, OPTIONAL, FILTER and *concatenation* via a point symbol (.) to construct graph patterns. We consider G_P as a 4-tuple $\langle \mathcal{T}, f, OPT, U \rangle$. $\mathcal{T}$ is a set of *triple patterns*, that are just like triples, except that any of the parts of a triple can be replaced with a *variable*, *i.e.* a variable starts with a ? and can match any node (resource or literal) in the RDF dataset. Given two triples patterns t_1 and t_2, they present an *intersection* if they have a node in common (*i.e.* subject or object), and we will denote it as $t_1 \leftrightarrow t_2$. $\mathcal{T}$ is organized in disjunctive sets of triple patterns, *i.e.*, $\mathcal{T} = \{T_1, T_2, \ldots, T_n\}$ where $\forall i, j : T_i \cap T_j = \emptyset$, with $i \neq j$. f is a FILTER constraint using boolean conditions to filter out unwanted query results. OPT is a set of OPTIONAL patterns trying to match a graph pattern, but does not fail the whole query if the optional match fails. U is a set of UNION statements modeled as a graph pattern. The *result clause* identifies which information to return from the query. It returns a table of variables (occurring in G_P) and values that satisfy the query. The *dataset definition* is optional and specifies the input RDF dataset to use during pattern matching. If it is absent, the query processor itself determines the dataset to use. The optional *solution-modifier* allows sorting of the mappings obtained from the pattern matching, as well as returning only a specific window of mappings (e.g., mappings 1 to 10). The result is a list L of mappings. The actual output of the SPARQL query is then determined by the *query-type*: (i) SELECT queries return projections of mappings from L (in order); (ii) ASK queries return a boolean: true if the graph pattern G_P could be matched in the input RDF dataset, and false otherwise; (iii) CONSTRUCT queries construct a new set of RDF triples based on the mappings in L; and (iv) DESCRIBE queries return a set of RDF triples that describes the IRIs and blank nodes found in L. The exact content of this description is implementation-dependent. We will not further describe the syntax and semantics of all constructs to use into a query, but refer instead to the SPARQL recommendation [12] for those components.

Example 1. We show three different SELECT queries.

Q1: SELECT ?x	Q2: SELECT ?y ?w	Q3: SELECT *
WHERE a ?y ?x. c ?z ?x.	WHERE a ?z ?x. ?y ?p ?x.	WHERE a ?z ?x
c ?p f. b ?w c.	FILTER (?y != a).	UNION
	OPTIONAL ?x ?w e	c ?p ?y

3   RDF Data Modeling

This section provides a general model capable of representing all aspects of a given RDF graph. Getting inspiration from [6], such representation is mapped within a standard tensorial framework.

A General Model. Let us define a set $\mathcal{E}$, with $\mathcal{E}$ being finite. A *property* of an element $e \in \mathcal{E}$ is defined as an application $\pi : \mathcal{E} \rightarrow \Pi$, where Π represents a suitable property codomain. Therefore, we define the application of a property $\pi(e) := \langle \pi, e \rangle$, *i.e.*, a property related to an element $e \in \mathcal{E}$ is defined by means of the pairing element-property; a property is a surjective mapping between e and its corresponding property value. Formally, let us introduce the family of properties π_i, $i = 1, \ldots, n < \infty$, and their corresponding sets Π_i; we may therefore model the product set: $\mathcal{E} \times \Pi_1 \times \ldots \times \Pi_n$.

Tensorial Representation of a RDF Graph. Given a RDF graph G, let us define the set S as the set of all resources (nodes) in G, with S being finite. A *property* of a resource $s \in S$ is defined as an application $\pi : S \to \Pi$, where Π represents a suitable property codomain. Therefore, we define the application of a property $\pi(s) := \langle \pi, s \rangle$, *i.e.*, a property related to a resource s is defined by means of the pairing resource-property; a property is a surjective mapping between a resource and its corresponding property value. Following the definition of our general model, a *RDF graph* $(\mathcal{G})$ is defined as the product set of all resources, and all the associated properties. Formally, let us introduce the family of properties π_i, $i = 1, \ldots, k + d < \infty$, and their corresponding sets Π_i; we may therefore model a RDF graph as the product set

$$\mathcal{G} = \underbrace{S \times \Pi_1 \times \ldots \times \Pi_{k-1}}_{\mathbb{B}} \times \underbrace{\Pi_k \times \ldots \times \Pi_{k+d}}_{\mathbb{H}} . \tag{1}$$

We highlight the indices employed in the definition of $\mathcal{G}$. The reader should notice as we divided explicitly the first k spaces $S, \Pi_1, \ldots, \Pi_{k-1}$, from the remaining ones. We have two different categories: *body* ($\mathbb{B}$) and *head* ($\mathbb{H}$). It should be hence straightforward to recognize the tensorial representation of $\mathcal{G}$:

Definition 1 (Tensorial Representation). *The* tensorial representation *of a RDF graph* $\mathcal{G}$, *as introduced in equation* (1), *is a multilinear form:* $\mathcal{G} : \mathbb{B} \longrightarrow \mathbb{H}$.

A RDF graph can be therefore rigorously denoted as a rank-k tensor with values in $\mathbb{H}$. Now we can model some codomains Π with respect to the conceptual modeling provided above. In particular we define the property Π_1 as the set $\mathcal{O}$ of all resources in G (*i.e.* $S \equiv \mathcal{O}$), Π_2 as the set of RDF properties $\mathcal{P}$ connecting a resource $s \in S$ (*i.e.* the subject) with a resource $o \in \mathcal{O}$ (*i.e.* the object). In the next paragraph we will illustrate how we organize such properties in $\mathbb{B}$ and $\mathbb{H}$. As a matter of fact, properties may *countable* or *uncountable*. By definition, a set A is *countable* if there exists a function $f : A \to \mathbb{N}$, with f being injective. For instance the sets $S, \mathcal{O}, \mathcal{P}$ are countable. Therefore countable spaces can be represented with natural numbers and we can introduce a family of injective functions called *indexes*, defined as $\text{idx}_i : \Pi_i \longrightarrow \mathbb{N}$, $i \in [1, k + d]$. When considering the set S, we additionally define a supplemental index, the *subject index function* $\text{idx}_0 : S \to \mathbb{N}$, consequently completing the map of all countable sets of $\mathcal{G}$ to natural numbers.

Implementation. Preliminary to describing an implementation of our model, based on tensorial algebra, we pose our attention on the practical nature of a RDF graph. Our treatment is general, representing a RDF graph with a tensor, *i.e.*, with a multidimensional matrix, due to the well known mapping between graphs and matrices (cf. [2]). However, a matrix-based representation need not to be *complete*. RDF graphs rarely exhibit completion: it is difficult to have an edge for each pairs of nodes (*i.e.* only if the graph is strongly connected). Hence, our matrix effectively requires to store only the information regarding connected nodes in the graph: as a consequence, we are considering sparse matrices [5], *i.e.*, matrices storing only non-zero elements. A sparse matrix may be indicated with different notations (cf. [5]), however, for simplicity's sake, we adopt the *tuple notation*. This particular notation declares the components of a tensor $M_{i_1 i_2 \ldots i_k}(\mathbb{F})$ in the form $\mathcal{M} = \left\{ \{i_1 i_2 \ldots i_k\} \to f \neq 0, f \in \mathbb{F} \right\}$, where we implicitly intended $f \neq 0 \equiv 0_{\mathbb{F}}$. As a clarifying example, consider a Kroneker vector $\delta_4 \in \mathbb{R}^5$: its sparse representation will be therefore constituted by a single tuple of one component with value 1, *i.e.*, $\{ \{4\} \rightarrow 1 \}$. In our case we have the notation $\mathbb{B} \longrightarrow \mathbb{H} = \mathbb{B}(\mathbb{H})$ where we consider $\mathbb{B} = \mathcal{S} \times \mathcal{O}$ and $\mathbb{H} = \mathcal{P}$.

Example 2. Let us consider Fig. 1, whose representative matrix is as follows:

$$\begin{pmatrix} \cdot & p1 & \cdot & p2 & \cdot & \cdot \\ \cdot & \cdot & p3 & \cdot & \cdot & \cdot \\ \cdot & \cdot & \cdot & p5 & p6 & p4 \\ \cdot & \cdot & \cdot & \cdot & p7 & \cdot \\ \cdot & \cdot & \cdot & \cdot & \cdot & \cdot \\ \cdot & \cdot & \cdot & \cdot & \cdot & \cdot \end{pmatrix}$$

where, for typographical simplicity, we omitted $0_{\mathbb{H}} = (0)$, denoted with a dot. For clarity's sake, we outline the fact that $\mathcal{G}$ is a rank-2 tensor with dimensions 6×6 (i.e. the rows and columns corresponding to the nodes). In fact, we have six nodes in the graph. Then we map the elements of $\mathcal{S}$ and $\mathcal{O}$ by using the index functions idx_0, and idx_1 (*i.e.* $\mathrm{idx}_0(a) = \mathrm{idx}_1(a) = 1$, $\mathrm{idx}_0(b) = \mathrm{idx}_1(b) = 2, \ldots, \mathrm{idx}_0(f) = \mathrm{idx}_1(f) = 6$). In the sparse representation we have $\{\{1,2\} \rightarrow (p1), \{1,4\} \rightarrow (p2), \ldots, \{4,5\} \rightarrow (p7)\}$. In this way we mean that if $\{i, j\} \rightarrow (0)$ then there does not exist a triple between the subject with index i and the object with index j otherwise $\{i, j\} \rightarrow (p)$ means that the subject with index i is connected to the object with index j by the RDF property p. In other terms, this implementation represents the adjacency matrix of the RDF graph in Fig. 1, where rows are subjects and columns are objects in the triples. Our model allows to express more information about triples (*e.g.* cardinality, reification, roles, properties such as symmetry or transitivity and so on). In the real implementation we use also the index function idx_2 for the properties; however for the sake of readability we will maintain the labels for properties in the rest of the paper.

4   SPARQL Answering

This section exemplifies our conceptual method, with the applicative objective of analyzing RDF data. With reference to the example provided in Section 3 (*i.e.* Fig. 1 and *Example 2*), in the following we simplify our notation, employing a matrix M_{ij}: i referring to the index of subjects, and j being the index of objects. In [1,7] the authors analyzed a log of SPARQL queries harvested from the DBPedia SPARQL Endpoint. The log contains 1343922 queries in total (after removal of syntactically invalid queries) posed by both humans and software robots. The log analysis produced interesting statistics on the usage of the different components in a query. Such statistics allow us to simplify the features of a SPARQL query. In the following we will consider a SPARQL query Q as a 2-tuple of the form $\langle RC, G_P \rangle$, *i.e.* only SELECT queries with *result clause* and *graph pattern*, employing the operators $\{$AND, FILTER, OPTIONAL, UNION$\}$. This simplification does not compromise the feasibility and generality of the approach.

SPARQL Tensor Applications. A triple-pattern searches for specific subgraphs in the input RDF graph. We can comfortably perform the search efficiently, using the model described in Section 3, by applying the tensor application. Therefore, given a triple-pattern $t_1 = \langle s_1, p_1, o_1 \rangle$, we can process t_1 by subject or by object. In the former case we search all subjects s connected to the object o_1, in the latter case we search all objects o connected to the subject s_1. If we process t_1 by the subject, given $j = \mathrm{idx}_1(o_1)$, we build a Kroneker vector as a vector δ_j, with $|\delta_j| = |\mathcal{O}|$, and finally apply of the rank-2 tensor represented by M_{ij} to δ_j, i.e.: $r = M_{ij}\delta_j$. If o_1 is a variable then we build the vector δ_j with all values equal to 1. Then if s_1 and/or p_1 are constant, we have to filter r with the map function between properties of $\mathcal{G}$ and a suitable space $\mathbb{F}$, e.g., natural numbers for a boolean result. Therefore $\tilde{r} = \mathrm{map}(p(\cdot) == p_1 \wedge i == \mathrm{idx}_0(s_1), r)$. The reader should notice a shorthand notation for an implicit boolean function: as for all descendant of the C programming language, such definition yields 1 if the condition is met, 0 otherwise. If we process t_1 by the object, given $i = \mathrm{idx}_1(s_1)$, we build a Kroneker vector as a vector δ_i, with $|\delta_i| = |\mathcal{S}|$, and finally apply of the rank-2 tensor represented by M_{ij} to δ_i, i.e.: $r = M_{ij}\delta_i$. If s_1 is a variable then we build the vector δ_i with all values equal to 1. Similarly to the processing by subject, if o_1 and/or p_1 are constant, we have to filter r with the map function: $\tilde{r} = \mathrm{map}(p(\cdot) == p_1 \wedge j == \mathrm{idx}_1(o_1), r)$. In the following, given a triple-pattern t, we will denote the processing by subject and the processing by object with $\mathcal{F}_s(t)$ and $\mathcal{F}_o(t)$, respectively. For instance, referring to the example pictured in Fig. 1, let us consider the triple pattern $t_1 = \langle a, ?x, b \rangle$. If we process t_1 by subject, we have $j = \mathrm{idx}_1(b) = 2$ and $\delta_{j=2} = \{\{2\} \to 1\}$. Therefore we result: $\mathcal{F}_s(t_1) = \mathrm{map}(i == \mathrm{idx}_0(a), M_{ij}\delta_j) = \{1,2\} \to (p_1)$. The result $\{1,2\} \to (p1)$ corresponds to the triple $\langle a, p_1, b \rangle$. Analogously, if we process t_1 by object, we have $i = \mathrm{idx}_0(a) = 1$ and $\delta_{i=1} = \{\{1\} \to 1\}$. Therefore we have $\mathcal{F}_o(t_1) = \mathrm{map}(j == \mathrm{idx}_1(b), M_{ij}\delta_i) = \{1,2\} \to (p_1)$. We highlight the processing $\mathcal{F}$ instantiates the triple patterns, i.e. $\mathcal{F}_s(t_1) = \{1,2\} \to (p_1)$ means that the triple $\langle a, p_1, b \rangle$ is an instance of the triple pattern t_1. Then, an important task is to process a graph pattern with two (or more) triple patterns in a SPARQL query. Let us consider two triple patterns $t_1 = \langle s_1, p_1, o_1 \rangle$ and $t_2 = \langle s_2, p_2, o_2 \rangle$, such that $t_1 \leftrightarrow t_2$. Now we process each pattern by the role (i.e. subject or object) of the node that is in common. Then given $\tilde{r}_1$ and $\tilde{r}_2$ the results from the processing of t_1 and t_2, respectively, we combine them by executing the Hadamard product: $r = \tilde{r}_1 \circ \tilde{r}_2$. For instance, referring to the example pictured in Fig. 1, let us consider the triple patterns $t_1 = \langle a, ?x, ?y \rangle$ and $t_2 = \langle ?y, ?z, c \rangle$. Such patterns have the variable $?y$ in common, that is object in t_1 and subject in t_2, i.e. we have to process t_1 by object and t_2 by subject. In this case we have

$$\tilde{r}_1 = \mathcal{F}_o(t_1) = M_{ij}\delta_{i=\mathrm{idx}_0(a)} = \{\{1,2\} \to (p1), \{1,4\} \to (p_2)\}$$

$$\tilde{r}_2 = \mathcal{F}_s(t_2) = M_{ij}\delta_{j=\mathrm{idx}_1(c)} = \{\{2,3\} \to (p_3)\}.$$

Finally $r = \tilde{r}_1 \circ \tilde{r}_2 = \{\{1,2\} \to (p_1), \{2,3\} \to (p_3)\}$. In particular we organize the result r as a map, $\langle key\text{-}value \rangle$, where the key is a triple pattern and the value is the set of corresponding instances. In the example we have $r = \{\langle t_1, \{\{1,2\} \to (p_1)\}\rangle, \langle t_2, \{\{2,3\} \to (p_3)\}\rangle\}$. Now let us consider a graph pattern with three triple patterns $t_1 = \langle s_1, p_1, o_1 \rangle$, $t_2 = \langle s_2, p_2, o_2 \rangle$ and $t_3 = \langle s_3, p_3, o_3 \rangle$, such that $t_1 \leftrightarrow t_2$ and $t_2 \leftrightarrow t_3$. Therefore we have to process t_1 with t_2 and t_2 with t_3. In this case t_2 may be processed in two different ways, i.e. both by subject and by object. Then we calculate r' $= \tilde{r}_1 \circ \tilde{r}_2' = \{\langle t_1, I_{t_1} \rangle, \langle t_2, I'_{t_2} \rangle\}$ and $r'' = \tilde{r}_2'' \circ \tilde{r}_3 = \{\langle t_2, I''_{t_2} \rangle, \langle t_3, I_{t_3} \rangle\}$, where $\tilde{r}_1$ and $\tilde{r}_3$ come from processing t_1 and t_3, respectively, while $\tilde{r}_2'$ and $\tilde{r}_2''$ come from processing t_2 in two different ways. Finally to perform the graph pattern, composed by t_1, t_2 and

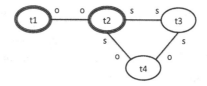

Fig. 2. An example of Execution Graph

t_3, we execute a *composition*, denoted by $\oplus$, that is $r = r' \oplus r'' = \{\langle t_1, I_{t_1}\rangle, \langle t_2, \{I'_{t_2} \cap I''_{t_2}\}\rangle, \langle t_3, I_{t_3}\rangle\}$. Such operation executes the union between the two processing but filtering r'_2 and r''_2 from all instances of t_2 that do not satisfy both r'_2 and r''_2 (*i.e.* we keep the intersection between the instances of t_2). If t_2 is processed in the same way as with t_1 as with t_2, then $\tilde{r}_2 = \tilde{r}_2{}' = \tilde{r}_2{}''$, *i.e.* $r = \tilde{r}_1 \circ \tilde{r}_2 \circ \tilde{r}_3$. For instance, let us refer again to Fig. 1. Consider three triple patterns $t_1 = \langle a, ?y, ?x\rangle$, $t_2 = \langle c, ?z, ?x\rangle$ and $t_3 = \langle c, ?p, f\rangle$, *i.e.* $t_1 \leftrightarrow t_2$ and $t_2 \leftrightarrow t_3$. In this case t_2 has to be processed by object due to the intersection with t_1 and by subject due to the intersection with t_3. We have

$$
\begin{bmatrix}
\mathcal{F}_o(t_1) \circ \mathcal{F}_o(t_2): &
\begin{array}{l}
t_1 : 0\ 0\ 0\ p_2\ 0\ 0 \\
t_2 : 0\ 0\ 0\ p_5\ 0\ 0
\end{array} \\
\hline
\mathcal{F}_s(t_2) \circ \mathcal{F}_s(t_3): &
\begin{array}{l}
t_2 : (0\ 0\ p_5\ 0\ 0\ 0)^t \\
\ \ \ \ \ \ (0\ 0\ p_6\ 0\ 0\ 0)^t \\
\ \ \ \ \ \ (0\ 0\ p_4\ 0\ 0\ 0)^t \\
t_3 : (0\ 0\ p_4\ 0\ 0\ 0)^t
\end{array}
\end{bmatrix}
$$

The execution of the graph pattern $\{t_1, t_2, t_3\}$ provides $r = \{\mathcal{F}_o(t_1) \circ \mathcal{F}_o(t_2)\} \oplus \{\mathcal{F}_s(t_2) \circ \mathcal{F}_s(t_3)\}$, where $I'_{t_2} \cap I''_{t_2} = \{3,4\} \to (p_5)$, and $r = \{\langle t_1, \{\{1,4\} \to p_2\}\rangle, \langle t_2, \{\{3,4\} \to (p_5)\}\rangle, \langle t_3, \{\{3,6\} \to (p_4)\}\rangle\}$. It is straightforward to process graph patterns with more then three triple patterns, performing Hadarmard products and compositions, as we will show in the next paragraph. As final remark, we discuss the maintenance of such framework. In order to eliminate duplicates (or erroneous data) or obsolete information, or to update the RDF dataset with new data (as it is the nature of Web of data), state-of-the-art approaches have to provide insertion and deletion operations, often complex tasks to maintain the organization of the RDF dataset (*e.g.* see [10]). As a matter of fact, in our computational framework these procedures are reflected by *assigning values* in a given tensor. Removing a value is comfortably implemented by assigning the null value to an element, *i.e.*, $M_{ij} = (0)$, and analogously, inserting or modifying a value is attained via a simple operation $M_{ij} = (p)$. Similarly if we delete or insert a node, we add or remove a row or a column.

SPARQL Execution. Exploiting the tensor applications defined above, we can execute a SPARQL query Q in terms of Hadarmard products and compositions. In our framework we have to define an *Execution Graph* E_G, supporting the automatic translation of Q into tensor applications and then the execution of such applications on our model.

Definition 2 (Execution Graph). *Given a query* $Q = \langle RC, G_P\rangle$, *an Execution Graph* E_G *is a 3-tuple* $\langle V, E, H\rangle$, *where* V *is a set of nodes, that are triple patterns in* G_P, E *is a set of edges of the form* (t_1^i, t_2^j) *where* $t_1, t_2 \in V$, $i, j \in \{s, o\}$ *and* H *is a map* $\langle key, value\rangle$ *where the key is a variable in* RC *and the value is a set of triple patterns containing the key. In* E_G *an edge* (t_1^i, t_2^j) *means that* $t_1 \leftrightarrow t_2$ *and* i, j *indicate how to process* t_1 *and* t_2, *i.e., by subject (s), or by object (o).*

Algorithm 1: Building of the Execution Graph

    Input : Result clause RC and triple patterns set T

    Output: The corresponding Execution Graph E_G

 1 $V \leftarrow T$;

 2 $H \leftarrow \texttt{Create\_H}(RC,T)$;

 3 $E \leftarrow \emptyset$;

 4 **while** T *is not empty* **do**

 5 $T \leftarrow T \smallsetminus \{\mathsf{tp}\}$;

 6 $E \cup \texttt{FindCC\_AUX}(\mathsf{tp}, T)$

 7 **return** $E_G : \langle V, E, H \rangle$;

For instance, referring to Example 1, let us consider the query

$$\texttt{Q1} = \{RC = \{?x\}, G_P = \langle \mathcal{T} = \{\{t_1, t_2, t_3, t_4\}\}, \_, \_, \_ \rangle\}$$

where $t_1 = \langle a, ?y, ?x \rangle$, $t_2 = \langle c, ?z, ?x \rangle$, $t_3 = \langle c, ?p, f \rangle$ and $t_4 = \langle b, ?w, c \rangle$. The corresponding E_G is depicted in Fig. 2. The nodes of E_G are t_1, t_2, t_3 and t_4, the edges are (t_1^o, t_2^o), (t_2^s, t_3^s), (t_2^s, t_4^o) and (t_3^s, t_4^o), and the set H is $\{\langle ?x, \{t_1, t_2\}\rangle\}$. In the graph, the double marked line indicates that the nodes contain a variable in H, while each edge depicts closed to the connected node how it will be processed. In this example Q1 presents G_P with $\mathcal{T}$ containing a single set T of triple patterns. If $\mathcal{T}$ provides multiple disjunctive sets T_i, we will produce an execution graph for each T_i.

Algorithm 2: FindCC_AUX: a recursive function for building E_G

    Input : A triple pattern tp, a graph pattern T.

    Output: The connected component built from tp.

 1 $\mathsf{PQ} \leftarrow \emptyset$;

 2 **foreach** $tp_i \in T\colon tp_i \leftrightarrow \mathsf{tp}$ **do**

 3 $T \leftarrow T - tp_i$;

 4   **if** $tp_i.s == \mathsf{tp}.s$ **then**

 5 $\mathsf{PQ} \leftarrow \mathsf{PQ} \cup \{\mathsf{tp}^s, tp_i{}^s\}$;

 6   **else**

 7   **if** $tp_i.s == \mathsf{tp}.o$ **then**

 8 $\mathsf{PQ} \leftarrow \mathsf{PQ} \cup \{\mathsf{tp}^s, tp_i{}^o\}$;

 9   **else**

10   **if** $tp_i.o == \mathsf{tp}.s$ **then**

11 $\mathsf{PQ} \leftarrow \mathsf{PQ} \cup \{\mathsf{tp}^o, tp_i{}^s\}$;

12   **else**

13 $\mathsf{PQ} \leftarrow \mathsf{PQ} \cup \{\mathsf{tp}^o, tp_i{}^o\}$;

14 $\mathsf{PQ} \leftarrow \mathsf{PQ} \cup \texttt{FindCC\_AUX}(tp_i, T)$;

15 **return** PQ;

Algorithm 1 describes the function `BuildGraph` to build an execution graph E_G from a SPARQL query Q. In this case we assume $\mathcal{T}$ as a single set T of triple patterns

---

Algorithm 3: Execution of the graph pattern

Input　: An Execution Graph E_G
Output: The result from the query answering

1　HP $\leftarrow \emptyset$;
2　**while** E *is not empty* **do**
3　　　$t_i^x \leftarrow$ take(E);
4　　　$CL \leftarrow \{\mathcal{F}_x(t_i)\}$;
5　　　HP $\cup \{$CreateCluster$(E, t_i^x, CL)\}$;
6　**return** Composition(HP);

(otherwise we have to perform the algorithm more times for each T_i). The set of nodes V is the copy of the set T from G_P (line [1]). The set H is built by using the function Create_H that takes as input the set of selection variables and the triple patterns of T (we omit the pseudocode since it is straightforward). Then the algorithm iterates on T (lines [4-6]) and extracts a triple pattern tp until T is empty. For each tp it invokes a recursive function FindCC_AUX (line [6]) that generates the edges to insert into E by computing the connected component between tp and the triples patterns in T. As shown in Algorithm 2, FindCC_AUX searches all triple patterns tp_i such that tp_i and tp present an intersection (i.e. $tp_i \leftrightarrow$ tp). If there exists tp_i, then the algorithm extract it from T (line [3]) and calculates for both tp_i and tp which role the node in common presents, *i.e.* subject or object (lines [4-13]). Then we can add the new edge to the set PQ representing the connected component and we recall the recursion (line [14]).

Algorithm 4: Create the group of Hadamard products

Input　: A set of edges E of E_G, a triple pattern t_i^x, a group CL
Output: The updated Group CL

1　**if** $\nexists(t_i^x, t_j^y) \in E$ **then**
2　　　**return** CL;

3　**else**
4　　　**foreach** $(t_i^x, t_j^y) \in E$ **do**
5　　　　　$CL \leftarrow CL \circ \{\mathcal{F}_y(t_j)\}$;
6　　　　　$E \smallsetminus \{(t_i^x, t_j^y)\}$;
7　　　　　CreateCluster(E, t_j^y, CL);

Once generated the execution graph E_G, to execute the corresponding SPARQL query, we simply explore E_G by using a *Depth-First Search* (DFS) traversal. First of all we execute the triple patterns in $\mathcal{T}$ of G_P, as described by Algorithm 3. The algorithm traverses E_G to group all processing of triple patterns to execute as Hadamard products. Finally we perform the composition of all groups. The algorithm starts from an empty set HP, representing the set of all groups CL. Then it iterates on the set of edges E and extracts (casually) the first element t_i^x of a pair $(t_i^x, t_z^w) \in E$ (line [3]). Such operation is performed by the function take (we omit the pseudo code since it is straightforward). We initialize a group CL with the processing $\mathcal{F}_x(t_i)$ (line [4]) and invoke a recursive function `CreateCluster` to fill CL by traversing E_G in depth (line [5]). Each time `CreateCluster` traverses an edge, it removes that edge from E. Therefore the iteration will finish when E will be empty. At the end the function `Composition` executes the compositions ($\oplus$) between the multiple sets of HP. The function converts each set $CL_i \in$ HP in a map structure $\langle T, I \rangle$, where the keys T are the triple patterns while the values I are the corresponding instances. Then for each key t_j occurring in more than one cluster in HP, the function calculates the intersection between each value I_j and updates the value of the maps identified by the key t_j with the intersection. (*i.e.* for space constraints we do not include the pseudo code of the function, that is very simple). Algorithm 4 illustrates the code of `CreateCluster`. If there does not exist a pair $(t_i^x, t_j^y) \in E$ then the traversing is finished and we return the group CL (lines [1-2]). Otherwise, for each (t_i^x, t_j^y) we update CL adding the processing $\mathcal{F}_y(t_j)$ (line [5]), we remove the pair from E (line [6]) and we recall the recursion on t_j^y (line [7]). For instance, referring to the execution graph of Fig. 2, we obtain the following processing

$$
\begin{bmatrix}
\mathcal{F}_o(t_1): & 0 & p_1 & 0 & p_2 & 0 & 0 \\
\mathcal{F}_o(t_2): & 0 & 0 & 0 & p_5 & p_6 & p_4 \\
\mathcal{F}_s(t_2): & (0 & 0 & p_5 & 0 & 0 & 0)^t \\
 & (0 & 0 & p_6 & 0 & 0 & 0)^t \\
 & (0 & 0 & p_4 & 0 & 0 & 0)^t \\
\mathcal{F}_s(t_3): & (0 & 0 & p_4 & 0 & 0 & 0)^t \\
\mathcal{F}_o(t_4): & 0 & 0 & p_3 & 0 & 0 & 0
\end{bmatrix}
\rightarrow
\begin{bmatrix}
\mathcal{F}_o(t_1) \circ \mathcal{F}_o(t_2): & t_1: & 0 & 0 & 0 & p_2 & 0 & 0 \\
 & t_2: & 0 & 0 & 0 & p_5 & 0 & 0 \\
\mathcal{F}_s(t_2) \circ \mathcal{F}_s(t_3) \circ \mathcal{F}_o(t_4): & t_2: & (0 & 0 & p_5 & 0 & 0 & 0)^t \\
 & & (0 & 0 & p_6 & 0 & 0 & 0)^t \\
 & & (0 & 0 & p_4 & 0 & 0 & 0)^t \\
 & t_3: & (0 & 0 & p_4 & 0 & 0 & 0)^t \\
 & t_4: & 0 & 0 & p_3 & 0 & 0 & 0
\end{bmatrix}
$$

Performing the Algorithm 3, we obtain two groups: $CL_1 = \{\mathcal{F}_o(t_1) \circ \mathcal{F}_o(t_2)\}$ and $CL_2 = \{\mathcal{F}_s(t_2) \circ \mathcal{F}_s(t_3) \circ \mathcal{F}_o(t_4)\}$. The composition of this two groups results four elements: $\{1,4\} \rightarrow (p_2)$, $\{3,4\} \rightarrow (p_5)$, $\{3,6\} \rightarrow (p_4)$ and $\{2,3\} \rightarrow (p_3)$. They represent instances of the corresponding triple patterns, as follows

I_{t_1}	I_{t_2}	I_{t_3}	I_{t_4}
$\langle a, p_2, d \rangle$	$\langle c, p_5, d \rangle$	$\langle c, p_4, f \rangle$	$\langle b, p_3, c \rangle$

Now to filter the result following the *result clause*, we use H to extract all triple patterns where the variables in H occur. Then using the values coming from the processing of T we instantiate the variables thought the key-value links of H. In our example $RC = \{?x\}$. The set H indicates t_1 and t_2 linked to $?x$. Therefore since $?x$ is object, and the composition returned $\{3,4\} \rightarrow (p_5)$ for t_2, the value 4 is the instance of $?x$, *i.e.* it corresponds to the node d. Through our model, the processing of the operators {FILTER, OPTIONAL, UNION} is quite simple. Let us consider the query

$$Q2 = \{RC = \{?y, ?w\}, G_P = \langle T = \{\{t_1, t_2\}\}, f = \{?y! = a\}, OPT = \{t_3\}, \_ \rangle\}$$

from Example 1. In the query we have three triple patterns $t_1 = \langle a, ?z, ?x \rangle$, $t_2 \doteq \langle ?y, ?p, ?x \rangle$ and $t_3 = \langle ?x, ?w, e \rangle$, where t_3 is optional. In this case we build two execution graphs E_G' and E_G'' where E_G' is built on t_1 and t_2 while E_G'' is built on all the patterns t_1, t_2 and t_3. The idea is to execute two queries with and without t_3 and then to make the union of the results. Generalizing, given the set T of triple patterns and $OPT = \{opt_1, opt_2, \ldots, opt_m\}$ from a query Q we build $m + 1$ execution graphs, that is E_G on T and m execution graphs E_G^i on T and t_{opt_i} (*i.e.* t_{opt_i} is the triple pattern from opt_i). Recalling Q2, we have $r_1 = \mathcal{F}_o(t_1) \circ \mathcal{F}_o(t_2)$ from E_G' and $r_2 = \mathcal{F}_o(t_1) \circ \mathcal{F}_o(t_2) \circ \mathcal{F}_s(t_3)$ from E_G'', resulting as follows

	I_{t_1}	I_{t_2}	I_{t_3}
$r_1:$	$\langle a, p_2, d\rangle$	$\langle a, p_2, d\rangle$	
		$\langle c, p_5, d\rangle$	
$r_2:$	$\langle a, p_2, d\rangle$	$\langle a, p_2, d\rangle$	$\langle d, p_7, e\rangle$
		$\langle c, p_5, d\rangle$	

In this case the final result r is r_2 itself (*i.e.* r_1 is subsumed by r_2). Finally we have to apply the FILTER constraint to r before processing the *result clause*. In fact, the instance $\langle a, p_2, d\rangle$ of t_2 does not satisfy the boolean constraint $?y\ != a$. Supported by the set H of the execution graph we select the instances of the triple patterns where the variables in the filter constraint occur, *i.e.* $I_{t_2} = \{\{1,4\} \rightarrow (p_2), \{3,4\} \rightarrow (p_5)\}$, and we apply the boolean condition through the map function. In the example, given $\mathrm{idx}_0(a) = 1$, we have $\mathrm{map}(i\ != 1, I_{t_2})$, that is to delete $\{1,4\} \rightarrow (p_2)$ from I_{t_2}. Finally applying the *result clause*, the variables $?y$ and $?w$ are instantiated by the values 4 (*i.e.* d) and p_7. The UNION operator is processed quite similar to OPTIONAL in our model. Referring to Example 1, let us consider the query

$$\text{Q3} = \{RC = \{*\}, G_P = \langle \mathcal{T} = \{\{t_1\}\}, \lrcorner, \lrcorner, U = \langle \mathcal{T}_U = \{\{t_2\}\}, \lrcorner, \lrcorner, \lrcorner\rangle\}$$

where $t_1 = \langle ?a, ?z?x\rangle$ and $t_2 = \langle c, ?p?y\rangle$. In this case for each $T_i \in \mathcal{T}$ we build an execution graph E_G^i and for each $T_j \in \mathcal{T}_U$ we build an execution graph E_G^j. Then we perform E_G^i to return r_i and E_G^j to return r_j. Finally we simply make the union between each r_i and r_j. In our example, first of all, we have to process t_1 and t_2 separately. In this case, since E_G^i corresponds to $\{t_1\}$ and E_G^j to $\{t_2\}$, both the execution graphs consist of one node. Therefore we process t_1 and t_2 indifferently by subject or by object (*e.g.* in this case both by object): we have $\mathcal{F}_o(t_1)$ and $\mathcal{F}_o(t_2)$ resulting the two instances $I_{t_1} = \{\langle a, p_1, b\rangle, \langle a, p_2, d\rangle\}$ and $I_{t_2} = \{\langle c, p_5, d\rangle, \langle c, p_6, e\rangle, \langle c, p_4, f\rangle\}$. The final result is the union $I_{t_1} \cup I_{t_2}$.

5   Experimental Results

We implemented our framework into TENSORRDF, a Java system for answering SPARQL queries over RDF datasets. All procedures for processing and maintenance the RDF dataset are linked to the Mathematica 8.0 computational environment. Experiments in a centralized environment (e^C) were conducted on a dual quad core 2.66GHz Intel Xeon, with 8 GB of memory, 6 MB cache, and a 4-disk 1Tbyte striped RAID array, running RedHat Linux and Java 1.6 installed. Experiments on a decentralized environment (e^D) were conducted on an equivalent cluster compute instance with 4 compute units (2 × 2.66GHz Intel Xeon and 2GB of memory), very high I/O performance (*i.e.* 10 Gigabit Ethernet), 64-bit platform and we embed *grid*Mathematica. In the following we will use the terms C and D as superscripts to mean *centralized* and *decentralized*.

Benchmark Environment. We evaluated the performance of our system comparing with the triple stores Virtuoso, Sesame, Jena-TDB, and BigOWLIM, and the open-source system RDF-3X. In our experiments, we employed three datasets: DBPEDIA (version 3.6), *i.e.* 200 M triples loaded into the official SPARQL endpoint, UNIPROT, *i.e.* 500 M triples consisting of 57GB of protein information, and BILLION, the dataset of Billion Triples Challenge that is 1000 M triples. For each dataset we defined a set of test queries. For DBPEDIA we wrote 25 queries of increasing complexity[1]. For UNIPROT and BILLION, we exploit the queries defined in [9].

---

[1] Available at https://www.dropbox.com/sh/sv5eo3eck3ovvzz/vbfoO7NVw6

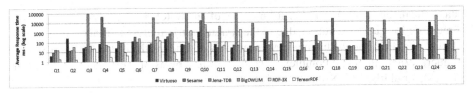

Fig. 3. Response Times on DBPEDIA (centralized environment): bars refer each system, using different gray scales, *i.e.* from Virtuoso, black bar, to TENSORRDF, light gray

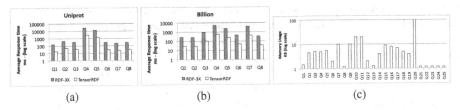

(a)   (b)   (c)

Fig. 4. Times on UNIPROT (a), and BILLION (b); Mem. Usage to query DBPEDIA (c)

Performance. The most important experiment is the comparative query execution on each dataset. We ran the queries ten times and measured the average response time (including the I/O times), *i.e.* ms and logarithmic scale. We performed *cold-cache* experiments (by dropping all file-system caches before restarting the systems and running the queries). Fig. 3 illustrates the response times on DBPEDIA. On the average Sesame and Jena-TDB perform poorly, BigOWLIM and Virtuoso better, and RDF-3X is really competitive.

TENSORRDF outperforms all competitors, in particular also RDF-3X by a large margin. The query performance of TENSORRDF is on the average 18 times better than that of RDF-3X, 128 times on the maximum, *i.e.*, Q21. In particular the queries involving OPTIONAL and UNION operators require the most complex computation: triple stores, *i.e.* Virtuoso, BigOWLIM, Sesame and Jena-TDB, depends on the physical organization of indexes, not always matching the joins between patterns. RDF-3X provides a permutation of all combinations of indexes on subject, property and object of a triple to improve efficiency. However queries, embedding OPTIONAL and UNION operators in a graph pattern with a considerable size, require complex joins between huge amount of triples (*i.e.* Q20) that compromises the performance. Due to the *sparse matrix* representation of tensors and vectors, TENSORRDF exploits parallel computations to perform efficiently map functions, reductions, Hadamard products and compositions ($\oplus$). In fact, in the Fig. 3, some queries (*e.g.* Q1, Q21 or Q22) do not report the white column of TENSORRDF because the response time is less than 1 ms. Similarly, we executed the test queries on UNIPROT and BILLION. Since RDF-3X is the most performing with respect to the other competitors, we compared TENSORRDF directly with RDF-3X. Fig. 4 presents the results. Also in this case TENSORRDF outperforms RDF-3X by a large margin, due to the fact our system has to perform only Hadamard products and compositions without any filtering or union operations. For space constraints, we do not report the response times for decentralized environment: briefly for few hundred of triples e^D is equally efficient as e^C while for thousand of triples the performance of e^D is comparable (*i.e.* slightly less efficient) to that of e^C (*i.e.* due to the large amount of data traffic in the cluster). Another strong point of our system is a very low consumption of memory for query execution, due to the *sparse matrix* representation of tensors and vectors. Fig. 4.(c) shows the memory usage (KB) to query DBPEDIA. On the average, all queries (also the most complex) require very few bytes of memory (*i.e.* dozens of KBytes); similarly for UNIPROT and BILLION.

6 Conclusions and Future Work

We have presented an abstract algebraic framework for the efficient and effective analysis of RDF data. Our approach leverages tensorial calculus, proposing a general model that exhibits a great flexibility with queries, at diverse granularity and complexity levels (*i.e.* both *conjunctive* and *non-conjunctive* patterns with filters). Experimental results proved our method efficient when compared to recent approaches, yielding the requested outcomes in memory constrained architectures. For future developments we are investigating the introduction of reasoning capabilities, along with a thorough deployment in highly distributed Grid environments. In addition, we are about to test our model on mobile devices, comprising more complex properties and queries.

References

1. Arias, M., Fernández, J.D., Martínez-Prieto, M.A., de la Fuente, P.: An empirical study of real-world sparql queries. CoRR abs/1103.5043 (2011)
2. Bondy, A., Murty, U.S.R.: Graph Theory. LNCS. Springer (2010)
3. Chong, E., Das, S., Eadon, G., Srinivasan, J.: An efficient SQL-based RDF querying scheme. In: VLDB, pp. 1216–1227. ACM (2005)
4. Abadi, D.J., Marcus, A., Madden, S.R., Hollenbach, K.: SW-Store: a vertically partitioned DBMS for semantic web data management. VLDB J. 18(2), 385–406 (2009)
5. Davis, T.A.: Direct Methods for Sparse Linear Systems. SIAM (2006)
6. De Virgilio, R., Milicchio, F.: RFID data analysis using tensor calculus for supply chain management. In: CIKM, pp. 1743–1748. ACM (2011)
7. Möller, K., Havsenbles, M., Cyganiak, R., Grimnes, G.: Learning from linked open data usage: Patterns & metrics. In: WebSci. Web Science Trust, pp. 1–8 (2010)
8. Neumann, T., Weikum, G.: The RDF-3X engine for scalable management of RDF data. VLDB J. 19(1), 91–113 (2010)
9. Neumann, T., Weikum, G.: Scalable join processing on very large rdf graphs. In: SIGMOD, pp. 627–640. ACM (2009)
10. Neumann, T., Weikum, G.: x-rdf-3x: Fast querying, high update rates, and consistency for rdf databases. PVLDB 3(1-2), 256–263 (2010)
11. Pérez, J., Arenas, M., Gutierrez, C.: Semantics and complexity of SPARQL. Trans. Database Syst. 34(3), 16:1–16:45 (2009)
12. Prud'hommeaux, E., Seaborne, A.: SPARQL query language for RDF (2008)
13. Schmidt, M., Meier, M., Lausen, G.: Foundations of SPARQL query optimization. In: ICDT, pp. 4–33. ACM (2010)
14. Vidal, M.E., Ruckhaus, E., Lampo, T., Martínez, A., Sierra, J., Polleres, A.: Efficiently Joining Group Patterns in SPARQL Queries. In: Aroyo, L., Antoniou, G., Hyvönen, E., ten Teije, A., Stuckenschmidt, H., Cabral, L., Tudorache, T. (eds.) ESWC 2010, Part I. LNCS, vol. 6088, pp. 228–242. Springer, Heidelberg (2010)

Non-binary Evaluation for Schema Matching

Tomer Sagi and Avigdor Gal

Technion – Israel Institute of Technology, Haifa, Israel

Abstract. In this work we extend the commonly used binary evaluation of schema matching to support evaluation methods for non-binary matching results as well. We motivate our work with some new applications of schema matching. Non-binary evaluation is formally defined together with two new, non-binary evaluation measures using a vector-space representation of schema matching outcome. We provide an empirical evaluation to support the usefulness of non-binary evaluation and show its superiority to its binary counterpart.

1 Introduction

Schema matching is the task of providing correspondences between concepts describing the meaning of data in various heterogeneous, distributed data sources (*e.g.* attributes in database schemata). In its origin, schema matching was conceived to be a preliminary process to schema mapping. A basic assumption that accompanied this research field from its inception is that schema matching provides a set of definite (true or false) correspondences to be then validated by some human expert before mapping expressions are generated. Schema matching evaluation follows this assumption closely, making extensive use of *Precision* and *Recall* [1], borrowed from the field of Information Retrieval. These measures provide a common-sense interpretation to our intuition of what is a "correct" matching by comparing a selected set of attribute correspondences against a set of attribute correspondences that is compiled by some domain expert.

Over the years, schema matching research has expanded and specialized to answer research and application questions in a variety of domains. Recent research is shifting focus from designing new matching algorithms to utilizing the various matchers to efficiently and effectively solve a specific problem at hand. Thus, recent work focuses on selecting appropriate matchers [2], evaluating matchers with respect to a specific schema pair at hand [3,4], tuning matcher parameters [5,4], and ensembling results from different matchers [6,7]. These changes create a major focus shift for schema matching evaluation as well: from a final judgement of quality at the end of the matching process to an intermediate assessment during the construction and application of matching tasks.

Existing measures no longer suffice when the need arises to evaluate non-binary intermediate results, with limited expert input. In particular, since existing evaluation measures are defined over binary results, system designers are forced to evaluate the final outcome with no insight towards the impact of intermediate results on performance.

P. Atzeni, D. Cheung, and R. Sudha (Eds.): ER 2012, LNCS 7532, pp. 477–486, 2012.
© Springer-Verlag Berlin Heidelberg 2012

In this work we propose new evaluation measures for *non-binary* schema matching outcomes. We devise these measures using an extension of the existing similarity matrix representation [8] onto a finite real vector space. The measures extend, in a natural manner, the traditional Precision and Recall measures. We have experimented with the proposed measures using real-world data and report on their performance. We thereby make the following contributions:

- At the conceptual level we introduce a framework for assessing the quality of a non-binary matching result.
- We propose two new non-binary evaluation measures for schema matching, formally analysing their properties.
- We provide an empirical evaluation, showing the benefits of using non-binary measures in common schema matching tasks.

The rest of the paper is organized as follows. Background on schema matching (Sect. 2) is followed by an evaluation model (Sect. 3). We introduce new measures in Sect. 4, followed by empirical evaluation (Sect. 5), related work (Sect. 6) and summary (Sect. 7).

2   Preliminaries

The following schema matching model is based on [8]. Let schema $S = \{a_1, a_2, \ldots, a_n\}$ be a finite set of attributes. Attributes can be both simple and compound, compound attributes should not necessarily be disjoint, *etc.* For any schema pair $\{S, S'\}$, let $\mathcal{S} = S \times S'$ be the set of all possible attribute correspondences between S and S'.

Let $M(S, S')$ be an $n \times n'$ *similarity matrix* over $\mathcal{S}$, where $M_{i,j}$ (typically a real number in $[0, 1]$) represents a degree of similarity between the i-th attribute of S and the j-th attribute of S'. $M(S, S')$ is a *binary* similarity matrix if for all $1 \leq i \leq n$ and $1 \leq j \leq n'$, $M_{i,j} \in \{0, 1\}$. A (possibly binary) similarity matrix is the output of the matching process. For any schema pair (S, S'), let the power-set $\Sigma = 2^{\mathcal{S}}$ be the set of all possible *schema matches* between this pair of schemata.

Example 1. Table 1 presents two similarity matrices between two simplified schemata, with four and three attributes, respectively. The similarity matrix on the left is the outcome of a matching process, using some matcher. The similarity matrix on the right is a binary similarity matrix, generated as the outcome of a decision-maker matcher [8] . This matcher enforces a binary decision while requiring participation of each attribute in at most one correspondence.   □

Matching schemas is often a stepped process in which different algorithms, rules, and constraints are applied. Several classifications of schema matching steps have been proposed over the years (see *e.g.*, [5,9]). Following Gal and Sagi [7], we separate matchers into those that are applied directly to the problem (*first-line matchers (1LM)*) and those that are applied to the outcome of other matchers (*second-line matchers (2LM)*). 1LMs receive two schemata and return a similarity matrix. 2LMs, which are often decision makers, receive a similarity matrix and return a (usually binary) similarity matrix.

Table 1. A Similarity Matrix Example

$S_1 \longrightarrow$ $\downarrow S_2$	1 cardNum	2 city	3 arrivalDay	4 checkIn Time
1 clientNum	0.84	0.32	0.32	0.30
2 city	0.29	1.00	0.33	0.30
3 checkInDate	0.34	0.33	0.35	0.64

$S_1 \longrightarrow$ $\downarrow S_2$	1 cardNum	2 city	3 arrivalDay	4 checkIn Time
1 clientNum	1	0	0	0
2 city	0	1	0	0
3 checkInDate	0	0	0	1

3   Schema Matching Evaluation

In this section we formally define similarity spaces (Sect. 3.1) and schema matching evaluation (Sect. 3.2).

3.1   Similarity Spaces

We propose a vector space representation of schema matching outcome named *similarity space*.[1] For convenience, we maintain matrix notation when referring to a dimension, marking a dimension as an (i, j) coordinate.

Definition 1. *Given schemata S and S', a similarity space $\mathcal{V}_S(S, S') = [0, 1]^{|S|}$ is an $|S|$-dimension vector space such that each dimension (i, j) in $\mathcal{V}_S(S, S')$ corresponds to the attribute pair (a_i, a_j) in S.*   □

Whenever the referenced schemata S and S' and the similarity matrix M are clear from the context we use $\mathcal{V}$ as a shorthand notation of $\mathcal{V}_S(S, S')$.

Definition 2. *For each dimension (i, j) in $\mathcal{V}_S(S, S')$ let $v^{i,j} = (0, 0, ..., 1, ..., 0)$ be a vector with all 0 values except the (i, j) element, assigned with a 1 value. Given a similarity matrix M over $\mathcal{S} = S \times S'$ and a similarity space $\mathcal{V}_S(S, S')$, a similarity vector $\mathbf{v}(M)$ from the space $\mathcal{S}$ is the vector:*

$$\mathbf{v}(M) = \sum_{(i,j) \ dimension \ in \ \mathcal{V}_S(S,S')} M_{i,j} v^{i,j} \qquad (1)$$

□

Whenever the similarity matrix M is clear from the context we use $\mathbf{v}$ as a shorthand notation of $\mathbf{v}(M)$. It is worth noting that each entry in a matrix M over $\mathcal{S}$ is represented as a similarity vector in $\mathcal{V}$. Therefore, a similarity vector represents the similarity of **pairs** of attributes.

Example 2. Consider Example 1. Let $\mathcal{V}$ be the vector space representation of the schema pair in Table 1. We present a simplified space defined over a single attribute from S_1, {checkInDay} and two attributes from S_2, {arrivalDay, CheckInTime}. Tables 2(a) and 2(b) show the relevant part of the

---

[1] The proposed term *similarity space* should not be confused with the one proposed by Zobel and Moffat [10] in the context of document vector spaces.

Table 2. Partial Similarity Matrix Examples

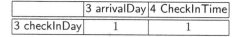

	3 arrivalDay	4 CheckInTime			3 arrivalDay	4 CheckInTime
3 checkInDay	0.35	0.64		3 checkInDay	0	1

     (a) Similarity Matrix   (b) Binary Similarity Matrix

	3 arrivalDay	4 CheckInTime
3 checkInDay	1	1

(c) Exact Match Matrix

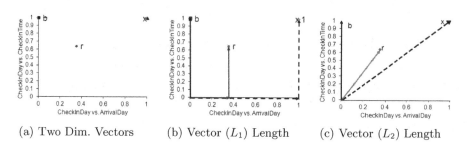

  (a) Two Dim. Vectors   (b) Vector (L_1) Length   (c) Vector (L_2) Length

Fig. 1. 2-Dimensional example

similarity matrices in Table 1, respectively. Table 2(c) illustrates the relevant part of an exact match for the matching of checkInDay with arrivalDay and CheckInTime, as a binary similarity matrix. It is worth noting that while the decision maker chose to match checkInDay only with arrivalDay, the exact match matches checkInDay with both arrivalDay and CheckInTime. We can now visualize the 2-dimensional similarity vectors of the three matrices. In Fig. 1 Each similarity vector is represented as a single point where vectors **r**, **b**, and **x** represent the matrices in Tables 2(a), 2(b) and 2(c) respectively. Figures 1b and 1c visualize vector length using Manhattan (L_1) and Euclidean (L_2) norms, respectively. □

3.2 Schema Matching Evaluation

Schema matching evaluation is a task aiming at assessing the quality of a matching result.

Definition 3. *Let* $\mathcal{M} = \{M_1, M_2, ..., M_n\}$ *be a set of similarity matrices over* $\mathcal{S} = S \times S'$. *A schema matching evaluation method is a function*

$$g : \mathcal{M} \times \mathcal{M} \times ... \times \mathcal{M} \to \mathbb{R}$$

g receives a set of similarity matrices representing matching results and evaluates them to return a single real value. We can designate one of the matrices to be an exact match, representing some "correct" match, typically provided by an expert. Such a match can be encoded as a binary similarity matrix, where "correct" correspondences are assigned the value of 1 and incorrect correspondences a value of 0. As an example, consider the well-known *Recall* evaluation method. Let $\mathcal{M} = \{M, M^e\}$ be a pair of two similarity matrices over $\mathcal{S} = S \times S'$. M is a binary matrix, representing the outcome of a decision maker schema matcher. In vector space representation, Recall is computed as follows:

$$g^{BR}(M, M^e) = \frac{\mathbf{v}(M) \cdot \mathbf{v}(M^e)}{\|\mathbf{v}(M^e)\|_1} \tag{2}$$

where $\mathbf{v}(\cdot)$ is a binary similarity vector over $\mathcal{V}_\mathcal{S}(S, S')$ and $\|\cdot\|_1$ represents the Manhatten (L_1) norm.

4   Non-binary Evaluation Measures

Our non-binary evaluators extend common schema matching evaluators, namely Precision and Recall, to support non-binary values as well. A natural extension is achieved by simply removing the requirement of $\mathbf{v}$ to be a binary vector. This simple extension enables evaluation of non-binary similarity vectors generated by 1LMs without applying any match selection rule or constraint and thus allowing for independent evaluation of 1LMs. We define these measures as follows:

Definition 4 (Non-binary Precision and Recall). *Let $\mathcal{V}_\mathcal{S}(S, S')$ be a similarity space and let $\mathbf{v}(M^e)$ be a similarity vector over $\mathcal{V}_\mathcal{S}$, where M^e represents an exact match.*
NBPrecision *is defined to be*

$$g^{NBP}(M, M^e) = \frac{\mathbf{v}(M) \cdot \mathbf{v}(M^e)}{\|\mathbf{v}(M)\|_1} \tag{3}$$

NBRecall *is defined to be*

$$g^{NBR}(M, M^e) = \frac{\mathbf{v}(M) \cdot \mathbf{v}(M^e)}{\|\mathbf{v}(M^e)\|_1} \tag{4}$$

*In both cases, $\mathbf{v}(M)$ is a **non** binary similarity vector over $\mathcal{V}$.*

Note that Eq. 2 and Eq. 4 differ in the type of their input vector $\mathbf{v}(M)$. A further relaxation of the evaluation measure can be achieved by allowing $\mathbf{v}(M^e)$ to be *non-binary* as well. Such relaxation can support probabilistic schema matching [11,12], where probabilistically correct correspondences are assigned non-0 values in the similarity matrix and the matrix value as a whole represents a probability space over the set of attribute correspondences. We defer a discussion of this relaxation to future work.

To investigate the relationship between the binary and non-binary variations we consider next a special type of 2LMs we term *filters*. Such matchers decide for each entry of the similarity matrix whether its value should remain unchanged or reduced to 0. Filters are the non-binary equivalent to decision maker 2LMs in the binary world. Formally,

Definition 5 (Filters). *Let* $\mathbf{v}(M)$ *be a similarity vector, and let* $\mathbf{v}(M^b)$ *be a similarity vector returned by a 2LM.* $\mathbf{v}(M^b)$ *is filtered if*

$$\mathbf{v}(M^b)_{i,j} = \begin{cases} \mathbf{v}(M)_{i,j} & \text{decision: unchanged} \\ 0 & \text{decision: filtered} \end{cases}$$

The following proposition shows that for NBRecall, just like Recall, removing correspondences (by setting their similarity to 0) cannot improve performance.

Proposition 6. *Let* M *be some similarity matrix resulting from the application of some 1LM and let* M^b *be the result of a filter 2LM. Let* M^e *be an exact match. Then, for* $\mathbf{v}(M)$ *and* $\mathbf{v}(M^e)$ *the following holds:*

$$g^{NBR}(M, M^e) \geq g^{NBR}(M^b, M^e)$$

Proof. $g^{NBR}(M, M^e) = \frac{\mathbf{v}(M) \cdot \mathbf{v}(M^e)}{\|\mathbf{v}(M^e)\|_1}$ by Definition 4. By way of contradiction, assume that

$$\mathbf{v}(M^b) \cdot \mathbf{v}(M^e) > \mathbf{v}(M) \cdot \mathbf{v}(M^e)$$

$$\Rightarrow \exists(i,j) \mid \mathbf{v}(M^e)_{i,j} = 1 \wedge \mathbf{v}(M)_{i,j} = 0 \wedge \mathbf{v}(M^b)_{i,j} > 0$$

contradicting the assumption that $\mathbf{v}(M^b)$ is zero biased. Therefore,

$$\mathbf{v}(M^b) \cdot \mathbf{v}(M^e) \leq \mathbf{v}(M) \cdot \mathbf{v}(M^e)$$

$$\Rightarrow \frac{\mathbf{v}(M^b) \cdot \mathbf{v}(M^e)}{\|\mathbf{v}(M^e)\|_1} \leq \frac{\mathbf{v}(M) \cdot \mathbf{v}(M^e)}{\|\mathbf{v}(M^e)\|_1}$$

since $\|\mathbf{x}\| > 1$

5   Empirical Evaluation

To better understand the impact of using NBPrecision and NBRecall on the assessment of 1LMs we look into the tuning of a 1LM called *Term*, which is part of the OntoBuilder matching system. *OntoBuilder* is a research prototype, developed for matching web-form based ontologies from the deep Web (see [13]). Schema pairs are drawn from the *webform* dataset[2] containing 249 schema pairs, automatically extracted from Web forms using the OntoBuilder Extractor [13] and matched into pairs. For each pair, an exact match was defined manually.

Term calculates string similarity between fields of Web forms. Each score is based on two string elements, the field name (*name score*) and the field label (*label score*). The weight of each score is a tunable parameter. Current approaches, such as eTuner [5] use machine learning on test datasets to learn the optimal values for such parameters. Learning requires the matcher to return a binary similarity matrix and therefore, a 2LM of type decision maker is applied. In our case, we use Stable Marriage (SM) [14] and *Threshold* (a simple threshold rule) to perform match selection.

---

[2] https://bitbucket.org/tomers77/ontobuilder/wiki/Downloads

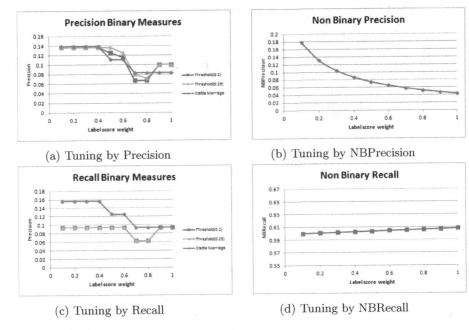

(a) Tuning by Precision

(b) Tuning by NBPrecision

(c) Tuning by Recall

(d) Tuning by NBRecall

Fig. 2. Using various measures to tune *term algorithm*

Figure 2 illustrates how binary and non-binary measures change in one schema pair when the *label* weight (X axis) is varied between 0 and 1. Figures 2a and 2c show how application of 2LMs SM and Threshold with 0.2 and 0.25 paint a different picture w.r.t. the effect of using label score than their non-binary counterparts (Figs. 2b and 2d). The difference between line shapes is due to the fact that results of the 2LMs are evaluated using binary Precision and Recall and therefore jump between values as correspondences are either added or removed. Analysis revealed that differences between 2LMs were due to one or two borderline correspondences, affected by the arbitrary cut-off of threshold or the combined score of SM. Direct analysis of the 1LMs using non-binary measures, is insensitive to these small scale interactions, therefore providing a clear and consistent view of result quality.

Matcher tuning is further complicated by the requirement for an exact match. Generating an exact match between two schemata of considerable size is a daunting task. Furthermore, parameters being tuned may vary between datasets, requiring re-tuning for each dataset anew. These realities often cause designers to have very few exact pairs to work with. Measure used are therefore required to be robust to small sample sizes and their behavior stable even when only few pairs are available. In Figs. 3a and 3b we present results of an experiment testing this stability. In the experiment we again varied the label score and compared Precision and NBPrecision over an increasing number of pairs. An outlier result in the second pair threw Precision completely off the trend and not until we

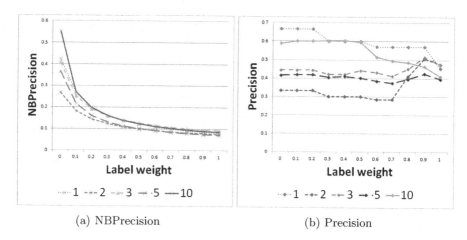

(a) NBPrecision   (b) Precision

Fig. 3. Impact of number of pairs on Precision vs. NBPrecision

reached 10 pairs did the Precision line converge back to the actual behavior. In contrast, NBPrecision maintained a stable line shape, varying in scale but not in form. Results demonstrate how binary measures mask the actual effect of using *label score* due to noise from small scale interactions between scores of different attribute correspondences. When compared to the smooth curve obtained from using NBPrecision and NBRecall, one immediately identifies the strength of these measures for tuning. Unsupervised learning as well as other automated tuning procedures benefit substantially from smooth curves creating a topology of performance which can be optimized.

6   Related Work

The overwhelming majority of current schema matching evaluation measures are based upon *Precision* and *Recall*, first introduced to schema matching by Li and Clifton [1]. These measures and their derivatives have served the classic role of schema matching as a preceding step to mapping expression generation and served well in evaluating various matching systems. The emergence of new applications and research domains challenged their universal adequacy and spawned attempts to provide additional measures for additional needs. F-measure, borrowed from IR by Berlin and Motro [15], Error, borrowed from IR by Modica et. al. [16], and Information Loss, suggested by Mena et. al. [17] all provide methods to aggregate the results of Precision and Recall and evaluate algorithms on a single measure. Overall/Accuracy [18] and HSR [19] assume that schema matching is followed by a manual effort to validate correspondences and therefore are suggested as a better measure for *post-match effort*. A thorough comparison of the use of Precision, Recall, and their derivatives appears in [8], Ch. 3.4.

The above-mentioned measures all assume that evaluating schema matching entails comparing an absolute correspondence list with a similarly absolute *exact match.* Work done by Euzenat [20] progresses towards relaxing this assumption by proposing alternative semantics to the set-inclusion semantics at the basis of classic Precision and Recall definitions. For example, to facilitate subsumption semantics, tree distances between two terms (one matched, the other an exact match) are used to partially recognize matches previously deemed invalid. However, these semantics still incur valuation of binary relationships between attributes matched. To the best of our knowledge, our work is the first to suggest a method for the assessment of non-binary matching results.

7   Conclusions

We have introduced non-binary schema matching evaluation, a tool in the hands of a matching system designer that can asses the quality of interim match results with no need for arbitrary match selection. We provide a formal model for schema matching evaluation using similarity spaces, offer two new non-binary measures and demonstrate how these new measures provide designers with capabilities that were previously unavailable.

In terms of future research we aim to continue and investigate different measures on various normed vector spaces. In addition, we intend to investigate the performance of different evaluation methods with respect to additional schema matching tasks. Finally, we shall also investigate non-binary reference vectors and the impact of using our new measures on schema matching problems where these vectors occur.

Acknowledgement. The research leading to these results has received funding from the European Union's Seventh Framework Programme (FP7/2007-2013) under the NisB[3] project, grant agreement number 256955.

References

1. Li, W., Clifton, C.: Semint: A tool for identifying attribute correspondences in heterogeneous databases using neural networks. Data and Knowledge Engineering 33(1), 49–84 (2000)
2. Duchateau, F., Bellahsene, Z., Coletta, R.: A Flexible Approach for Planning Schema Matching Algorithms. In: Meersman, R., Tari, Z. (eds.) OTM 2008, Part I. LNCS, vol. 5331, pp. 249–264. Springer, Heidelberg (2008)
3. Tu, K., Yu, Y.: CMC: Combining Multiple Schema-Matching Strategies Based on Credibility Prediction. In: Zhou, L.-Z., Ooi, B.-C., Meng, X. (eds.) DASFAA 2005. LNCS, vol. 3453, pp. 888–893. Springer, Heidelberg (2005)
4. Mao, M., Peng, Y., Spring, M.: A harmony based adaptive ontology mapping approach. In: Proc. of SWWS (2008)
5. Lee, Y., Sayyadian, M., Doan, A.H., Rosenthal, A.S.: eTuner: tuning schema matching software using synthetic scenarios. The VLDB Journal 16(1), 97–122 (2007)

---

[3] http://nisb-project.eu/

6. Algergawy, A., Nayak, R., Saake, G.: XML Schema Element Similarity Measures: A Schema Matching Context. In: Meersman, R., Dillon, T., Herrero, P. (eds.) OTM 2009, Part II. LNCS, vol. 5871, pp. 1246–1253. Springer, Heidelberg (2009)
7. Gal, A., Sagi, T.: Tuning the ensemble selection process of schema matchers. Information Systems 35(8), 845–859 (2010)
8. Gal, A.: Uncertain schema matching. Synthesis Lectures on Data Management 3(1), 1–97 (2011)
9. Do, H.H., Rahm, E.: Coma: a system for flexible combination of schema matching approaches. In: Proceedings of VLDB, VLDB Endowment, pp. 610–621 (2002)
10. Zobel, J., Moffat, A.: Exploring the similarity space. SIGIR Forum 32, 18–34 (1998)
11. Dong, X., Halevy, A., Yu, C.: Data integration with uncertainty. The VLDB Journal 18, 469–500 (2009)
12. Gal, A., Martinez, M., Simari, G., Subrahmanian, V.: Aggregate query answering under uncertain schema mappings. In: ICDE, pp. 940–951 (2009)
13. Gal, A., Modica, G., Jamil, H., Eyal, A.: Automatic ontology matching using application semantics. AI Magazine 26(1), 21 (2005)
14. Marie, A., Gal, A.: On the stable marriage of maximum weight royal couples. In: Proceedings of AAAI Workshop on Information Integration on the Web (2007)
15. Berlin, J., Motro, A.: Autoplex: Automated Discovery of Content for Virtual Databases. In: Batini, C., Giunchiglia, F., Giorgini, P., Mecella, M. (eds.) CoopIS 2001. LNCS, vol. 2172, pp. 108–122. Springer, Heidelberg (2001)
16. Modica, G., Gal, A., Jamil, H.: The Use of Machine-Generated Ontologies in Dynamic Information Seeking. In: Batini, C., Giunchiglia, F., Giorgini, P., Mecella, M. (eds.) CoopIS 2001. LNCS, vol. 2172, pp. 433–447. Springer, Heidelberg (2001)
17. Mena, E., Kashyap, V., Illarramendi, A., Sheth, A.P.: Imprecise answers in distributed environments: Estimation of information loss for multi-ontology based query processing. Int. J. Cooperative Inf. Syst. 9(4), 403–425 (2000)
18. Melnik, S., Garcia-Molina, H., Rahm, E.: Similarity flooding: A versatile graph matching algorithm and its application to schema matching. In: ICDE, pp. 117–128. IEEE (2002)
19. Duchateau, F., Bellahsene, Z., Coletta, R.: Matching and Alignment: What Is the Cost of User Post-Match Effort? In: Meersman, R., Dillon, T., Herrero, P., Kumar, A., Reichert, M., Qing, L., Ooi, B.-C., Damiani, E., Schmidt, D.C., White, J., Hauswirth, M., Hitzler, P., Mohania, M. (eds.) OTM 2011, Part I. LNCS, vol. 7044, pp. 421–428. Springer, Heidelberg (2011)
20. Euzenat, J.: Semantic precision and recall for ontology alignment evaluation. In: Proc. IJCAI, pp. 348–353 (2007)

An Integrated Conceptual Model to Incorporate Information Tasks in Workflow Models

Sandeep Purao[1], Wolfgang Maass[2], Veda C. Storey[3],
Bernard J. Jansen[1], and Madhu Reddy[1]

[1] College of IST, Penn State University, University Park, PA, USA
[2] Saarland University, 66123 Saarbrücken, Germany
[3] University Plaza, Georgia State University, Atlanta, GA, USA
spurao@ist.psu.edu

Abstract. In information-rich environments, participants can, and often must, access and use diverse sources of data to support their decision-making tasks. Modeling such environments is important, but cannot be done without effective conceptual models. A problem in information-rich environments is a disconnect between the control-flow across tasks and the information flow that must accompany these tasks. This can pose a challenge for supporting workflows in such environments. Micro-level concerns such as information seeking, sharing, recording, interpreting and hand-offs are not captured in existing workflow models. Without these information-related tasks, the control flows depicted appear to occur magically. We propose an integrated conceptual modeling technique that allows modeling both, control-flows and information-flows. The technique overloads some constructs while retaining their semantic origins, obviating the need to learn new constructs. We elaborate on the model with authentic examples drawn from ethnographic studies of healthcare practices in intensive care units. The paper demonstrates how: the proposed overloading can model information-related tasks; and help bring together conceptual modeling of control and information-flows in information-rich environments. The technique is evaluated with the help of multiple real-world examples.

Keywords: workflow, conceptual model, information flow, micro-behaviors, information-rich environments, healthcare.

1   Introduction

Workflow models and technologies hold significant promise for capturing and coordinating organizational processes (Wang et al, 2003; Ledermann and Morrison, 2002). Conceptual modeling of workflows attempts to achieve these goals by emphasizing tasks and dependencies across tasks. In spite of the rigidities this might impose (Glance et al, 1996), the underlying formalisms such as Petri Nets allow monitoring of organizational processes effectively. The emphasis on 'tasks' and 'dependencies across tasks' because of the formalisms, however, inadvertently introduces a separation between the control flow and the information flow (necessary in information-rich environments). For decision makers and organizational actors in

P. Atzeni, D. Cheung, and R. Sudha (Eds.): ER 2012, LNCS 7532, pp. 487–500, 2012.

information-rich domains, this separation poses significant challenges. An information-rich environment is one in which organizational actors can, and often must, access and use multiple sources of data in support of their tasks (Simon 1971). Consider, for example, the domain of health services delivery. Many 'tasks' that are performed in this domain require access to, and use of, a great deal of information from diverse data sources. Further, some tasks require action, not only from an individual, but also from a team of specialists (e.g. doctors, nurses, pharmacists) who must collaborate contemporaneously while making sense of diverse data sources (Reddy and Jansen 2008) and performing tasks that are distributed in time and space. Prior work on conceptual models for workflows has largely ignored such concerns related to information seeking, sharing and interpretation.

This disconnect between 'control-flow' and 'information-flow' can be traced separate streams of research. Research related to workflows has focused on dependencies across tasks and their enactment, and monitoring for concerns such as soundness (Aalst, 1996) and reachability (Verbeek et al, 1999). Conceptual modeling of workflows builds on workflow-nets (Aalst, 1997), which in turn are based upon Petri nets (Murata, 1989). Second, research on information seeking focuses on human behaviors surrounding information-related tasks. This stream rarely connects the findings to the task context, relying instead on conceptualizing information-related tasks such as needs and actions from individuals for retrieving data sources, and finding relevant information. One perspective that suggests connecting the two is offered by Ellis and Haugan (1997) who argue for 'task-oriented information seeking,' which subsumes information-related behaviors as part of task performance.

The consequences of disconnect between control-flow and information flow can be severe in information-rich environments. For example, poor information flow accompanying well-specified control flow can result in non-trivial performance bottlenecks (Lederman and Morrison, 2002). Healthcare organizations present an important domain in which to study these concerns (Gaudioso 2010) because they represent an information-rich context where the lack of integration between control flow and information flow can pose severe problems. Examples of these problems have been documented, including repeating information seeking efforts (for example, when physicians attempt to reconstruct the rationale for prior decisions) or incomplete information due to a lack of inference-sharing among multiple members of a healthcare team (Lederman and Morrison, 2002).

The *objective* of this research, therefore, is to develop and evaluate an integrated conceptual model that captures (a) information-related tasks such as information seeking and information use behaviors of participants (aka information flow), and (b) normal tasks and dependencies across tasks (aka control flow). The motivation is derived from both: a review of prior work that shows this disconnect across the two streams; and ethnographic studies of care delivery in healthcare organizations that highlight the problem. In this paper, we identify modeling concerns for workflows in information-rich domains, demonstrate how constructs may be overloaded to address these while remaining faithful to their original semantic intent, and demonstrate potential benefits with additional episodes drawn from the ethnographic studies. Key *contributions* of the research include: an extension of an established conceptual modeling technique to integrate control and information flows in information-rich environments; and a first evaluation based on application to authentic examples.

2   Background and Prior Work

Consider the following example of a routinized workflow in a surgical intensive care unit in a healthcare organization, an information-rich environment. The example draws on an ethnographic study[1] that reveals how healthcare professionals interact carry out long-running routines for patient care (Reddy and Jansen 2008).

Illustrative Scenario. The routinized workflow for a healthcare facility starts when the patient is admitted to the surgical intensive care unit (SICU), and continues with recurring episodes that are often similar to the following. The SICU team visits the patient during morning rounds. The resident reports that the patient is in the unit to recover from an operation to fix a perforation in the abdomen. The patient is in acute renal failure with small bowel obstruction and is intubated. The SICU team is interested in discussing the patient's condition and care plan for the day. They begin by declaring the need to find out what is happening with the patient. They ask for the notes from the patient's latest operation. The fellow and two of the residents start looking for the notes. After searching, they only find notes for an operation performed two months ago and determine that that was the latest operation. The SICU team wants to determine whether the patient really has a perforation. Without this information, the team cannot determine the type of treatment. In this instance, it turns out that the resident had incorrect information and the patient did not have a perforation. The team needs to examine whether there is infection in the abdomen because that might be the first thing to treat. They find that the patient has high white blood cell count, even through he is receiving the full spectrum of antibiotics. The team decides that additional information is necessary and schedule a CT scan to check the abdomen. To ensure that the patient is receiving the antibiotics, they change all the PICC lines (providing long-term antibiotic access) in the patient. The change is a response to the observation that the patient has a high white blood count. The following day, the resident and the SICU team examine the results of the CT scan and find them to be negative. These results, together with a check of the patient's current status allow the team to decide that the patient be intubated to put food directly into the patient's stomach.

The example shows how contemporary healthcare delivery takes place in an information-rich environment (see also Gaudioso 2010). For example, in the description above, note the following phrases: "the resident reports that...," "team is interested in discussing the patient's condition...," "they begin by declaring the need to find out what is happening with the patient," "resident start looking for notes," "they find that the patient has high white blood cell count," and so on. Each is an indication of how integral information seeking, interpretation and use are to the actual tasks. This episode, like others, can be parsed to understand the different problems related to information-related tasks. Our analysis appears in Table 1.

---

[1] An ethnographic study is a detailed field study of work practices that can show the unique perspectives of participants to reveal their particular, situated actions with contextual rationale reflecting of their system of meanings (see also Hammersley and Atkinson 2007).

Table 1. Problems related to information seeking tasks drawn from the ethnographic data

Problem	Example
Incorrect information	Resident has wrong information that the patient had perforation
Incomplete information	The team does not know whether there was infection in the abdomen.
Untimely information	The team has access only to information from an operation done two-months ago.
Repeated Information Seeking Efforts	Multiple members of the team review notes on prior operations the patient has undergone.
Inferences not shared	The team does not record the rationale for ordering the CT scan as the need to find out about the elevated white blood cell count
Limited Information channels	Multiple members of the team searching for the same information may need to access the same channel.

The problems are not addressed by the conventional focus on control flows in conceptual models of workflows.

2.1 Workflows and Workflow Modeling

A workflow may be described as a set of coordinated tasks undertaken to fulfill a specific goal (WFMC, 2007, Aalst and Hee, 2002). Seminal work by Zisman (1977) and van der Aalst (2002) has established Petri nets (Murata, 1989) as a key technique for modeling workflows, proposing workflow-nets (van der Aalst, 1997), a variation of Petri nets, to capture workflow concepts. The fundamental concepts in workflow-nets include tasks, places, and arcs for connecting the two. A Petri Net $N = (T, P, A)$, where T is the set of tasks, P is the set of places and A is the set of arcs, is a Workflow Net iff N has two special places, $p1$ and $p2$, where $\bullet p1 = $ null and $p2\bullet = $ null (i.e. the pre-set of place $p1$ and the post-set of place $p2$ are null). With the net in place, individual cases are denoted by tokens that traverse different paths through the net. The dynamic behavior of workflows is captured by firing rules that dictate how tokens move from one place to the next. The firing rules enable tasks e.g. task $t1$ is enabled when places in the pre-set of $t1$ contain tokens. When task $t1$ is fired, the tokens from these places are deposited in the post-set of $t1$. Figure 1 shows a workflow net representation of the example described above, with little emphasis on information seeking tasks relevant in an information-rich environment.

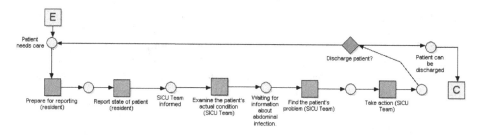

Fig. 1. Modeling the example with a simple Workflow net

2.2   Task-Oriented Information Seeking

In contrast, information-seeking has been investigated as an important concern elsewhere. Theories of information seeking have focused on understanding human behaviors; they have, however, often ignored the context of organizational work (Dourish et al, 1996). An exception is task-oriented information seeking theories, which place human 'information seeking behaviors' within the context of a single task. These efforts, however, still ignore the set of surrounding tasks.

Elsewhere, prior work suggests that information seeking within an organizational task setting exhibits characteristics that differentiate it from information searching in other contexts, such as the home (Rieh, 2003). Ellis and Haugan (1997) present a six-stage general model of information seeking behaviors based on empirical studies of the information seeking patterns of social scientists, research physicists and chemists, and engineers and research scientists in an industrial firm. The Ellis-Haugan model describes six categories of information seeking activities as generic: starting, chaining, browsing, differentiating, monitoring, and extracting. Examining thirty-four users from seven IT companies over a two week period, Choo et al. (1998) report that a behavioral framework of information seeking within an organization that relates motivations (Aguilar, 1967) and moves (Ellis and Haugan, 1997) may be a fruitful approach for understanding information seeking within an organization.

One issue with these models of information seeking is that they are separated from the task itself, with the major focus on the people and their behavior. In one exception, Reddy and Jansen (2008) present a model of collaborative information seeking that addresses the information problem, and the agents involved in the interaction. These agents can be other people or systems. However, in their work, the information need is isolated and not located within an overall workflow model. This is the challenge we address in the current paper.

2.3   Challenges in Combining the Two Perspectives

A key challenge for integrating the two points of view is the difference and apparent mismatch between the two perspectives. The workflow perspective refers to control-flows to ensure temporal dependencies between tasks, while paying little attention to the information gathering, generating and hand-offs. The information seeking perspective, on the other hand, is agnostic about the sequencing among tasks. Instead, it focuses on problems related to individual motives for information gathering and implicitly, to use of information to accomplish individual tasks. Without the context provided by the surrounding tasks in the workflow perspective, the information seeking perspective, therefore, fails to account for problems and opportunities related to merging, splitting, refining, embedding or otherwise manipulating or retaining information as it passes from one perspective to another. Combining the two, therefore, presents an interesting dilemma: whether to devise a technique *de novo* that elicits and captures both perspectives or to extend an established perspective. The first alternative has been the focus of standardization efforts related to business process modeling notation (BPMN, 2007), which contains artifacts as data objects and message flows (White, 2004), which may either be coupled with the control flows or shown as additions. This choice, however, cannot reflect the micro-behaviors related to information seeking identified in prior work related to information behaviors (outlined in the previous section). Neither does it take advantage of the significant research that has been undertaken on analyses of workflow-nets (van der Aalst, 1997). Although transformations from BPMN to Petri Nets are possible, they can result in loss of content. Finally, a modeling technique such as BPMN can make demands on designers to learn new constructs beyond the foundational ones available in Petri Nets. One might also argue that techniques such as conventional data flow diagrams allow incorporating data flows. They, however, do provide neither the granularity nor the task dependencies that we require. To address these concerns, we extend the existing conceptual model underlying workflow-nets. Our intent is to develop a conceptual modeling technique for an integrated model that can account for both, dependencies across tasks and information behaviors and flows.

3   An Integrated Conceptual Model

To overcome the above challenges we develop an integrated conceptual modeling technique while retaining the foundational formalisms of workflow-nets. We introduce specialized constructs to deal with 'information tasks' and 'information arcs' corresponding to the constructs of 'tasks' and 'directed arcs,' effectively overloading the constructs while remaining consistent with the original intent. The model uses the following notation:

$t_1..t_k \in T$ set of tasks
$p_1..p_m \in P$ set of places
$a_1..a_n \in A$ set of directed arcs
$i_1..i_x \in I$ set of information tasks (overloading the task construct)
$f_1..f_y \in F$ set of information arcs (overloading the arc construct)
$\forall f_y \in F$ $f_y =$ generation $\vee$ flow $\vee$ consumption

An information arc can be one of three kinds: generation, flow or consumption.

$\exists p_1, p_2 \in P$ such that $\bullet p_1 =$ null $\wedge p_2 \bullet =$ null
        i.e. the pre-set of place p_1 and the post-set of place p_2 are null.

The workflow net has a source and a sink element.

$\forall f_y \in F$ such that $f_y =$ generation $\Rightarrow \bullet f_y \in p_m$
$\forall f_y \in F$ such that $f_y =$ flow $\Rightarrow \bullet f_y \in i_w \wedge f_y \bullet \in i_x$
$\forall f_y \in F$ such that $f_y =$ consumption $\Rightarrow f_y \bullet \in p_m$

An information generation arc must have a task as its originator, an information flow arc must have an information task as its originator and receiver, and an information consumption arc must have a task as its receiver.

$\forall i_l \in I$ such that $\bullet i_l = t_k \vee (i_x \wedge t_k)$

        i.e. the pre-set of information task i_l must be either a task t_k or a task t_k and an information task i_x.

An information task must be preceded by either a simple task or a simple task along with an information task.

$N = (T, P, A, I, F)$ integrated conceptual model of workflows

Together, the above elements describe an enhanced workflow-net. With the integrated conceptual model following the construct overloading suggested above, individual cases can continue to be denoted by tokens that traverse different paths through the net. The dynamic behavior of workflows can also be captured by firing rules that dictate how tokens move from one place to the next. For example, the same firing rule that enables tasks (e.g. task $t1$ is enabled when places in the pre-set of $t1$ contain tokens) captures how tasks may be enabled with the presence of information available by the information tasks. Non-completion of information tasks signals the absence of information available for firing the task. When the task $t1$ is fired, tokens from the pre-set of that task are removed and deposited in the post-set of $t1$, which may include other tasks as well as information tasks.

The integrated conceptual model can be used to explicitly recognize and capture problems related to information seeking, interpretation and hand-offs in workflows. These three categories will directly map to the three types of information arcs respectively: information generation, information flow, and information consumption.

Table 2. Addressing information-related problems with an integrated conceptual model

Problem	Solution offered by the integrated conceptual model
Incorrect information	Information consumption arcs ensure that appropriate information sources are tapped to compile the needed information
Incomplete information	Firing rules for enabling tasks ensure that all necessary sources of information are accounted for
Untimely information	Techniques from timed Petri Nets can be added to the enhanced workflow-nets to ensure on-time information delivery
Repeated Information Seeking Efforts	Information flow arcs ensure that information needed to from one information task to another is passed as needed.
Inferences not shared	Information generation arcs capture the need to share inferences with downstream participants in the workflow.
Limited Information channels	Simulations with varying token arrival and processing rates can reveal problems related to information channels capacity

To demonstrate key elements of the integrated conceptual model, we revert to the earlier scenario and show how it can be represented with an integrated model that captures both, the control flow and the information flow. Figure 2 shows the outcome.

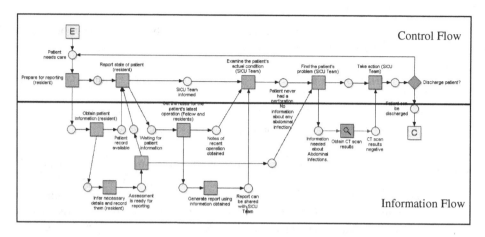

Fig. 2. Modeling the example as an Integrated Conceptual Model

The integrated conceptual modeling technique, thus, does both. It overloads the constructs of task and directed arc in a manner that allows for the capture and representation of information tasks and information flows. Yet, it retains the original semantic intent of the constructs. Both uses of the construct 'task' in the integrated model retain the meaning of an organizational participant performing an action. In the first use, it refers to an action in the organizational process (e.g. 'prescribe medication'); in the second, it refers to an action that relates to information behaviors such as generating, interpreting or handing-off information (e.g. 'access patient notes'). This is an important distinction that has been lost due to the separate trajectories of work in workflow modeling and information retrieval. Similarly, both uses of the construct 'arc' in the integrated model retain the meaning of a dependency across different kinds of tasks. In the first use, it refers to the dependencies across normal organizational tasks. In the second, it refers to the dependency across a normal task and an information task. This, in turn, can take two forms. It can capture the need to perform an information task before a normal task can be completed (e.g. 'interpret patient notes' before 'diagnose patient'), or it can capture how the completion of a normal task can lead to the emergence of an information task (e.g. 'diagnose patient' leading to 'notify nursing staff'). Each of these represents important distinctions that facilitate work in information-rich environments. Conventional workflow models either ignore these information-related behaviors, relying on job knowledge and expertise for their completion, or worse, assume that they do not exist, and instead, rely on only organizational tasks. This can present significant challenges in information rich environments, where such information-related tasks can often outnumber the normal organizational tasks (see, e.g. Figure 2 above).

Use of the integrated conceptual model also allows that the information generation, flow and consumption arcs as primitives can be used to validate that the generated conceptual model is feasible. During the process simulation, the workflow along with the information tasks can be converted into regular XML with a schema that allows correspondence to the Petri Net schema with underlying PNML. Process enactment can be simulated using cases (tokens). This simulation can be performed before deploying the process or for the purpose of evaluating the process, which in turn can lead to the design of data sources.

4   Application and Evaluation

To ensure that the integrated conceptual model of workflows (with overloaded the task and arc constructs for information elements and behaviors) provided adequate representation mechanisms, three additional cases were used. One was derived from the ethnographic study in the healthcare domain described earlier. The second and third were generated based on accounts of short processes in higher education.

Case 1. (*Domain: Healthcare; Source: Ethnographic study, Source: Reddy and Jansen 2008*) This routinized workflow starts when the patient is admitted to the surgical intensive care unit (SICU) for hepatic failure (liver failure) thought to be caused by hepatitis B. Similar to the first case, the workflow consists of recurring episodes. One of these begins with the SICU team discussing the patient's condition and plan of care for the day. The team members first debate whether the Swan line needs to be pulled back. The resident examines the patient's x-ray and notices for the Swan-Ganz line (type of IV tube used to give medication and monitor pressures in the heart and lungs). He wants the attending physician to verify that the line is too far in. The attending physician agrees and indicates that it needs to be pulled back. The resident passes along these instructions to the nurse. The next question discussed is why Ampheticin B (an antibiotic) is being given. One of the fellows asks the resident this because the patient is on two antibiotics (ampheticin B and fluconozal). He wants to take the patient off of fluconozal, but cannot find in the patient notes as to why the patient is on that drug. He asks the resident to ask the Liver Team that is responsible for the patient if there is a reason for giving that drug. The resident cannot seem to find a reason why he is on fluconozal. Lacking this information, the team decides to discontinue that drug. The team then tries to figure out whether they should get the patient's white blood count checked. It is dropping, which could mean that the patient is more susceptible to an infection. They decide to order a re-test to make sure that it is dropping before they do anything. It is time to decide why the patient is still in the unit. The fellow has completed a physical exam of the patient and is wondering why the patient has not been transferred out of the unit. He thinks that patient is ready to go. The resident tells him that the patient's white blood count is dropping. The fellow still thinks that the patient can be transferred out. They decide to keep the patient in the unit for another day. The next day's discussions begin with the same question: why didn't we transfer him out yesterday? The resident tells the fellow that the patient started to desaturate (have trouble breathing) so they had to intubate him during the night (put a tube in the lungs) to help him breath. They attempt to find the patient's blood count. His blood count is the lowest in 3 days. They give him medication to make sure he is not getting any infections. The resident notices by looking at the results from the electronic medical record that the patient's fluid level is much higher than it should be. The fellow tells him that they have to get him to drop his fluid level, so they give him lasix.

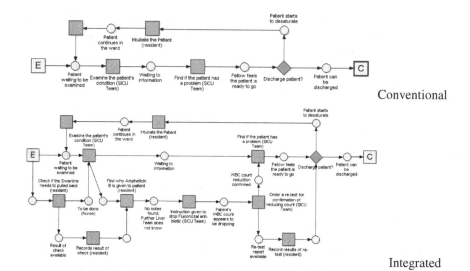

Fig. 3. Application to a new case in the healthcare domain

Figure 3 shows two versions of workflow models. The version labeled 'conventional' shows tasks and dependencies across tasks such as 'examine the patient' and 'intubate the patient'. The version labeled 'integrated' adds information-related tasks such as 'checking if a certain medication needs to be pulled back' etc. The difference between the two is apparent in the numerous information-related tasks that are depicted in the integrated conceptual model.

Case 2. (*Domain: Education; Source: User Interviews*). The process followed for a graduate student registering for courses contains several tasks. First, the student selects the course(s) by searching for information on available courses for the semester. The student may interact with related faculty members and colleagues to obtain additional information about the courses. The courses selected are approved by the graduate advisor, who may also perform a search on information about the courses completed by past students with similar profiles, to provide useful suggestions prior to approval. To complete the registration, a staff member at the registrar's office searches a database to check availability of seats and accepts the student's request.

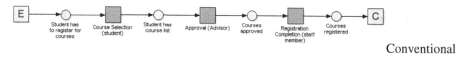

Fig. 4. Application to a new case in a the higher education domain

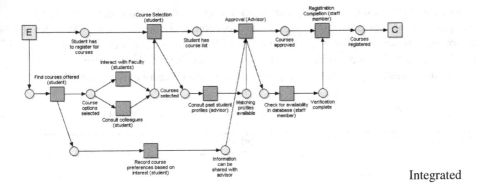

Integrated

Fig. 4. (*continued*)

Figure 4 allows us to appreciate the information-related tasks added to the model.

Case 3. (Domain: Education; Source: User Interviews). The process followed by a faculty member for ordering course texts contains several tasks. First, the faculty member locates book(s) and finds information about the book(s). For this, the faculty member searches for information about various text books in the subject and then identifies two or three that will suit his course requirements. Based on this information, the secretary uses the book information to search for the ISBN details and finds publisher contacts. The secretary then performs a catalog search for details about the publisher for ordering test copies, and finally places the order for the book copies.

Figure 5 also shows the difference between 'conventional' and 'integrated' models.

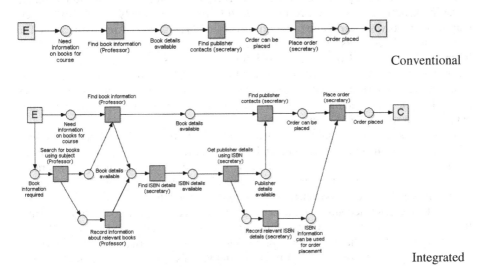

Conventional

Integrated

Fig. 5. Application to another new case in a the higher education domain

The three test cases show how the integrated conceptual model can be useful for the purpose of detecting and overcoming problems associated with micro-level information flows. In table 3, we bring together how the three text cases demonstrate the use of overloaded constructs, and more specifically, how these constructs contribute to detecting and addressing information related tasks in each workflow.

Table 3. Use of overloaded constructs and contribution to the test cases

Problem	Information Task	Information Generation Arc	Information Flow Arc	Information Consumption Arc
Incorrect information				
Incomplete information	Case 1 Case 3	Case 3	Case 1	Case 1 Case 3
Untimely information	Case 1			
Repeated Information Seeking Efforts	Case 1 Case 2 Case 3	Case 3	Case 2 Case 3	Case 3
Inferences not shared Limited Information channels	Case 1	Case 1		Case 1

The test cases, thus, demonstrate how the integrated conceptual model can help overcome concerns related to supporting information behaviors.

5   Discussion and Concluding Remarks

This research builds upon prior work on workflow modeling, retaining fundamental modeling primitives (and, as a result, the ability to draw on analyses in prior work). Our work involves overloading the constructs of 'Task' and 'Arc' to account for information-related tasks and behaviors in information-rich environments – resulting in an integrated model of control and information flows in workflow models. The theoretical basis for including the information-related tasks and behaviors is drawn from theories of task-oriented information seeking. We do not, however, negate the possibility of introducing additional diagrammatic notations but find it more persuasive to rely on existing notations.

The results of this research have clear implications for practice. Situations where micro-level information generating and sharing practices in routinized workflows must be managed abound. Examples include healthcare and education (as illustrated above), as well as national security, and applications such as repair and maintenance. These information-rich environments are poorly served by conventional workflow models that cannot account for the information-related tasks in such environments. Recent work related to collaborative information seeking (Reddy and Jansen, 2008) is beginning to surface these problems more clearly. Complex organizational activities, however, require coordination efforts from individuals who may participate in the workflow at different places and times. The enhancements presented here – in the form of an integrated conceptual model – are a first step in this direction.

There are a number of areas for future research. First, in this research, we have not highlighted differences in the nature of information sources. First, in this research, we have not highlighted differences in the nature of information sources. A broad interpretation of this concept (e.g. colleagues, informal documents, unstructured documents on the Web, search results, liveness of the results) might require a deeper investigation of how to represent and capture these. Second, our proposals do not account for issues of privacy and security, which can justifiably require filters and constraints on which information may be shared with other participants in a workflow. Third, patterns of workflows proposed in prior work (van der Aalst et al, 2000) could be investigated to ascertain whether they could be used to capture and enforce rules that govern the simultaneous use of tasks and information tasks as well as arcs and information arcs in the enhanced workflow-net model. Finally, our modeling efforts should be considered as a part of the overall set of technologies that help collaboration including but not limited to decision support, health records and other. As a result, integration across these remains an important concern..

References

1. Aguilar, F.J.: Scanning the Business Environment. Macmillan, New York (1967)
2. Becker, J., Uthmann, C.V., Muhlen, M., Rosemann, M.: Identifying the Workflow Potential of Business Processes. In: Proceedings of the 32nd Annual Hawaii International Conference on Systems Sciences, vol. 5. IEEE (1999)
3. BPMN. BPMN Information Home, http://www.bpmn.org/ (accessed on March 12, 2012)
4. Browne, E.D., Schrefl, M., Warren, J.R.: Goal-Focused Self-Modifying Workflow in the Healthcare Domain. In: Proceedings of the 37th Annual Hawaii International Conference on System Sciences, vol. 6. Track 6. IEEE (2004)
5. Choo, C., Detlor, B., Turnbull, D.: A Behavioral Model of Information Seeking on the Web: Preliminary Results of a Study of How Managers and IT Specialists Use the Web. In: The 61st Annual Meeting of the American Society for Information Science, pp. 290–302 (1998)
6. Dourish, P., Holmes, J., MacLean, A., Marqvardsen, P., Zbyslaw, A.: Freeflow: mediating between representation and action in workflow systems. In: Proceedings of the 1996 ACM Conference on Computer Supported Cooperative Work, pp. 190–198 (1996)
7. Ellis, C.A.: Workflow technology. In: Beaudouin-Lafon, M. (ed.) Computer Supported Co-operative Work. Trends in Software, vol. 7, pp. 29–54. John Wiley & Sons (1999)
8. Ellis, D., Haugan, M.: Modelling the Information Seeking Patterns of Engineers and Research Scientists in an Industrial Environment. Journal of Documentation 53, 384–403 (1997)
9. Gaudioso, C.: A practical approach to breast cancer knowledge management: A tumor board perspective. Ph.D. Dissertation, The University of Wisconsin - Milwaukee (2010)
10. Hammersley, M., Atkinson, P.: Ethnography: principles in practice. Taylor & Francis (2007)
11. Hassell, L., Holmes, J.: Modeling the workflow of prescription writing. In: Proceedings of the 2003 ACM Symposium on Applied Computing, pp. 235–239 (2003)
12. Glance, N.S., Pagani, D.S., Pareschi, R.: Generalized process structure grammars (GPSG) for flexible representations of work. In: Proceedings of the 1996 ACM Conference on Computer Supported Cooperative Work, pp. 180–189 (1996)
13. Lamb, R., Kling, R.: Reconceptualizing users as social actors in information systems research. MIS Quarterly 27(2), 197–235 (2003)

14. Lederman, R., Morrison, I.: Examining Quality of Care - How Poor Information Flow can Impact on Hospital Workflow and Affect Patient Outcomes. In: Proceedings of the 35th Annual Hawaii International Conference on System Sciences, vol. 6. IEEE (2002)

15. Ling, S., Loke, S.W.: Advanced Petri Nets for Modelling Mobile Agent Enabled Interorganizational Workflows. In: Proceedings of the 9th IEEE International Conference on Engineering of Computer-Based Systems, pp. 245–252. IEEE (2002)

16. Murrata, T.: Petri nets: Properties, analysis and applications. Proceedings of the IEEE 77(4), 541–580 (1989); 99/02, Department of Math. and Comp. Sci. Eindhoven University of Technology, The Netherlands (1999)

17. Reddy, M.C., Jansen, B.J.: A model for understanding collaborative information behavior in context: A study of two healthcare teams. Information Processing & Management 44(1), 256–273 (2008)

18. Rieh, S.Y.: Investigating Web Searching Behavior in Home Environments. In: The 66th Annual Meeting of the American Society for Information Science and Technology, pp. 255–264 (2003)

19. Simon, H.A.: Designing Organizations for an Information-Rich World, in Martin Greenberger. In: Computers, Communication, and the Public Interest. The Johns Hopkins Press, Baltimore, MD (1971)

20. Tan, J., Wen, H., Awad, N.: Healthcare and services delivery systems as complex adaptive systems. Communications of the ACM 48(5), 36–44 (2005)

21. Twidale, M., Marty, P.: Coping with errors: the importance of process data in robust sociotechnical systems. In: Proceedings of CSCW 2000, pp. 269–278 (2000)

22. van der Aalst, W.: Structural characterizations of sound workflow nets. Technical Report Computing Science Report 96/23, Eindhoven University of Technology. The Netherlands (1996)

23. van der Aalst, W.: Verification of workflow nets. In: 18th International Conference Proceedings Application and Theory of Petrinets 1997, pp. 407–426 (1997)

24. van der Aalst, W., Hofstede, A.T., Kiepuszewski, B., Barros, A.P.: Workflow patterns. Technical Report BETA Working Paper Series, WP47, Eindhoven University of Technology, The Netherlands (2000)

25. van der Aalst, W., Hee, K.: Workflow Management: Models, Methods, and Systems. The MIT Press (2002)

26. van der Aalst, W.M.P., Stahl, C.: Modeling Business Processes: A Petri Net Oriented Approach. MIT Press, Cambridge (2011)

27. Verbeek, H., Basten, T., van der Aalst, W.: Diagnosing workflow processes using Woflan. Technical Report

28. Wong, et al.: Workflow-enabled distributed component-based information architecture for digital medical imaging enterprises. IEEE Transactions on Information Technology in Biomedicine 7(3), 171–183 (2003)

29. WFMC. WFMC.org Homepage, http://www.wfmc.org/ (accessed on April 12, 2007)

30. Weick, K.E., Sutcliffe, K.M.: Managing the Unexpected: Assuring High Performance in an Age of Complexity. Jossey-Bass, San Francisco (2001)

31. White, S.A.: Introduction to BPMN. IBM (May 2004), http://www.bpmn.org/Documents/Introduction%20to%20BPMN.pdf (accessed on March 12, 2012)

32. Zisman, M.: Representation, Specification, and Automation of Office Procedures, PhD Thesis, University of Pennsylvania (1977)

A Method for the Definition and Treatment of Conceptual Schema Quality Issues

David Aguilera, Cristina Gómez, and Antoni Olivé

Department of Service and Information System Engineering
Universitat Politécnica de Catalunya — BarcelonaTech
Barcelona, Spain
{daguilera,cristina,olive}@essi.upc.edu

Abstract. In the literature, there are many proposals of quality properties of conceptual schemas, but only a few of them (mainly those related to syntax) have been integrated into the development environments used by professionals and students. A possible explanation of this unfortunate fact may be that the proposals have been defined in disparate ways, which makes it difficult to integrate them into those environments. In this paper we define quality properties in terms of quality issues, which essentially are conditions that should not happen, and we propose a unified method for their definition and treatment. We show that our method is able to define most of the existing quality properties in a uniform way and makes it possible to integrate quality issues into development environments. The method can be adapted to several languages. We present a prototype implementation of our method as an Eclipse plugin. We have evaluated the potential usefulness of our method by analyzing the presence of a set of quality issues in a set of conceptual schemas developed by students as part of their projects.

Keywords: Quality Issues, Conceptual Modeling, Methodologies & Tools.

1 Introduction

A conceptual schema defines the general knowledge that an information system needs to know in order to perform its functions [1]. The increasingly important role conceptual schemas play in information system development requires that they must be of high-quality [2, 3]. This quality can be analyzed in terms of properties (or dimensions). All conceptual schemas should have the fundamental properties of syntactic and semantic correctness, relevance and completeness, but other quality properties have been proposed in the literature [4–11], and may be required or recommended in particular projects. In this paper we focus on the quality of conceptual schemas.

It is a fact that only a few quality properties (mainly those related to syntax) have been integrated into the development environments used by professionals and students, and thus enforced in the conceptual schemas developed by them. As a consequence, the support offered by conceptual modeling tools is

P. Atzeni, D. Cheung, and R. Sudha (Eds.): ER 2012, LNCS 7532, pp. 501–514, 2012.

significantly poorer than the support offered by modern IDEs [12]. A possible explanation of this unfortunate fact may be that the proposals have been defined in the literature in disparate ways, which makes it difficult to integrate them into those environments. In this paper we propose a unified method for defining conceptual schema quality properties.

The method proposed here is based on the notion of conceptual schema quality issue that, according to the dictionary, we understand as "an important [*quality*] topic or problem for debate or discussion"[1]. In essence, an issue is a condition. The condition may be an integrity constraint a schema must satisfy to be syntactically correct, a necessary condition for a schema to be satisfiable, a condition for a schema element to be relevant, a best practice defined as a condition that must be satisfied, and so on.

In our method, quality issues are defined by engineers of conceptual modeling methods, and included in a catalog. A given development environment may choose from that catalog the quality issues that it may enforce in the projects developed using that environment. In a similar way, an organization using a development environment for a given project may choose the quality issues included in that environment that it wants to enforce in that project. The idea is that if all chosen issues are satisfied by a project, then the schema developed will have the level of quality aimed at by the organization in that project.

Of course, our method requires that the development environments must allow the addition (or the removal) at any time of new quality issues developed by method engineers. A necessary requirement for this to be possible is that quality issues are defined in a uniform way, as proposed here.

In general, the quality of a conceptual schema can be evaluated when requested by the conceptual modeler (usually, when it is complete) or continuously, that is, every time that the conceptual modeler makes a change to that schema. In the evaluation on request, a process, which can be manual or automated, analyzes the whole schema and reports all quality issues that have been found. In the continuous evaluation, an automated process analyzes the changes to a schema and points out the quality issues that have arisen due to those changes. The quality issues proposed in this paper can be evaluated in both ways.

The overall framework of our research is that of design science [13]. The problem we try to solve is to increase the quality of conceptual schemas. The problem is significant because each information system development project requires the development of its conceptual schema, and that conceptual schema must be correct, complete and follow a set of best practices. In this paper we formalize the notion of conceptual schema issue, propose a method for defining quality issues, and describe how to automatically detect the issues present in a given conceptual schema. As far as we know, this is the first work that proposes a unified method for defining conceptual schema quality properties. We have evaluated the expressiveness and usefulness of our method by means of its application to a set of issue types and to a set of conceptual schemas developed by students of advanced courses as part of their projects.

---

[1] Oxford dictionaries (`http://oxforddictionaries.com/`)

The paper is structured as follows. In Sect. 2 we formalize the notions of conceptual schema issue and issue type. Section 3 presents an algorithm that can be used to compute the issues raised in a schema, either under request or continuously. Section 4 briefly describes the implementation of our method in the Eclipse Platform. Section 5 presents the experimental evaluation we have done of the expressiveness and usefulness of our method. Section 6 discusses related work. Section 7 summarizes the conclusions and points out future work.

2   Conceptual Schema Issues

In this section, we first introduce a simple motivating example that serves as an illustration of the quality issues we deal with in this paper. We then formalize the notions of conceptual schema issue type and issue instance, issue lifecycle, and issue precedence.

2.1   A Motivating Example

Consider the structural conceptual schema of Fig. 1. There is a 4-level hierarchy of *IsA* (*Car* and *motorcycle IsA MotorVehicle IsA LandVehicle IsA Vehicle*). A person may own any number of motor vehicles. On the other hand, there is the typical parent-child relationship type.

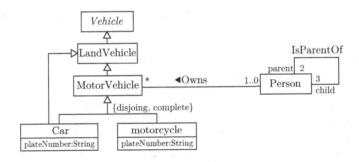

Fig. 1. Motivating example

Even if the example is very small, we can detect several quality problems:

a) The cardinality constraint [1..0] of the participant *person* in association *Owns* is syntactically incorrect.
b) Entity type *motorcycle* does not start with a capital letter. Several naming guidelines recommend that entity types start with a capital letter (e.g. [4]).
c) Abstract entity type *Vehicle* has only one subtype (*LandVehicle*). Either *Vehicle* is not abstract, or both type and subtype have the same population.
d) The cardinality constraints of association *IsParentOf* are not satisfiable.
e) The specialization *Car IsA LandVehicle* is redundant.
f) Attribute *plateNumber* is repeated in all subtypes of *MotorVehicle*.

On the other hand, there are a few issues that should be checked. For example:

g) The name of the association *IsParentOf* (as well as the name of the association *Owns*) must make sense.
h) The recursive association *IsParentOf* could be *symmetric, asymmetric* or *antisymmetric*. If so, the schema must include the corresponding constraint.

In the following, we describe how these quality problems can be uniformly defined as instances of issue types.

2.2   Formalization of Issue and Issue Types

Let S be a schema that consists of n schema elements $e_1, \ldots, e_n$, which are an instance of the corresponding schema metatypes. The most important schema elements are entity, relationship and event types, *IsA* relations, constraints, derivation rules and pre/post conditions. Among the auxiliary schema elements there are strings (for example, names of relationship types) and integers (for example, minimum values of cardinality constraints).

We define a conceptual schema quality issue instance (for short, issue) of type I_x as a fact $I_x(e_1, \ldots, e_m)$ where $e_1, \ldots, e_m$ are schema elements, $m \geq 1$. In a schema there may be several distinct issues of the same issue type, and there may be several issues for the same tuple $\langle e_1, \ldots, e_m \rangle$.

For example, consider the issue (e) of Fig. 1, and assume that g is the schema element corresponding to *Car IsA LandVehicle*. Then, issue (e) can be formalized as an issue $I_e(g)$ of issue type $I_e = $ "The specialization is redundant".

Each issue is an instance of an issue type I_x. Formally, I_x is a tuple:

$$I_x = \langle S_x, \phi_x, \rho_x, \mathcal{K}_x, \mathcal{A}_x, \mathcal{O}_x, \mathcal{P}_x \rangle \qquad (1)$$

where

- S_x is the *scope* of the issue type
- ϕ_x is the *applicability condition*
- ρ_x is the *issue condition*
- $\mathcal{K}_x$ is the *kind* of the issue type
- $\mathcal{A}_x$ is the *acceptability* of the issue type
- $\mathcal{O}_x$ is a set of *issue actions*
- $\mathcal{P}_x$ is a set of *precedents*.

The scope S_x of an issue type I_x is a tuple $S_x = \langle T_1, \ldots, T_m \rangle$ of m ($m \geq 1$) schema metatypes. At a given time, there could be an instance of I_x for each element of the Cartesian product of $T_1 \times \ldots \times T_m$. In the previous example, assuming that T_{IsA} is the schema metatype corresponding to the *IsA* relations of the schema S, then $S_e = \langle T_{\text{IsA}} \rangle$. In principle, there could be an instance of issue type I_e for each instance of T_{IsA} in schema S.

In practice, often not all elements of $T_1 \times \ldots \times T_m$ may raise an issue of type I_x, but only a subset of them. Therefore, we find it convenient to define for each issue type I_x an applicability condition $\phi_x(e_1, \ldots, e_m)$ such that the potential set $Pot(I_x)$ of elements of $T_1 \times \ldots \times T_m$ that may raise an issue of type I_x is:

$$Pot(I_x) = \{\langle e_1, \ldots, e_m \rangle \mid \langle e_1, \ldots, e_m \rangle \in T_1 \times \ldots \times T_m \wedge \phi_x(e_1, \ldots, e_m)\} \quad (2)$$

When $Pot(I_x) = T_1 \times \ldots \times T_m$ then we define $\phi_x(e_1, \ldots, e_m) = True$. In the above example, if we take into account that an *IsA* specialization can be redundant only if its subtype is a subtype of another specialization and its supertype is a supertype of another specialization, then we could define $\phi_e(g) =$ "the subtype of g is a subtype of another specialization and the supertype of g is a supertype of another specialization"

An instance of issue type I_x at a given time is an element of $Pot(I_x)$ that satisfies the issue condition $\rho_x(e_1, \ldots, e_m)$ at that time. The set $Raised(I_x)$ of issues of type I_x raised at a given time is:

$$Raised(I_x) = \{\langle e_1, \ldots, e_m \rangle \mid \langle e_1, \ldots, e_m \rangle \in Pot(I_x) \wedge \rho_x(e_1, \ldots, e_m)\} \quad (3)$$

In the above example, the issue condition would be (written in the appropriate language) $\rho_e(g) =$ "there is an indirect specialization between the subtype and supertype of g".

As before, when $Raised(I_x) = Pot(I_x)$ then we define $\rho_x = True$. In the example of Fig. 1, the type of issue (g) requires checking that the name of an association makes sense. In this case, we have:

$$S_g = \langle Association, String \rangle$$
$$\phi_g(a, s) = \text{"Association } a \text{ has name } s\text{"}$$
$$\rho_g(a, s) = True$$

That is, there is a raised issue for each tuple $\langle a, s \rangle$ such that $a \in Association$ and $s \in String$, and s is the name of a. The issue requires the conceptual modeler checking that s makes sense.

In our method, issue types have an issue kind $\mathcal{K}_x$, which may be either *problem* or *checking* issue. All issues of an issue type are of the same kind. A *problem issue* $I_x(e_1, \ldots, e_m)$ is an issue that in principle (we will see later on that there may be exceptions) should not happen in a schema. Once raised, the issue should be solved (and thus, it ceases to exist), which can only be done by changing the schema in a way such that:

- $\langle e_1, \ldots, e_m \rangle$ is not an element of $T_1 \times \ldots \times T_m$, or
- $\langle e_1, \ldots, e_m \rangle$ does not satisfy $\phi_x(e_1, \ldots, e_m)$, or
- $\langle e_1, \ldots, e_m \rangle$ does not satisfy $\rho_x(e_1, \ldots, e_m)$

In the running example, $\mathcal{K}_e = problem\ issue$, because it is not considered good practice to have redundant specializations in a schema. A possible solution to this problem issue could be to remove the redundant specialization from the schema.

A *checking issue* $I_x(e_1, \ldots, e_m)$ is an issue that requires the conceptual modeler to check something that cannot be automatically checked, or—in general—to perform some action that cannot be automatically performed. Once the checking has been done, or the action has been performed, the issue usually remains raised, but in a different state (as we will describe shortly).

In the schema of Fig. 1, an example of checking issue could be I_h = "The symmetric property of the association is well defined". In this case, we have:

$$\mathcal{S}_h = \langle Association \rangle$$
$$\phi_h(a) = \text{"Association } a \text{ is binary and recursive"}$$
$$\rho_h(a) = True$$

The issue requires the conceptual modeler to check whether or not *IsParentOf* is *symmetric, asymmetric* or *antisymmetric* and, if so, to check that there is a constraint (invariant) that enforces it [1, p. 203]. Once checked, the issue becomes checked, although it is still a raised issue.

On the other hand, the issues of an issue type I_x may or may not be acceptable, $\mathcal{A}_x = \{True, False\}$. An issue type may be defined as acceptable if the method engineer believes that some of its instances are acceptable in some circumstances. The exact meaning of the acceptability depends on the issue kind.

If $\mathcal{K}_x$ = *problem issue*, then:
- $\mathcal{A}_x = True$ means that a conceptual modeler may find it reasonable that there are some instances of I_x in a particular schema.
- $\mathcal{A}_x = False$ means that all issues of type I_x must be solved.

If $\mathcal{K}_x$ = *checking issue*, then:
- $\mathcal{A}_x = True$ means that a conceptual modeler may find it reasonable not to check some instances of I_x in a particular schema.
- $\mathcal{A}_x = False$ means that all issues of type I_x must be checked.

In the example of Fig. 1, the issue I_c = "Abstract entity type has only one subtype" could be an instance of an *acceptable* issue type, because there may be situations in which the conceptual modeler can accept issues of this type.

Table 1 shows the classification according to *kind* and *acceptability* of the issues raised in the motivating example of Fig. 1.

An issue type I_x has a set $\mathcal{O}_x$ of one or more *issue actions*. Each issue action of I_x with $\mathcal{S}_x = \langle T_1, \ldots, T_m \rangle$ is an operation $op(p_1{:}T_1, \ldots, p_m{:}T_m)$ whose effect depends on $\mathcal{K}_x$. If it is a problem issue, then the execution of the operation solves the issue $I_x(e_1, \ldots, e_m)$ (that is, $\langle e_1, \ldots, e_m \rangle \notin Raised(I_x)$), and the issue ceases to exist. If I_x is a checking issue, then the execution sets the state of the issue to *Checked*.

Table 1. Classification of some issues of the schema shown in Fig. 1

	Acceptable	Non-acceptable
Problem issue	I_b: Entity type *motorcycle* does not start with a capital letter.	I_a: Cardinality constraint of participant *person* in association *Owns* is syntactically incorrect.
	I_c: Abstract entity type *Vehicle* has only one subtype.	I_d: Cardinality constraints of association *IsParentOf* are not satisfiable.
	I_f: Attribute *plateNumber* is repeated in all subtypes of *MotorVehicle*.	I_e: The specialization *Car IsA LandVehicle* is redundant.
Checking issue	I_g: The name of the association *IsParentOf* must make sense.	I_h: The *symmetric* property of *IsParentOf* is well defined.

Issue actions can be automated or manual. For example, consider $I_f =$ "An attribute is repeated in all subtypes of a complete generalization set". One possible automated issue action to solve issues of this type could be the refactoring operation *pullUpAttribute* (see [10]), which moves the attributes up to the supertype automatically. Now consider, for instance, $I_a =$ "The cardinality constraint of a participant in an association is syntactically incorrect". In order to solve an issue of this type, a modeler should, in general, perform a manual operation that requires her to change the multiplicities so that they are syntactically correct. For the particular case where she set the values upside down, she could execute an automated operation *swapLowerAndUpperValues* instead.

2.3   The Lifecycle of Issues

An issue $I_x(e_1, \ldots, e_m)$ is automatically created at the time when $\langle e_1, \ldots, e_m \rangle \in Raised(I_x)$. Similarly, an existing issue $I_x(e_1, \ldots, e_m)$ is automatically deleted at the time when $\langle e_1, \ldots, e_m \rangle \notin Raised(I_x)$. The initial state of an issue is *Pending* but, depending on its kind and acceptability, it can change its state as indicated below and summarized in Fig. 2.

The simplest case is when I_x is a non-acceptable problem issue (Fig. 2 (a)). Issues of this type are created in the initial state of *Pending*, and they remain in this state until they cease to exist. An example is $I_a =$ "The cardinality constraints of an association participant are syntactically incorrect". When an issue of this type is raised, it becomes *Pending* and it remains in this state until the conceptual modeler changes the schema in a way that the issue is not raised.

Problem issues that are acceptable may be in the states of $\{Pending, Accepted\}$ as shown in Fig. 2 (b). If the conceptual modeler accepts (event *Acceptance*) one of these issues, then the issue changes to the state of *Accepted*. The transition can be reversed if the conceptual model reconsiders the acceptance (event *Reconsideration*). An example is $I_b =$ "The name of entity type does not start with a capital letter". When an issue of this type is raised, it becomes *Pending*. In most cases, the conceptual modeler will change the name and then the issue will be automatically deleted. However, given that I_b has been defined as acceptable, the conceptual modeler may choose to accept issues of this type.

Non-acceptable checking issues can be in the states of $\{Pending, Checked\}$ as shown in Fig. 2 (c). Once the conceptual modeler checks (event *Checking*) the

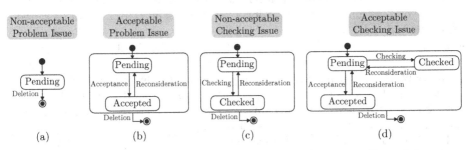

Fig. 2. The lifecycle of an issue of type I_x, depending on its $\mathcal{K}_x$ and $\mathcal{A}_x$

issue, it changes to the state of *Checked*. The transition can be reversed if the conceptual model reconsiders the checking (event *Reconsideration*). An example is I_h = "The symmetric property of the association is well defined" When an issue of this type is raised, it becomes *Pending* and it remains in this state until the conceptual modeler performs (event *Checking*) one of the issue actions, which checks the issue.

Finally, acceptable checking issues can be in the states of {*Pending, Accepted, Checked*} as shown in Fig. 2 (d). The semantics of *Checked* is the same as the previous case, but now an issue can be accepted (event *Acceptance*), meaning that the conceptual modeler, for whatever reason, decides not to check the issue. The transition to *Accepted* can be reversed. An example is I_g = "The name of the association must make sense". If the conceptual modeler performs (event *Checking*) one of the issue actions, then the issue becomes *Checked*. However, the conceptual modeler may decide not to check the issue and accept the name of the association as it is.

2.4   Issue Precedence

In some cases, the issues of a given issue type I_x should only be considered if there are no other unsolved issues of some specific types. For example, the issues of type I_d = "The cardinality constraints of an association are not satisfiable" should only be considered if there are no unsolved issues of type I_a = "The cardinality constraints of an association participant are not syntactically correct". Clearly, it makes no sense to check satisfiability if some cardinality constraints are incorrect. We formalize these issue relationships by means of precedence relationships, which are obviously acyclical. For each issue type I_x we can define two sets of issue type precedents: *global* and *instance*.

On the one hand, if an issue type I_y is a *global precedent* of another issue type I_x, then there cannot be issues of type I_x as long as there are any unsolved issues of type I_y. For instance, in the above example we would define I_a as a global precedent of I_d. On the other hand, if I_y is an *instance precedent* of I_x, then I_x should only be considered *for those instances* that are not involved in an issue of type I_y. For example, consider the issue type I_g = "The name of the association must make sense" and a new problem issue type based on a certain naming guideline I_n = "The name of the association is not a verb phrase in third person singular". If there is a schema with two associations a_1 and a_2 and the name of a_1 is not in the correct form, then the issue $I_n(a_1)$ would be raised. Clearly, the modeler has to fix the name of a_1 prior to checking whether it makes sense or not. However, if $I_n(a_2) \notin Raised(I_n)$, then she should be able to check whether the name of a_2 makes sense regardless the issue $I_n(a_1)$ is raised or not.

Formally, the *global precedents* of an issue type I_x, denoted $\mathcal{GP}_x = \{I_1, \ldots, I_p\}$, is a set of issue types $I_1, \ldots, I_p$ such that any issue of type I_x should be considered only if there is no any unsolved issue of types $I_1, \ldots, I_p$. Then, the first example would be formalized as $\mathcal{GP}_d = \{I_a\}$.

The *instance precedents* of an issue type I_x, denoted $\mathcal{IP}_x = \{\langle I_1, \mathcal{M}_{x:1}\rangle, \ldots, \langle I_q, \mathcal{M}_{x:q}\rangle\}$, is a set of tuples of the form $\langle I_y, \mathcal{M}_{x:y}\rangle$ where I_y is an issue type whose scope $S_y \subseteq S_x$ and $\mathcal{M}_{x:y}$ is a set of pairs $\langle i, j \rangle$ ($1 \leq i \leq |S_y|$ and $1 \leq j \leq |S_x|$) that map each metatype in S_y to one, and only one, metatype in S_x. Therefore, an issue $I_x(e_{x_1}, \ldots, e_{x_m})$ should not be considered if there is an unsolved issue $I_y(e_{y_1}, \ldots, e_{y_l})$ such that $\forall i, j \mid \langle i, j \rangle \in \mathcal{M}_{x:y} \Rightarrow e_{y_i} = e_{x_j}$. Then, the second example would be formalized as $\mathcal{IP}_g = \{\langle I_n, \{\langle 1, 1 \rangle\}\rangle\}$ (assuming that the scopes of I_g and I_n are $S_g = \langle Association, String \rangle$ and $S_n = \langle Association \rangle$).

3   Computing Issue Instances

In this section we propose an algorithm for computing the issues that are present in a conceptual schema S, given a concrete set of issue types $\mathcal{J}$. The algorithm can be used to compute raised issues either under request or continuously.

In general, given any I_x and I_y issue types, the computation of the respective issue instances are independent processes that do not interfere each other. However, if I_y is a precedent of I_x, it is a necessary condition that the issues of type I_y are determined *before* the issues of type I_x, since the existence of issues of the former type may modify the set of raised issues of the latter type. In order to compute the issues of any type $I_x \in \mathcal{J}$ (whose precedents have already been computed), we first have to determine the set $Pot(I_x)$. As we have already seen, the elements of a tuple $\langle e_1, \ldots, e_m \rangle \in Pot(I_x)$ must (1) be instances of the corresponding metatypes in the scope S_x (that is, $T_1(e_1) \wedge \ldots \wedge T_m(e_m)$), (2) satisfy the applicability condition $\phi_x(e_1, \ldots, e_m)$, and (3) there must be no unsolved issues of a type $I_y \in \mathcal{P}_x$ (see Sect. 2.4).

Next, we have to compute the set $Raised(I_x)$. This can be easily achieved by simply selecting those tuples $\langle e_1, \ldots, e_m \rangle \in Pot(I_x)$ that satisfy the issue condition $\rho_x(e_1, \ldots, e_m)$. As a result, we have an issue $I_x(e_1, \ldots, e_m)$ for each tuple $\langle e_1, \ldots, e_m \rangle$ in $Raised(I_x)$.

Finally, we have to properly set the state ε of each issue instance we have just created. As we have seen in Sect. 2.3, when an issue instance is created, its state ε is automatically set to *Pending*. During a continuous evaluation, however, this process is iteratively executed each time a change occurs, and thus it may be the case that some of the issues of any type I_x we have just created ($Raised(I_x)$) already existed from a previous iteration ($Raised_{\text{prev}}(I_x)$). Therefore, for each issue instance $I_x(e_1, \ldots, e_m)$ in $Raised(I_x)$, we have to check whether the issue is also in $Raised_{\text{prev}}(I_x)$. If that is the case, then the state ε of the issue instance $I_x(e_1, \ldots, e_m)$ must be $\varepsilon_{\text{prev}}$ instead of *Pending*.

Algorithm 1 iteratively computes the raised issues found in the schema. In each iteration, an issue type $I_c \in \mathcal{J}$ (whose issue precedents have already been computed in previous iterations) is randomly selected (line 5). Then, for each tuple $\langle e_1, \ldots, e_m \rangle$ of the schema to which I_c may apply (that is, $\langle e_1, \ldots, e_m \rangle \in Pot(I_x)$), the algorithm checks that (a) its issue precedents are satisfied (as described in Sect. 2.4), and (b) it satisfies the issue condition ρ_c (lines 7 to 13). Finally, for those tuples $\langle e_1, \ldots, e_m \rangle \in CandidateIssues$, the algorithm generates the associated issues in the proper state. If there was a pair $i = \langle I_c(e_1, \ldots, e_m), \varepsilon \rangle$ in Ψ_{prev}, i is added to the result Ψ_{new}; otherwise, a new pair i such that $i = \langle I_c(e_1, \ldots, e_m), Pending \rangle$ is created and added to Ψ_{new} (lines 14 to 18).

Algorithm 1. Computing Issue Instances

Input	– S: a set of n schema elements $e_1, \ldots, e_n$,
	– $\mathfrak{I}$: the set of issue types $I_1, \ldots, I_p$, and
	– Ψ_{prev}: the set of pairs $\langle I_x(e_1, \ldots, e_m), \varepsilon \rangle$ computed in the previous execution, where $I_x(e_1, \ldots, e_m)$ was a raised issue and ε was its state.
Output	– Ψ_{new}: the set of pairs $\langle I_x(e_1, \ldots, e_m), \varepsilon \rangle$ computed in this execution, where $I_x(e_1, \ldots, e_m)$ is a raised issue and ε is its state.

1: **procedure** $updateIssues(S, \mathfrak{I}, \Psi_{\text{prev}}) : \Psi_{\text{new}}$
2:     $\Psi_{\text{new}} \leftarrow \emptyset$
3:     $\mathfrak{I}_{\text{pending}} \leftarrow \mathfrak{I}$
4:     **while** $\mathfrak{I}_{\text{pending}} \neq \emptyset$ **do**
5:         $I_c \leftarrow I_x \mid (I_x \in \mathfrak{I}_{\text{pending}} \land (\nexists I_y \mid I_y \in \mathfrak{I}_{\text{pending}} \land I_y \in \mathcal{P}_x))$
6:         $CandidateIssues \leftarrow \emptyset$
7:         **for all** $\langle e_1, \ldots, e_m \rangle \in Pot(I_c)$ **do**
8:             **if** $arePrecedentsSatisfied(\langle e_1, \ldots, e_m \rangle, \mathcal{P}_c, \Psi_{\text{new}})$ **then**
9:                 **if** $\rho_c(e_1, \ldots, e_m)$ **then**
10:                     $CandidateIssues \leftarrow CandidateIssues \cup \langle e_1, \ldots, e_m \rangle$
11:                 **end if**
12:             **end if**
13:         **end for**
14:         $IssuesToKeep \leftarrow \{ \langle e_1, \ldots, e_m \rangle \mid \langle I_c(e_1, \ldots, e_m), \varepsilon \rangle \in \Psi_{\text{prev}} \land$
$\langle e_1, \ldots, e_m \rangle \in CandidateIssues \}$
15:         $IssuesToCreate \leftarrow CandidateIssues - IssuesToKeep$
16:         $\Psi_{\text{new}} \leftarrow \Psi_{\text{new}} \cup$
$\{ i = \langle I_c(e_1, \ldots, e_m), \varepsilon \rangle \mid i \in \Psi_{\text{prev}} \land \langle e_1, \ldots, e_m \rangle \in IssuesToKeep \} \cup$
$\{ \langle I_c(e_1, \ldots, e_m), Pending \rangle \mid \langle e_1, \ldots, e_m \rangle \in IssuesToCreate \}$
17:         $\mathfrak{I}_{\text{pending}} \leftarrow \mathfrak{I}_{\text{pending}} - \{ I_c \}$
18:     **end while**
19:     **return** Ψ_{new}
20: **end procedure**

---

4   Prototype Implementation

As a proof of concept for our method, we have built a prototype tool in the Eclipse Platform [14]. Eclipse is an extensible IDE that can be used as a UML modeling environment thanks to the UML2 Tools plugin. Figure 3 shows a screenshot of our prototype tool with the example introduced in Sect. 2.1.

Figure 4 depicts the architecture of our prototype. Currently, our tool detects and treats a set of issue types included in an initial catalog (see next section for more details on how this catalog was defined). Issue types are defined for the UML using OCL expressions [15]. Each issue type includes the information described in Sect. 2, as well as some additional information that is necessary to provide meaningful feedback to modelers (such as the *label* to be shown when an issue instance is raised or a *description*). Since the UML2 Tools plugin integrates an OCL interpreter, our tool is able to load a set of issue types defined in OCL and evaluate the applicability and issue conditions against the conceptual

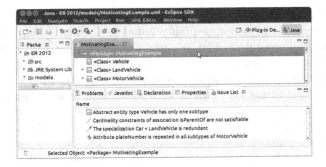

Fig. 3. Screenshot of our prototype tool running the example from Fig. 1.

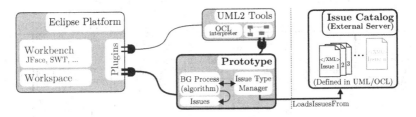

Fig. 4. Architecture of our prototype tool in the Eclipse Platform.

schema using OCL. This evaluation is implemented as a background process that executes the algorithm presented in Sect. 3.

In our implementation, the type of the applicability condition ϕ depends on the size of the scope. When the scope of an issue includes only one metatype T, then ϕ is a Boolean condition. Thus, the background process is responsible for selecting the instances of T in the schema and, for each instance, checking whether ϕ is satisfied or not. On the other hand, when the scope of an issue is more than one metatype $\langle T_1, \ldots, T_m \rangle$, the applicability condition returns the set of tuples to which the issue may apply. In these (less common) cases, the OCL expression is targetted at T_1.

5   Experimental Evaluation

This section describes two experiments conducted to evaluate our method in terms of its *expressiveness* and *usefulness* when used to define conceptual schemas. The first experiment evaluates the *expressiveness* of our method analyzing its capability to define most of the existing quality properties (i.e. issue types) we may find in the literature. In order to create an initial catalog, we randomly selected a set of 112 issue types (98 problem issues and 14 checking issues) for conceptual schemas expressed in UML/OCL and we classified them into three categories: 72 *syntactic issue types*, 29 *best practices* and 11 *naming guidelines*. As expected, we were able to define all of them using our method. The complete description and definition of those issue types may be found in [15].

The second experiment evaluates the *usefulness* of our method. For that purpose, we analyzed the presence of quality issues in a set of conceptual schemas, using the prototype tool described in Sect. 4. The starting point of the experiment was the random selection of 13 conceptual schemas that were developed by students as part of their final projects during the last year of their Computer Science degree at the *Universitat Politècnica de Catalunya*. The conceptual schemas were defined using different modeling environments. Table 2 summarizes, for each conceptual schema, the number of classes, association classes, associations, attributes, and invariants present in the conceptual schema, as well as the syntactic issues detected, and best practices and naming guidelines not followed. Note we did not execute any issue actions because the schemas were already "definitive".

Table 2. Summary of the results of the experiment

	P1	P2	P3	P4	P5	P6	P7	P8	P9	P10	P11	P12	P13	Total
Classes	10	16	31	18	15	11	14	44	15	18	20	28	366	606
AssClasses	1	3	3	0	4	1	0	6	1	0	6	6	55	86
Assocations	8	9	23	10	6	5	19	18	15	27	12	7	264	423
Attributes	16	54	109	75	33	91	11	117	50	73	42	56	1144	1871
Invariants	3	0	5	0	6	2	15	9	16	19	23	23	386	507
Syntactic Issues	2	0	6	0	1	2	4	7	2	6	0	0	0	30
Best Practices	13	24	54	34	10	21	7	64	5	10	6	21	57	326
Naming Issues	4	13	47	19	16	6	16	25	16	38	7	4	53	264

When we defined the schemas using our tool, we found that all projects had a significant number of unsolved quality issues. The results reveal that syntactic issues (as for instance, "member ends of a class are not distinguishable") are less frequent than unfollowed best practices or naming guidelines (as for instance, "classes without identifier" or "attributes whose name is the same as an existing class"), probably because of the support offered by development environments.

6   Related Work

To our knowledge, this is the first work that proposes a unified method for the definition of conceptual schema quality properties. There is plenty of proposals in the literature (naming guidelines [5, 6, 16–18]; presentation guidelines [4], best practices—such as using stereotyped constraints to define general constraints [8], making implicit constraints explicit [7], or writing tests [19]—, or applying refactorings [10]) that would improve the quality of conceptual schemas if adopted. Some approaches discuss the importance of efficiently detecting inconsistencies (i.e. violations of metamodel constraints) as they arise, providing useful feedback to modelers in a non disruptive manner. Blanc et al. [20] propose an incremental consistency checker based on the idea of representing models as sequences of primitive construction operations. Their approach infers that an inconsistency appears when a concrete sequence of operations is detected. In [21], Egyed proposes an instant consistency checking for the UML. His approach defines a few consistency rules for UML 1.3 and checks whether they hold or not each time a change is performed. His main contribution involves the *detection of scope*, which is performed in terms of instances instead of types.

Despite some of these works have been implemented into their own prototype tools (for example, Boger et al. implemented refactorings into Poseidon for UML in [22]), it is a fact that the default support offered by most commercial tools only includes violations of metamodel constraints. A notable exception is *ArgoUML* [23], which integrates a set of "design critics". Each design critic implements a best practice the conceptual schema should follow. Its approach, however, does not provide a formal background on how to specify a design critic, which means the tool cannot be easily extended nor the design critics can be easily reused.

Altogether, applying these proposals in a project becomes unfeasible, either because of the difficulty of using several different tools to individually check each property, or because there is no tool at all. By using our method, however, development environments could easily integrate issue types and modelers could take advantage of them.

7 Conclusions

In this paper, we have formalized the concepts of quality issue and issue type, proposed a method for their definition, and described an algorithm that computes the quality issues present in a schema at a given moment of time. The method should be powerful enough to define in a uniform way most (if not all) of the quality properties proposed in the literature, which is a necessary prerequisite for fostering their integration and use in the diverse development environments used by professionals and students. The overall goal is that of improving the quality of the resultant conceptual schemas.

We have implemented our method in a development environment and we also have made a preliminary evaluation of its expresiveness and usefulness. As expected, we were able to define all the issue types we considered using our method, and we found that all the conceptual schemas we tested had a significant number of unsolved quality issues.

The work reported here can be continued in several directions. First, we plan to build a comprehensive catalog with as many as possible issue types, all taken from the public literature. Each issue type would be defined using the method proposed here, and implemented in the UML/OCL. A second direction is then the implementation of the method and the full catalog in a development environment used by professionals and/or students, and the analysis of the results.

Acknowledgements. This work has been partly supported by the *Ministerio de Ciencia y Tecnología* and *FEDER* under project TIN2008-00444/TIN, *Grupo Consolidado*, and by *Universitat Politècnica de Catalunya* under FPI-UPC program.

References

1. Olivé, A.: Conceptual Modeling of Information Systems. Springer (2007)
2. Lindland, O.I., Sindre, G., Sølvberg, A.: Understanding quality in conceptual modeling. IEEE Softw. 11(2), 42–49 (1994)

3. Maes, A., Poels, G.: Evaluating Quality of Conceptual Models Based on User Perceptions. In: Embley, D.W., Olivé, A., Ram, S. (eds.) ER 2006. LNCS, vol. 4215, pp. 54–67. Springer, Heidelberg (2006)
4. Ambler, S.W.: The Elements of UML 2.0 Style. Cambridge University Press (2005)
5. Becker, J., Delfmann, P., Herwig, S., Lis, L., Stein, A.: Formalizing Linguistic Conventions for Conceptual Models. In: Laender, A.H.F., Castano, S., Dayal, U., Casati, F., de Oliveira, J.P.M. (eds.) ER 2009. LNCS, vol. 5829, pp. 70–83. Springer, Heidelberg (2009)
6. Chen, P.: English sentence structure and entity-relationship diagrams. Inf. Sci. (2-3), 127–149 (1983)
7. Costal, D., Gómez, C.: On the Use of Association Redefinition in UML Class Diagrams. In: Embley, D.W., Olivé, A., Ram, S. (eds.) ER 2006. LNCS, vol. 4215, pp. 513–527. Springer, Heidelberg (2006)
8. Costal, D., Gómez, C., Queralt, A., Raventós, R., Teniente, E.: Facilitating the Definition of General Constraints in UML. In: Wang, J., Whittle, J., Harel, D., Reggio, G. (eds.) MoDELS 2006. LNCS, vol. 4199, pp. 260–274. Springer, Heidelberg (2006)
9. Deissenboeck, F., Pizka, M.: Concise and consistent naming. Softw. Qual. Control, 261–282 (2006)
10. Fowler, M.: Refactoring: Improving the Design of Existing Code. Addison-Wesley (1999)
11. Cherfi, S.S.S., Comyn-Wattiau, I., Akoka, J.: Quality Patterns for Conceptual Modelling. In: Li, Q., Spaccapietra, S., Yu, E., Olivé, A. (eds.) ER 2008. LNCS, vol. 5231, pp. 142–153. Springer, Heidelberg (2008)
12. Wohed, P.: Tool Support for Reuse of Analysis Patterns - A Case Study. In: Laender, A.H.F., Liddle, S.W., Storey, V.C. (eds.) ER 2000. LNCS, vol. 1920, pp. 196–209. Springer, Heidelberg (2000)
13. Hevner, A.R., March, S.T., Park, J., Ram, S.: Design science in information systems research. MIS Quarterly (1), 75–105 (2004)
14. Eclipse Foundation: Eclipse project, http://www.eclipse.org
15. Aguilera, D., Gómez, C., Olivé, A.: Issue catalog, http://helios.lsi.upc.edu/phd/catalog/issues.php
16. Barker, R.: CASE Method: Entity Relationship Modelling, 1st edn. Addison-Wesley Longman Publishing (1990)
17. Embley, D.W., Kurtz, B., Woodfield, S.: Object-Oriented Systems Analysis: A Model-Driven Approach. Yourdon Press (1992)
18. Meyer, B.: Reusable Software: the Base object-oriented component libraries. Prentice-Hall (1994)
19. Tort, A., Olivé, A., Sancho, M.-R.: The CSTL Processor: A Tool for Automated Conceptual Schema Testing. In: De Troyer, O., Bauzer Medeiros, C., Billen, R., Hallot, P., Simitsis, A., Van Mingroot, H. (eds.) ER Workshops 2011. LNCS, vol. 6999, pp. 349–352. Springer, Heidelberg (2011)
20. Blanc, X., Mougenot, A., Mounier, I., Mens, T.: Incremental Detection of Model Inconsistencies Based on Model Operations. In: van Eck, P., Gordijn, J., Wieringa, R. (eds.) CAiSE 2009. LNCS, vol. 5565, pp. 32–46. Springer, Heidelberg (2009)
21. Egyed, A.: Instant consistency checking for the UML. In: ICSE Proceedings, pp. 381–390. ACM (2006)
22. Boger, M., Sturm, T., Fragemann, P.: Refactoring Browser for UML. In: Aksit, M., Awasthi, P., Unland, R. (eds.) NODe 2002. LNCS, vol. 2591, pp. 366–377. Springer, Heidelberg (2003)
23. ArgoUML: ArgoUML, http://argouml.tigris.org

EERMM: A Metamodel
for the Enhanced Entity-Relationship Model

Robson Do Nascimento Fidalgo[1], Elvis Maranhão De Souza[1], Sergio España[2],
Jaelson Brelaz De Castro[1], and Oscar Pastor[2]

[1] Center for Informatics, Federal University of Pernambuco, Recife(PE), Brazil
`{rdnf,ems7,jbc}@cin.ufpe.br`
[2] Centro de Investigación ProS, Universitat Politècnica de València, València, España
`{sergio.espana,opastor}@pros.upv.es`

Abstract. A metamodel describes the elements of a model, the relationships between them, and the structuring rules that constraint the model elements and the way they are arranged/related in order to respect the domain rules. That is, a metamodel provides an abstract syntax to distinguish between valid and invalid models. Although the Enhanced Entity-Relationship (EER) model has been extensively researched, and various extensions and enhancements have been proposed, to the best of our knowledge, a metamodel for the EER model, based on the classical notation of Chen, has not been proposed yet. That is, we have found no evidence of a metamodel that gives a precise and expressive definition of constructors and constructions needed to create, interchange or transform valid EER models. With aim of overcoming these shortcoming, in this paper we propose an expressive metamodel for EER modeling, named EER MetaModel (EERMM), which provides a novel perspective for scientific researches and industrial applications that need an EER metamodel as a starting point. As a proof of concept, we have implemented a CASE tool (EERCASE) according to our metamodel and, by exploiting this tool, we have designed an EER schema that makes use of all constructors of the EER model.

Keywords: Metamodels, Conceptual Data Models, Entity Relationship Model.

1   Introduction

The Enhanced Entity-Relationship (EER) model [1] is one of the mostly used modeling languages for database conceptual design. That is, there are a numerous CASE tools that aim to support EER modeling, most computer science university curricula include it [2] and most database textbooks [3][4][5][6] dedicate one or more chapters to present it. However, although much research related to the EER model has been conducted, to the best of our knowledge, a metamodel for the EER model, based on the classical notation of Chen (see Figure 1), has not been proposed yet. That is, we have found no evidence of a metamodel that: (*i*) defines the constructors and the constructions rules required to create syntactically valid EER schemas, (*ii*) can used as a reference for works that aim to extends (or reuse part of) the classic EER model; (*iii*) allows interchange or transformation of models between EER tools; and

P. Atzeni, D. Cheung, and R. Sudha (Eds.): ER 2012, LNCS 7532, pp. 515–524, 2012.
© Springer-Verlag Berlin Heidelberg 2012

(*iv*) supports the automatic SQL/DDL code generation. For these reasons, we propose in this paper an expressive metamodel for EER modeling named EER MetaModel (EERMM). As a proof of concept, we implemented the CASE tool named EERCASE and, by exploiting this CASE tool, we have constructed an EER schema that uses of all constructors of EER model.

The remainder of this paper is organized as follows. In Section 2 we briefly describe the main concepts of the EER model, then we make some reflection regarding the use of EER model and the Class Diagram of UML for database conceptual modeling and, at last, we discuss some works that propose metamodels for the EER model. Next, in Section 3, we present our metamodel (EERMM). In Section 4 we give an overview of our CASE tool (EERCASE) and present a full example of the EER model that demonstrates the correctness and usefulness of the EERMM. Finally, in Section 5, we provide some conclusions and indications for future work.

2   Models and Metamodels for Database Conceptual Design

There are many notations for the EER model [11], where some of them allow at most binary relationships, while others allow n-ary relationships. In general, n-ary notations are semantic and conceptually more expressive than binary notations [11][12], because, different from binary notations, n-ary notations allow attributes in relationships and three or more entities connected into the same relationship. Regardless of the notation used, there are two different conventions to refer to the place where roles, cardinalities and participations are specified on a relationship, namely: look-across (opposite-side of entity) and look-here (same-side of entity) [11].

In Figure 1 we show the constructors of the EER model according to the Elmasri-Navathe's textbook [3]. Our work is based on this notation because it is very close to the classical notation of Chen, is widely used in database courses and is well accepted by the database community [13]. We highlight that Elmasri-Navathe's notation is n-ary, uses look-here for specifying roles and participations and look-across for specifying cardinalities.

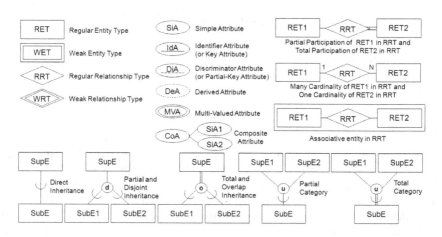

Fig. 1. Constructors of the EER model

Besides the EER model, other notations also can be used to model a database, such as the Class Diagram (CD) of the Unified Modeling Language (UML) [14]. Three works [15][16][17] have performed comparisons between the EER model and the CD and report interesting results. According to the first two works [15][16], the authors suggest that the CD notation is generally more comprehensible than the EER notation. However, in the third work [17], the authors conclude that the understanding of EER model is significantly higher than CD if "composite attribute", "multi-value attribute" and "weak entity" are required in a modeling. To summarize this discussion, we think that any comparison between the EER model and the CD can provide results to debate and arguing, and the choice for a particular notation rather than other is a matter of preference or project contingency.

Despite the weaknesses of CD, many UML vendors have expressed a desire to use CD for data modeling and ended up defining their own tool-specific profiles for each. As a result, there is neither an accepted standard nor interoperability of models developed using such profiles/tools [19]. With aim of overcoming such limitation, the Object Management Group (OMG) is revising its Common Warehouse Metamodel (CWM) [20] in order to provide more normative metamodels (including a metamodel for the EER model) which they call Information Management Metamodel (IMM) [21]. These metamodels are briefly presented as follows.

The CWM metamodel extends some UML metaclasses and specifies the metaclasses for the main concepts of the EER model (i.e., Entity, Relationship and Attribute). However, considering the EER constructors presented in Figure 1, CWM does not include metaclasses nor meta-attributes for: *Associative Entity, Weak Entity, Discriminator Attribute, Composite Attribute, Attribute on Relationship, Category, Inheritance, Generalization Relationship* and *Specialization Relationship*. Furthermore, the CWM metamodel defines the metaclass *ForeignKey*, which is a constructor for relational database modeling. That is, this metaclass is not a constructor of the EER model.

Unlike the CWM metamodel, the EER metamodel of IMM does not extend UML metaclasses for specifying its metamodel, nor mixes EER constructors with relational database ones. However, also taking in account the EER constructors presented in Figure 1, the IMM does not specify metaclasses or meta-attributes for: *Associative Entity, Category, Discriminator Attribute, Composite Attribute, Multi-valued Attribute, Attribute on Relationship* and *Specialization Relationship*. With aim of overcoming the CWM and IMM shortcomings, in the next section we propose a more expressive metamodel for EER modeling.

3   EERMM - The Enhanced Entity Relationship MetaModel

The Enhanced Entity Relationship MetaModel (EERMM) is designed considering that the EER model is essentially a node-link diagram [22][23], where an EER schema is composed of nodes representing: *Entities, Associative Entities, Attributes, Relationships, Inheritances* and *Categories* and links representing: *Attribute Links, Relationship Links, Generalization Links* and *Specialization Links*. In Figure 2 we depict some nodes and links of the EER model. A conceptual view of our metamodel is presented in Figure 3.

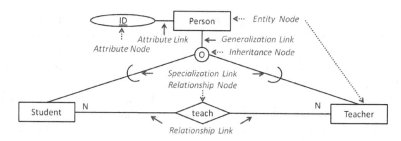

Fig. 2. Some nodes and links of the EER model

In Figure 3 there are three main meta-entities: *Schema, Node* and *Link. Schema* is the root meta-entity that corresponds to the drawing area of an EER schema. For this reason, *Schema* can have many instances of *Node* and many instances of *Link*, which cannot exist without a *Schema* (i.e., they have a total participation with *Schema*). Besides the main meta-entities, our metamodel has three specialized meta-entities for *Node*: *Element, Inheritance* and *Category,* which are presented as follows.

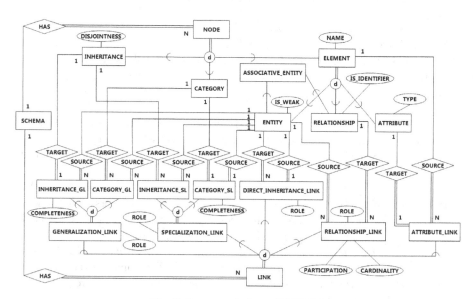

Fig. 3. Conceptual view of EERMM

The *Element* meta-entity has a meta-attribute called *name* and it is specialized in three meta-entities: *Entity, Relationship* and *Attribute*. Note that: (*i*) the *Entity* meta-entity has a meta-attribute called *is_Weak* that is a Boolean value; (*ii*) the *Relationship* meta-entity has a meta-attribute called *is_Identifier* that is also a Boolean value; (*iii*) the *Attribute* meta-entity has a meta-attribute called *type* that specifies whether an instance of *Attribute* is "*common*", "*identifier*", "*discriminator*", "*derived*" or "*multivalued*", exclusively; and (*iv*) the *Associative_Entity* meta-entity is a specialization of the *Entity* and a *Relationship* meta-entities.

The *Inheritance* meta-entity captures the "inheritance" concept and has the *disjointness* meta-attribute, which defines whether an inheritance is mutually exclusive (*"disjoint"*) or not (*"overlap"*). Finally, the *Category* meta-entity captures the "category" concept (also called "union type") that, by definition [3], is disjoint. Then, for this reason, the EERMM does not model the *disjointness* meta-attribute for a category.

Besides the *Node* meta-entities, our metamodel has five specialized meta-entities for *Link*: *Attribute_Link, Relationship_Link, Specialization_Link, Generalization_Link* and *Direct_Inheritance_Link*. The *Attribute_Link* meta-entity has a *source* relationship with the *Element* meta-entity, where an instance of *Attribute_Link* has only one instance of *Element* as source, but an instance of *Element* can be source for many instances of *Attribute_Link*. Furthermore, the *Attribute_Link* meta-entity has a *target* relationship with the *Attribute* meta-entity, where an instance of *Attribute_Link* has only one instance of *Attribute* as target and an instance of *Attribute* can only be target for one instance of *Attribute_Link*. That is, an instance of *Element* can be source for many instances of *Attribute_Link* (many attributes on entities, many attributes on relationships and many composite attributes), but an instance of *Attribute* can only be target for one instance of *Attribute_Link* (an *Attribute* cannot be linked to more than one *Element*).

The *Relationship_Link* meta-entity has a *source* relationship with the *Entity* meta-entity, where an instance of *Relationship_Link* has only one instance of *Entity* as source, but an instance of *Entity* can be source for many instances of *Relationship_Link*. Moreover, the *Relationship_Link* meta-entity has a *target* relationship with the *Relationship* meta-entity, where an instance of *Relationship_Link* has only one instance of *Relationship* as target, but an instance of *Relationship* can be target for many instances of *Relationship_Link*. That is, an instance of *Relationship_Link* connects one instance of *Entity* with one instance of *Relationship*, but an instance of *Entity* or an instance of *Relationship* can be connected to many instances of *Relationship_Link* (an entity can be linked to many relationships and a relationship can be n-ary). Furthermore, the *Relationship_Link* meta-entity has three meta-attributes: *participation* (i.e., *"total"* or *"partial"*); *cardinality* (i.e., *"one"* or *"many"*) and *role* (i.e., a text that describes the function of an entity in a relationship).

The *Specialization_Link* meta-entity also has the *role* meta-attribute, but it describes the function of a sub-entity in a specialization. The *Specialization_Link* is specialized in two meta-entities: *Inheritance_SL* (*Inheritance Specialization Link*) and *Category_SL* (*Category Specialization Link*), which are used to define a specialization link between a sub-entity and an inheritance or between a sub-entity and a category, respectively. Note that: (*i*) an instance of *Inheritance_SL* or *Category_SL* has only one instance of *Entity* as source. However, while an instance of *Entity* can be source for many instances of *Inheritance_SL* (i.e., a sub-entity can be a specialization of many inheritances – a multiple inheritance), an instance of *Entity* can only be source for one instance of *Category_SL* (i.e., a sub-entity can only be a specialization of one category); (*ii*) an instance of *Inheritance_SL* or *Category_SL* has only one instance of *Inheritance* or *Category* as target, respectively. However, while an instance of *Inheritance* can be target for many instances of *Inheritance_SL* (i.e., a simple inheritance with many sub-entities), an instance of *Category* can only be target for one instance of *Category_SL* (i.e., a category can only have one sub-entity); and (*iii*), by notation definition [3], only *Category_SL* can have the *completeness* meta-attribute, which specifies whether a category is completely defined (*"total"*) or not (*"partial"*).

The *Generalization_Link* meta-entity has also the *role* meta-attribute, but it describes the function of a super-entity in a generalization. The *Generalization_Link* is specialized in two meta-entities: *Inheritance_GL* (*Inheritance Generalization Link*) and *Category_GL* (*Category Generalization Link*), which are used to define a generalization link between a super-entity and an inheritance, or between a super-entity and a category, respectively. Note that: (*i*) an instance of *Inheritance_GL* or *Category_GL* has only one instance of *Entity* as source, but an instance of *Entity* can be source for many instances of *Inheritance_GL* (i.e., a super-entity can be a generalization of many inheritances) or *Category_GL* (i.e., a super-entity can participate in many categories); (*ii*) an instance of *Inheritance_GL* or *Category_GL* has only one instance of *Inheritance* or *Category* as target, respectively. However, while an instance of *Inheritance* can only be target for one instance of *Inheritance_GL* (i.e., an inheritance can have only one super-entity), an instance of *Category* can be target for many instance of *Category_GL* (i.e., a category can have many super-entities); and (*iii*), by notation definition [3], only *Inheritance_GL* can have the *completeness* meta-attribute, which specifies whether a inheritance is completely defined (*"total"*) or not (*"partial"*).

The *Direct_Inheritance_Link* meta-entity connects two entities without to use an inheritance node. A *Direct_Inheritance_Link* is important, because sometimes it necessary to model a multiple inheritance with one single sub-entity that directly inherits from two or more super-entities or model an inheritance with only one sub-entity. Note that in both cases, it is a mistake to define the disjointness and completeness properties of an inheritance. For this reason, a *Direct_Inheritance_Link* does not have these properties. The *Direct_Inheritance_Link* meta-entity has a *source* relationship with the *Entity* meta-entity, where an instance of *Direct_Inheritance_Link* has only one instance of *Entity* as source and an instance of *Entity* can be source for only one instance of *Direct_Inheritance_Link*. That is, an *Entity* (super-entity) can only has one *Direct_Inheritance_Link* and a *Direct_Inheritance_Link* can only be connected with one super-entity. Furthermore, the *Direct_Inheritance_Link* meta-entity also has a *target* relationship with the *Entity* meta-entity, where an instance of *Direct_Inheritance_Link* has only one instance of *Entity* as target, but an instance of *Entity* can be target for many instance of *Direct_Inheritance_Link* (a multiple inheritance). At last, a *Direct_Inheritance_Link* also has the *role* meta-attribute.

Note that a *Link* cannot exist without a *Node* (total participation), but a *Node* can be instantiated without a connection to a *Link* (partial participation), because we understand that a designer first models nodes and after models links to connected the nodes. Furthermore, all inheritances are disjoint and total, and the EERMM meta-attributes, except the *role* meta-attribute, are mandatory.

4   A CASE Tool Based on EERMM

In order to demonstrate the feasibility, the expressiveness and usefulness of our metamodel, we have developed the EERCASE CASE tool, which is based on the notation presented in Figure 1 and was used to design our metamodel (Figure 3). Our CASE tool is implemented using Java/Eclipse technologies (i.e., Eclipse Modeling Framework (EMF)

[7], Graphical Modeling Framework (GMF) [8] and Epsilon Framework [9]), which are conform to the Essential Meta Object Facility (EMOF) [10] standard and, in its current version, it generates code only for PostgreSQL. In Figure 4 we show the metamodel used to develop the EERCASE tool. This metamodel is specified in Ecore [7] and extends EERMM in order to (*i*) add eight meta-attributes: (*datatype, size, isNull, isUnique, check, defaultValue, comment* and *cardinality*) in the *Attribute* metaclass, because these meta-attributes are useful for code generation or annotation; (*ii*) add the *isIdentifier* meta-attribute in the *RelationshipLink* meta-entity (this allows to set *Entity.isWeak = True* and *Relationship.isIdentifier = True* with only one click to set *RelationshipLink.isIdentifier = True*); (*iii*) incorporate the enumerations that specify the valid values for an attribute; and (*iv*) remodel the *Associative_Entity* metaclass as an inheritance of an *Entity* and a composition of *Relationship*, because Java does not support multiple inheritance.

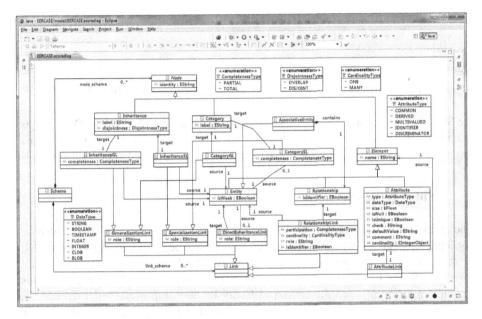

Fig. 4. Implementation view of EERMM

In order to show the expressiveness of EERMM, in Figure 5 we show the Graphical User Interface (GUI) of EERCASE with a hypothetical schema for a University that covers all constructors of the EER model. Note that the EERCASE GUI has a palette (area "B" in Figure 5) with all constructors of the EER Model, a drawing region (area "A" in Figure 5) and a property palette (area "C" in Figure 5). Once finished the EER modeling, the designer may validate its schema using our CASE tool (Menu Edit → Validate). For example, it can check whether: (*i*) there are two entities or two attributes (in the same entity) with the same name; (*ii*) there are isolated nodes; and (*iii*) a relationship has a identifier attribute. The validations are implemented using the Epsilon Validation Language (EVL)[9]. Once validated, the designer may automatically generate

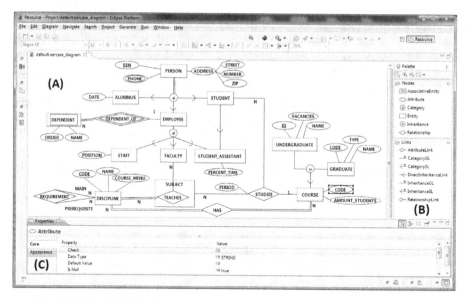

Fig. 5. EERCASE GUI with an EER schema using all constructors of the EER model

the SQL/DLL code for the modeled schema. The generation code for SQL/DDL scripts is implemented using the Epsilon Generation Language (EGL) [9]. Due to scope and space limitations, a detailed specification of EERCASE is not included in this paper.

5   Final Remarks

The EER model is one of the mostly used modeling languages for database conceptual design. However, although much work has been done on the EER model, to the best of our knowledge, a metamodel for the EER model, according to all constructors of Chen's notation, had not been proposed yet (see Section 2). For this reason, we have proposed the Enhanced Entity Relationship MetaModel (EERMM), which describes the constructors and the restrictions needed to design EER schemas that are syntactically consistent. Our metamodel cover the classical notation of Chen, but it can be extended, reused or transformed (i.e., model-to-model (M2M) transformation [24]) to support others EER notation (e.g., Higher-order Entity-Relationship Model - HERM [25]). That is, EERMM provides a novel perspective for scientific researches and industrial applications that need an EER metamodel as a starting point.

Our metamodel is conceptually defined by means of the EER notation (see Figure 3) and is implemented by means of the Ecore/EMOF specifications (see Figure 4). Furthermore, our metamodel (*i*) is more expressive than others (see related works in Section 2), because it explicitly support all constructors of the EER model (see Figure 1) and (*ii*) automatically assures well-formed rules for creating syntactically correct models, for instance: (*i*) a *Node* cannot be connected to another *Node* without a *Link*, and vice-versa; (*ii*) a *Link* cannot be depicted isolated in the drawing area; (*iii*) a

Relationship_Link cannot connect an *Entity* to another nor a *Relationship* to another; (*iv*) an *Inheritance* or a *Category* cannot have an *Attribute* and an *Attribute* cannot be linked to more than one *Entity, Associative_Entity, Relationship* or *Attribute;* (*v*) an *Inheritance* cannot have more than one *Generalization_Link;* (*vi*) a *Category* cannot have overlapping nor more than one *Specialization_Link;* and (*vii*) an *Entity* cannot be source for more than one *Direct_Inheritance_Link*.

In order to evaluate our proposal, we have implemented a CASE tool (EERCASE), which was used to build a full application of EERMM, demonstrating the feasibility, the expressiveness and usefulness of our metamodel. In its current version, EERCASE generates SQL/DDL code only for PostgreSQL, but in future work, other DBMS will also be covered. Others directions for future work are extending EERMM to support spatial and temporal data modeling and developing modules for EERCASE that provide reverse engineering and structural validity [18] of modeled schemas.

Acknowledgments. (*i*) This work was supported by the process BEX 5026/11 CAPES/DGU from the CAPES Brazilian foundation and (*ii*) the authors wish to thank the anonymous reviewers for their helpful comments, especially for the reviewer who sent your comments in an attachment file.

References

1. Chen, P.P.: The Entity-Relationship Model - Toward a Unified View of Data. ACM Trans. Database Syst., 9–36 (1976)
2. ACM & IEEE, Computer Science Curriculum 2008: An Interim Revision of CS 2001. Computing Curriculum Series, http://www.acm.org/education/curricula/ ComputerScience2008.pdf (accessed March 2012)
3. Elmasri, R., Navathe, S.: Fundamentals of Database Systems, 6th edn., p. 1200. Addison Wesley (2010)
4. Silberschatz, A., Korth, H., Sudarshan, S.: Database System Concepts, 6th edn., p. 1376. McGraw-Hill Science/Engineering/Math (2010)
5. Connolly, T.M., Begg, C.E.: Database Systems: A Practical Approach to Design, Implementation and Management, 5th edn., p. 1400. Addison Wesley (2009)
6. Garcia-Molina, H., Ullman, J.D., Widom, J.: Database Systems: The Complete Book, 2nd edn., p. 1248. Prentice Hall (2008)
7. Steinberg, D., Budinsky, F., Paternostro, M., Merks, E.: EMF: Eclipse Modeling Framework, 2nd edn., p. 744. Addison-Wesley Professional (2008)
8. ECLIPSE, Graphical Modeling Framework, http://www.eclipse.org/gmf/ (accessed March 2012)
9. Kolovos, D., Rose, L., Paige, R.: The Epsilon Book, http://www.eclipse.org/epsilon/doc/book/ (accessed March 2012)
10. OMG, EMOF - Essential Meta-Object Facility, http://www.omg.org/spec/MOF/2.4.1/PDF (accessed March 2012)
11. Song, I., Evans, M., Park, E.K.: A comparative analysis of entity-relationship diagrams. J. Comput. Softw. Eng. 3(4), 427–459 (1995)
12. Jones, T.H., Song, I.: Analysis of Binary/Ternary Cardinality Combinations in Entity-Relationship Modeling. Data Knowl. Eng., 39–64 (1996)

13. Song, I., Chen, P.P.: Entity Relationship Model. Encyclopedia of Database Systems, 1003–1009 (2009)

14. Rumbaugh, J., Jacobson, I., Booch, G.: The Unified Modeling Language Reference Manual, 2nd edn. The Addison-Wesley Object Technology Series. Addison-Wesley Professional

15. Lucia, A.D., Gravino, C., Oliveto, R., Tortora, G.: Data Model Comprehension: An Empirical Comparison of EER and UML Class Diagrams. In: ICPC, pp. 93–102 (2008)

16. Lucia, A.D., Gravino, C., Oliveto, R., Tortora, G.: An experimental comparison of EER and UML class diagrams for data modelling. Empirical Software Engineering, 455–492 (2010)

17. Bavota, G., Gravino, C., Oliveto, R., De Lucia, A., Tortora, G., Genero, M., Cruz-Lemus, J.A.: Identifying the Weaknesses of UML Class Diagrams during Data Model Comprehension. In: Whittle, J., Clark, T., Kühne, T. (eds.) MODELS 2011. LNCS, vol. 6981, pp. 168–182. Springer, Heidelberg (2011)

18. Dullea, J., Song, I., Lamprou, I.: An analysis of structural validity in entity-relationship modeling. Data Knowl. Eng., 167–205 (2003)

19. OMG, The IMM Story, http://portals.omg.org/imm/doku.php?id=motivation_behind_imm_and_what_is_imm (accessed March 2012)

20. OMG, Common Warehouse Metamodel (CWM) Specification Volume 2. Extensions, p. 218 (2001)

21. OMG, Information Management Metamodel (IMM) Specification Volume II - Business Modeling, p. 54 (2009)

22. Irani, P., Tingley, M., Ware, C.: Using Perceptual Syntax to Enhance Semantic Content in Diagrams. IEEE Computer Graphics and Applications, 76–85 (2001)

23. Ware, C., Bobrow, R.J.: Motion to support rapid interactive queries on node–link diagrams. In: TAP, pp. 3–18 (2004)

24. Stahl, T., Völter, M., Bettin, J., Haase, A., Helsen, S.: Model-Driven Software Development: Technology, Engineering, Management, p. 444. John Wiley (2006)

25. Thalheim, B.: The Enhanced Entity-Relationship Model. In: Handbook of Conceptual Modeling, pp. 165–206. Springer, Heidelberg (2011)

Introducing Usability in a Conceptual Modeling-Based Software Development Process[*]

Jose Ignacio Panach Navarrete[1], Natalia Juristo Juzgado[2], and Óscar Pastor[3]

[1] Universitat de València
Escola Tècnica Superior d'Enginyeria, Departament d'Informàtica
Vicent Andrés Estellés, s/n 46100 Burjassot, València, Spain
joigpana@uv.es
[2] Universidad Politécnica de Madrid
Campus de Montegancedo, 28660 Boadilla del Monte, Madrid, Spain
natalia@fi.upm.es
[3] Universitat Politècnica de València
Centro de Investigación en Métodos de Producción de Software,
Camino de Vera s/n, 46022 Valencia, Spain
opastor@pros.upv.es

Abstract. Usability plays an important role to satisfy users' needs. There are many recommendations in the HCI literature on how to improve software usability. Our research focuses on such recommendations that affect the system architecture rather than just the interface. However, improving software usability in aspects that affect architecture increases the analyst's workload and development complexity. This paper proposes a solution based on model-driven development. We propose representing functional usability mechanisms abstractly by means of conceptual primitives. The analyst will use these primitives to incorporate functional usability features at the early stages of the development process. Following the model-driven development paradigm, these features are then automatically transformed into subsequent steps of development, a practice that is hidden from the analyst.

Keywords: Model-Driven-Development, Usability, Conceptual Model.

1 Introduction

Historically, many SE authors have considered usability as a non-functional requirement. Recently, however, some authors have identified several usability features strongly related to functionality [4]. These features do not only affect interfaces but also architecture, and are hard to deal with if they are not considered from the early stages of development. Incorporating usability from requirements elicitation is not for

---

[*] This work has been developed with the support of, MICINN (TIN2011-23216, PROS-Req TIN2010-19130-C02-02), GVA (ORCA PROMETEO/2009/015), and co-financed with ERDF. We also acknowledge the support of the ITEA2 Call 3 UsiXML (20080026) and financed by the MITYC under the project TSI-020400-2011-20 .

P. Atzeni, D. Cheung, and R. Sudha (Eds.): ER 2012, LNCS 7532, pp. 525–530, 2012.

free. Generating usable software has a number of unwanted collateral impacts: increased complexity [4]; increased cost; increased maintenance difficulty [5].

To mellow these effects we propose including functional usability features in a model-driven development (MDD) software process. If usability is considered from the early stages of development, it can be included in an MDD method and benefit from the advantages of the MDD paradigm [11]. MDD claims that developers should focus their efforts on building a conceptual model, then the system is implemented by means of transformation rules that can be automated [8]. If we study other works in the literature, we find very few proposals to deal with usability features in an MDD method. Moreover, when it is discussed, very few precise details are given, which makes it difficult to understand how these approaches could work correctly in practical settings. Examples of these works have been developed by Tao [12] and Raneburger [10]. From our point of view, usability is as important as functional requirements, and therefore MDD methods should provide a mechanism to abstractly represent usability. In the following, we explain how we propose including usability features with functional involvement in an MDD method.

The paper is structured as follows. Second section describes our proposal for adding usability features to an MDD method. Third section discusses an experiment to evaluate user satisfaction improvement applying our proposal. Finally, Fourth section presents some conclusions.

2   Incorporating Usability Functionalities in a Model-Driven Development Method

Human-Computer Interaction literature provides many different recommendations to improve the usability of a software system. In [4], authors present three groups of recommendations: (1) Usability recommendations with impact on the user interface; (2) Usability recommendations with impact on the development process; (3) Usability recommendations with impact on the architectural design. These last recommendations involve building certain functionalities into the software to improve user-system interaction. This set of usability recommendations is referred to as functional usability features. Examples of these features are providing undo and feedback facilities. A big amount of rework is needed to include these features in a software system unless they are considered from the first stages of the software development process [1]. Moreover, their inclusion from the first steps of the development process increases the complexity of the software development. To minimize the problems of including usability features with impact on the architecture, we aim to incorporate them in an MDD method. This way, we benefit from the advantages of the MDD paradigm [11]. Our approach is divided into four steps:

1. Identify the possible use ways of each usability functionality.
2. Identify the properties that configure each use way.
3. Define conceptual primitives to abstractly represent the use way properties.
4. Describe the changes that must be made to the model compiler to generate code.

First and second steps are based on interaction patterns and usability guidelines that define how to deal with functional usability features. From all the existing works in the literature, we have chosen a list of usability recommendations called *Functional Usability Features (FUFs)* [4] as illustrative example to apply our proposal. Each FUF was defined with a main objective that can be specialized into more detailed goals we named mechanisms. The list of FUFs and their mechanisms are shown in [4, 7]. We focus our example on the usability mechanism called *Structured Text Entry*, which belongs to *User Input Error Prevention* FUF. We select this mechanism because its goal --*help the user when the system only accepts inputs in a specific format*- - is simple enough to allow its presentation in a couple of pages. As MDD method to include Structured Text Entry, we have selected OO-Method [9]. The OO-Method has been successfully implemented in industry (INTEGRANOVA [1]). The analyst does not have to implement any code because all the code is automatically generated from a conceptual model by means of a model compiler. Next, we explain the steps of our proposal to include Structured Text Entry in OO-Method.

2.1   Identification of Use Ways

Each functional usability feature can achieve its goal in different means. We have called each such mean Use Way (UW).   Each UW has a specific target to achieve as part of the overall goal of the usability feature. We propose deriving UWs from existing works in the literature (interaction patterns and usability guidelines), such as FUFs. In FUFs list, each mechanism is defined with a set of questions to identify usability requirements. Regarding our proposal, these questions can be used to identify UWs. If we focus on the Structured Text Entry mechanism, we identify three UWs:

- **Specify the input widget visualization type (UW1):** This UW aims to specify the format of the input widget to help the user and it is derived from the usability mechanism question, *Which is the format of input arguments?*
- **Mask definition (UW2):** This UW aims to stop the user from entering data that is not in a valid format and it is derived from the usability mechanism question, *What guidance should the user receive to enter the input in the required format?*
- **Default values (UW3):** This UW aims to provide the user with guidance on which format to use to enter data and it is derived from the same question as UW2.

2.2   Identification of Properties

We have called the different UW configuration options to satisfy usability functionalities as **Properties**. In this second step, we identify properties also from the questions used in the usability mechanism definition. Focusing on the UWs derived from *Structured Text Entry*, we have identified the following properties:

- UW1 has only one property: **Type of input widget (UW1_P1)**. This property aims to define how the user will visualize input arguments and it has been derived from the mechanism question, *which is the format of input arguments?*

- UW2 has two properties: **Widget selection (UW2_P1):** This property aims to specify the widgets that need a mask and it has been derived from the mechanism question, *which widgets require a specific format for their data?.* **Regular expression (UW2_P2):** This property aims to define the regular expression that specifies the mask and it has been derived from the mechanism question, *which is the required format for the widget?*
- UW3 has two properties: **Widget selection (UW3_P1):** This property aims to specify the widgets that need a default value and it has been derived from the mechanism question, *which widgets require a default value?* **Definition of the default value (UW3_P2):** This property aims to define the default value and it has been derived from the mechanism question, *which is the required default value?*

2.3   Definition of Conceptual Primitives

The conceptual model of the MDD method needs to be enriched to support the UWs. This step involves verifying whether or not there are already conceptual primitives in the MDD method that represent a property. If there is no conceptual primitive already to represent a property, the conceptual model needs to be expanded to ensure the required expressiveness. In order to include *UW1_P1* property in OO-Method, we need to enrich the OO-Method conceptual model with new primitives that represent the different widget types. For example, a numbered list can be represented with a ComboBox or with a RadioButton. *UW2* and *UW3* are already supported by the OO-Method conceptual model and they do not involve new conceptual primitives.

2.4   Changes in the Model Compiler

The changes in the model compiler aim to implement the code derived from new conceptual primitives. These changes involve adding new attributes, services and classes in the generated code. In our example, the only change to be made to the model compiler is to include *Specify the input widgets visualization type (UW1).* This change has the aim of generating the code that implements the type of widget specified by means of conceptual primitives.

3   A Lab Evaluation

We have carried out an empirical evaluation with 66 subjects using a Web application for car rental. The users of this system are the employees of offices all over the world. This Web application has been fully developed using INTEGRANOVA [1]. The UWs not supported by INTEGRANOVA were manually included in the generated code. We included a total of 7 UWs. The aim of this evaluation is to study whether or not the end-user perceives the benefits of including UWs in the system. If the answer is positive, the effort to include UWs in an MDD method is justified, since UWs will improve the end-users' approval of the software. We divided the experimental subjects into two groups: subjects that interact with the system without UWs and subjects that interact with the system including several UWs.

We identified the following null hypotheses:

- $H1_0$: The satisfaction for the users that interact with UWs is the same as the satisfaction for users that interact without UWs.
- $H2_0$: The time for the users that interact with UWs is the same as the time for users that interact without UWs.

There are two response variables [3] in the experiment. One variable is called *user satisfaction level*. This variable measures whether or not the user is satisfied with the interaction and it is measured by means of a five-point Likert-scale questionnaire. The other variable is *time to finish* the tasks. This variable measures how long it takes the user to perform the experimental tasks. This is measured timing the seconds needed to finish the tasks. There are two factors [3] in the experiment. One factor is *use of UWs*. This factor involves studying the Web application with UWs and without them. The other factor is *previous experience of applications generated with INTEGRANOVA*. We combined both factors across the subjects to see how they affect the response variables.

We had 22 subjects with experience in INTEGRANOVA; 11 interacted with UWs and other 11 without UWs. We had 44 subjects without experience in INTEGRANOVA; 22 interacted with UWs and other 22 interacted without UWs. We analyzed the data using two methods: ANOVA and box and whisker plots. First, the ANOVA results show that the satisfaction of subjects strongly depends on using UWs (p-value=0,001 for most studied UWs). Moreover, the ANOVA analysis shows that there is no relationship between: (1) *User satisfaction level* and *previous experience of applications generated with INTEGRANOVA* (p-value=0.558 for most studied UWs); (2) *Time to finish the tasks* and *use of UWs* (p-value=0.628); (3) *Time to finish the tasks* and *previous experience of applications generated with INTEGRANOVA* (p-value=0.057). Second, the box and whiskers plots illustrate the median and quartile for both response variables (*user satisfaction level* and *time to finish tasks*). Figure 1a shows the plot that compares *user satisfaction level* when using and not using the Use Way called *Warning message (UW_W1)*. Figure 1b shows the box and whisker plot for *time to finish the tasks* with reference to *the use of UWs factor*.

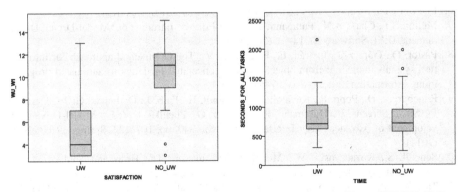

Fig. 1. a) Box and whisker plot for *user satisfaction level* with and without UW_W1 **b)** Box and whisker plot for *time to finish the tasks* with and without UWs

According to our analysis, we state that UWs generally improve user satisfaction. Moreover, satisfaction does not depend on experience in the use of INTEGRANOVA applications. So, we reject the hypothesis $H1_0$. With regard to the time hypothesis ($H2_0$), the analysis shows that time is independent of interacting with or without UWs. Moreover, there is no difference between the time of the experts in INTEGRANOVA applications and beginners.

4   Conclusions

In this paper, our aim is to incorporate functional usability features into an MDD process, benefiting from the MDD advantages and minimizing manual implementations. The method we propose can be easily applied to other MDD methods than OO-Method. Use ways and properties can be directly applied to other methods; while the conceptual primitives and the changes in the model compiler depend on a specific MDD method, since conceptual model and model compiler are exclusive of a MDD method. The difficulty of applying our proposal to other MDD methods depends on the expressiveness of their conceptual models. OO-Method has an Interaction Model, which facilitates the inclusion of new conceptual primitives to represent interaction features. However, MDD methods with less expressiveness to deal with interaction would require adding more conceptual primitives to represent UWs.

References

1. Bass, L., John, B.: Linking usability to software architecture patterns through general scenarios. The Journal of Systems and Software 66, 187–197 (2003)
2. CARE Technologies S.A, http://www.care-t.com
3. Juristo, N., Moreno, A.: Basics of Software Engineering Experimentation. Springer (2001)
4. Juristo, N., Moreno, A., Sánchez, M.I.: Analysing the impact of usability on software design. Journal of Systems and Software 80, 1506–1516 (2007)
5. Lawrence, B., Wiegers, K., Ebert, C.: The top risk of requirements engineering. IEEE Software 18, 62–63 (2001)
6. List of FUFs, http://hci.dsic.upv.es/FUF/FUFList.html
7. Mellor, S.J., Clark, A.N., Futagami, T.: Guest Editors' Introduction: Model-Driven Development. IEEE Software 20, 14–18 (2003)
8. Pastor, O., Gómez, J., Insfrán, E., Pelechano, V.: The OO-method approach for information systems modeling: from object-oriented conceptual modeling to automated programming. Information Systems 26, 507–534 (2001)
9. Raneburger, D., Popp, R., Kavaldjian, S., Kaindl, H., Falb, J.: Optimized GUI Generation for Small Screens. In: Hussmann, H., Meixner, G., Zuehlke, D. (eds.) Model-Driven Development of Advanced User Interfaces. SCI, vol. 340, pp. 107–122. Springer, Heidelberg (2011)
10. Sendall, S., Kozaczynski, W.: Model Transformation: The Heart and Soul of Model-Driven Software Development. IEEE Software 20, 42–45 (2003)
11. Tao, Y.: An Adaptive Approach to Obtaining Usability Information for Early Usability Evaluation. In: Proc. of IMECS (2007)

Conceptualizing Data in Multinational Enterprises: Model Design and Application

Verena Ebner [*], Boris Otto, and Hubert Österle

Institute of Information Management, University of St. Gallen, St. Gallen, Switzerland
{verena.ebner,boris.otto,hubert.oesterle}@unisg.ch

Abstract. Collaboration and coordination within multinational enterprises need unambiguous semantics of data across business units, legal contexts, cultures etc. Therefore data management has to provide enterprise-wide data ownership, an unambiguous distinction between "global" and "local" data, business-driven data quality specifications, and data consistency across multiple applications. Data architecture design aims at addressing these challenges. Particularly multinational enterprises, however, encounter difficulties in identifying, describing and designing the complex set of data architectural dimensions. The paper responds to the research question of what concepts need to be involved to support comprehensive data architecture design in multinational enterprises. It develops a conceptual model, which covers all requirements for defining, governing, using, and storing data. The conceptual model is applied in a case study conducted at a multinational corporation. Well-grounded in the existing body of knowledge, the paper contributes by identifying, describing, and aggregating a set of concepts enabling multinational enterprises to meet business requirements.

Keywords: Enterprise data architecture design, enterprise data architecture management, data quality management, data modeling, data classification.

1   Introduction

Consumer-centric business models pose the demand for a 360° perspective of the customer, increasing value chain integration creates the need for business collaboration and information sharing [1, 2], and a fast-growing number of legal regulations and contractual obligations requests consolidated and integrated data across the enterprise. Especially in multinational enterprises, these needs for enterprise-wide collaboration, coordination and interoperability are faced by ambiguous definitions of enterprise data[1] across multiple business units, legal contexts, as well as geographical regions, numerous stakeholders and missing responsibilities for enterprise data, multiple distributed, heterogeneous, internal and external applications storing and managing enterprise data in a redundant and inconsistent manner, and a variety of business processes using and managing enterprise data with different goals.

---

[1] Enterprise data refers to what in literature is commonly named master data. To distinguish from the definitions of master data in the practitioners' community, the term enterprise data is used here.

P. Atzeni, D. Cheung, and R. Sudha (Eds.): ER 2012, LNCS 7532, pp. 531–536, 2012.
© Springer-Verlag Berlin Heidelberg 2012

Enterprise data in multinational enterprises can be described with regard to the characteristics time reference, change frequency, volume volatility and existential independence [3]. Enterprise data stores and describes characteristics of a company's core business entities, e.g. customers, suppliers, or products [1]. Shared access, replication, and flow of enterprise data in order to ensure data quality is controlled in the enterprise data architecture [4]. Enterprise data architecture aims to support collaborative use and management of enterprise data by providing an enterprise data model, enterprise data applications and a description of data flow between applications [3, 5]. Therefore, it is necessary to assess, describe and document business requirements to be met by enterprise data on an attribute level.

The paper takes up on the research question as to what are the enterprise data architecture design decisions and which criteria, also called classifiers, best support the assessment of alternative design options. More precisely, the paper aims to answer the question as to what are the components a conceptual model for enterprise data needs to involve, so that all criteria to define, govern, store and use data in the environment of a multinational enterprise are taken into consideration. A conceptual model is designed, that aims to identify, describe and aggregate a complex set of classifiers enabling multinational enterprises to design an enterprise data architecture that meets business requirements of pressing relevance. The conceptual model was designed by the analysis of existing conceptual data models, and a literature review identifying enterprise data conceptualizations. The results were reflected in two multiple expert interviews. Subsequently, the applicability of the model was demonstrated in a real-world context at a multinational electronic and electrical engineering corporation [6].

2   Model Design

The conceptual model for enterprise data shown in Fig. 1 contains both structural and behavioral components [7]. The structural components reflect the *enterprise data elements,* e.g. classes, attributes, or their relationships. The behavioral component describes the enterprise data elements. The behavior may differ for each enterprise data element. An element can be described from four different perspectives, so called views, namely *Administration*, *Governance*, *Storage* and *Usage* [8].

Administration is concerned with the definition, description and instantiation of an enterprise data element [9]. Governance focuses on data ownership and data management processes. Storage handles the distribution of data between applications. Usage refers to the use of data in business processes [10]. For each view, a set of characteristics is specified called *enterprise data classifiers* which can be operationalized by value sets. Tab. 1 shows a reference list of enterprise data classifiers. The list is exemplary and by no means complete.

The Administration view supports an unambiguous definition of enterprise data within the enterprise with regard to its meaning and semantics [1]. *Validity* refers to the reach of the definition to be compulsory within the organization. It can be valid for the whole enterprise (global), for some parts of the enterprises (inter-divisional), or in one single enterprise division only (local). *Data definition autonomy* describes how independent data can be defined among various business units. For some enterprise data elements, e.g. unique identifiers, a distinct instantiation of data values is necessary [2], i.e. *Uniqueness*. The concept of *Structural constraints* describes structural guidelines. There can be no structure given at all, some elements of the structure, or the complete structure can be given. *Data volume* denotes the number of entities handled for the data element. Further classifiers are *Change frequency*, i.e. the occurrence of modifications of data values, *Versioning*, i.e. the ability to track changes, and *Historiography*, i.e. archiving of values. The two latter classifiers are often derived from regulatory specifications.

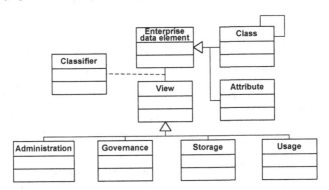

Fig. 1. Conceptual model for enterprise data

The Governance view relates to organizational aspects of managing data instances. Governance classifiers refer to organizational roles, i.e. *Data ownership*, and *Data stewardship* [11]. While the former role is responsible for the definition, the latter is responsible for the creation and instantiation of an enterprise data element. A distinction is made between the *scope* and the *distribution* of these roles. While scope defines the area of responsibility and accountability, distribution defines the location, which can be central, in one division, or distributed over multiple divisions [12].

Classifiers concerning data stored in applications and data exchange among applications are enclosed in the Storage view. Data storage applications distinguish the dimensions *distribution, autonomy* and *heterogeneity* [13]. *Application distribution* refers to physical data storage, being centrally in one application or distributed over many applications. *Application autonomy* denotes the ability of applications to independently process data. With regard to heterogeneity it can be distinguished between *Application heterogeneity,* and *Data model heterogeneity* [13]. *Application integration* concerns the physical merging of data into one application, or the logical mapping of data from distributed applications. *Redundancy* relates to data being stored in multiple applications. Thereby, identical copies of the same instance (duplicates), different attributes (consolidation), or different instances of the same class (aggregation) can be distributed among multiple applications. With regard to the exchange of data between applications and provisioning to data consumers the following classifiers were identified: *Application accessibility, Data provision, Processing, Distribution type, Distribution direction,* and *Distribution initiation* [14].

In the Usage view focuses on data in business processes. As enterprise data quality is often described as "fitness for use" [15], enterprise architecture design highly depends on the usage of data in business processes. It is distinguished between core and management processes, and support processes [16]. Data *Usage types* can be analytical or transactional scenarios [2]. *Process diversity* provides an indicator for data being used differently in various processes. *Process dependency* refers to multiple processes that can be interrelated. They may depend on one another along the value chain (horizontal) or between divisions (vertical). The *Transaction rate* provides information about the frequency in which data is exchanged between processes.

Table 1. Enterprise data classifiers

Administration	Governance	Storage	Usage
-Data definition validity	-Data ownership scope	-Data distribution	-Usage type
-Data definition autonomy	-Data ownership distribution	-Application autonomy	-Process heterogeneity
-Uniqueness	-Data stewardship scope	-Application heterogeneity	-Process dependency
-Structural constraints	-Data stewardship distribution	-Data model heterogeneity	-Process types
-Data volume		-Application integration	-Transaction rate
-Change frequency		-Redundancy	
-Versioning		-Application accessibility	
-Historiography		-Data provision	
		-Processing	
		-Distribution type	
		-Distribution direction	
		-Distribution initiation	

3   Model Application

The conceptual model was applied at a multinational electronic and electrical engineering enterprise (hereinafter called EEE). When setting up an enterprise wide data management organization, a framework for managing enterprise data was created and an enterprise wide set of goals was specified. Country specific regulations and enterprise wide consolidation of enterprise supplier data to gain strategic benefits for negotiations with vendors and for external e-business processes resulted in the need for world-wide transparency, consolidation, clear and unambiguous responsibilities on an enterprise wide level and standardized processes to maintain and ensure the quality of enterprise supplier data. To identify similarities and differences in handling supplier data, a conceptualization seems a promising approach. Fig. 2 shows the structural components of and relationships between enterprise supplier data. For reason of clarity, attributes were omitted in the figure. As the classifiers are enterprise specific, nine classifiers were selected. For each classifier a company specific value set was defined represented in a morphological field [17]. Tab. 2 shows the classifiers, the value sets, and instantiates them for the enterprise data element Identifier: VAT registration number. Selected values for the elements are colored in grey.

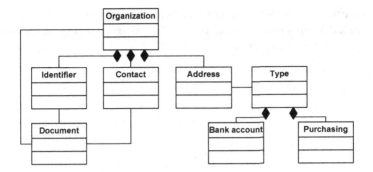

Fig. 2. Conceptual model for enterprise supplier data at EEE

Table 2. Instantiation of classifiers for *Identifier: VAT registration number*

View	Classifier	Value sets		
Administration	Data definition	inter-divisional	intra-divisional	local
	Structural constraints	none	partial	complete
	Data volume	high	medium	low
	Change frequency	high	low	none
Governance	Data ownership scope	inter-divisional	intra-divisional	local
Storage	Distribution type	broadcast	individual	none
	Data provision	central	hybrid	distributed
Usage	Process diversity	high	medium	low
	Business processes	support processes	core & management processes	

4 Summary and Outlook

The conceptual model for enterprise data describes structural and behavioral components of enterprise data elements and their relationships. Enterprise data classifiers represent business requirements to be met by enterprise data and need to be taken into account when designing enterprise data architecture. Interdependencies between enterprise data elements, between the views, or between individual enterprise data classifiers were not taken into account. Relations between enterprise data elements may involve concepts from entity relationship modeling and object oriented modeling like associations or inheritance. Concepts like multiplicity, direction, aggregation, and composition can be of relevance. Future research should analyze these interdependencies.

The conceptual model for enterprise data is well-grounded in theory and practice and is applied to a single case study. In order to assess proper model design, an evaluation of the conceptual model against predefined research goals should be performed [18, 19]. The research contributes to the scientific body of knowledge by identifying, structuring and aggregating concepts of enterprise data architecture management. Future research on requirements analysis and strategic design of enterprise data architecture can build on the conceptual model. In the practitioners' community the conceptual model supports enterprise data architects in gathering and analyzing requirements to be met by enterprise data.

References

1. Loshin, D.: Master Data Management. Elsevier Science & Technology Books, Burlington (2009)
2. Dreibelbis, A., Hechler, E., Milman, I., Oberhofer, M., van Run, P., Wolfson, D.: Enterprise Master Data Management: An SOA Approach to Managing Core Information. Pearson Education, Boston (2008)
3. Otto, B., Schmidt, A.: Enterprise Master Data Architecture: Design Decisions and Options. In: 15th International Conference on Information Quality (ICIQ 2010), Little Rock (2010)
4. Dama: The DAMA Guide to the Data Management of Knowledge. In: Dama (ed.), 1st edn. Technics Publications, Bradley Beach (2009)
5. Periasamy, K.P.: The State and Status of Information Architecture: An Empirical Investigation, Orlando, FL (1993)
6. Yin, R.K.: Case Study Research. Design and Methods, 3rd edn. Applied Social Research Methods Series, vol. 5. Sage Publications, London (2002)
7. Schütte, R.: Grundsätze ordnungsmässiger Referenzmodellierung: Konstruktion konfigurations- und anpassungsorientierter Modelle. Gabler, Wiesbaden (1998)
8. Sowa, J.F., Zachman, J.A.: Extending and formalizing the framework for information systems architecture. IBM Systems Journal 31(3), 590–616 (1992)
9. Shankaranarayanan, G., Even, A.: Managing Metadata in Data Warehouses: Pitfalls and Possibilities. Communications of AIS 14, 247–274 (2004)
10. Levitin, A., Redman, T.: Quality Dimensions of a Conceptual View. Information Processing & Management 31(1), 81–88 (1995)
11. Bitterer, A., Newman, D.: Organizing for Data Quality. Gartner Research, Stamford (2007)
12. Khatri, V., Brown, C.V.: Designing Data Governance. Communications of the ACM 53(01), 148–152 (2010)
13. Leser, U., Naumann, F.: Informationsintegration - Architekturen und Methoden zur Integration verteilter und heterogener Datenquellen. Dpunkt. verlag, Heidelberg (2007)
14. Jung, R.: Architekturen zur Datenintegration. Gestaltungsempfehlungen auf der Basis fachkonzeptueller Anforderungen. Deutscher Universitäts-Verlag, Wiesbaden (2006)
15. English, L.P.: Improving Data Warehouse and Business Information Quality, 1st edn. John Wiley & Sons, Inc., New York (1999)
16. Porter, M.E.: Competitive Advantage: Creating and Sustaining Superior Performance. Free Press, New York (1998)
17. Ritchey, T.: Problem structuring using computer-aided morphological analysis. Journal of the Operational Research Society, 792–801 (2006)
18. Becker, J., Rosemann, M., Schütte, R.: Grundsätze ordnungsmäßiger Modellierung. Wirtschaftsinformatik 37(5), 435–445 (1995)
19. Gregor, S.: The Nature of Theory in Information Systems. MIS Quarterly 30(3), 611–642 (2006)

A Metadata-based Approach to Leveraging the Information Supply of Business Intelligence Systems

Benjamin Mosig and Maximilian Röglinger

FIM Research Center, University of Augsburg,
Universitätsstraße 12, 86159 Augsburg, Germany
{benjamin.mosig,maximilian.roeglinger}@wiwi.uni-augsburg.de

Abstract. Ensuring adequate information provision continues to be a key challenge of corporate decision making and the usage of business intelligence systems. As a matter of fact, the situation becomes increasingly paradox: Whereas decision makers struggle to specify their information requirements and spend much time on obtaining the information they believe to require, the amount of information supplied by business intelligence systems grows at a speed that makes it hard to keep track. Thus, it is very likely that the required information or suitable alternatives are available, but neither found nor used. Instead, manual searching causes considerable opportunity cost. Existing approaches to information requirements analysis pay attention to incorporate information supply, but do not provide means for leveraging it in a systematic and IT supported manner. As a first step to close this research gap, we propose a metadata-based approach consisting of a procedure model and formalism that help identify a suitable subset of the information supplied by an existing business intelligence system. The formalism is specified using set theory and first-order logic to provide a general foundation that may be integrated into different conceptual modelling approaches.

Keywords: Business intelligence, Data warehouse, Metadata, Information provision, Information supply.

1 Motivation

Ensuring adequate information provision has been a recurring topic over the last decades. Numerous approaches to information requirements analysis and various types of management support systems – most recently business intelligence (BI) systems – have been proposed to increase the clarity in corporate decision making [5]. Despite these efforts, decision makers still face information overload with negative consequences such as mental stress, loss of clarity, and reduced decision quality [1]. A particular problem in recent times is that the convenient access to BI systems and the high storage capacity of underlying data warehouses entice companies into accumulating large amounts of information [4]. As BI systems are historically grown and have been subject to uncontrolled growth in many organizations, it is hard to keep track with the information they supply. Academics approvingly report that "not

P. Atzeni, D. Cheung, and R. Sudha (Eds.): ER 2012, LNCS 7532, pp. 537–542, 2012.

missing information [is] the primary problem" and that "all information is available somewhere" [6] in most companies. Against this backdrop, it is very likely that the required information or suitable alternatives are available within an organization, but neither found nor used. The potential of existing information supply to satisfy information requirements is not sufficiently tapped [6].

As mentioned, literature contains numerous approaches dedicated to the elicitation and specification of information requirements particularly for the development of data warehouses and BI systems [3, 5]. Apart from few exceptions, the proposed approaches pay attention to incorporating existing information supply. The approaches share several characteristics: First, most activities related to leveraging existing information supply require manual effort and are hardly IT supported. Second, the approaches center on informal or semi-formal concepts, which makes it difficult to cover large amounts of existing information supply systematically. Third, some approaches deal with the initial development of a data warehouse or BI system and thus focus on the information supply of operational information systems. Fourth, the approaches provide no explicit means for coping with decision makers' struggles when specifying information requirements. Despite the value of the presented approaches, there is a need for additional support to leverage the information supply of existing BI systems. This leads to the following research question: *How can the information supply of existing BI systems be leveraged in a systematic and IT supported manner?*

We address the research question by proposing a metadata-based approach consisting of a procedure model and formalism that complement the approaches discussed above and help identify a suitable subset of the information supplied by an existing BI system. We rely on metadata because they play an important role in BI systems and have the potential to structure large amounts of data [2]. Following an axiomatic and deductive research approach, we first sketch the general setting and explicate our assumptions (section 2). We then derive the procedure model on this foundation (section 3). The paper concludes with a discussion (section 4). The formalism is specified using set theory and first-order logic. It has been omitted due to space restrictions, but can be requested from the authors.

2   General Setting

Our unit of analysis is a single historically grown BI system that is based on a data warehouse as informational infrastructure. The data warehouse is based on a multi-dimensional data schema whose core elements on schema level are measures and dimensions. We treat all dimensions as orthogonal and abstract from structural abnormalities such as parallel hierarchies [3]. While an examination on the schema level is reasonable in the context conceptual modeling, information requirements analysis extends to the instance level because information requirements typically relate to the actual values of measures and hierarchic levels. We assume:

(A.1) The multi-dimensional data schema consists of measures and dimensions. Each dimension includes hierarchic levels.

To incorporate metadata into the procedure model and the formalism, information requirements need to be split into two parts where the first part includes requirements that directly relate to the core elements of the multi-dimensional data schema and the second part comprises requirements that relate to meta-attributes (see Table 1).

Table 1. Considered components of the information requirements

	Related to the core elements of the multi-dimensional data schema	Related to additional meta-attributes
Schema level	• Requirements regarding measures • Requirements regarding dimensions • Requirements regarding hierarchic levels	
Instance level	• Requirements regarding the domain of selected measures • Requirements regarding the domain of selected hierarchic levels	• Requirements regarding the value of a meta-attribute <u>for each single</u> selected measure • Requirements regarding the value of a meta-attribute <u>for all</u> selected measures

The first part of the information requirements helps specify requirements where the decision makers know precisely which combinations of measures and dimensions they need. These requirements can be elicited using the existing approaches to information requirements analysis. As known from conceptual modeling, there is a dependency between requirements on schema level and on instance level. That is, requirements regarding the instance level of measures or hierarchic levels relate to the domains of the measures or hierarchic levels selected on schema level.

Requirements belonging to the second part of the information requirements relate to additional meta-attributes. What is special about using meta-attributes is that usually multiple subsets of the information supply exist that meet the related requirements. Requirements can be defined at two distinct reference levels (see Table 1). Either each single selected measure has to fulfill a requirement individually (reference level: each single measure) or all selected measures together have to fulfill a requirement (reference level: all measures). Moreover, the collection effort of all selected measures must not exceed a defined limit (meta-attribute: 'collection effort'). We assume:

(A.2) Each measure features the same meta-attributes. Meta-attributes are only assigned to measures.

(A.3) The information requirements $I^{req} = \{F^{model,schema}, F^{model,instance}, F^{meta}\}$ comprise requirements related to core elements of the multi-dimensional data schema on schema level, $F^{model,schema} = f_1^{model,schema} \wedge \dots \wedge f_s^{model,schema}$, requirements related to the core elements of the multi-dimensional data schema on instance level $F^{model,instance} = f_1^{model,instance} \wedge \dots \wedge f_t^{model,instance}$, and requirements related to meta-attributes $F^{meta} = f_1^{meta} \wedge \dots \wedge f_u^{meta}$. All requirements are specified in first-order logic. $F^{\text{model,schema}}$ only contains requirements that can be covered by the information supply.

(A.4) The subset of I^{supply} that meets all requirements related to the core elements of the multi-dimensional data schema on schema level ($F^{\text{model,schema}} = \text{T}$) is denoted by $I^{\text{selected,model,schema}}$. The subsets of I^{supply} that meet all requirements related to meta-attributes ($F^{\text{meta}} = \text{T}$) are denoted as set family $(I^{\text{selected,meta}})_v$ where V is an index set, $|V|$ is the number of different sets, and $v \in V$.

Due to the logical AND operator ($\wedge$) in (A.3), $F^{\text{model,schema}}$, $F^{\text{model,instance}}$, and F^{meta} only evaluate to *true* (T) if all respective requirements are met. Each of the v set unions $I^{\text{selected,model,schema}} \cup (I^{\text{selected,meta}})_v$ – subsequently referred to as I_v – is a feasible alternative containing the required information on schema level. $F^{\text{model,instance}}$ has not been considered so far as the enclosed requirements can partly be formulated after one of the I_v has been selected. In order to determine which of the I_v should be selected, we assume that the decision makers assess the utility and disutility of each alternative.

(A.5) Decision makers strive to maximize the net benefit they receive from the selected subset of the information supply. Each measure m_p features a subjective utility value $u(m_p)$ and disutility value $d(m_p)$. The utility and disutility values of a particular subset of the information supply I_v are calculated as follows:
$$U(I_v) = \sum_{m_p \in \{m_z | \exists (m_z, \ldots) \in I_v\}} u(m_p) \quad \text{and} \quad D(I_v) = \sum_{m_p \in \{m_z | \exists (m_z, \ldots) \in I_v\}} d(m_p).$$
The overall net benefit is calculated as $U^{\text{net}}(I_v) = U(I_v) - D(I_v)$.

Finally, we need to know how decision makers specify their information requirements. Based on our experience from related industry projects, we assume:

(A.6) Decision makers base their information requirements primarily on measures. Moreover, decision makers are able to specify information requirements related to the core elements of the multi-dimensional data schema and requirements related to meta-attributes independent of one another.

3   Procedure Model

Based on the elaborations concerning the general setting, we are able to derive properties of a procedure model for leveraging the information supply of existing BI systems. The overall procedure model is shown in Fig. 1. First, the procedure model can start with the simultaneous specification of requirements from $F^{\text{model,schema}}$ regarding measures as well as F^{meta}. This is because decision makers base their information requirements primarily on measures (see A.6) and meta-attributes are only assigned to measures (see A.2). This results in steps ❶ and ❷ of the procedure model. Second, the utility and disutility of the selected measures can be assessed directly afterwards as only measures are assessed (see A.5). This results in steps ❸ and ❹ of the procedure model. Due to the interdependency of requirements on schema level and on instance level, the requirements regarding dimensions and hierarchic levels from $F^{\text{model,schema}}$ have to be specified first. After that, $F^{\text{model,instance}}$ can be formulated when it comes to report parameterization. This results in steps ❺ and ❻ of the procedure model. The position of steps ❸ and ❹ is also reasonable because labour-intense effort is reduced as decision makers would otherwise have to assess the (dis-) utility of measures, dimensions, and dimensional hierarchy levels that are not implemented.

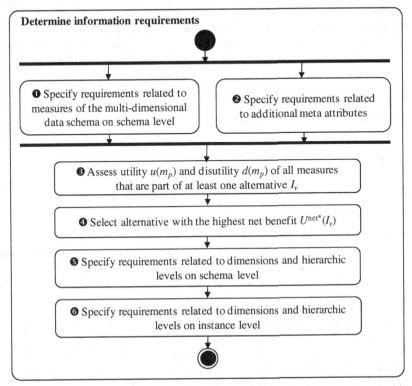

Fig. 1. Procedure model for leveraging the information supply of existing BI systems

4   Discussion

The proposed approach is beset with limitations that need to be taken into account when applying it in industry settings. Other limitations motivate future research endeavors:

1. Prior to application, appropriate meta-attributes have to be identified and – if not already available – filled with values. While this may be quite costly for a single use case, it is worth the effort in case of repeated applications for multiple groups of decision makers. As Stroh et al. [5] point out, information requirements analysis is not a one-time project, but a continual process. The proposed formalization based on metadata is a first step in this direction since it enables the realization of meaningful automation potential.

2. The approach restricts itself to consider existing information supply – which will in general not fully satisfy a decision maker's information requirements. In this case,

the remaining parts of the information requirements have to be covered using existing approaches to information requirements analysis.

3. Although IT support plays an important role, an implementation is pending. This is already part of on-going research and will shape up useful for practical application and evaluation issues.

4. The information requirements are currently treated as constant. While this is approximately appropriate for standard reporting and well-structured problems, it is not always the case in a complex and disruptive business environment.

Despite its limitations, the proposed approach is a first step to address the research gap of leveraging the information supply of existing BI systems and towards an enhanced usage of metadata in the context of BI systems.

References

1. Bawden, D., Robinson, L.: The dark side of information: overload, anxiety and other paradoxes and pathologies. Journal of Information Science 35(2), 180–191 (2009)
2. Foshay, N., Mukherjee, A., Taylor, A.: Does Data Warehouse End-User Metadata Add Value? Communications of the ACM 50(11), 70–77 (2007)
3. Kimball, R., Ross, M., Thornthwaite, W., Mundy, J., Becker, B.: The Data Warehouse Lifecycle Toolkit, 2nd edn. John Wiley & Sons, Indianapolis (2008)
4. Oppenheim, C.: Managers' use and handling of information. International Journal of Information Management 17(4), 239–248 (1997)
5. Stroh, F., Winter, R., Wortmann, F.: Method Support of Information Requirements Analysis for Analytical Information Systems. State of the Art, Practice Requirements, and Research Agenda. Business & Information Systems Engineering 3(1), 33–43 (2011)
6. Winter, R., Strauch, B.: A Method for Demand-driven Information Requirements Analysis in Data Warehousing Projects. In: Proceedings of the 36th Hawaii International Conference on System Sciences (HICSS 2003), p. 231.1 (2003)

SPIDER-Graph: A Model for Heterogeneous Graphs Extracted from a Relational Database

Rania Soussi

Ecole Centrale Paris, MAS Laboratory, Business Intelligence Team,
Grande Voie des Vignes 92295 Chatenay-Malabry
rania.soussi@ecp.fr

Abstract. An adapted modeling and visualization technique of links and interactions between several objects,e.g. customers and products, social network and etc,is a precious mean to permit a good understanding of a lot of situations in the enterprise's context. In this latter context, most of the time, these objects and their relations are stored in relational databases. But extracting and modeling heterogeneous graphs from databases are outside of the classical graph models possibilities, moreover when each node contains a set of values. On the other hand, graph models can be a natural way to present these graphs and facilitates their query. In this way, we propose a graph model named SPIDER-Graph which is adapted to represent interactions between complex heterogeneous objects extracted from a relational database. This model is used in our approach of heterogeneous object graph extraction from a relational database which is detailed in this paper.

Keywords: SPIDER-Graph model, graph extraction, relational database.

1 Introduction

In the enterprise context, people need to visualize different types of interactions between heterogeneous objects (e.g. people interactions like social networks...) in order to improve data analysis and to make the good decision. Indeed, these different graphs can help enterprises sending product recommendations (using the graph of Products and Clients), finding experts (using a social network), etc. Nevertheless, in a business context, important expertise information is stored in files, databases and especially relational databases (RD). Indeed, databases contain information about all objects and In order to analyze these interactions and to facilitate their querying, it is relevant to model such interaction by using a graph structure. Graphs are a natural way of representing and modeling heterogeneous data in a unified manner. The main advantage of such structure relies on its dynamic aspect and its capability to represent relations, even multiple ones, between objects. It also facilitates data query and analysis using graph operations. In this work, we propose a new graph model adapted to support complex objects interaction which is called SPIDER-Graph (Structure Providing Information for Data whith Edge or Relations) and inspired from an existing model.

P. Atzeni, D. Cheung, and R. Sudha (Eds.): ER 2012, LNCS 7532, pp. 543–552, 2012.

Then, we use it to model graphs extracted from RDs. Then, the structure of this chapter is the following. First, we present the different definitions of graph and graph database (Gdb) model related to our work. The SPIDER-Graph model is introduced on the section 3. In section 4, we detail our approach of heterogeneous graph extraction from a RD using the SPIDER-Graph model. Finally, we conclude and give some perspectives to our work.

2 Graph and Graph Database Models

Graphs are the natural way to model entities interconnectivity and many applications are based on graph model like social network, hypertext, etc. Some graph definitions are presented below:

Definition 1 (Graph). A *graph* $G := (V, E)$ consists of a set V of vertices (also called nodes), a set E of edges where $E \subseteq V \times V$.

Definition 2 (Labeled Graph). Given two alphabets Σ_V and Σ_E a *labeled graph* is a triple $G := (V, E, l_V)$ where V is a finite set of nodes, $E \subseteq V \times \Sigma_E \times V$ is a ternary relation describing the edges (including the labeling of the edges) and $l_V : V \to \Sigma_V$ is a function describing the labeling of the nodes.

A graph can contain more information by adding attributes to nodes or edges.

Definition 3 (Attributed graph). If nodes in a graph have attributes, the graph is an *attributed graph* denoted by $AG := (V, E, \alpha)$, where α is a labeling function: $\alpha : V \to L_N$. If both nodes and edges in a graph have attributes, the graph is an attributed graph denoted by $AG := (V, E, \alpha, \beta)$where $\beta : E \to L_E$edge labeling functions. α and β are restricted to labels consisting of fixed-size tuples, that is, $L_N = R^p$, $L_E = R^q$, $p, q \in N \bigcup \{0\}$.

Simple graphs are not sufficient to model real life applications. Moreover, they cannot represent heterogeneous graphs based on complex objects. Then, the basic structure of a graph is complemented with the use of hypernodes and hypergraphs extensions that provide support for nested structures.

Definition 4. (Hypernode). A *hypernode*[1] is defined by $H := (V, E)$ where $H \in L$ is termed the defining label of the hypernode (L is a st of labels) and *(V, E)* is a digraph such that $V \subset (P \bigcup L)$ (P is the set of primitive node). *(V,E)* is termed the digraph of the hypernode H.

Definition 5 (Hypergraph). An *hypergraph*[2] G is a tuple (V, E, μ), where V is a finite set of nodes, E is a finite set of edges, $\mu : E \to V^*$ is a connection function where V^* means multiple nodes.

In order to model a more complex structures and connections using the different proposed graph aspects, Gdb model has been proposed. It leads to a more natural modeling (graph structures). It can keep all the information about an entity in a single node and show related information by arcs connected to it. This model also facilitates data query using graph operations. A Gdb is defined[3] as a "database where the data structures for the schema and/or instances are modeled as a (labeled) (directed) graph or generalizations of the graph data structure, where data manipulation is expressed by graph-oriented operations and type constructors, and has integrity constraints appropriate for the graph structure". There is a variety of models for Gdb (see [3]). All these models have their formal foundation as variations of the basic mathematical definition of a graph. These models can be divided in two categories:

1. models based on simple nodes data being represented in these models by a (directed or undirected) graph with simple nodes and edges (like GOOD [4], GMOD [5], etc.).
2. models based on hypernode: in these models, the basic structure of a graph (node and edge) and the presentation of entities and relations are based on hypernodes and hypergraphs (like Hypernode Model[1],GGL[6],etc.).

We have studied and compared these different models in [7]. From this comparison, models based on a simple graph are unsuitable for complex networks where entities have many attributes and multiple relations. However, models based on hypernodes can be very appropriate to represent complex and dynamic objects. In particular, the hypernode model (HM) with its nested graphs can provide an efficient support to represent every real-world object as a separate entity. Hypernode also encapsulates all the attributes related to each object in a same node, thus facilitating the visualization of objects with different and multiples attributes. However, the HM does not offer an explicit representation of labeled edges.

In order to have a better suited model to represent complex heterogeneous graphs, we propose a graph model based on complex-nodes which have a structure similar to a hypernode. The data model we propose here contains some characteristics of the HM. Indeed, both models encapsulate the object attributes, inside the complex-nodes or the hypernode. Attributes can also be multi-valued. The differences between a hypernode and our model are the following (see Fig.1):

1. Our model proposes an explicit representation of labeled edges between nodes to describe objects relations instead of encapsulating two hypernodes within a further hypernode in the case of the hypernode;
2. Attributes type can be a reference to another node instead of using the encapsulation in the HM in order not to have a multiple encapsulation leading to hidden interactions between objects.
3. the SPIDER-Graph contains a schema graph and instance level. However the HM merges the two levels.

3   Spider-Graph Model

In this section we present SPIDER-Graph which is designed to explicitly represent the interaction between objects having multiples attributes. SPIDER-Graph, our model is built on the traditional graph-based model. This model is

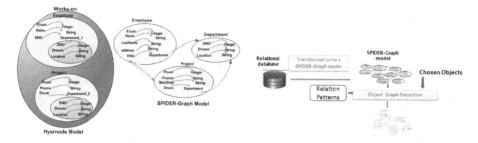

Fig. 1. SPIDER-Graph Model VS Hyper-node model   **Fig. 2.** The Graph extraction approach

designed to explicitly represent the interaction between objects having multiple attributes. The underlying data structure of the SPIDER-Graph model is the complex-node, which is used to represent objects. A schema graph in the SPIDER-Graph model is schematized by an attributed labeled graph: objects are represented by means of typed complex-nodes which may carry attributes. Attributes can be atomic, multi-valued, or a reference to another object (another complex-node). Relations between objects are modeled by labeled edges without attributes. Our model is defined as follows:

Definition 6 (Complex-node). A Complex-node CN is defined by $CN :=
(cn, A_{cn})$ where: cn denotes the name of CN and A_{cn} denotes a set of attributes $A_{cn} := \{ a_{cn} | a_{cn} := \langle n, t \rangle \}$ where n is the attribute name, t is the type. t is a basic type (such as: Integer, String,...) or a reference to another CN.

Now we can formally represent the graph schema (Example in Fig. 4) by:

Definition 7 (SPIDER-Graph schema). The schema is defined by $S =
N_{cn} \bigcup R$ where: N_{cn} is a set of complex nodes and R is the set of relations between N_{cn} defined by $R := \{ \langle r, CN_s, CN_d \rangle , r \in R, CN_s, CN_d \in N_{CN} \}$ where r denotes the name of R, CN_s denotes the node source name and CN_d denotes the node destination name.

In the SPIDER-Graph model, an instance graph is an attributed graph: concrete objects represented by an instance CN where values are added to the attributes and relations represented as edges labeled with the corresponding relation name according to the schema. The instance of a CN is defined by:

Definition 8 (Complex-node instance). An instance CN CN_I of CN is defined by $CN_I := (cni, cn, A_{cni})$ where: cni denotes the name of the instance, cn denotes the name of CN and A_{cni} denotes a set of valuated attributes $A_{cni} := \{ a_{cni} | a_{cni} := \langle n, V \rangle \}$ where n is the attribute name (according to the schema), V is the set of values which can be atomic values or a CN_I (having the same type mentioned in the schema).

Definition 9 (SPIDER-Graph instance). The SPIDER-Graph instance is defined by $IS := N_{Icn} \bigcup R_I$ where: N_{Icn} is a set of instance CN and R_I is the set of relations between N_{Icn} defined by $R_I := \{\ \langle r, CN_{Is}, CN_{Id}\rangle, r \in R, and CN_{Is},$ $CN_{Id} \in N_{cn}\}$.

4   Graph Extraction Using the SPIDER-Graph Model

A heterogeneous object graph, which describes the interactions between different objects like social network graph will facilitate the analysis of the interactions and will help make better decisions in enterprises. In order to extract object graphs from RD we have proposed an approach based on two main steps (Fig.4): (1)converting the RD model to the SPIDER-Graph model and(2) Extracting the heterogeneous object graph (with the chosen entities) from the graph model.

4.1   Converting Relational Model into SPIDER-Graph Model

The transformation of a relational model into SPIDER-Graph model includes schema translation and data conversion [7]. The schema translation can turn the source schema into the target schema by applying a set of mapping rules. In our work, we propose a translation process which directly transforms the relational schema into a SPIDER-Graph schema:Each table is transformed to a complex-node and Four types of relation can be detected using the foreign key.

From the RD tables (Fig. 3), we extract six *CN* (objects) (Fig. 4): *Employee, Manager, Department, Product, Project, Project-Product*. The table *Works-on* is transformed into a relation because it contains only foreign keys linking the tables *Employee* and *Project*. The CN *Employee* = (*"Employee"*,<*Enum, Integer*>,<*Name, String*>,<*LastName, String*>, <*address, String*>, <*DNO, Department*>); has four attributes (*Enum, Name, LastName, address*) with predefined types and one node *DNO* having *Department* as type because it represents a foreign key exported from the CN *Department*.

Once the schema level is extracted, data contained in the relational schema will be used for getting the instance level. Data stored as tuples (rows) in

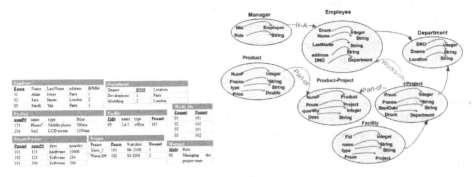

Fig. 3. Relational database   **Fig. 4.** The extracted SPIDER-Graph schema level

the relational model are converted into nodes and edges in the graph model. This involves unloading and restructuring relational data, and then reloading them into a target model in order to populate the schema generated during the translation process. Then, for each *CN* in the SPIDER-graph schema, a set of *CN* instances is extracted from the relational tuples. For example, *Employee_3* is an instance of *Employee* and is defined by $Employee\_3 =<$ $Employee, Employee\_3, A_{Employee\_3} >$ where $A_{Employee\_3} = \{ < Enum,$ $Integer, 03 >, < Name, String, "Smith" >, < LastName, String, "Yan" >, <$ $address, String," Paris" >, < DNO, Departments, Department\_1 > \}$.

Finally, the transformed data are loaded into the SPIDER-Graph schema (An excerpt is shown in Fig. 5). The main advantages of this transformation are (1) to discover underlying graphs of objects from RD, taking into account the implicit relations expressed by the means of primary and foreign keys and (2) to model data in a more flexible way (objects can easily be added or removed in a graph). The reader interested by details about this approach can refer to [7].

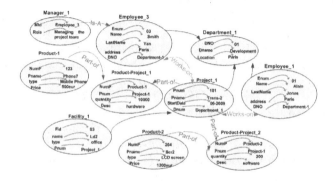

Fig. 5. The Spider-Graph instance level

4.2   Graph Extraction According to the User Chosen Objects

From the graph depicted in the previous section, we apply extraction rules according to the user chosen objects . The objects interaction graph is defined by $GO = (O_I, R_O)$ where:

- O_I is a finite set of object such $O_I : = \{o_I | o_I \in N_{CN}\}$.
- R_O is a finite set of relations between objects such $R_O := \{r | r := \langle l, o_{I1}, o_{I2} \rangle ,$ $o_{I1}, o_{I2} \in O_I\}$ where l is the relation name.

As a first step, the user select from the schema graph the objects that he would like to see their interactions. Then, the relation construction algorithm search the hidden relations between these objects. The identified objects can already share relations on the SPIDER-Graph. The objective here is to put these relations on the object interaction graph and find hidden ones. In our process to transform the RD into a graph database model, we have defined four types of relations: IS-A, Part-of, dependency with known name (using the initial relational tables containing only foreign keys), dependency with unknown name. Then, in order to detect all the relations, we have proposed an approach based on these existing relations. From the different existing relations and by using the SPIDER-Graph schema, the proposed approach creates a set of pattern relations. After that, it seeks these patterns in the SPIDER-Graph instance and creates the set of relations between the identified objects. A pattern relation P_r is defined by $P_r = \langle n, o_{I1}, o_{I2}, o_m \rangle$ such as: n is the name of the relation, o_{I1} and o_{I2} are the two types of the two objects that should share a relation. o_{I1} and $o_{I2} \in O_I$, o_m is a CN mediator (a connector) for the relation. In other word, this CN is used to build this relation.

The relation pattern is able not only to find the existing relations between the objects that should be added to the final graph but also to add new attributes on the objects and find new objects that cannot be identified using the previous process.

In order to create these patterns, we use the SPIDER-Graph schema which mentions the type of CN and all the relations between them. Then, these patterns are searched for the SPIDER-Graph instance. In the following part, we describe the extraction process of each pattern and how it is used to find the relations between the objects.

Patterns to identify new objects. The relation *IS-A* allows to find hidden objects which may not be identified in the entities identification process (Table 1). In the relation construction process, we start by analyzing this kind of relation to find in the next steps the relations related to the hidden objects.

Patterns based on relation between chosen objects. The chosen objects can be directly linked by an existing relation in the graph. We put these relations directly on the object graph. In table 2 the pattern Pr_1 finds all the existing relations between the instances of two chosen objects, while the pattern Pr_2 creates a new relation between the instances of the same object. In some cases, the chosen objects are not directly linked. Then, we search whether there is a path between them or not. The relations between the CNs are directed or bidirectional relations and we assume that each edge has the weight 1. Then, we can apply the Dijkstra's Algorithm. Then, we can find three cases:

-There is no path between the two objects,
-There is a direct path and it is denoted by the path :$< CN_s, CN_j, \ldots, CN_d >$ where $\{CN_j\}$ is the set of CNs between the two chosen objects CN_s and CN_d

Table 1. Patterns to identify new objects

	Initial Relation	Process and description
Definition	$R_1 \quad := \langle "IS - A", CN_s, CN_d \rangle,$ CN_s or CN_d instances$\in O_I$	CN_s or CN_d instances are added to O_I
Example	< "IS-A", Employee, Manager>	Manager instances are added to O_I

Table 2. Pattern used to extract relations between chosen objects

Initial Relation	Pattern	Process and description
R_2 := $\langle r, CN_s, CN_d\rangle$ where CN_s and CN_d instances$\in O_I$	$Pr_1 := \langle r, CN_{si}, CN_{di}, null\rangle$ where CN_{si} and CN_{di} are CN_S and CN_d instances.	Find all the existing relations between CN_s and CN_d instances then add them to R_O
	Pr_2 :=< $Same_{(}CN_d.name), CN_{SIi},$ $CN_{SIj}, CN_d >$ where CN_{SIi} and CN_{SIj} are CN_s instances and i!=j	Find all the CN_s instances which have relations with a same CN_d instances and link them with a new relation "Same_(CN_d.name)"
CN_s and CN_d instances$\in O_I$ there is no relation between them	$Pr_{path}:=<"", CN_{si},$ $CN_{di},~ P>$ where P is a semi-path or a path between CNs and CN_d	See description above

-There is a semi-directed path. The semi directed path is denoted by the semi-path :< CN_s, CN_j ... CN_k, CN_d> where $\{CN_j\}$ is the set of CNs between the first chosen objects CN_s and CN_k and $\exists R_i \in R$ such as $R_i := \langle "Part - of",$ $CN_d, CN_k\rangle$

This kind of relation is represented by a dotted edge (see Fig.8).

Patterns based on relations between a chosen object and another CN (Table 3). Naturally, chosen objects may also have some relations with other CNs (which are not chosen to be added on the object graph). These relations can reveal hidden relations between objects or enrich the attributes of a CN. The pattern Pr_3 finds the object instances which are in relation with the same CN_d instance. The pattern Pr_4 is more complex and corresponds to the following case: if an object has a *"part-of"* relation with another CN CN_d and other CNs have a *"part-of"* relation with CN_d, then the chosen object has a relation with this CN. Pr_4 detects these relations. This set of patterns is applied in order to extract the set of relations between chosen objects. After having identified relations and objects (of interest), we build the transformed graph corresponding to the chosen objects. In the next section, we present a use case experimentation corresponding to the different chosen objects extracted from an actual RD.

4.3   Example of Extracted Graphs

In this section, we will follow the previous steps to extract the interaction graph between *Project* and *Employee* from the RD shown in Fig.3.

First, the object identification process is performed. In this case, the identification is simple because we can detect the chosen objects directly from their name. Then all the instances of Employee and Project are added to the set O_I ($O_I = \{Employee\_1, Employee\_2, Employee\_3, Project\_1, Project\_2\}$). The

Table 3. Patterns based on relations between a chosen object and another CN

Initial Relation	Pattern	Process and description
$R_3 := \langle r, CN_s, CN_d \rangle$ where $CN_s \in O_I$ and $CN_d \notin O_I$	$\text{Pr}_3 := <Same\_(CN_d.name),$ $CN_{SIi}, CN_{SIj}, CN_d >$ where CN_{SIi}, CN_{SIj} are CN_S instances, i!=j and r != "Part-of"	Find all the CN_s instances which have relations with a same CN_d instances and link them with a new relation
	$\text{Pr}_4 := <Same\_CN_k.name,$ $CN_{SIi}, CN_{SIj}, CN_k >$ where CN_{SIi}, CN_{SIj} are CN_S instances, i!=j, r="Part-of" and CN_k $\in\{CN\backslash CN$ has the relation $R:=<"Part\text{-}of",$ $CN, CN_d>\}$	1. Find all the CN_k having a "Part-of" relation with CN_d such as $R:= <"Part\text{-}of", CN_k,$ $CN_d >$ 2. add the new attribute to CN_s containing the name of CN_j 3. Then the pattern Pr_4 is applied : find all the CN_d instances which have relations with a same CN_j instances and link them with a new relation
$R_4 := \langle r, CN_s, CN_d \rangle$ where $CN_s \notin O_I$ and $CN_d \in O_I$	$\text{Pr}_5 := <Same\_CN_s.name,$ $CN_{dI}, CN_{jI}, CN_s >$ where CN_{dI} is CN_d instance and CN_{jI} is instance of CN_j where $CN_j \in O_I \bigcap\{CN\backslash CN$ has the relation $R:=<"",$ $CN, CN_s>\}.$	1. add a new attributes in CN_d containing CN_s reference. 2. Then the pattern Pr_5 is applied : if CN_s has relations with other objects, link each object instance with a CN_d instance if they are in relation with the same CN_s instance.

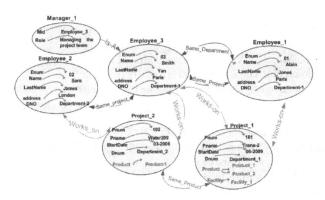

Fig. 6. The Employee-Project graph

relation construction process is then performed using the O_I set. We apply the patterns on the set of existing CNs relations using the following steps:

Step 1. Identify new objects using the *"IS-A"* relation. In our example, the process identifies *"Manager"* as a new object to add in the set O_I due the relation $R_1 = <$ *"IS-A", Manager, Employee*$>$.

Step 2. Add the existing relations between identified objects by applying Pr_1. In the database schema, *Employee* and *Project* share the relation $R :=< "", Employee, Project >$. Then, R is added to the set O_I.

Step 3. Create new relations and new attributes using the patterns. From the existing relation between *Project, Employee* and other *CN*s, we have detected many new relations. For example, from the relation $R :=< "", Employee, Department >$ and using the pattern Pr_3, the new relation $Rnew :=< "Same_Department", Employee_i, Employee_j >$. The result graph is presented in Fig. 6.

5   Conclusion

In this paper, we have presented a new graph data model named SPIDER-Graph which can represent complex objects and their interactions. Indeed, SPIDER-Graph encapsulates the object attributes inside the *CN*s, where attributes can be multi-valued or a reference to other CN, and offers labeled edge between *CN*s. We have used this model to represent heterogeneous object graph extracted from RD. Then, we have detailed the objects graph extraction from RD based on SPIDER-Graph. This approach allows having different graphs using as input the same RD and the enterprise concepts chosen by the user. The extraction approach is based on two steps: (1) transformation of a relational model into the SPIDER-Graph model, and (2) objects graph extraction. In our future work, we will focus on how to improve this extraction method by adding other enterprise data resource to enrich the extracted graphs. Then, we will define graph query language better suited to the heterogeneous object graph.

References

1. Levene, M., Loizou, G.: A Graph-Based Data Model and its Ramifications. IEEE Trans. on Knowl. and Data Eng., 809–823 (1995)
2. Berge, C.: Hypergraphs. North-Holland, Amsterdam (1989)
3. Angles, R., Gutierrez, C.: Survey of graph database models. ACM Comput. Surv. 40(1), 1–39 (2008)
4. Gyssens, M., Paredaens, J., Gucht, D.V.: A graph-oriented object model for database end-user interfaces. SIGMOD Rec. 19(2), 24–33 (1990)
5. Andries, M., Gemis, M., Paredaens, J., Thyssens, I., Bussche, J.D.: Concepts for Graph-Oriented Object Manipulation. In: Pirotte, A., Delobel, C., Gottlob, G. (eds.) EDBT 1992. LNCS, vol. 580, pp. 21–38. Springer, Heidelberg (1992)
6. Graves, M., Bergeman, E.R., Lawrence, C.B.: Graph database systems for genomics. IEEE Eng. Medicine Biol. 14, 737–745 (1995)
7. Soussi, R., Aufaure, M.A., Baazaoui, H.: Graph Database For collaborative Communities. In: Pardede, E. (ed.) Community-Built Databases: Research and Development, 1st edn., p. 400 (May 2011)

Exploiting Data Dependencies with Null Values for Ontology Extraction

Maria Amalfi

KRDB Research Centre, Free University of Bozen-Bolzano, Bolzano, Italy
mamalfi@inf.unibz.it

Abstract. I present a novel methodology to automatically extract ontologies in the form of conceptual schemes characterizing *raw data* and then generate mappings linking the obtained ontology to raw data with the aim to extend *Ontology Based Data Access* to *raw data*. Additionally, I propose new definitions of *functional* and *multivalued dependencies* under SQL (in short, resp., nFDs and nMVDs) to model the semantics of raw data and to achieve raw data normalization. The main novelty in this work is to adapt existing techniques to deal with the case of data with missing or incomplete information that is witnessed through the presence of null values.

1 Introduction

Describing data by means of conceptual models is becoming a crucial task for several data access and data integration tools. To supply this need, in recent years, researchers have came up with systems for *Ontology Based Data Access* (OBDA) (e.g., in [1, 2]) giving rise to methods for constructing some knowledge on top of structured data (e.g., relational schemes).

Nonetheless, more and more often data needed to be classified by ontologies is *raw data*, i.e., unstructured and incomplete data, where incompleteness is witnessed by null values. A suitable example is represented by Semantic Desktop tools from which raw data related to the file systems can be extracted.

The extension of ontology extraction systems to generate ontologies directly from raw data, would be a powerful means to classify more information. On the other hand the automatization of the whole process would be a means to overcome most of the disadvantages arising from manual procedures. Hence, to extend the activity of OBDA to unstructured and incomplete data, and to make the activity of ontology extraction more flexible and more reliable, I introduce a novel automatic methodology (Section 3) which starting from raw data devises (i) ontologies in the form of conceptual models, and (ii) mappings from the ontology to raw data expressed in the form of views. The process of conceptual model generation is based on the extraction of a richer vocabulary coming from the elicitation of new potential concepts exploited to derive (Extended)ER schemes which will have the properties of being dependency and information preserving.

The procedure I propose, results in an extension of the ontology extraction system shown by Lubyte el al. in [3] enriched to handle raw data. It is based

P. Atzeni, D. Cheung, and R. Sudha (Eds.): ER 2012, LNCS 7532, pp. 553–562, 2012.

on known techniques of database design and normalization [4–6] coping with incomplete relations (i.e., relations with null values) where raw data is stored. The extraction of the ontology concepts is based on a raw data normalization process which exploits new definitions of *functional* and *multivalued dependencies* for incomplete relations under SQL (Section 4), which are used to identify the semantics of raw data. The existing notions of FDs and MVDs with null values for incomplete relations [7–12] (defined under the *unknown* [9, 12] and the *no-information semantics* [13] of null values) are not suitable to identify the meaning of raw data where every null value is assumed (i) not to be equivalent to any other value, and (ii) to identify either existent or inapplicable values. The properties (i) and (ii) are instead well modeled in SQL [14] which results to be the best standard to classify raw data. Since the existing definitions of FDs and MVDs with null values on incomplete relations have problems when related with the SQL behavior, I have revised the notions of functional and multivalued dependencies with null values defined in [7, 8, 11, 15] to cope with the SQL standard (Section 4). Additionally, I have redefined the sets of inference rules for both kinds of introduced constraints, which have been proved to be sound and complete (Section 4.3). From now on, the revised definitions of FDs and MVDs I propose will be respectively referred as nFDs and nMVDs.

To the best of my knowledge, this work is the first attempt to generate ontologies from raw data. The paper presents the overall ontology-mapping methodology and then it focuses on the formalization of nFDs and nMVDs introduced (i) to state the semantics of raw data, and (ii) to allow the raw data normalization which is the basis of the ontology extraction procedure.[1]

Due to lack of space, in this paper I shall assume the reader is familiar with the basic notions of relational database theory [17], semantics of null values [13, 18] and data constraints with null values [7–9, 11, 12, 15].

2   Related Works

The process of extracting conceptual models from given relational schemes, known as Database Reverse Engineering (DBRE), has been widely studied over the years (e.g., in [19]). However, the existing approaches are usually based on normalized relational schemes on which constraints are stated. To the best of my knowledge, the only approach dealing with not normalized schemes is presented in [20], where after the derivation of functional and inclusion dependencies, data is normalized into a 3NF schema which is finally transformed into an ER schema.

Even though there is an obvious connection between this area and the problem I am tackling, the crucial difference is that DBRE approaches usually produce just a graphical representation of a conceptual model, without generating any mapping linking the conceptual model to the database. My approach instead, together with the ontology, generates also ontology-to-data mappings which enable the ontology to be directly employed for OBDA on raw data.

---

[1] Details are available in [16].

My methodology performs ontology extraction after a process bound to achieve normalization of raw data with null values. The works coming closer to mine are those that arose (i) in the context of ontology extraction to support OBDA; and (ii) in the context of normalization of data with null values.

In [3] is presented an automatic technique to perform ontology extraction provided with mappings linking the extracted ontology to the underlying data.

Even though there is a close connection between my results and the ones produced by the procedure proposed in [3], the two methodologies differ in the input. While the technique in [3] relies on structured data as relational data sources where none null value appears, on the other hand my procedure relies on the more general raw data, i.e., unstructured and incomplete data, with the consequence that my work results in an extension of the one in [3].

In the area of relational normalization, the works coming close to mine are those concerning the normalization of incomplete relations. Even though, approaches to achieve lossless decompositions of incomplete relations have been formalized [11, 21], they exploit NFDs and NMVDs [7, 8, 11, 15] defined for *subsumption free* relations, which do not properly model raw data with null values; and do not always guarantee lossless decompositions. On the contrary, my methodology, which does not require that relations are *subsumption free*, is based on the new notions of nFDs and nMVDs (Section 4) which have the advantage of being able to model raw data; and can be used to normalize incomplete relation such that lossless decompositions are always achievable.

3   The Ontology Extraction Methodology

My ontology extraction methodology is based on the notions of data constraints, database design and normalization techniques exploited to retrieve ontologies from raw data. By raw data I mean any data structured just as a set of attributes, without the preliminary knowledge of the different relationships the may belong to. Typically raw data is defined by any information not based on database technology, such as an API of an application, a data stream, a web service, etc. Here I abstract the notion of raw data as an incomplete universal relation.

The ontology generation is fundamentally based on a relational decomposition process which starts from raw data stored into an *incomplete universal relation* and exploits the revised notions of nFDs and nMVDs (which are formalized in Section 4) to identify the semantics of raw data.

Given as input raw data, the whole process consists of three main phases: (1) the universal relation decomposition; (2) the derivation of relational constraints; and (3) the ontology and mappings extraction (Fig. 1).

Phase (1): Universal Relation Decomposition. In phase (1) the raw data normalization is performed. nFDs are treated as nMVDs [5] which are automatically checked to satisfy the properties such as conflict-freedom and contention-freedom [11]. These properties are needed to guarantee the data dependency preservation. The obtained set of nMVDs is exploited to decompose the universal relation into a family of 4NF relations, whose data is

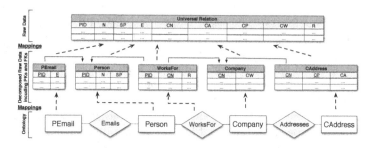

Fig. 1. The Ontology Extraction Methodology

represented by views (identifying mappings from the decomposed relations to raw data) over the original relation. At the end of phase (1), we have (i) a set of relations; and (ii) a set of mappings that link the decomposed relations to raw data.

Phase (2): Derivation of Relational Constraints. To perform the ontology-mapping extraction, my methodology relies on Lubyte et al.'s process [3], whose input consists of a relational structure and a set of relational constraints such as Primary Keys (PKs) and Foreign Keys (FKs). Hence, in phase (2), the translation of nFDs and nMVDs into relational constraints is performed. To convert nFDs and nMVDs into PKs and FKs, I exploit the equivalence between the data structure known as Bachman Diagram (BD) [22] and the relational schema obtained by decomposing raw data by means of MVDs from phase (1). Exploiting paths among the nodes in the BD, I am able to deduce the set of Inclusion Dependencies embedded into the decomposed relational schema from which PKs and FKs are derived. At the end of phase (2), we remain with (i) and (ii) as before, plus (iii) a set of PKs and FKs holding in the decomposed relations.

Phase (3): Ontology and Mappings Extraction. In the last phase, the ontology (ER schema) and the mappings from the ontology to the decomposed relational schema will be generated by means of Lubyte et al.'s process [3]. Actually, Lubyte et al.'s procedure handles relational structures where none null value appears. Hence it must be customized to process also relational structures with null values.

Finally, the ontology and the ontology-to-data mappings, which are obtained by composing (i) mappings from the ontology to the decomposed relational schema extracted in phase(3), and (ii) mappings from the decomposed relational schema to raw data extracted in phase(1), are returned.

3.1 An Example

Assume U is a universal relation retrieved from an address book tool by a Semantic Desktop system. The attributes in U are: *PersonID, Name, SPouse, Email, CompanyName, CompanyAddress, CompanyPhone, CompanyWebsite*, and *Role*, while the following set of nFDs and nMVDs identifies the semantics of U:

$PID \to N, SP$ (A person has one name and at most one spouse)
$PID, CN \to R$ (Every person may work for a company with one role)
$CN, CP \to CA$ (A company has a phone number associated with any factory)
$CN \to CW$ (A company may have a web site)
$PID \twoheadrightarrow E$ (A person may have several emails)
$PID, CN \twoheadrightarrow R$ (A person may work for several companies with some role)
$CN \twoheadrightarrow CA, CP$ (A company may have different factories and phones).

The execution of my ontology extraction procedure produces the results shown (phase by phase) in Fig. 2.

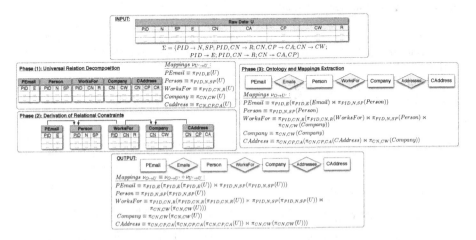

Fig. 2. The Ontology and Mappings associated to the given Raw Data

4   FDs and MVDs with **Null** Values

The existing interpretations of FDs and MVDs with null values for incomplete relations [7–12] defined under the *unknown* [9, 12] and the *no-information semantics* [13] of null values, are not appropriate to represent the structure of raw data, where every null value is assumed (i) not to be equivalent to any other value, and (ii) to identify either an existent or an inapplicable value. Indeed, under the unknown interpretation, a null value always identifies an existing value, but that is not always true in the case of raw data. Under the no-information interpretation, even if the interpretation is more general than the unknown one, it is possible that the comparison of two null values evaluates to true, while in the case of raw data, we shall never assume that two null values are equivalent. Properties (i) and (ii) are well modeled in SQL [14] by which the comparison of a null value with any other value is never evaluated to *true*.

Hence, starting from the more general definitions of NFDs and NMVDs formalized in [10, 11], to identify the semantic of raw data, I introduce revised notions of FDs and MVDs with null values (resp., nFDs and nMVDs) that cope better with SQL.

4.1   NFDs and NMVDs Under the No-Information Semantics

In database theory, given an incomplete relation R defined on a certain set of attributes U, a tuple $t \in R$ is said to be *total* if it does not contain any null value and it is said to be X-total if for any attribute $A \in X$, $t[A]$ is not null. The notation $R_X[XZ]$ denotes the X-total projections of R on XZ where XZ is a subset of U. Given two tuples t, $t' \in R$, it is said that t *subsumes* t' if for every attribute $A \in U$ either $t[A] = t'[A]$ or $t[A] = $ null. Consequently, a relation with no subsumed tuples is said to be *subsumption-free*.

Given an incomplete relation R, which is assumed to be *subsumption-free*, defined over a set of attributes U

- A *Functional dependency with* null *values* (NFDs) [11] $X \to Y$ is satisfied in R if and only if for every X-total tuples t, t' in R, it is true that if $t[X] = t'[X]$ then $t[Y] = t'[Y]$.
- A *Multivalued Dependency with* null *values* (NMVDs) [10, 11] $X \twoheadrightarrow Y$ is satisfied in R if and only if for every tuple t in $R_X[XZ]$ (where $Z = U - XY$) it is true that $R_{XZ=t}[Y] = R_{X=t[X]}[Y]^2$.

Example 1. Let R_1 and R_2 be two incomplete relations defined as follows.

R_1:

A	B	C
a_1	b_1	null
a_2	b_1	null
a_3	null	c_1

R_2:

A	B	C	D
a_1	null	c_1	d_1
a_1	null	c_2	d_1
a_1	null	c_1	d_2
a_1	null	c_2	d_2
null	b_1	c_3	d_3

According to the definitions of NFD and NMVD, R_1 satisfies the two NFDs $\{A \to B, B \to C\}$, and R_2 satisfies the two NMVDs $\{A \twoheadrightarrow BC, A \twoheadrightarrow D\}$.

We notice that by the definitions of NFD and NMVD, it is possible that the comparison of two null values in a relation evaluates to true (e.g., the null values in attribute C in the first two tuples in R_1), but in the case of raw data, we shall never assume that two null values are equivalent. Moreover, even if NFDs and NMVDs are based on the the definition of FD and MVD over complete relations [23], because of null values appearing in attributes in Left Hand Sides (LHSs) of dependencies, they do not preserve the *Join Dependency* (JD)[3] which is instead implicitly derivable from FD and MVD. The consequence is that the direct application of normalization techniques (which are known to produce lossless

---

[2] Let R be a relation over $U = XYZ$ and t a tuple in $R_X[XZ]$, then $R_{XZ=t}[Y]$ and $R_{X=t[X]}[Y]$ denote, respectively, the projections on Y of those tuples whose values in XZ are equivalent to values in the tuple t; and the projections on Y of those tuples whose values in X are equivalent to values in $t[X]$.

[3] Given a total relation R defined over a set of attributes U, a *Join Dependency* (JD) $\bowtie [X_1, \ldots, X_n]$, where $X_1, \ldots, X_n \subseteq U$ and $\bigcup_{i=1}^{n} X_i = U$, is satisfied in R if $R = \bowtie [\pi_{X_1}(R), \ldots, \pi_{X_n}(R)]$.

decompositions) on NFDs and NMVDs could be affected by loss of information, e.g., the last tuple in R_1, where for the attribute B there is a null value, cannot be retrieved after a decomposition of R_1 by the NFDs in Example 1.

4.2   FDs and MVDs with Null Values and SQL: nFDs and nMVDs

The semantics of FDs and MVDs with null values under the SQL semantics (in short, resp., nFDs and nMVDs), that I propose, specializes the notions of NFDs and NMVDs [8, 10, 11] in a way that (i) the comparison of values obeys the SQL semantics by which any null value is never equivalent to any other value; and (ii) the strong relationship between the definitions of nFD and nMVD and the definition of JD is preserved by the assumption that attributes in LHSs of dependencies identify a Null Free Subeschema (NFS) [7]. A NFS is a constraint requiring that a certain subset U_S of a relation schema $R(U)$ does not contain null values (written as $R(U, U_S)$).

Formally, let R be an incomplete relation (possibly not *subsumption free*) over the schema $R(U, U_S)$ where $U = XYZ$

Definition 1. *A Functional dependency with* null *values under SQL (nFD) $X \to Y$, where $X, Y \subseteq U$, is satisfied in R if (i) each tuple in R is X-total, and (ii) for each pairs of tuples $t, t' \in R$, $t[X] = t'[X]$ implies $t[Y] = t'[Y]$.*

Definition 2. *A Multivalued Dependency with* null *values under SQL (nMVD) $X \twoheadrightarrow Y$, where $X, Y \subseteq U$ and $Z = U - XY$, is satisfied in R if $R = \pi_{XY}(R) \bowtie \pi_{XZ}(R)$[4].*

In the case of total tuples, nFDs and nMVDs assume the same semantics as for FDs and MVDs over complete relations [23].

Referring to Example 1, by the new definitions of nFD and nMVD, the relation R_1 satisfies only the nFD $A \to B$, however there are two subrelations of R_1 (say R_{1_1} and R_{1_2}) that satisfy both nFDs $\{A \to B, B \to C\}$. Instead, the relation R_2 does not satisfy any nMVD in $\{A \twoheadrightarrow BC, A \twoheadrightarrow D\}$, while there are two subrelations of R_2 (say R_{2_1} and R_{2_2}) that satisfy both nMVDs.

R_{1_1}:

A	B	C
a_1	b_1	null

R_{1_2}:

A	B	C
a_2	b_1	null

R_{2_1}:

A	B	C	D
a_1	null	c_1	d_1
a_1	null	c_2	d_1

R_{2_2}:

A	B	C	D
a_1	null	c_1	d_2
a_1	null	c_2	d_2

From Definitions 1 and 2 and by the SQL semantics of null values follows that if a nFDs $X \to Y$ (resp. nMVD $X \twoheadrightarrow Y$) holds on a relation $R(U, U_S)$, then it is that $X \subseteq U_S$.

On the other hand, if a relation R satisfies a nFD $X \to Y$, it is that for every tuple $t \in R$ s.t. $t[Y]$ is not total, then there is none tuple $t' \in R$ s.t. $t'[X] = t[X]$. While, in the case of nMVDs, if a relation R satisfies a nMVD $X \twoheadrightarrow Y$, for every tuple $t \in R$ s.t. $t[Y]$ is not total, it is that for each tuple $t' \in R$ s.t. $t'[X] = t[X]$ then it is also that $t'[U - XY] = t[U - XY]$, more generally, $t'[U - Y] = t[U - Y]$.

---

[4] Due to the preservation of the natural join property, each tuple in R is X-total.

4.3   Inference Rules for nFDs and nMVDs

Let X, Y, Z, and W be subsets of U and let U_S a NFS s.t. $U_S \subseteq U$. The following are sound and complete inference rules for nFDs with null values.

nFD1 *(Reflexivity)* If $X \subseteq U_S$ and $Y \subseteq X$, then $X \to Y$.
nFD2 *(Augmentation)* If $X \to Y$ and $Z \subseteq W \subseteq U_S$, then $XW \to YZ$.
nFD3 *(Transitivity)* If $X \to Y$ and $Y \to Z$, then $X \to Z$.
nFD4 *(Union)* If $X \to Y$ and $X \to Z$, then $X \to YZ$.
nFD5 *(Decomposition)* If $X \to YZ$, then $X \to Y$ and $X \to Z$.

Theorem 1. *Axioms $nFD1-nFD5$ form a sound and complete set of inference rules for nFDs on incomplete relations with NFS.*

Proof. The proof is shown in [16].

Given an incomplete relation R and the nMVD $g : \emptyset \twoheadrightarrow V$, from Definition 2 follows that, g is valid in R iff and only iff $V = U$. If the LHS of a nMVD is $\emptyset$ then from Definition 2, the relation R must be equivalent to the cartesian product of $V \times U - V$, but in the presence of null values, by the SQL semantics for null values, if some null value is present for some attribute in V then it is unique and it is never possible to duplicate it to generate the cartesian product.

The following are sound and complete inference rules for nMVDs where if the nMVD $\emptyset \twoheadrightarrow V$ is given then $V = U$.

nMVD0 *(Complementation)* Let X, Y and Z be sets s.t. their union is U and $Y \cap Z \subseteq X$ then $X \twoheadrightarrow Y$ if and only if $X \twoheadrightarrow Z$.
nMVD1 *(Reflexivity)* If $X \subseteq U_S$ and $Y \subseteq X$, then $X \twoheadrightarrow Y$.
nMVD2 *(Augmentation)* If $X \twoheadrightarrow Y$ and $Z \subseteq W \subseteq U_S$, then $XW \twoheadrightarrow YZ$.
nMVD3 *(Transitivity)* If $X \twoheadrightarrow Y$ and $Y \twoheadrightarrow Z$, then $X \twoheadrightarrow Z - Y$.
nMVD4 *(Union)* If $X \twoheadrightarrow Y$ and $X \twoheadrightarrow Z$, then $X \twoheadrightarrow YZ$.
nMVD5 *(Decomposition)* If $X \twoheadrightarrow Y$ and $X \twoheadrightarrow Z$, then $X \twoheadrightarrow Y \cap Z$ and $X \twoheadrightarrow Y - Z$.

Theorem 2. *Axioms $nMVD1 - nMVD5$ form a sound and complete set of inference rules for nMVDs on incomplete relations with NFS.*

Proof. The proof is shown in [16].

The following are inference rules for sets of nFDs and nMVDs.

nFD-nMVD1 *(Conversion)* If $X \to Y$ then $X \twoheadrightarrow Y$.
nFD-nMVD2 *(Interaction)* If $X \twoheadrightarrow Z$ and $Y \to Z'$ with $Z' \subseteq Z$, where Y and Z are disjoint, then $X \to Z'$.

Theorem 3. *Axioms $nFD-nMVD1$ and $nFD-nMVD2$ are sound inference rules for nFDs and nMVDs on incomplete relations with NFS.*

Proof. The proof is shown in [16].

5   Conclusions and Future Work

I have described the current state of my work, showing a novel methodology to extract ontologies and mappings from raw data whose information and data dependencies are preserved. I have introduced FDs and MVDs with null values under the SQL standard (resp., nFDs and nMVDs) in the context of incomplete relation and NFS together with the reformulation of sound and complete sets of inference rules for both kinds of revised constraints.

Actually, my methodology preserves data dependencies but does not always fully preserve information. Hence the preliminary work done to define nFDs and nMVDs has to be enriched to obtain data decompositions which will always be lossless. On the other hand, the preservation of data dependencies has been accomplished exploiting properties which bound the selection of data constraints, hence, more investigations are needed to overcome these limitations.

For the ontology extraction, my methodology relies on Lubyte et al.'s process [3] which was designed to manage complete relational structures. Hence, to exploit it, I will customize it to also process incomplete relational structures.

Furthermore, some investigation will be done (i) in the area of horizontal decomposition trying to derive ontologies where every possible embedded concept, also those that cannot be directly discovered by means of vertical decompositions techniques, is elicited; and (ii) in the area of data structure as hierarchical schemata [15], since I believe that the classification of concepts could be directly performed during the decomposition process.

Finally, I aim to automate every process of which my procedure consists and as a last step I shall provide experimentation on real data to evaluate and to show the validity of my methodology.

Acknowledgements. I thank Sergio Tessaris and Enrico Franconi for their helpful comments.

References

1. Calvanese, D., Giacomo, G.D., Lembo, D., Lenzerini, M., Poggi, A., Rosati, R.: Ontology-based database access. In: SEBD, pp. 324–331 (2007)
2. Rodriguez-Muro, M., Lubyte, L., Calvanese, D.: Realizing ontology based data access: A plug-in for protege. In: ICDE Workshops, pp. 286–289 (2008)
3. Lubyte, L., Tessaris, S.: Automatic Extraction of Ontologies Wrapping Relational Data Sources. In: Bhowmick, S.S., Küng, J., Wagner, R. (eds.) DEXA 2009. LNCS, vol. 5690, pp. 128–142. Springer, Heidelberg (2009)
4. Bernstein, P.A.: Synthesizing third normal form relations from functional dependencies. ACM Trans. Database Syst. 1(4), 277–298 (1976)
5. Fagin, R.: Multivalued dependencies and a new normal form for relational databases. ACM Trans. Database Syst. 2(3), 262–278 (1977)
6. Fagin, R.: Normal forms and relational database operators. In: SIGMOD Conference, pp. 153–160 (1979)

7. Atzeni, P., Morfuni, N.M.: Functional dependencies and constraints on null values in database relations. Information and Control 70(1), 1–31 (1986)
8. Hartmann, S., Link, S.: When data dependencies over sql tables meet the logics of paradox and s-3. In: PODS, pp. 317–326 (2010)
9. Imielinski, T., Lipski Jr., W.: Incomplete information and dependencies in relational databases. In: SIGMOD Conference, pp. 178–184 (1983)
10. Lien, Y.E.: Multivalued dependencies with null values in relational data bases. In: VLDB, pp. 61–66 (1979)
11. Lien, Y.E.: On the equivalence of database models. J. ACM 29(2), 333–362 (1982)
12. Vassiliou, Y.: Functional dependencies and incomplete information. In: VLDB, pp. 260–269 (1980)
13. Zaniolo, C.: Database relations with null values. J. Comput. Syst. Sci. 28(1), 142–166 (1984)
14. Türker, C., Gertz, M.: Semantic integrity support in sql: 1999 and commercial (object-) relational database management systems. VLDB J. 10(4), 241–269 (2001)
15. Lien, Y.E.: Hierarchical schemata for relational databases. ACM Trans. Database Syst. 6(1), 48–69 (1981)
16. Amalfi, M.: Technical report: A methodology to devise conceptual schemes from raw data with null values (2012),
    http://www.inf.unibz.it/~mamalfi/TechnicalReport-v02.pdf
17. Abiteboul, S., Hull, R., Vianu, V.: Foundations of Databases. Addison-Wesley (1995)
18. Codd, E.F.: Extending the database relational model to capture more meaning. ACM Trans. Database Syst. 4(4), 397–434 (1979)
19. Andersson, M.: Extracting an Entity Relationship Schema from a Relational Database Through Reverse Engineering. In: Loucopoulos, P. (ed.) ER 1994. LNCS, vol. 881, pp. 403–419. Springer, Heidelberg (1994)
20. Petit, J.M., Toumani, F., Boulicaut, J.F., Kouloumdjian, J.: Towards the reverse engineering of denormalized relational databases. In: ICDE, pp. 218–227 (1996)
21. Jajodia, S., Springsteel, F.N.: Lossless outer joins with incomplete information. BIT 30(1), 34–41 (1990)
22. Bachman, W.: Data structure diagrams. DataBase, 4–10 (1969)
23. Beeri, C., Fagin, R., Howard, J.H.: A complete axiomatization for functional and multivalued dependencies in database relations. In: SIGMOD Conference, pp. 47–61 (1977)

Partial Multi-dimensional Schema Merging in Heterogeneous Data Warehouses

Marius-Octavian Olaru

Department of Information Engineering (DII),
University of Modena and Reggio Emilia
Modena, Italy
mariusoctavian.olaru@unimore.it

Abstract. In recent years, novel approaches and architectures to support Data Warehouse integration (like the P2P Data Warehouse[1], also called *Business Intelligence Network, or BIN*) have been proposed. Although the concept has been defined from a theoretical point of view, until now no *automatic* or *semi-automatic* method that allows the integration of multidimensional information coming from heterogeneous Data Warehouses has been proposed. This paper proposes a method capable of exploiting dimension mappings between two or more multidimensional schemas in order to import and integrate heterogeneous DW instances subsets. The proposed method is based on a *semi-automatic* mapping generation technique previously developed, and on the *dimension-chase* algorithm defined in literature for integration purposes.

1 Introduction

During the last two decades, Data Warehousing has been the main instrument that permits the analysis of large amounts of operational data. It allows managers to take strategic decisions based on valuable, aggregated information synthesized from the companys transactional data.

The classic approach in Data Warehousing has been that in which a Data Warehouse is built from one or more data sources belonging to the same company or organization through a process called *Extract-Transform-Load* (ETL). When companies require to unify two or more heterogeneous DWs (for example when a company acquires another company and needs to create a new DW) the solution used so far was to create a new DW by combining the data sources of the initial DWs into an unique repository, in the *data staging* area, and subsequently build a new DW based on the unified data repository. This kind of integration is, however, considered a *low-level* solutions, as the actual integration is being performed at the initial phase of the DW building procedure.

In recent years, however, new approaches with a different perspective over Data Warehouse integration have been proposed in the context of the Business Intelligence 2.0 scenario. These new architectures, like the Peer-to-Peer Data Warehouse[1] or the Federated Data Warehouse(e.g. [2]), suggest a collaborative solution to information management, which means integrating information from

P. Atzeni, D. Cheung, and R. Sudha (Eds.): ER 2012, LNCS 7532, pp. 563–571, 2012.
© Springer-Verlag Berlin Heidelberg 2012

a network of collaborative companies/associations in order to offer participants a broader view of the operating context. In such kind of scenario, when a relatively elevated number of DWs must be combined, the classical solution through the use of ETL procedures is hardly feasible, mainly because of the amount of work required. That is why the author believes that a *high-end* integration of heterogeneous DWs (i.e., integrating the final multidimensional information) has the possibility of greatly reducing the resources needed for the implementation of the solution.

Such approach, however, presents various challenges, like the identification of common facts and dimensions and the resolution of conflicts between common dimensions[3]. In fact, a common problem to Data Warehouse integration is that heterogeneous DWs contain similar information, but usually structured differently (for example with different dimensions), that makes the simultaneous interrogation difficult. In this context, this paper presents a method for the integration and merging of similar dimensions from heterogeneous DWs based on the *dimension chase* procedure, that permits the *uniformization* of the dimensions in order to allow a more easily DW integration.

The rest of this paper is structured as follows: Section 2 provides an overview of the related work; Section 3 describes the proposed method in detail, while Section 4 draws the conclusions and presents the future work.

2   Related Work

Although it can be seen as a special case of data integration, until now, DW integration has received little attention from the research community[3]. Some solutions proposed so far suggest the use of classical data integration approaches, likes the use of semantics. For example, [4] makes use of *similarity* functions (based on earlier work in data integration [5,6,7]) to identify pairs of *related* concepts (facts, attributes, measures, etc.) that may be used in the process of DW integration, however the solution lacks the presentation of any integration strategy.

Other researchers consider that the common structure of DWs along the acknowledged concepts of *facts* and *dimensions* may suggest the use of specific solutions [3,8]. For example, the authors in [8] propose a dimension mapping strategy based on the *topology* of dimensions and on *cardinality-based* properties, whereas they make use of semantics only as an optional validation tool. One interesting aspect of the semantic validation step, however, is the use of the *Combined Wordsense Disambiguation* (CWSD) [9,10] technique developed inside the MOMIS [11] data integration project, that is capable of doing both wordsense disambiguation and context specific abbreviation expansion. The proposed method is a DW specific solution and is suited even in cases where a classic solution (like one based purely on semantics) would fail.

The first mention of *dimension integration* can be found in [12]. Although the authors don't provide a real integration strategy, they define a design methodology to *facilitate* the integration of autonomous *data marts* by providing guidelines to create and maintain *conformed dimensions* (dimensions with consistent dimension keys, consistent column names, consistent attribute definitions and consistent attribute values). A more detailed formalization can be found in [13], where the author defines the term of *compatible dimensions* through the use of *Dimension Algebra* (DA) expressions. The paper also defines *matchings* among compatible dimensions, and properties that the matchings may have, namely *coherence, soundness* and *consistency*. The author also introduces the *dimensions chase* procedure, which is a special case of the *chase* procedure [14], defined for integration purposes and for reasoning on functional dependencies. The work in [13] is different from the current work because it uses the *dimension chase* only as a way of identifying properties of the matchings among dimensions defined through the use of DA expressions, whiles the present paper suggests the use of the procedure for integration purposes.

3   Dimension Integration Strategy

The work presented in the current paper is the continuation of the method proposed in [8], where the authors present a method for the *semi-automatic* discovery of mappings between dimensional attributes of two multi-dimensional schemas. The method uses topological properties of dimension hierarchies to *semi-automatically* generate a set of candidate mapping predicates that express the relations among different dimension hierarchies of two DW schemas. Each mapping in the set is then validated using CWSD. The result of the process is a subset of mappings that with a high probability represent semantic relations among dimensional attributes. A DW integration strategy may use the generated mappings to identify and integrate relevant information retrieved from compatible information sources.

The method presented in the current paper is divided in two steps:

- **Dimensional attributes insertion**: starting from two heterogeneous dimensions, compatible attributes are inserted from a remote dimension in a local dimension;
- **Attributes values insertion**: the newly inserted attributes are populated with relevant information from the remote dimensions, by using the *dimension chase* procedure;

3.1   Motivating Example

Consider the two example schemas in Figure 1. Suppose that the first (S_1) represents the REVENUE fact table in a Data Warehouse (call it DW_1 from now on) and that the second(S_2) represent the SALE fact table in a second Data Warehouse (DW_2). Using the method proposed in [8] it is possible to discover a set of mapping predicates between the two dimensions. For example, $\omega_1 : S_1.city < equi - level > S_2.city$, as they both refer to the same concept, at the same granularity, and $\omega_2 : S_1.region < roll - up > S_2.city$ that is used to state that the dimensional attribute $S_1.region$ has the same semantic meaning

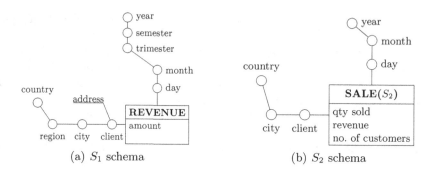

(a) S_1 schema   (b) S_2 schema

Fig. 1. Example

as the dimensional attribute $S_2.city$, but at a higher aggregation level. In fact, a *region* is an aggregation of *cities* as among them there is a *many − to − one* relation.

The entire set of mappings can be used to integrate information coming from the two different DWs. For example, a user having the possibility to pose queries over the two DWs may obtain the total *revenue* by combining the revenue coming from the two DWs, divided by *city* and *month*. This is possible as the requested information is available in both DWs at the required granularity level. On the other hand, a query requesting the total revenue divided by *month* and *region* cannot be executed over the second DW as the available information can only be summarized by *city* and *country*, but not by *region*, which is not a dimensional attribute in the schema S_2.

The heterogeneity of the two schemas makes the integration difficult as for some queries the required information is available in both DWs at the same granularity level, meanwhile, for other queries the requested information may be found at different granularity. This can be a major limit when the goal is to interrogate both DWs contemporaneously.

One way of overcoming this limit is to *import*, where possible, parts of the remote schema into the local schema, augmenting this way the capabilities of the local schema itself. For example, if the dimensional attribute *region* could be imported from schema S_1 to schema S_2, this would allow users to have the same unified view over both DWs, and would offer new interrogation possibilities to users that have access to DW_2.

3.2   Preliminaries

Although several methodologies for designing a Data Warehouse at conceptual level exist, the method presented in this paper relies on the *Dimensional Fact Model* described in [15]. The DFM describes a fact schema as a sextuple $f = (M, A, N, R, O, S)$, where:

- M is a set of *measures* defined by a numeric or Boolean value
- A is a set of *dimensional attributes*

- N is a set of *non-dimensional attributes*
- R is a set of ordered couples that define the $quasi - tree$ representing the dimension hierarchy
- $O \subseteq R$ is a set of *optional* relationships.
- S is a set of aggregation statements

In order to express the semantic relations among multidimensional schema elements, we used the mapping predicates presented in [16]. In particular, the authors of [16] define five mapping predicates, namely:

- **same** predicate: used to state that a measure in the first schema is semantically equivalent to a measure or a set of measures in the second schema, using, if needed, an encoding function;
- **equi-level** predicate: used to state that two dimensional attributes in two different schemas have the same granularity and meaning;
- **roll-up** predicate: used to indicate that a dimensional attribute (or set of attributes) of the first schema aggregates an attribute (or set of attributes) of the second schema;
- **drill-down** predicate: used to indicate that an attribute (or set of attributes) of the first schema disaggregates an attribute (or set of attributes) of the second schema;
- **related** predicate: indicates that between two dimensional attributes there is a $many - to - many$ relation;

3.3 Schema Importation Rules

The first step of our method consists in the identification of parts of remote dimensions that are compatible and can be inserted in the local dimension. Suppose that by using the method described in [8] we obtain the mappings set $\mathcal{M}$ containing the following mappings:

- $\omega_1 : S_2.client < equi - level > S_1.client,$
- $\omega_2 : S_2.city < equi - level > S_1.city,$
- $\omega_3 : S_2.country < equi - level > S_1.country,$
- $\omega_4 : S_2.country < roll - up > S_1.region,$
- $\omega_5 : S_1.region < drill - down > S_2.country,$
- $\omega_6 : S_1.region < roll - up > S_2.city,$
- $\omega_7 : S_2.city < drill - down > S_1.region$

The presence of at least one $equi - level$ mapping suggests that the dimensions have common information, so their schemas are overlapped (the schema intersection is not *null*). The key idea is to use this common schema information as a starting point for importing other dimensional attributes. The attributes are first inserted as optional attributes, and then, if sufficient information is available in the two DWs, the attributes are *promoted* to mandatory attributes. The first step is based on the following rule:

Rule 1. *Given two fact schemes* $f_1 = (M', A', N', R', O', S')$ *and* $f_2 = (M'', A'', N'', R'', O'', S'')$, *let* $a'_i, a''_i \in A'$ *such that* $(a'_i, a''_i) \in R'$, *and* $a'_j \in A''$. *If* $\{(a'_i < equi - level > a'_j), (a''_i < roll - up > a'_j)\} \subseteq \mathcal{M}$, *then:*

$$A'' := A'' \cup \{a''_i\} \qquad (1)$$
$$O'' := O'' \cup \{(a'_j, a''_i)\}$$

If $\exists\, a''_j \in A''$ *such that* $\{(a'_j, a''_j)\} \subseteq R''$, *and* $\{(a''_i < drill - down > a''_j)\} \subseteq \mathcal{M}$, *then:*

$$R'' := R'' \cup \{(a''_i, a''_j)\} \qquad (2)$$

Figure 2 contains a graphical example of the importation Rule. Dimensions d_1 and d_2 are the corresponding dimensions of DW1 and DW2 as defined in Fig. 1. As $region, city \in A'$ and $(city, region) \in R'$ and $city \in A''$, and $\{(S_1.region < roll - up > S_2.city)\} \subseteq \mathcal{M}$, then, according to (1), the attribute $region$ is inserted among the attributes of S_2 (more precisely, in the set A'') and the ordered tuple $(city, region)$ is inserted in O''. Next, for every dimensional attribute in A'' that is a $roll - up$ of the newly inserted attribute, add an ordered couple in R'', in order to express the given semantic relation. As $S_2.country < roll - up > S_1.region$, the ordered couple $(region, country)$ is inserted in the set R'' to express the given semantic relation.

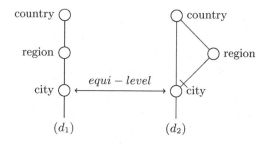

Fig. 2. Graphical example of the importation rule

3.4   Data Importation

In order to import relevant information from remote dimensions, the *dimension-chase*[13] (or *d-chase*) algorithm will be used. The procedure consists in creating an initial *tableau* from the dimensions and applying a *chase step* recursively until the completion of the tableau.

Table 1. Tableau

client	city	region	country	dimension
M.ROSSI	ROME	LAZIO	ITALY	d1
P.BIANCHI	TURIN	PIEDMONT	ITALY	d1
A.RENZO	MILAN	LOMBARDY	ITALY	d1
A.MANCINO	ROME	$v1$	ITALY	d2
S.RUSSO	TURIN	$v2$	ITALY	d2
T.CONTI	MODENA	$v3$	ITALY	d2

With a little abuse of notation, the information contained in the two dimensions will be represented as a table. Table 1 represents the initial tableau, that was built by adding all the attributes of the two dimensions (the couples of attributes in the two dimensions that are connected by an $< equi - level >$ relation are inserted only once, as they represent the same concept), and a final column containing the label of the dimension. The tuples in the table are the tuples of the dimensions. For every column representing an attribute not contained in a given dimension, the value is replaced by a variable (see last three rows). Suppose that the first three rows represent the information contained in the first dimensions (d_1), while the last three rows represent the information from the second considered dimension (d_2). The *chase-step* consists in recursively applying the following rule:

Rule 2. $\forall a_1, a_2 \in A'$ such that $(a_1, a_2) \in R'$, or $a_1, a_2 \in A''$ such that $(a_1, a_2) \in R''$, if $\exists t_1, t_2 \in T$ such that $t_1[a_1] = t_2[a_1]$ and $t_1[a_2] \neq t_2[a_2]$, then if $t_1[a_2]$ is a variable, assign it the value $t_1[a_2] := t_2[a_2]$ (vice-versa if $t_2[a_2]$ is a variable).

The chase ends after no possible assignment can be made using the information available in the tableau. In the given example, the procedure successfully assigns v_1 the value "LAZIO" and v_2 the value "PIEDMONT". However, no value is assigned to variable v_3, because the required information is not contained in the first dimension. The final tableau (Table 2) can be used to import the required information in the second dimension. First of all, the tableau needs to be projected on the final schema of the dimension in which to import the information (d_2 in the given example), in order to import only the values of the dimension levels of interest. Secondly, the obtained information is inserted in the considered instance of schema S_2.

There are two important aspects of the information importation step. First of all, the newly inserted dimensional attribute (*region*) is populated with compatible values contained in the other dimension. This gives analysts the opportunity of formulating queries previously unavailable on schema S_2 (for example, the total revenue for a specific month for the region "PIEDMONT"), although this kind of interrogations may provide partial results, if the newly inserted attribute remains optional. Secondly, the importation step increases the information previously available in the initial dimension, due to the possibility of importing the tuples from the tableau that originated from the remote dimension (d_1 in our case). For example, the information regarding the clients contained in the

Table 2. Final Tableau

client	city	region	country	dimension
M.ROSSI	ROME	LAZIO	ITALY	d1
P.BIANCHI	TURIN	PIEDMONT	ITALY	d1
A.RENZO	MILAN	LOMBARDY	ITALY	d1
A.MANCINO	ROME	LAZIO	ITALY	d2
S.RUSSO	TURIN	PIEDMONT	ITALY	d2
T.CONTI	MODENA	$v3$	ITALY	d2

instance of schema S_1 (the first three tuples) are also inserted in the instance of schema S_2. This new information may be later used for interrogation/analysis purposes.

As stated earlier, if sufficient information is contained in the two dimensions, then the attribute originally inserted (*region*) can be *promoted* to a normal attribute. This is possible only if the values of the newly inserted attribute had all been populated. Using the tableau, this is true if and only if all variables in the column have been assigned a correct value. This is not the case in the given example, as variable v_3 has been assigned no value after the execution of the *d-chase* algorithm.

4   Conclusions and Future Work

The work presented in this paper describes a method for integrating compatible dimensions levels between heterogeneous dimension hierarchies. The method is based on a set of mapping predicates previously generated. The mappings may be manually generated, or one may use semi-automatic methods, like the one proposed in [8]. In both cases, the accuracy of the mappings directly influences the efficiency of the proposed method. If the mappings are accurate, then the proposed method, thanks to the use of the *dimension chase* procedure, is able to correctly integrate the available information in the compatible levels.

It is the author's believe that Data Warehouse Integration will be a growing challenge in the near future. That is why there is a pressing need to develop complete methods that are able to start from identifying common concepts in different Data Warehouses and continue by correctly integrating the required information both at logical level and at physical level.

One important challenge that still needs to be addressed is the identification of identical information that for design strategies, or other reasons, is represented in a different form and makes the integration difficult. This particular issue, together with the management and integration of common (or partially overlapped) derived summary information [17,18] may constitute the final step in designing a complete methodology for heterogeneous Data Warehouse integration.

References

1. Golfarelli, M., Mandreoli, F., Penzo, W., Rizzi, S., Turricchia, E.: OLAP query reformulation in peer-to-peer data warehousing. Information Systems 37(5), 393–411 (2012)
2. Berger, S., Schrefl, M.: From Federated Databases to a Federated Data Warehouse System. In: Proceedings of the 41st Hawaii International International Conference on Systems Science, HICS 2008, p. 394 (2008)
3. Özsu, M.T., Liu, L.: Encyclopedia of Database Systems. Springer US (2009)
4. Banek, M., Vrdoljak, B., Tjoa, A.M., Skocir, Z.: Automated Integration of Heterogeneous Data Warehouse Schemas. IJDWM 4(4), 1–21 (2008)
5. Bergamaschi, S., Castano, S., Vincini, M.: Semantic Integration of Semistructured and Structured Data Sources. SIGMOD Record 28(1), 54–59 (1999)
6. Madhavan, J., Bernstein, P.A., Rahm, E.: Generic Schema Matching with Cupid. In: Apers, P.M.G., Atzeni, P., Ceri, S., Paraboschi, S., Ramamohanarao, K., Snodgrass, R.T. (eds.) VLDB, pp. 49–58. Morgan Kaufmann (2001)
7. Melnik, S., Garcia-Molina, H., Rahm, E.: Similarity Flooding: A Versatile Graph Matching Algorithm and Its Application to Schema Matching. In: ICDE, pp. 117–128 (2002)
8. Bergamaschi, S., Olaru, M.O., Sorrentino, S., Vincini, M.: Dimension matching in Peer-to-Peer Data Warehousing. In: 16th IFIP WG8.3 International Conference on Decision Support Systems, DSS 2012, Anávissos, Greece (2012)
9. Sorrentino, S., Bergamaschi, S., Gawinecki, M.: NORMS: An automatic tool to perform schema label normalization. In: ICDE, pp. 1344–1347 (2011)
10. Sorrentino, S., Bergamaschi, S., Gawinecki, M., Po, L.: Schema label normalization for improving schema matching. Data & Knowledge Engineering 69(12), 1254–1273 (2010)
11. Beneventano, D., Bergamaschi, S., Guerra, F., Vincini, M.: The MOMIS Approach to Information Integration. ICEIS (1), 194–198 (2001)
12. Kimball, R., Ross, M.: The Data Warehouse Toolkit: The Complete Guide to Dimensional Modeling, vol. 32. John Wiley & Sons, Inc., New York (2002)
13. Torlone, R.: Two approaches to the integration of heterogeneous data warehouses. Distributed and Parallel Databases 23(1), 69–97 (2008)
14. Abiteboul, S., Hull, R., Vianu, V.: Foundations of Databases. Addison-Wesley (1995)
15. Golfarelli, M., Maio, D., Rizzi, S.: The Dimensional Fact Model: A Conceptual Model for Data Warehouses. Int. J. Cooperative Inf. Syst. 7(2-3), 215–247 (1998)
16. Golfarelli, M., Mandreoli, F., Penzo, W., Rizzi, S., Turricchia, E.: Towards OLAP query reformulation in Peer-to-Peer Data Warehousing. In: Song, I.Y., Ordonez, C. (eds.) DOLAP, pp. 37–44. ACM (2010)
17. Sato, H.: Handling summary information in a database. In: Proceedings of the 1981 ACM SIGMOD International Conference on Management of Data, SIGMOD 1981, p. 98. ACM Press, New York (1981)
18. Malvestuto, F.M., Zuffada, C.: Statistical and Scientific Database Management. LNCS, vol. 339. Springer, Heidelberg (1989)

Knowlog: A Declarative Language for Reasoning about Knowledge in Distributed Systems

Matteo Interlandi

University of Modena and Reggio Emilia
matteo.interlandi@unimore.it

Abstract. In the last few years, researchers started to investigate how recursive queries and deductive languages can be applied to find solutions to the new emerging trends in distributed computing. We conjecture that a missing piece in the current state-of-the-art in logic programming is the capability to express statements about the knowledge state of distributed nodes. In fact, reasoning about the state of remote nodes is fundamental in distributed contexts in order to design and analyze protocols behavior. To reach this goal, we leveraged Datalog¬ with an epistemic modal operator, allowing the programmer to directly express nodes' state of knowledge instead of low level communication details. To support the effectiveness of our proposal, we introduce, as example, the declarative implementation of the two phase commit protocol.

1   Introduction

Pushed by the new interest that Datalog is acquiring in the database community, the goal of this paper is to open a new direction in the investigation on how Datalog could be adopted to program distributed systems. Many authors have stated how logic programming in general [6] and Datalog in particular [5] seems to particularly suited for the representation of distributed programs implementation and properties.

We think that a missing point is the possibility to express statements about the knowledge state of distributed nodes in Datalog. In fact, the ability to reason about the knowledge state of remote nodes has been demonstrated [4] to be a fundamental tool in multi-agent systems in order to specify global behaviors and properties of protocols. Motivated by all these facts, we leveraged Datalog¬ with an epistemic modal operator, allowing the programmer to express directly nodes' state of knowledge instead of low level communication details. To support our assertions, we describe our implementation of the two phase commit protocol.

The remainder of the paper is organized as follow: Section 2 contains some preliminary notations about Datalog¬. Section 3 describes our distributed system model and introduces some concepts such as *global state*, *run* and the modal operator K. Section 4 introduces Knowlog and the implementation of the two phase commit protocol. The last section contains future works and conclusions.

P. Atzeni, D. Cheung, and R. Sudha (Eds.): ER 2012, LNCS 7532, pp. 572–577, 2012.

2 Preliminaries

In order to define Knowlog, we first introduce some principles of Datalog¬ [1], and Datalog¬ augmented with temporal constructs [8,3]. A Datalog¬ rule is an expression in the form:

$$H(\bar{u}) \leftarrow B_1(\bar{u}_1), ..., B_n(\bar{u}_n), \neg C_1(\bar{v}_1), ..., \neg C_m(\bar{v}_m)$$

where $n, m \geq 0$, H, B_i, C_j are relation names, $i = 0, ..., n$ and $j = 0, ..., m$ and $\bar{u}, \bar{u}_i, \bar{v}_j$ are tuples of appropriate arities. Tuples are composed by *terms* and each term can be a constant in the domain **dom** or a variable in the **var** set. We will use interchangeably the terms *predicates* and *relations*. As usual $H(\bar{u})$ is referred as the *head*, $B_i(\bar{u}_i)$, $C_j(\bar{v}_j)$ as the *body*, and in general $H(\bar{u})$, $B_i(\bar{u}_i)$ and $C_j(\bar{v}_j)$ as *atoms*. A *literal* is an atom (in this case we refer to it as *positive*) or the negation of an atom. If $m = 0$, the rule is in Datalog form and express a *definite clause*, while if $m = n = 0$ the rule is expressing a *fact clause*. We refer to a fact clause directly as a *fact* or equivalently as a *groud atom* if it does not contain variable terms.

In this paper we assume that each rule is *safe*, i.e. every variable occurring in a rule-head appears in at least one positive literal in the rule body. Then, a *Datalog¬ program* Π is a set of safe rules. For a database schema **R**, a *database instance* is a finite set **I** constructed by the union of the relation instances over R, with $R \in \mathbf{R}$ a relation name and where each relation instance is a finite set of facts.

As introductory example, we use the program depicted in Listing 1.1 where we employed a relation `link`, containing tuples in the form (S,D), to specify the existence of a link between a source node S and a destination node D. In addition, we used the `path` relation, which is computed starting from the `link` relation (r1) and recursively adding a new path when, roughly speaking, there is a link from A to B and already exists a path from B to C (r2).

```
r1: path(X,Y):-link(X,Y)
r2: path(X,Z):-link(X,Y),path(Y,Z)
```

Listing 1.1. Simple Recursive Datalog Program

3 System Model

Before starting the discussion on how we leverage the language with epistemic operators, we first introduce our model of distributed system and how communication among nodes is performed. We define a distributed message-passing system to be a non empty finite set N of share-nothing nodes joined by bidirectional communication links. Each node identifier has a value in the domain **dom** but here we consider the set $N = \{1, ..., n\}$ of node identifiers, where n is the total number of nodes in the system. We denote with *adb* a new set of *accessible* relations encompassing all the tables that are horizontally partitioned among nodes and through which nodes are able to communicate. Each relation $R \in adb$ contains a *location specifier* term [7]. This term maintains the identifier of the remote node to which every new fact inserted into the relation R must be issued.

Since distributed systems are not static but evolving with time, following [3,8] we incorporate in the language a notion of time. Thus, each relation is labeled with a *time-step identifier* having values in $\mathbb{N}$ and which specifies at what time-step a given instance has been derived and is true. A consequence of this approach is that tuples by default are considered *ephemeral*, i.e., they are valid only for one single time-step. Obviously, tuples can became *persistent* - once derived, for example at time s, they last for every time $t \geq s$ - if they are stored in *persistent* relations. We embrace the Dedalus [3] notation, thus programs rules are divided in two sets: *inductive* and *deductive*. The former set contains all the rules employed for transferring tuples among time-steps i.e., persistency rules, while the latter encompasses all rules that are local in a single time-step. Deductive rules appear as usual Datalog¬ rules, while a **next** suffix is introduced in head relations to characterize inductive rules.

Continuing with the example introduced in the previous section, in order to describe Listing 1.2 we can imagine a real network configuration where each node has locally installed the program, and where each **link** relation reflect the actual state of the connection between nodes. For example, we will have the fact **link(A,B)** in node A's instance if a communication link between A and node B exists. The location specifier term is identified by the prefix **#**. To note that rule **r1** is an inductive rule defining **link** as a persistent relation, while **r2** is a rule for the modification of the **link** relation if a tuple is inserted into the ephemeral relation **link_down**.

```
r1: link(X,Y)@next:-link(X,Y),¬del_link(X,Y)
r2: del_link(X,Y):-link_down(X,Y)
r3: path(#X,Y):-link(X,Y)
r4: path(#X,Z):-link(X,Y),path(#Y,Z)
```

Listing 1.2. Distributed Program

The semantics of the program in Listing 1.2 is the same of the one in the previous section, even though operationally they substantially differ. In fact, in this new version, computation is performed in parallel on multiple distributed nodes. Communication is achieved through rule **r4** which, informally, specify that a path from nodes A to C exists if there is a link from A to another node B and this last knows that a path from B to C exists.

3.1   The Knowledge Model

In every point in time, each node is in some particular *local state* encapsulating all the information the node possesses. We use s_i to denote the local state of node i . We define the *global state* of a distributed system as a tuple $(s_1, ..., s_n)$ where s_i is the node i's state. We define how global states may change over time through the notion of *run*, which binds time values to global states, i.e., $r : \mathbb{N} \to \mathcal{G}$ where $\mathcal{G} = \{S_1 \times ... \times S_n\}$ and S_i is the set of possible local states for node $i \in N$. Following [4] we define a *system* as a set of runs. In knowledge-based systems, nodes are able to accomplish actions not only based on their local state, but also on the knowledge that the node has, i.e., information about the state of the system. If we consider two runs of a system, with global states respectively $g = (s_1, ..., s_n)$ and $g' = (s'_1, ..., s'_n)$, g and g' are *indistinguishable* for node i, and we will write $g \sim_i g'$ if i has the same local state both in g and g', i.e., $s_i = s'_i$. We use the modal operator K_i and we write $K_i\psi$ to express that a node i "*knows*" sentence ψ: in every global state that i considers possible - i.e., the set of all global states that are indistinguishable for i - the sentence ψ is true. This definition of knowledge follows the set of axioms called *S5*. We refer the reader to the technical report [9] for a detailed discussion about the modal operator K.

4   Incorporating Knowledge: Knowlog

We employ $\square$ to denote a (possibly empty) sequence of modal operators K and we use the following statement to express the sentence ψ in rule form:

$$\square(H \leftarrow B_1, ..., B_n, \neg C_1, ..., \neg C_m) \tag{1}$$

with $n, m \geq 0$ and each positive literal is in the form $\square R$, while negative literals are in the form $K_i\square R$ where K_i is equal to the *model context*. From [10] we adopt the term *modal context* to refer to the sequence - with the maximum length of one - of modal operators appearing in front of a rule. We put some restriction on the sequence of operators permitted in $\square$.

Definition 1. *Given a (possibly empty) sequence of operators* $\square$, $\square$ *is in restricted form if it does not contain* K_iK_i *subsequences, with* i *specifying a node identifier.*

Definition 2. *A Knowlog program is a set of rules in the form (1), containing only (possibly empty) sequences of modal operators in the restricted form and where the subscript* i *of each modal operator* K_i *can be a constant or a variable.*

Informally speaking, given a Knowlog program, using modal context we are able to assign to each node the rules the node is responsible for, while atoms and facts residing in the node i are in the form $K_i\square R$. We define communication rules as follow:

Definition 3. *A communication rule in Knowlog is a rule where no modal context is set and body atoms have the form* $K_i\square R$ - *namely they are prefixed with modal operators related to the same node - while the head atom has the form* $K_j\square R'$, *with* $i \neq j$.

In this way, we are able to abstract away all the low level details about how information is exchanged, leaving to the programmer just the task to specify *what* a node should know, and not *how*.

The Two-Phase-Commit Protocol. Inspired by [2], we implemented the two-phase-commit protocol (2PC) using the epistemic operator K. 2PC is used to execute distributed transaction and can be divided in two phases: in the first phase, called the *voting phase*, a coordinator node submits to all the transaction's participants the willingness to perform a distributed commit. Consequently, each participant sends a vote to the coordinator, expressing its intention. In the second phase - namely the *decision phase* - the coordinator collects all votes and decides if performing a global *commit* or an *abort*. The decision is then issued to all the participants, which act accordingly. In the 2PC implementation of Listing 1.3, we assume that our system is composed by three nodes: one coordinator and two participants. Due to the lack of space, we considerably simplify the 2PC protocol.

```
\\Initialization at coordinator
  r1: Kc(part_cnt(count<N>):-nodes(N))
  r2: Kc(transaction(Tx_id,State):-log(Tx_id,State))
\\Decision Phase at coordinator
  r3: Kc(yes_cnt(Tx_id,count<part>):-vote(Vote,Tx_id,part),Vote == "yes"))
  r4: Kc(log(Tx_id,"commit")@next:-part_cnt(C),yes_cnt(Tx_id,C1),C==C1,
        State=="vote-req",transaction(Tx_id,State))
  r5: Kc(log(Tx_id,"abort"):-vote(Vote,Tx_id,part),Vote == "no",
        transaction(Tx_id,State),State =="vote-req")
\\Voting Phase at participants
  r6: Kp(log(Tx_id,"prepare"):-State=="vote-req",Kctransaction(Tx_id,State))
  r7: Kp(log("abort",Tx_id):-log(Tx_id,State),State=="prepare",
        db_status(Vote),Vote=="no")
\\Decision Phase at participants
  r8: Kp(log(Tx_id,"commit"):-log(Tx_id,State_1),State_1=="prepare",
        State_t=="commit",Kctransaction(Tx_id,State_t))
  r9: Kp(log(Tx_id,"abort"):-log(Tx_id,State_1),State_1=="prepare",
        State_t=="abort",Kctransaction(Tx_id,State_t))
\\Communication
  r10:Kxtransaction(Tx_id, State):-Kcsubs(X),
        Kctransaction(Tx_id,State),Kcpath(#Y,X)
  r11:Kcvote(Vote,Tx_id,"sub1"):-Kp1log(Tx_id,State),
        State=="prepare",Kp1db_status(Vote),Kp1path(#P1,C)
  r12:Kcvote(Vote,Tx_id,"sub2"):-Kp2log(Tx_id,State),
        State=="prepare",Kp2db_status(Vote),Kp2path(#P2,c)
```

Listing 1.3. Two Phase Commit Protocol

In the above example, for simplicity we wrote K_P as a modal context instead of K_{P1} and K_{P2}. If the reader is interested in the Knowlog semantics, we refer to the technical report [9].

5   Conclusion

In this paper we present Knowlog, a programming language based on Datalog¬ leveraged with a notion of time and the modal operator K. Through Knowlog reasonings about state of knowledge in distributed systems can be performed, therefore lighten the programmer's burden of expressing low level communication details. Our work in particular is motivated by [5] where the author discusses how Datalog¬ programs are interestingly suited to express, logically, distributed systems and their properties. For what concern the notion of knowledge applied to distributed systems, we have taken inspiration from the comprehensive exposition of [4]. In literature many approaches for enhancing logic programming with modal operators have been developed. A survey is presented in [11].

What we discuss here is a first step towards the definition of a comprehensive logical framework able to define a declarative as well as operational semantics, and generic enough to be adopted in multiple contexts. To this purpose, we will incorporate in Knowlog the *distributed knowledge*, and overall the *common knowledge* operator that has been proven to be linked to concepts such as coordination, agreement and consistency [4]. The following step will be the definition in Knowlog of weaken forms of common knowledge such as *eventual* common knowledge. Other interesting steps that will be pursued is the definition of the model-theoretic semantics of Knowlog starting from the work of [10].

References

1. Abiteboul, S., Hull, R., Vianu, V.: Foundations of Databases. Addison-Wesley (1995)
2. Alvaro, P., Condie, T., Conway, N., Hellerstein, J.M., Sears, R.: I do declare: consensus in a logic language. Operating Systems Review 43(4), 25–30 (2009)
3. Alvaro, P., Marczak, W.R., Conway, N., Hellerstein, J.M., Maier, D., Sears, R.: DEDALUS: Datalog in Time and Space. In: de Moor, O., Gottlob, G., Furche, T., Sellers, A. (eds.) Datalog 2010. LNCS, vol. 6702, pp. 262–281. Springer, Heidelberg (2011)
4. Fagin, R., Halpern, J.Y., Moses, Y., Vardi, M.Y.: Reasoning About Knowledge. MIT Press, Cambridge (2003)
5. Hellerstein, J.M.: The declarative imperative: experiences and conjectures in distributed logic. SIGMOD Rec. 39, 5–19 (2010)
6. Lamport, L.: The temporal logic of actions. ACM Trans. Program. Lang. Syst. 16, 872–923 (1994)
7. Loo, B.T., Condie, T., Garofalakis, M., Gay, et al.: Declarative networking: language, execution and optimization. In: SIGMOD 2006, pp. 97–108. ACM, New York (2006)
8. Ludäscher, B.: Integration of Active and Deductive Database Rules, Infix Verlag, St. Augustin, Germany. DISDBIS, vol. 45 (1998)
9. Interlandi, M.: Knowlog^K: A Declarative Language for Reasoning about Knowledge in Distributed Systems. Technical Report, DBGroup, University of Modena and Reggio Emilia (March 2011),
http://www.dbgroup.unimo.it/TechnicalReport/interlandi2012.pdf
10. Nguyen, L.A.: Foundations of modal deductive databases. Fundam. Inf. 79, 85–135 (2007)
11. Orgun, M.A., Ma, W.: An Overview of Temporal and Modal Logic Programming. In: Gabbay, D.M., Ohlbach, H.J. (eds.) ICTL 1994. LNCS, vol. 827, pp. 445–479. Springer, Heidelberg (1994)

A Framework for Populating Ontological Models from Semi-structured Web Documents[*]

Hassan A. Sleiman and Inma Hernández

University of Sevilla, Spain
{hassansleiman,inmahernandez}@us.es

Abstract. The Web is the largest repository of information that has ever existed. This information is presented in a human friendly format using HTML, which complicates the consumption of this information by automatic processes. Solutions to this problem are the Semantic Web and Web Services, but the lack of such services in the majority of web sites has increased the interest on information extraction, which allow extracting and structuring information from web documents in ontological models. Despite the high number of proposals on information extraction, there does not exist a universally applicable information extractor. As a consequence, when populating an ontology model automatically from a web site, it is not unusual to need more than one information extractor. We propose a framework that allows the development, training, and the application of information extractors on semi-structured web documents to produce semantic data. We have developed a version of the framework and verified it by means of experiments on 15 web sites. Experimental results are very promising.

Keywords: Populating Ontological Models, Web Information Extraction, Framework.

1 Introduction

The Web contains a huge amount of information and is a still growing data container. This unlimited repository aroused enterprises' interests in exploiting web information, so new applications that consume and analyse this information have emerged. Unfortunately, the information in the Web is embedded in HTML tags and in other contents that in many cases are not interesting. Ontological models provide structured information and provide many services that allow their integration avoiding high costs. Information extraction can be used in these cases to obtain the information in which user is interested from current web sites, and structure it into ontological models.

An Information Extractor is a general algorithm that can be configured by means of rules so that it extracts the information of interest from a web document and returns it according to a structured model. Information extractors

---

[*] Supported by the European Commission (FEDER), the Spanish and the Andalusian R&D&I programmes (grants grants TIN2010-21744-C02-01, TIN2007-64119, P07-TIC-2602, P08-TIC-4100, TIN2008-04718-E, and TIN2010-09988-E).

P. Atzeni, D. Cheung, and R. Sudha (Eds.): ER 2012, LNCS 7532, pp. 578–583, 2012.
© Springer-Verlag Berlin Heidelberg 2012

can be used to extract and structure information from free text web documents, such as news, or from result and detail web documents that contain information about products or services [2]. We focus on information extractors used for semi-structured web documents.

The majority of information extractors are configurable by means of rules. Beyond handcrafted information extraction rules, there are many proposals in the literature to learn them in a supervised and in an unsupervised manner. Supervised techniques require user intervention to learn these rules. Generally, it needs the user to annotate a set of web documents by selecting and assigning a type for the relevant information in a set of web documents used then in the learning process [5,7]. This latter group also contains a number of unsupervised techniques, which do not require the user to provide annotations. Input web documents are analysed to detect repeating patterns or templates used at the server side to generate these pages [6,4].

Developing and maintaining information extractors is still a tedious process because of the lack of development support tools. Relying on a framework in the domain of information extractors has many benefits. Using a framework in developing proposals reduces development and testing costs. Furthermore, it allows producing structured data to populate the same model, which reduces information integration costs. The framework should allow the development and integration of several web information extractors in order to allow obtaining data from one or more web sites, and to structure this data according to a given ontology.

We propose an information extraction framework for semi-structured web documents. The framework is composed of several layers, and aims at creating semantic data (OWL) to populate ontological models. The populated ontology models can be then used for reasoning or other tasks. We present an overview of the framework architecture which has been validated by developing several techniques and used to populate ontological models in several fields.

This paper is organised as follows: First, Section 2 classifies and lists related work briefly and then, Section 3 describes the architecture of the framework. The framework is then used to develop a set of proposals and to populate different ontological models. The the experimental results are reported in Section 4. We conclude our work in Section 5.

2   Related Work

The literature provides some information extraction frameworks for semi-structured web documents, namely NoDoSe [1], DEByE [8], WL2 [3]. These tool do not provide any support to develop new rule learners and its components are not documented for their usage as an API in devising different rule learning techniques. These tools do not offer any services to populate models and are no longer maintained. The literature also provides some information

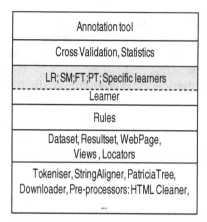

Fig. 1. Framework layers

extraction frameworks for free-text documents, such as UIMA[1] and Gate[2]. They focus on free text documents and offers a set of components to extract and manage extracted information. There are several proposals on platforms and toolkits for information extraction from free texts [9], which is not our focus in this paper.

3   Architecture of Our Framework

The framework is composed of six layers, namely: An annotation tool, cross validator, Learners, Rules, Datasets, and Utilities, c.f. Figure1. These layers can be used to annotate training datasets, learn extraction rules, to download a web document, extract information, populate a given ontology model, and to assess the information extractor. In the following we give more details on each layer.

Annotation tool: A tool that helps users download and annotate web documents to create Datasets according to an OWL ontology in which he or she describes classes, properties and their relationships. Ontology classes are used to represent records of information, object properties represent nested records, and data properties represent attributes. First, the user shall select the ontology URI which is used to assign a type and a relation between the annotations. Then, he or she navigate to web documents and add them to the created Dataset. Once added, contents from this web document can be selected, dragged and dropped into the ontology. This allows the creation of individuals of a certain class and assigning properties to them, saving their locators too. When an information extractor is applied on a collection of web documents, it creates Datasets, which can be then loaded and modified in the tool.

---

[1] http://uima.apache.org/
[2] http://gate.ac.uk/ie/

Cross Validator: Our framework provides a cross validation layer to evaluate information extractors. The purpose of this layer is to assess information extractors on web sites and produce comparable results. These results can be used then to select which information extraction technique best performs on a given web site. This layer includes the following classes: CrossValidation and Statistics. CrossValidation implements a k-folds cross-validator, where k is typically 10. Statistics is a class that collects information during the cross validation process. It provides methods to collect the following statistics: true positives, false positives, true negatives, false negatives, the total number of annotations at each iteration and the weight of each class and property in each dataset. The provided methods in this class to calculate effectiveness and efficiency values use these attributes.

Learners: The Learners layer provides classes used during rule learning process. The key of this layer is that it is open, i.e., is intended to provide an extension point software engineers should use to create their own rule learners, cf. the gray band in Figure1. This layer contains the following classes: LearnerFactory, Learner, and SkeletonLearner. LearnerFactory is a class that is intended to provide a method to create a Learner of a given type. Learner is an interface that provides a number of template methods software engineers must provide to implement their techniques. SkeletonLearner transforms a set of annotations into a Transducer in which each state identifies a type of data to be extracted, and the transitions between them maintain the separators found in the input web document for future use by the learning algorithms.

Rules: This layer contains the rules created by learners and used for information extraction. Note that this layer has the responsibility of assigning a type to the extracted information, and structure them according to a given ontology. The main classes in this layer are the following: Rules, Transducer, and Transition. Rules is an interface that should be implemented by the different types of extraction rules learnt by the Learners in the upper layer. Transducer models a type of rules some information extraction techniques learn. It contains a collection of States and Transitions. Inside each Transition resides its conditions which are regular expressions. A transducer extracts information of a certain class in the ontological model, such as Book, Car, Doctor. Inside the transducer, a State extracts a type of data and assigns it to the individual extracted by the transducer. Each State contains a URI that indicates what type of information is extracted inside this state. If the URI is of a basic type, then it assigns a DataProperty to the individual being extracted by the individual. If the URI is of type class C, then the state invokes the transducer that extracts information of type C over the extracted text by the state, and adds an ObjectProperty from the current individual to the individual extracted by the invoked transducer. Transition This class models the conditions that should be fulfilled to start or end the extraction of a certain type of information.

Datasets: This layer contains all the information annotated by user during the annotation process or extracted by an information extractor during an extraction process and the populated models. The classes in this layer are Dataset, This is a map-like structure that associates a set of annotations to a number of web documents. These annotations represent the relevant data in this web document and are represented by a Resultset. Resultset: is a class that contains the annotations that mark the relevant information in a web document. Each annotation has its description besides the relations between these annotations. An annotation is an individual of a given class in the ontological model, and its properties. We used reification to add Locators to the properties of each individual. Locators are pointers to each annotated fragment in a web document. They are of two types: TreeLocators which contain an XPath that points to annotation's node or TextLocators which points to the beginning offset and the length of annotation in the web document. The classes TextView and TreeView offer working with the text contained inside a web document and with the HTML tree and its nodes, respectively.

Utilities: This layers contains the utility classes that are used by the upper layers. These utilities include a configurable tokeniser, a multiple string alignment algorithm, a web page downloader, and preprocessors to clean and fix HTML code inside downloaded web pages.

4 Experimental Results

Table 1. Comparing NLR, SM, and FT techniques.

Dataset	NLR			SM			FT		
	P	R	T	P	R	T	P	R	T
awesomebooks.com	1.00	0.94	15.54	1.00	0.93	7.50	1.00	0.67	4.01
betterworldbooks.com	0.99	0.91	88.37	0.87	0.89	17.85	0.92	0.51	9.40
manybooks.net	0.97	0.82	13.82	0.97	0.82	8.31	0.77	0.53	5.07
waterstones.com	1.00	1.00	138.46	1.00	0.94	9.54	0.92	0.54	5.64
citwf.com	0.91	0.91	61.56	0.98	0.87	5.06	0.99	0.89	3.17
allmovie.com	0.64	0.69	26.90	0.92	0.28	13.62	0.73	0.54	9.87
allconferences.com	0.61	0.59	20.28	1.00	0.82	11.03	0.69	0.65	6.09
mbendi.com	1.00	1.00	9.37	1.00	0.90	1.95	0.86	0.44	1.59
netlib.org	0.43	0.39	2.31	0.85	0.53	7.76	0.81	0.76	3.76
rdlearning.org.uk	0.64	0.28	5.00	0.80	0.41	5.03	0.78	0.58	2.14
doctor.webmd.com	0.62	0.62	989.78	0.98	0.95	11.14	0.83	0.61	4.29
extapps.ama-assn.org	0.61	0.61	384.84	0.79	0.38	4.46	0.70	0.58	3.65
dentists.com	1.00	1.00	18.82	0.64	0.62	2.32	1.00	0.30	1.84
steadyhealth.com	0.75	0.72	265.75	1.00	0.96	11.68	1.00	0.78	6.21
en.uefa.com	1.00	1.00	39.79	1.00	0.94	8.17	1.00	1.00	4.59

To validate our framework, we have implemented a number of proposals in the literature and populated ontological models from 15 web sites. We provide a toolkit with the following learners NLR, SM and FT which are inspired by [7], [5], and [6] respectively.

Table 1 reports on the results of applying these techniques in practice on several datasets using different ontological models. The first column contains the web site on which we have applied each technique. Other columns contain precision (P), recall (R), and the time that was necessary to learn extraction rules by each technique for each web site. The results regarding precision and recall were calculated using 10-fold cross validation. Note that the table allows us to select the information extractor that best performs on each web site to extract its information and populate the ontological model. For instance, if we are interested in populating an ontological model on Books from the site `www.waterstones.com`, we should apply the NLR technique since it achieves very high precision and recall. On the other hand, if the web site is `www.abebooks.com`, it is better to use the SM technique since the other techniques have achieved very low results.

5   Conclusions

In this paper we have described our information extraction framework that allows populating ontological models from semi-structured web documents. We also performed our first experimental results which confirms the fact that using the framework allows the development of several web information extraction techniques and their adaptation to populate ontological models. Our future work will focus on expanding the framework so that unsupervised information extraction techniques can be developed to automatically extract information, create an ontological model, and populate it. We also plan to add a validator that allows to check the correctness of the extracted data before saving them.

References

1. Adelberg, B., et al.: NoDoSE - a tool for semi-automatically extracting semi-structured data from text documents. In: SIGMOD (1998)
2. Chang, C.-H., et al.: A survey of web information extraction systems. IEEE Trans. Knowl. Data Eng. 18(10) (2006)
3. Cohen, W.W., et al.: A flexible learning system for wrapping tables and lists in HTML documents. In: WWW (2002)
4. Crescenzi, V., et al.: Roadrunner: Towards automatic data extraction from large web sites. In: VLDB (2001)
5. Hsu, C.-N., Dung, M.-T.: Generating finite-state transducers for semi-structured data extraction from the web. Inf. Syst. 23(8) (1998)
6. Kayed, M., Chang, C.-H.: FiVaTech: Page-level web data extraction from template pages. IEEE Trans. Knowl. Data Eng. (2010)
7. Kushmerick, N., et al.: Wrapper induction: Efficiency and expressiveness. Artif. Intell. 118(1-2) (2000)
8. Laender, A.H.F., et al.: DEByE - data extraction by example. Data Knowl. Eng. 40(2) (2002)
9. Suchanek, F.M., et al.: SOFIE: a self-organizing framework for information extraction. In: World Wide Web Conference Series (2009)

Data Management in a Modern ITS: Problems and Solutions

Luca Carafoli

University of Modena and Reggio Emilia
luca.carafoli@unimore.it

Abstract. The sustainable urban mobility of people and goods in a territory has become one of the major challenges which has recently gained much interest in several ICT research areas. In this context the Pegasus Project aims to provide reliable and timely information to improve the safety and the efficiency of vehicles' and goods' flows. To do this, it is necessary to collect and integrate large amounts of geo-located stream items coming from On Board Units (OBUs) installed on vehicles. The final aim is to exploit these information for providing end users with various services to enhance mobility.

We propose a data management solution where stream items coming from vehicles are processed and stored in a result container. The required data access functionalities are provided through an SQL-like query language enhanced with stream, event, spatial and temporal operators.

Finally, we propose several communication-saving techniques for Vehicle to Infrastructure (V2I) and Vehicle to Vehicle (V2V) communications in order to aggregate and reduce the data coming from vehicles.

1 Introduction

The support to sustainable urban mobility of people and goods in a territory has become one of the major challenges of modern society because of the increase of traffic congestion. Traffic congestion means increased fuel consumption and pollution (noise and emission) and in general a deterioration of people's life quality.

In this context, the PEGASUS project aims to provide mobility solutions for an efficient and effective traffic management with the employment of infotelematics systems. The reference scenario is shown in Figure 1. Vehicles are equipped with sensor-based devices called OBUs which have the goal to collect data from sensors and to send these to a data Control Centre. The Control Centre has the goal to collect, integrate and analyze the large amount of geo-located data coming from vehicles. Data coming from OBUs are analyzed to produce real-time maps, including traffic and crashes updates, which are distributed to the OBUs according to user's location. The OBUs' smart navigation engine exploits this information for providing end users with various services to enhance mobility.

In this context, our work concerns the management of the stream items coming from vehicles in order to efficiently solve both continuous and one-shot queries

P. Atzeni, D. Cheung, and R. Sudha (Eds.): ER 2012, LNCS 7532, pp. 584–589, 2012.

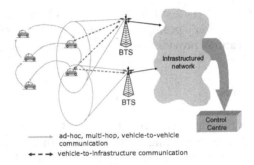

Fig. 1. The PEGASUS project reference scenario

as the backbone for services that retrieve useful information from raw data. In communication-saving area we focus both on the Vehicle to Infrastructure (V2I) and Vehicle to Vehicle(V2V) communications.

The paper is organized as follows. Section 2 presents the problem formulation, related works are dealt with in Section 3. In Section 4 we describe our preliminary ideas and solutions for the problems presented in Section 2. Finally Section 5 concludes.

2   Problem Formulation

From a data management point of view, the application scenario considered in the Pegasus project can be viewed as a data-intensive streaming application with temporal and spatial requirements. In fact, data coming from vehicles into the Control Centre are a continuous stream of items. The Pegasus scenario poses two main issues:

1. the Control Centre has to be in charge of storing real-time geo-located stream items which must be accessed and manipulated efficiently to promptly answer users' requests;
2. stream items must be retained beyond their real-time processing as the Control Centre should be able to process not only continuous queries but also any ad-hoc query which could be issued to the system, also after data acquisition;

Communication saving techniques in the Pegasus Project are essential to reduce the large cost of communications between vehicles and infrastructure and to optimize the data acquisition of the Control Centre. This research field is about both V2V and V2I communications. The issues that the project scenario poses in this field are the following:

1. to find the right trade-off between quality of data and rate of communications. We need to reduce cost without losing essential information;

2. to obtain the optimal trade-off between reactivity of system and communication saving. If we don't promptly communicate the events of interest, the Control Centre cannot provide useful information to users;

3 Related Work

In [1,2], with the same approach, are introduced Continuous Queries over unbounded streams of data, modeling a pure DSMS. In a pure DSMS items are interesting for a small time period and, at the end of its, they are deleted from the system. Instead our system is able to process in real-time stream of data and to store and to manage them. Similarly to [3], we present a different approach and architecture from the above papers, based on two layers: the Stream Processing Engine and the Storage Manager System. In order to obtain better performance than [3], where the authors, introducing a producer-consumer model with several access patterns to storage, de facto have built a DSMS with historical data, we manage data in different ways depending their position in memory hierarchy. Finally, differently from [4] and the other papers mentioned, our system not only stores historical data but also it allows users to submit traditional and continuous queries on these.

In the Communication Saving Techniques fields, [5,6,7] presents V2V systems with a no centralized structure. The traffic information are exchanged among vehicles, not only near ones, so each vehicle can get from the nearest vehicles information about all vehicles in the road network. This kind of system works optimally if there are many vehicles that share information. In fact, for example, in case of small traffic conditions some vehicles can have not updated information about other and some vehicles can not have other in the vicinity. In [8] and [9] are presented some communication reduction techniques. This techniques are independent techniques and do not support an integration with the V2V communications. In our system we exploit V2V communications to aggregate information about near vehicles and V2I communication reduction techniques to send this information only if they are useful to Control Centre. In this way we optimize the number of the connection and we avoid the fragmentation.

4 Preliminary Ideas

4.1 Control Centre

The Control Centre is composed of two modules, the Service Module and the Data Module (see Figure 2). The Service Module interacts with the Data Module to obtain information which is used for answering users' service requests through the Communication Manager.

In order to satisfy the scalability requirements needed by the Pegasus Scenario, the Data Module should employ a flexible mechanism for data storage to differently manage fresh data which should be timely available, thus needing

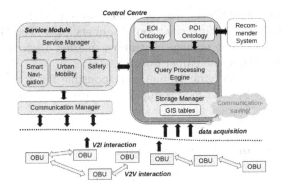

Fig. 2. The control centre architecture

main memory allocation, and historical data which can be stored on disks and cached when needed.

In fact one of the main data management constraints of the Pegasus project is that stream items must be retained beyond their real-time processing. Control Centre must be able to process not only continuous queries but also any ad-hoc query. Therefore, it is not possible to adopt the traditional DSMS data management solutions [1]. Instead, the system is based on a two-tiered architecture where stream items are pulled from the source input stream, processed and stored in a result container to be further pulled for query processing [3].

We propose a two-layers decoupled system like [3] formed by a Query Processing Engine (QPE) and a Storage Manager (SM) layer. The SM provides data access functionalities and implements a scalable storage mechanism which efficiently supports the wide variety of storage requirements emerging in the Pegasus project scenario. The data coming from OBUs are complex and particularly time consuming to manage, because they contain spatial and temporal attributes.

To meet all the constraints of the project, the system works on two distinct areas of memory. Similarly to other DSMSs, the tuples required by the windowed operations will be mainly maintained in main-memory data structures, such as circular queues [1] or linked lists [3]. In order to support high insert rates, we use a linked list of *blocks* of items, in which items are ordered by their insertion times. A *block* itself is implemented as an array of items and has a size equal to secondary memory disk blocks. When a *block* is expired [3] (that is it contains only expired items), it can be directly written on a disk block speeding up the transfer to secondary memory. By this lazy removal policy, that is moving multiple items from main-memory to secondary-memory then a *block* expires, the system gains efficiency compared to transferring item by item. Further over the main-memory structures there is a higher level index [10] to speed up query response time.

Whenever a tuple has been consumed by all the current sliding windows, it becomes an "historical" item. Historical items will be maintained in secondary-memory data structures. Secondary-memory storage is organized in files with traditional indexes.

The Query Processing Engine has the same goal of traditional DBMSs' QPE, that is the building and the optimization of a query plan. The QPE is fully decoupled from SM and it can manage continuous queries. To do this, we extended the SQL language with specific operators.

When a query is issued, the system is responsible for delivering the query results which consist in one single result set, in case of one-shot queries, or in a stream of result sets, in case of continuous queries.

4.2   Communication Saving Techniques

Each vehicle, equipped with an OBU device, can collect data through the GPS and Accelerometer units and perform real-time communications through the GPRS (V2I) and WIFI (V2V) units. The data which is acquired and managed by an OBU are: 1) position, velocity and other GPS-acquired data; 2) accelerations read from accelerometer (used for assisting emergency detection e.g. in case of an accident).

This item, of fixed dimension, is sent to Control Centre by each vehicle with fixed rate. In order to optimize communications, each OBU follows an innovative hybrid communication strategy, where V2V communications between vehicles do not try to completely replace V2I ([5],[11]) but are instead exploited to reduce V2I payload. In particular vehicles exploit the WiFi communication channel to organize themselves in cluster, similarly to [5]. Cluster Members (CMs) communicate to Cluster Heads (CHs) inside a cluster and CHs communicate the results to the Control Centre in V2I mode. CHs can adopt the best V2I techniques depending on the traffic and the kind of road. We provide two family of V2I techniques: the *sampling techniques* and the *information-need techniques*. The *sampling* techniques span from *simple sampling*, which sends data at regular time/travelled distance intervals, to *map-based sampling* that allows for smarter communication choices on the basis of the vehicle position on the map. In the *information-need* techniques the vehicle decides when to transmit on the basis of specific information made available by the Control Centre. For example, in the *deterministic information-need* policy, each OBU transmits, for each travelled segment, the measured v_m if and only if $|v_b - v_m|$ exceed a given threshold T, where v_b is the broadcasted velocity from the Control Centre for each segment of interest [11].

5   Conclusion

This paper presents a scalable storage mechanism to meet the Pegasus Project and, in general, the requirements of an advanced ITS scenario. The system is able to manage continuous queries as well as one-shot queries all, streams of items coming from vehicles and historical data in real-time. The main ideas are to exploit specific techniques for current and historical data management, strictly decoupling QPE and SM, and ultimately providing users with a consistent and uniform way of dealing with their data. The majority of other papers, instead, maintain a clear separation between stream and persistent data management. Furthermore only [3] decouples query processing and storage management.

On the side of communication saving techniques, several of them have been analysed and combined in order to optimize communication on the basis of the traffic situation. Communication costs are reduced by aggregating data with V2V policies and with several V2I techniques.

References

1. Abadi, D., Carney, D., Çetintemel, U., Cherniack, M., Convey, C., Lee, S., Stone-braker, M., Tatbul, N., Zdonik, S.: Aurora: a new model and architecture for data stream management. VLDB J. 12(2), 120–139 (2003)
2. Arasu, A., Babu, S., Widom, J.: The CQL continuous query language: semantic foundations and query execution. VLDB J. 15(2), 121–142 (2006)
3. Botan, I., Alonso, G., Fischer, P., Kossmann, D., Tatbul, N.: Flexible and scalable storage management for data-intensive stream processing. In: Proc. of EDBT Conf., pp. 934–945 (2009)
4. Tufte, K., Li, J., Maier, D., Papadimos, V., Bertini, R.L., Rucker, J.: Travel time estimation using niagarast and latte. In: Proceedings of the 2007 ACM SIGMOD International Conference on Management of Data, SIGMOD 2007, pp. 1091–1093. ACM, New York (2007)
5. Ding, R., Zeng, Q.A.: A clustering-based multi-channel Vehicle-to-Vehicle (V2V) communication system. In: First International Conference on Ubiquitous and Future Networks, ICUFN 2009, pp. 83–88 (June 2009)
6. Yang, X., Liu, L., Vaidya, N., Zhao, F.: A vehicle-to-vehicle communication protocol for cooperative collision warning. In: The First Annual International Conference on Mobile and Ubiquitous Systems: Networking and Services, MOBIQUITOUS 2004, pp. 114–123 (August 2004)
7. Shinkawa, T., Terauchi, T., Kitani, T., Shibata, N., Yasumoto, K., Ito, M., Higashino, T.: A technique for information sharing using inter-vehicle communication with message ferrying. In: 7th International Conference on Mobile Data Management, MDM 2006, p. 130 (May 2006)
8. Ayala, D., Lin, J., Wolfson, O., Rishe, N., Tanizaki, M.: Communication reduction for floating car data-based traffic information systems. In: 2010 Second International Conference on Advanced Geographic Information Systems, Applications, and Services (GEOPROCESSING), pp. 44–51 (February 2010)
9. Kerner, B., Demir, C., Herrtwich, R., Klenov, S., Rehborn, H., Aleksic, M., Haug, A.: Traffic state detection with floating car data in road networks. In: Proceedings of the IEEE Intelligent Transportation Systems, pp. 44–49 (Septemeber 2005)
10. Golab, L., Garg, S., Özsu, M.T.: On Indexing Sliding Windows over Online Data Streams. In: Bertino, E., Christodoulakis, S., Plexousakis, D., Christophides, V., Koubarakis, M., Böhm, K. (eds.) EDBT 2004. LNCS, vol. 2992, pp. 712–729. Springer, Heidelberg (2004)
11. Goel, S., Imielinski, T., Ozbay, K.: Ascertaining viability of wifi based vehicle-to-vehicle network for traffic information dissemination. In: Proceedings of the 7th International IEEE Conference on Intelligent Transportation Systems, pp. 1086–1091 (October 2004)

Author Index